Lecture Notes in Control and Information Sciences

Volume 452

For further volumes:
http://www.springer.com/series/642

About this Series

This series aims to report new developments in the fields of control and information sciences—quickly, informally and at a high level. The type of material considered for publication includes:

1. Preliminary drafts of monographs and advanced textbooks
2. Lectures on a new field, or presenting a new angle on a classical field
3. Research reports
4. Reports of meetings, provided they are
   (a) of exceptional interest and
   (b) devoted to a specific topic. The timeliness of subject material is very important.

Lei Liu · Quanmin Zhu · Lei Cheng
Yongji Wang · Dongya Zhao
Editors

Applied Methods and Techniques for Mechatronic Systems

Modelling, Identification and Control

Springer

Editors
Lei Liu, Yongji Wang
Huazhong University of Science and Technology
Wuhan
People's Republic of China

Lei Cheng
Wuhan University of Science and Technology
Wuhan
People's Republic of China

Quanmin Zhu
University of the West of England
Bristol
UK

Dongya Zhao
China University of Petroleum
Qingdao
People's Republic of China

ISSN 0170-8643 ISSN 1610-7411 (electronic)
ISBN 978-3-642-36384-9 ISBN 978-3-642-36385-6 (eBook)
DOI 10.1007/978-3-642-36385-6
Springer Heidelberg New York Dordrecht London

Library of Congress Control Number: 2013956325

Printed on acid-free paper

Springer is part of Springer Science+Business Media (www.springer.com)

Preface

Most of the research and experiments in the fields of science, engineering, and social studies have spent significant effort to find rules from various complicated phenomena by principles, observations, measured data, and logic derivations. The rules are normally summarized as concise and quantitative expressions or "models." "Identification" provides mechanisms to establish the models from measured data and "control" provides mechanisms to improve the system (represented by its model) performance. The book will reflect the relevant studies in mechatronic fields, with the latest research from interdisciplinary theoretical studies, computational algorithm development to exemplary applications.

Mechatronics system analysis and design include a combination of mechanical engineering, electrical engineering, control engineering, computer engineering, and so on. Mechatronics is a multidisciplinary field of engineering and technology, that is to say, it rejects splitting engineering into separate disciplines. Originally, mechatronics just included the combination of mechanics and electronics, hence the word is a combination of mechanics and electronics; however, as technical systems have become more and more complex to provide high-level functions, the word has been "updated" during recent years to include more technical areas. Research and applications in the synergistic integration of mechanical engineering, electronic control and systems concepts have contributed significantly to the design of systems, devices, processes, and products in modern technology development.

It has been deemed that such fast and highly demanded development and applications should be reported in book, journals, and other media regularly and systematically. So does this book. There are 22 chapters from nine countries in this edited volume presenting a clear route/unique format of each paper, background → motivation → quantitative development (equations) → case studies/illustration/tutorial (curve, table, etc.), which is compiled as an application-oriented reference (Applied Methods and Techniques for Mechatronic Systems—Modelling, Identification, and Control) for readers/users easily tailoring the techniques to accommodate their ad hoc applications. Primary readers could be postgraduate students (M.Sc. and Ph.D.), researchers (post-doc research fellows), and teachers (lecturers and professors) in academia. The secondary readers could be engineers (technical engineers, managers, and directors) in company R&D departments.

The best way to introduce the major contribution from this book is to include a brief abstract of each chapter, which gives readers a clear picture about the book contents.

Chapter 1 *Synchronized Control of Mechanical Systems: A Tutorial*. This study summarizes the work on synchronized control mainly developed from the authors and their colleagues. In the presentation, it tries to use several classical mechanical systems to show the design philosophy of the synchronized control systems, consequently establishes a reader/user-friendly framework with tutorial, survey, applications, and potential research expansion.

Chapter 2 *Control Reconfiguration on Deadlocked Gimballed Thrust of Launch Vehicle*. This study investigates actuator failure compensation for new generation launch vehicle control. A control reconfiguration scheme of fault-tolerant control is developed to enhance the reliability of launch vehicle attitude control systems and prevent the control invalidation caused by the deadlock of the oscillating actuator, based on the congruity of the composite moment before and after the failure happens. This reconfiguration scheme is capable of utilizing the remaining control authority to achieve the desired performance in the presence of deadlock at certain detectable angle occurring in one or several actuators at unknown time instants.

Chapter 3 *Synthesis of an Advanced State Feedback Control for Continuous Nonlinear Polynomial Systems*. This study considers the problem of approximate linearization of affine nonlinear control systems by a static state feedback. First of all, it proposes an analytical method, based on the development into generalized Taylor series expansions and the Kronecker product tools, in order to simplify the complex implementation of the input-state feedback linearization formalism. Next, to improve the synthesized polynomial feedback control, the genetic algorithm, as an optimization method, is used. Finally, the new approach presented in this work is applied to investigate the control problem of a chemical reactor. Moreover, it proves that the controlled process is locally asymptotically stable in a wide region around the operating point, in the Lyapunov sense.

Chapter 4 *Recent Advances on Nonsingular Terminal Sliding Mode Control Method*. This study reviews the research history of the singularity and introduces the recent advance on nonsingular and fast terminal sliding mode (NFTSM) control method. The synthesis of NFTSM controller synthesis is based on a newly proposed nonsingular fast terminal function and a terminal attractor with non-negative exponential coefficient. Both theoretical analyses and computer simulations have proved its effectiveness under the condition that plant uncertainties are bounded.

Chapter 5 *Flocking Behavior via Leader's Backstepping on Nonholonomic Robot Group*. This study aims to improve flocking control for group of nonholonomic robots. It introduces a new flocking control algorithm with potential-based flocking being its foundation. By incorporating Leader's Backstepping algorithm into the flocking strategy, an improved flocking performance is obtained, which leads the flock to the target point swiftly in a smoothed trajectory. Simulations in this paper test and verify the effectiveness of the algorithm, in which key parameters' influences on system performance are discussed.

Chapter 6 *Performance Comparison Between NCTF and PV Techniques for the Control of Linear Motion Servo System*. This study considers two types of controllers for Proportional-Velocity (PV) and Nominal Characteristic Trajectory Following (NCTF). The experimental results showed that the PV was successfully implemented which controlled the settling time, rise time, and steady-state error of the desired position. However, the overshoot performances show its disadvantages. Additionally, the PV controller design is time-consuming process, since model and parameters of the linear motion servo system are needed. Therefore, the needs for higher performance controller become important for the simplicity of the controller design. Hence, the investigation proceed with the nonmodel-based NCTF controller was to control the cart position of the linear motion servo system. The NCTF controller consists of a Nominal Characteristic Trajectory (NCT) and PI compensator. The NCTF controller was designed based on a simple open-loop experiment of the object. The experimental results showed that the NCTF controller is more effective for controlling position of linear motion servo system than the PV controller.

Chapter 7 *A High-Order PID-Sliding Mode Control: Simulation on a Torpedo*. This study deals with the basic concepts, mathematics, and design aspects of a control for nonlinear systems that make the chattering effect lower. As solution to this problem it adopts as a starting point the high order sliding mode approaches then the PID sliding surface. Simulation results show that this control strategy can attain excellent control performance with no chattering problem.

Chapter 8 *Sliding Mode Control with Self-Turning Law for Uncertain Nonlinear Systems with Time-Delay and External Disturbances*. This study proposes a novel sliding mode control with self-turning law for nonlinear systems with time-delay and external disturbances possessing uncertain parameters. The adjustable control gain and a bipolar sigmoid function are online tuned to force the tracking error to approach zero. The proposed control scheme provides good transient and steady-state performance. Moreover as the proposed controller, the chatting phenomenon can be avoided and the problem of the time-delay and the external disturbances are solved for a class of nonlinear systems. The closed-loop control system stability is proved to use the Lyapunov method. Steady-state system performance and the chattering are considerably improved. Numerical simulation results are given to illustrate the effectiveness of the proposed procedure.

Chapter 9 *Applied Methods and Techniques for Modelling and Control on Micro-Blog Data Crawler*. This study presents an algorithm on modelling and control on microblog data crawler based on simulating browsers' behaviors. This needs to analyze the simulated browsers' behaviors in order to obtain the requesting URLs, to simulate and parse and analyze the sending URLs requests according to the order of data sequence. The experimental results and the analysis show the feasible of the approach. Further works are also presented in the end.

Chapter 10 *Development of an Improved Genetic Algorithm for Resolving Inverse Kinematics of Virtual Human's Upper Limb Kinematics Chain*. This study presents an Improved Genetic Algorithm (IGA) is proposed to resolve the inverse kinematics problem in upper limb kinematics chain (ULKC). First, the joint units of ULKC and its mathematical models are constructed by using D–H method; then population diversity and population initialization are accomplished by simulating human being population, and the adaptive operators for mutation are designed. The simulation results show that compared with the Standard Genetic Algorithm (SGA), the IGA can provide higher precise solutions in searching process and avoid "premature" stop or inefficient searching in later stage with high probability.

Chapter 11 *Sliding Mode Control for Nonlinear Discrete Time Systems with Matching Perturbations*. This study considers sliding mode control of nonlinear discrete time systems with matching perturbations. The nonlinear sliding mode controller, whose parameters assure the closed-loop system stable, is designed in order to drive the state trajectories toward a small-bounded region. The controller is approximated by a polynomial equation in current control term according to Taylor series expansion. The algebraic solutions can be obtained by resolving a polynomial equation in the latest control term. The integrated procedure provides a straightforward methodology to apply sliding mode control design technique for nonlinear systems. The simulation results are provided to illustrate the effectiveness of the proposed scheme.

Chapter 12 *Type-2 Fuzzy Wavelet Neural Network Controller Design Based on an Adaptive Gradient Descent Method for Nonlinear Dynamic Systems*. This study develops a novel structure of Type-2 Fuzzy Wavelet Neural Networks (T2FWNN) to control a nonlinear system. This has been performed by invoking some of the specific advantages of wavelets, such as dynamic compatibility, compression, and step parameter adaptation along with a combination to type-2 fuzzy concepts regarding the neural networks abilities. The proposed network is constructed based on a set of TSK fuzzy rules that includes a wavelet function in the consequent part of each rule. This can provide appropriate tools on adaptation of plant output signal to follow a desired one. In this regard, the merits of utilizing wavelets and type-2 FLS simultaneously have been discussed and explored to efficiently handle the uncertainties. It is worth mentioning that the stability of the system is effectively dependent on the learning procedure and the initial values of the network parameters. Here, an adaptive gradient descent strategy is used to adjust the unknown parameters. Furthermore, the performance of the proposed T2FWNN is compared with the type-1 FLS networks. As it is investigated, this method has considerably gained high levels of accuracy with the reasonable number of the parameters. Finally, the efficiency of the proposed approach is demonstrated via the simulation results of two nonlinear case studies.

Chapter 13 *Multivariable Closed-Loop Identification and Its Application to Boiler–Turbine System of Power Unit*. This study presents a new technique for multivariable closed-loop identification. On the basis of process input and output data in the control loops, the process frequency–response matrix is estimated with signal decomposition and frequency spectrum analysis, and then a transform function matrix is identified by least square method. The required input and output data are obtained while the processes are still in normal closed-loop operation. The closed-loop identification is applied to the boiler–turbine coordinated control system of power unit. Simulation example is given to show both effectiveness and accuracy of the identification method for boiler–turbine unit.

Chapter 14 *Tracking and Statistics Method Based on LBTM for Traffic Car Flow*. This study proposes a novel tracking method based on local block graphs targets matching. Car flow tracking is applied to prove the effectiveness of the method. First, the images containing the targets are captured by the video frames to get the targets block graphs. Second, the block graphs can be used to achieve targets matching, and the detection process can be achieved by targets matching to get the optimal targets set. Finally, the targets set can be used to achieve the minimum deviation forecasting of all frames, and the targets tracking can be achieved. A junction video is selected as the experimental data, and a large number of experiments have been done. The results show that the proposed method not only has a great effect to track cars, but also has a better detection rate and tracking accuracy. The car flow statistics can be completed effectively.

Chapter 15 *Object Manipulation Strategy Analysis and Realization for a Humanoid Robot*. This study proposes humanoid robot object operation strategy that integrates the visual feedforward control strategy and the visual feedback control strategy. Using the visual feedforward control strategy, the humanoid robot can walk approaching the target object and make its arm close to the target object; using the visual feedback control strategy, the humanoid robot achieves the robot hand fine alignment with the target object. Based on the visual feedforward control strategy the time of the humanoid robot hand approaching the target object have been reduced; based on the visual feedback control strategy the accuracy of the operation has been improved. Based on the control strategy of this paper proposed, the humanoid robot will be able to walk and operate objects independently.

Chapter 16 *Low Intensity Laser Irradiation Influence Proliferation of Mesenchymal Stem Cells: Comparison of Experimental Data to an Intelligent Agent-based Model Predictions*. This study provides some of the experiments employed to measure influence of low intensity laser on proliferation of mesenchymal stem cells which can vary considerably according to many parameters and biological conditions such as laser nature of emission, irradiation time, wavelength, and energy density. These experiments are compared to an intelligent agent-based model predictions and detailed information about the model description and comparison results is provided. The model is capable of predicting the data for the scenarios fairly well although a few are somewhat problematic.

Chapter 17 *Motor Soft Starter Based on Variable Reactance*. This study establishes the mathematical model of the soft starter which is based on variable reactance, and the relationship between the thyristor conduction angle and the motor current is analyzed. The purpose of the motor and equipment's protection is achieved by selecting the appropriate conduction angle to control the starting current amplitude.

Chapter 18 *Mixed H_2/H_∞ Robust Controller Design-Based LMI Techniques*. This study introduces the work of design of improved LMI-based robust output feedback controller and related simulations.

Chapter 19 *Advanced Control of Atomic Force Microscope for Faster Image Scanning*. This study discusses the design and experimental implementation of an observer-based model predictive control (OMPC) scheme which aims to compensate for the effects of creep, hysteresis, cross-coupling, and vibration in piezoactuators in order to improve the nanopositioning of an AFM. The controller design is based on an identified model of the piezoelectric tube scanner (PTS) for which the control scheme achieves significant compensation of its creep, hysteresis, cross-coupling, and vibration effects and ensures better tracking of the reference signal. A Kalman filter is used to obtain full-state information of the plant. The experimental results illustrate the use of this proposed control scheme.

Chapter 20 *Design of the State Estimation System in the Advanced Drive Assistance System*. This study designs a novel loose coupling senor fusion, which uses the Extended Kalman filtering to fuse the sensor measurements from odometers, accelerometers, gyroscope, and GPS. By using a novel four-wheel vehicle model, the EKF is able to conduct a multioutput rate sensor fusion, compensate the latency for GPS signals, and increase the accuracy of vehicle state estimation even if there exist sensor errors, such as GPS outage, odometer reading error due to wheel slippage. From the road test, it is proved that the designed EKF has achieved good results for vehicle state estimation.

Chapter 21 *A Collaborative Learning Optimization Strategy for Shared Control of Walking-Aid Robot*. This study develops a collaborative learning optimization strategy for shared control of an intelligent walking-aid robot for the purpose of assisting elderly and disabled people. The proposed architecture can adjust two user control weights dynamically by a learning algorithm according to user control habit and walking environment, allowing both human and robot to maintain control of the walking-aid robot. Finally, the experiment results illustrate the validity of the collaborative learning optimization strategy as part of a shared control algorithm.

Chapter 22 *Vibration Suppression of Deformable Linear Object Based on Vision Feedback*. This study proposes an approach based on robot vision to suppress the vibration of a DLO, by means of real-time image processing of high-speed visual feedback to obtain the geometric coordinates and posture of the DLO. A PID controller verifies the validity of the theory along with the expected posture of the DLO. The effectiveness of the proposed strategy is confirmed by experiment. Results show that the proposed method can damp vibrations effectively.

Acknowledgment

The editors wish to express their sincere gratitude to the supports from The National Natural Science Foundation of China (Grant No. 61273188, 61004080, 60705035); Shandong Provincial Natural Science Foundation under Grant ZR2011FM003; China and the Fundamental Research Funds for the Central Universities of China; Taishan Scholar Construction Engineering Special Funding; Key Program of Open Foundation of Hubei Province Key Laboratory of Systems Science in Metallurgical Process of China (Grant No. Z201102). Finally, the editors wish to show their sincere thanks to the reviewers who helped to provide constructive comments and corrections.

Contents

Chapter 1
Synchronized Control of Mechanical Systems: A Tutorial

Dongya Zhao, Quanmin Zhu, Shaoyuan Li and Feng Gao

Abstract Mechanical systems with synchronization functionality have been used for many complex production tasks that cannot be carried out by a single machine and/or individual mechanism. Synchronized control approach is one of the key issues to achieve the synchronized motion of mechanical systems, which attracts more and more attention from academic research to applications. There are many types of mechanical systems that require synchronized control, which have different kinematics and dynamics. It is not an easy job to use a unified synchronized control algorithm for all mechanical systems. This chapter summarizes the work on synchronized control mainly developed by the authors and their colleagues. In the presentation it tries to use several classical mechanical systems to show the design philosophy of the synchronized control systems, and consequently establishes a reader/user-friendly framework with tutorial, survey, applications, and potential research expansion.

Keywords Mechanical systems · Synchronized control · Cross-coupling error

D. Zhao (✉) · Q. Zhu
College of Chemical Engineering, China University of Petroleum, Qingdao, 266580, China
e-mail: dongyazhao@gmail.com

Q. Zhu
Department of Engineering Design and Mathematics, University of the West of England, Coldharbour Lane, Bristol, BS 16 1QY, UK

S. Li
Institute of Automation, Shanghai JiaoTong University, Shanghai, 200240, China

F. Gao
School of Mechanical Engineering, Shanghai JiaoTong University, Shanghai, 200240, China

L. Liu et al. (eds.), *Applied Methods and Techniques for Mechatronic Systems*,
Lecture Notes in Control and Information Sciences 452,
DOI: 10.1007/978-3-642-36385-6_1,

1.1 Introduction

Multiple machine and/or mechanism cooperation is increasingly in demand for use in large-scale modern manufacturing processes such as assembling, transporting, painting, and welding, to name a few [1–3]. These applications require large maneuverability and manipulability. Most of them cannot be achieved by a single machine and/or manipulability. In these cases, the use of multiple machine and/or mechanism is an effective option [4, 5]. Note that coordinated or cooperative control algorithms are required to effectively achieve such capacity and functionality. In the main, conventional control algorithms such as centralized and/or decentralized control algorithms have not addressed coordination/cooperation tasks [6, 7]. It has also been noticed that most of the existing coordinated control algorithms, such as cooperative control and master–slave control, require internal force measurement in implementation, which is difficult to be obtained or estimated.

From a practical point of view, synchronization, coordination, and cooperation are intimately linked subjects and have been used as synonyms to describe such system characteristics [8]. Further research has found that position synchronized control can coordinate multiple mechanical systems without using internal force [9–12]. Because of their easy and efficient implementation capability, synchronization and synchronized control have attracted much attention from research and practice [1–20].

It is useful to summarize the design philosophy of synchronized control of mechanical systems. One of the key points of synchronized control design is the synchronization error that directly relates to the kinematics of the mechanical systems. Note that different types of mechanical systems have entirely different kinematics and lead to varied synchronized control algorithms. It is difficult to describe the control algorithms with a unified format. Despite the diversiform kinematics, there are some common properties in synchronized control algorithm development. The purpose of this chapter is to overview the synchronized control by using the work mainly developed by the authors and their colleagues. For easy reference and applications, the study also lists the corresponding step-by-step control design procedures with illustrative examples.

There are many types of mechanical systems that require synchronized control, such as multiple robotic manipulators systems, parallel robotic manipulators, multiple mobile robots, and multi-agent systems. For each type of mechanical system, the synchronized control should be designed by addressing the kinematics and dynamics of the concrete-type mechanical systems. Three main principles have been used in the author developed synchronized controller design, the synchronization function [21], cross-coupling error [22], and the control algorithms (such as adaptive control, robust control, sliding mode control, and so on). In light of these principles, an effective synchronized control procedure will be presented systematically.

The rest of this chapter is organized as follows: Sect. 1.2 defines position error, synchronization error, and cross-coupling error, and then explains the synchronized control design procedure. Section 1.3 presents the synchronized control of multiple robotic manipulators systems. Section 1.4 discusses the synchronized control of par-

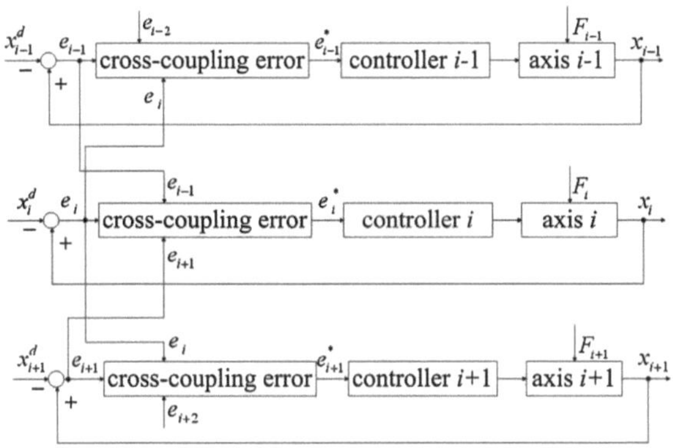

Fig. 1.1 Block diagram of multiple motion axes position synchronized control system

allel robotic manipulators. Section 1.5 describes the formation control of multiple mobile robots by using synchronization definition. Section 1.6 describes adaptive backstepping sliding mode synchronized control for multi-agent systems. Finally, Sect. 1.7 draws the concluding remarks.

1.2 Preliminaries

In this section, multiple axes systems are selected to formulate the synchronization issues (position error, synchronization error, cross-coupling error, and control) of a mechanical system.

1.2.1 Synchronization Error

Consider a motion control system with n axes, the dynamics of the ith-axis is described as [13]

$$M_i(x_i)\ddot{x}_i(t) + C_i(x_i, \dot{x}_i)\dot{x}_i(t) + G_i(x_i) = \tau_i(t) + F_i(x_i, \dot{x}_i, t), \qquad (1.1)$$

where $x_i(t)$ denotes the coordinate of motion in the ith-axis, $M_i(x_i)$ and $C_i(x_i, \dot{x}_i)$ are terms representing inertia and nonlinear effect (such as induced by centripetal and Coriolis), respectively, $G_i(x_i)$ is the gravitational force, $F_i(x_i, \dot{x}_i, t)$ denotes the effect caused by friction force, interconnection force from other axes and unexpected disturbances, τ_i denotes the input torque. Figure 1.1 shows the block diagram of multiple motion axes position synchronized control system.

Define the position tracking error or simply called position error of the ith-axis as

$$e_i(t) = x_i(t) - x_i^d(t), \tag{1.2}$$

where the $x_i^d(t)$ denotes the desired position of the ith-axis.

To synchronize multiple motion axes, the key problem is to maintain some special kinematic relationship among axes. Suppose that these axes are subject to the following synchronization function [6]:

$$\mathbb{R}\{(x_{1\sim n}(t)) \mid f(x_1(t), x_2(t), \cdots, x_n(t)) = 0\} \tag{1.3}$$

Function (1.3) is valid for all desired coordinates $x_i^d(t)$, namely

$$\mathbb{R}\left\{\left(x_{1\sim n}^d(t)\right) \middle| f\left(x_1^d(t), x_2^d(t), \cdots, x_n^d(t)\right) = 0\right\} \tag{1.4}$$

$f(x_1(t), x_2(t), \cdots, x_n(t)) = 0$ can be expanded at the desired coordinates $x_{1\sim n}^d(t)$ by using Taylor series expansion:

$$\begin{aligned} f(x_1(t), x_2(t), \cdots, x_n(t)) &= f\left(x_1^d(t), x_2^d(t), \cdots, x_n^d(t)\right). \\ &+ \sum_{i=1}^{n}\left[\left.\frac{\partial f(x_1(t), x_2(t), \cdots, x_n(t))}{\partial x_i(t)}\right|_{x_i^d}\left(x_i(t) - x_i^d(t)\right) + o(x_i(t))\right] \\ &= \sum_{i=1}^{n}\left[\left.\frac{\partial f(x_1(t), x_2(t), \cdots, x_n(t))}{\partial x_i(t)}\right|_{x_i^d} e_i(t) + o(x_i(t))\right], \end{aligned} \tag{1.5}$$

where $o(x_i(t))$ are the higher order terms.

Defining $c_i(t) = \left.\frac{\partial f\left(x_1(t), x_2(t), \cdots, x_n(t)\right)}{\partial x_i(t)}\right|_{x_i^d}$, function (1.5) can be written as

$$f(x_1(t), x_2(t), \cdots, x_n(t)) = \sum_{i=1}^{n}[c_i(t)e_i(t) + o(x_i(t))]. \tag{1.6}$$

Function (1.6) must be satisfied to guarantee the goal of synchronization. For a multiple motion axes control system, the synchronization function must be subjected to the following expression [6, 7]:

$$f(x_1(t), x_2(t) \cdots x_n(t)) = x_1(t) - x_2(t) - C_1 = \cdots = x_n - x_1 - C_n, \tag{1.7}$$

where C_i is a constant. The expression means that the difference between positions/orientations of each two axes must be a constant so as not to damage the axes.

According to function (1.7), the following expression can be derived as

$$e_1(t) - e_2(t) = e_2(t) - e_3(t)$$
$$= \cdots = e_{n-1}(t) - e_n(t) = e_n(t) - e_1(t) = 0. \tag{1.8}$$

If expression (1.8) is satisfied the following expression can also be derived:

$$e_1(t) = e_2(t) = \cdots = e_n(t). \tag{1.9}$$

Synchronization errors of multiple axes are defined as differential position errors among multiple motion axes. It will result in heavy online computational burden if all synchronization errors are considered in each axis control. Based on the assumption that motion of all axes are synchronized if every pair of axes is synchronized, one can design the synchronization controller to stabilize synchronization errors between each axis and the other two axes to zero. The synchronization error has been defined as [6, 7]

$$\begin{cases} \varepsilon_1(t) = e_1(t) - e_2(t) \\ \varepsilon_2(t) = e_2(t) - e_3(t) \\ \vdots \\ \varepsilon_n(t) = e_n(t) - e_1(t) \end{cases} \tag{1.10}$$

1.2.2 Cross-Coupling Error

Under the synchronization error definition (1.10), the control torque τ_i is to control the position error $e_i(t) \to 0$ and at the same time to synchronize the motions of the $(i-1)$th-axis, ith-axis, and $(i+1)$th-axis, i.e., when $e_i(t) = 0$, $\varepsilon_{i-1}(t) = 0$, and $\varepsilon_i(t) = 0$ at the same time.

To accommodate both position error and synchronization error, cross-coupling error insight has been introduced to provide an effective method to eliminate the interconnections among multiple motion axes systems. This means that the interconnections will vanish after the motion axes work in a synchronous manner. Accordingly, cross-coupling technique lays a foundation for synchronized control design. Mathematically, the cross-coupling error in terms of position error and synchronization error has been properly defined as follows [6, 7, 22]:

$$\begin{cases} e_1^*(t) = e_1(t) + \beta \int_0^t (\varepsilon_1(\omega) - \varepsilon_n(\omega))d\omega \\ e_2^*(t) = e_2(t) + \beta \int_0^t (\varepsilon_2(\omega) - \varepsilon_1(\omega))d\omega \\ \vdots \\ e_n^*(t) = e_n(t) + \beta \int_0^t (\varepsilon_n(\omega) - \varepsilon_{n-1}(\omega))d\omega \end{cases}, \tag{1.11}$$

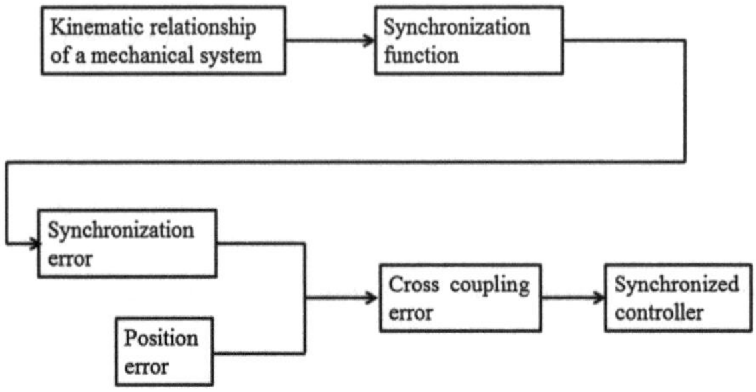

Fig. 1.2 Synchronized control design procedure

where $e_i^*(t)$ is the cross-coupling error, β is positive coupling parameter to balance the weight between the position error and synchronization error.

Including motion responses of all axes in the controller formulation of each axis may lead to heavy online computational work, especially when the axes number n is large. The synchronization error, covering two or three adjoining axes, makes the adjoining axes synchronous, and therefore all of the axes synchronous so as to dramatically reduce the design complexity and computation burden.

1.2.3 Synchronized Control Design Procedure

By using the synchronization error and cross-coupling error, the synchronized controller can be derived. In a real scenario, the synchronized controller should achieve the synchronization while overcoming the system uncertainty and external disturbance. Different types of mechanical systems have different kinematics which lead to different synchronized control algorithms. However, on the other hand, the synchronized control algorithms share a common design procedure. Figure 1.2 gives the controller design procedure which can help the readers to grasp the controller design routine easily.

1.2.4 Summary

In this section, the multiple axes system is selected as an example to demonstrate the definition/design of synchronization error, cross-coupling error, and the synchronized control. The synchronization error is a measure to evaluate the system synchronization fitness. The cross-coupling error presents a simple and effective technique for the synchronized controller design. The synchronized control objective is to guarantee that each agent tracks its own desired trajectory while synchronizing its motion with other agents. A commonly used control design procedure is summarized in Fig. 1.2.

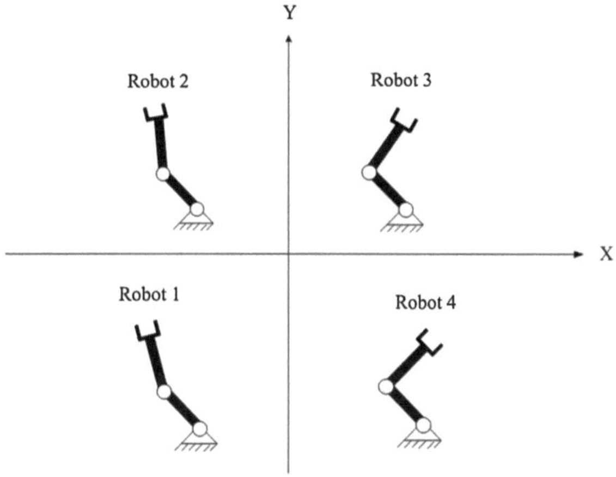

Fig. 1.3 An MRMS containing four planar robotic manipulators

These design methods can be extended to other types of mechanical systems easily and smoothly.

1.3 Chattering-Free Sliding Mode Synchronized Control for Multiple Robotic Manipulators Systems

This section is focused on the synchronized control of multiple robotic manipulators systems. The details can be found in paper [14].

1.3.1 Multiple Robotic Manipulators Systems

Assume that a multiple robotic manipulators system (MRMS) is composed of n robotic manipulators. An MRMS containing four robotic manipulators is shown in Fig. 1.3.

Joint space dynamic equation of the ith robotic manipulator can be described as [23]:

$$M_i(q_i)\ddot{q}_i + C_i(q_i, \dot{q}_i)\dot{q}_i + G_i(q_i) = \tau_i + d_i(t), \qquad (1.12)$$

where $q_i(t) \in R^m$ denotes the joint angular position of the ith robotic manipulator, $M_i(q_i) \in R^{m\times m}$ is symmetric positive definite inertia matrix, $C_i(q_i, \dot{q}_i) \in R^{m\times m}$ is the centripetal and Coriolis matrix, $G_i(q_i) \in R^m$ is the gravitational force, $d_i(t) \in R^m$ is the lumped system uncertainty caused by modeling error, backlash, friction, external disturbance etc., $\tau_i \in R^m$ denotes the input torque, $i = 1, \cdots, n$.

For industrial robots, operations are completed in task space. The trajectory of the end-effector is also planned in task space, such as Cartesian coordinate system. Suppose $x_i(t) \in R^m$ denotes task space coordinate vector. The joint coordinate can be transformed into task space coordinate by the following equation:

$$x_i = h_i(q_i), \tag{1.13}$$

where $h_i(q_i) \in R^m$ is transform function, $i = 1, \cdots, n$.

The task space velocity $\dot{x}_i$ is related to joint space velocity $\dot{q}_i$, the relationship can be expressed as:

$$\dot{x}_i = J_i(q_i)\dot{q}_i, \tag{1.14}$$

where $J_i(q_i) \in R^{m\times m}$ is Jacobian matrix from joint space to task space.

According to Jacobian matrix, task space dynamic equation of the ith robotic manipulator can be expressed as [24]

$$M_i(q_i)J_i^{-1}\ddot{x}_i + \left(C_i(q_i,\dot{q}_i) - M_i(q_i)J_i^{-1}\dot{J}_i\right)J_i^{-1}\dot{x}_i + G_i(q_i) = \tau_i + d_i(t). \tag{1.15}$$

Based on dynamic equation (1.14), the set of dynamic equation of n robotic manipulators system can be written as [25]

$$M(q)J^{-1}\ddot{x} + \left(C(q,\dot{q}) - M(q)J^{-1}\dot{J}\right)J^{-1}\dot{x} + G(q) = \tau + d(t), \tag{1.16}$$

where

$$q = \{q_i\} \in R^{mn}, \quad x = \{x_i\} \in R^{mn}, \quad M(q) = \{M_i(q_i)\} \in R^{mn\times mn},$$
$$C(q,\dot{q}) \in \{C_i(q_i,\dot{q}_i)\}\, R^{mn\times mn}, G(q) = \{G_i(q_i)\} \in R^{mn}, d(t) = \{d_i(t)\} \in R^{mn},$$
$$J = \{J_i\} \in R^{mn\times mn}, \tau(t) = \{\tau_i(t)\} \in R^{mn}, i = 1, \cdots, n.$$

1.3.2 Low Pass Filter-Based Position Synchronization Sliding Mode Control for MRMS

Figure 1.4 shows the control structure of the Low Pass Filter (LPF)-based position synchronization sliding mode control (LPFPSSMC), in which LPF_i denotes ith LPF.

LPFPSSMC control law is designed as

$$u = u_1 + u_2 \tag{1.17}$$

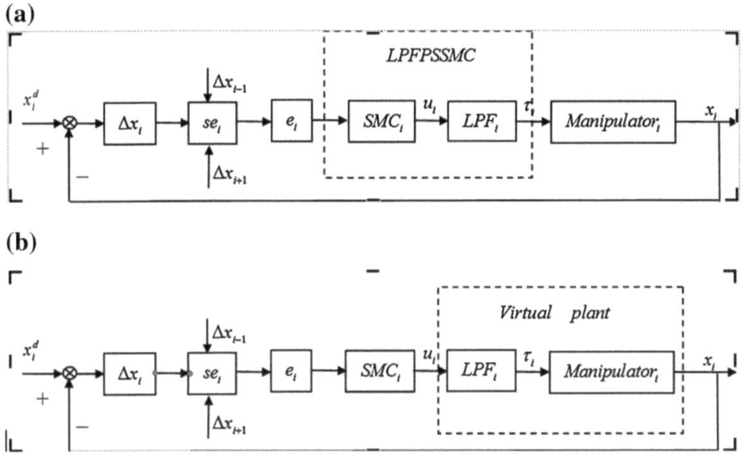

Fig. 1.4 **a** Block diagram of the real system of *i*th robotic manipulator. **b** Block diagram of the virtual system of *i*th robotic manipulator

$$\begin{aligned} u_1 =& \Lambda^{-1}\left(H_2\ddot{x} + H_3\dot{x} + H_4\right) + H_1\Lambda^{-1}\dddot{x}^{d} \\ & - \left(I + \gamma T\right)^{-1} H_1\Lambda^{-1}\left[K_1\left(I + \gamma T\right)\Delta\ddot{x} + K_2\left(I + \gamma T\right)\Delta\dot{x}\right]. \end{aligned} \tag{1.18}$$

$$u_2 = \begin{cases} -\Lambda^{-1}\dfrac{\left(s^T(I+\gamma T)H_1^{-1}\right)^T}{\left\|s^T(I+\gamma T)H_1^{-1}\right\|^2}\left\|s^T\right\|\left\|\left(I+\gamma T\right)H_1^{-1}\right\|\eta & \|s\|^1 0 \\ 0 & \|s\| = 0, \end{cases} \tag{1.19}$$

where $\eta > D + \bar{\varepsilon}$, $\bar{\varepsilon} > 0$ is a positive number.

u_1 is a continuous control law used to compensate the known part of system dynamics. Note that u_1 can make $\dot{s}$ to be equal to 0 in the absence of the lumped system uncertainty. In the principle of sliding mode control [26], u_1 is equivalent control indeed. u_2 is a discontinuous control law used to make cross-coupling error reach to the sliding mode in finite-time and cope with the lumped system uncertainty. Although control law includes a discontinuous term (1.20) it will not stimulate chattering. u is the virtual control signal which is not applied to robotic manipulators directly. τ is the real control signal. Through the LPF, the high frequency discontinuous part of u is eliminated. Hence the real control signal τ is smooth and chattering free. Because the proposed approach considers LPF and multiple robotic manipulators as a whole plant, the robustness of SMC will not be lost. This controller can guarantee that the robotic manipulators move in a synchronous manner.

1.3.3 Summary

A novel PSSMC has been constructed by integrating several efficient tools into a capable procedure. In light of the cross-coupling error technique the proposed approach can stabilize position errors and synchronization errors to zero asymptotically and simultaneously. Under this control approach, the position errors can converge to zero in a synchronous manner, which may achieve higher precision and more flexibility in manufacturing with multiple robotic manipulators. An LPF is applied to significantly reduce chattering effects, which makes the proposed approach more practical with smoother operation. It is worthwhile to note that the study has provided a good example to develop new solutions to the challenging issues from multiple robotic manipulators systems. The future work can be conducted in applying this new scheme to resolve some special problems encountered in MRMS.

1.4 Fully Adaptive Synchronized Tracking Control of Parallel Manipulators

This section illustrates the synchronized control of parallel robotic manipulator systems. The readers can find the details in paper [10].

1.4.1 A Brief Introduction to Parallel Manipulators

As a typical six degree of freedom (DOF) parallel manipulator, Stewart platform has been extensively studied [27–30]. This parallel manipulator can provide better accuracy, rigidity, load-to-weight ratio, and load distribution than serial manipulator. It has been used in the area of low speed and large payload conditions, such as motion base of flight simulator and motion bed of a machine tool [27, 31]. Stewart Platform is a general configuration of parallel manipulator. Its structure is shown in Fig. 1.5.

By employing synchronization error and cross-coupling error, the synchronized control approach can guarantee that both position error and synchronization error converge to zero asymptotically. Because the actuators work in a synchronous manner, the tracking performances can be improved greatly.

By using the natural orthogonal complement method, the dynamic model of Stewart platform can be derived in joint space [27, 31–33] as

$$M(q)\ddot{q} + C(q,\dot{q})\dot{q} + G(q) = \tau, \tag{1.20}$$

where $q \in R^{n\times 1}$ is the generalized coordinate, $M(q) = diag\{M_i(q)\} \in R^{n\times n}$ is the inertia matrix, $C(q,\dot{q}) = diag\{C_i(q,\dot{q})\} \in R^{n\times n}$ is the Coriolis and centrifugal

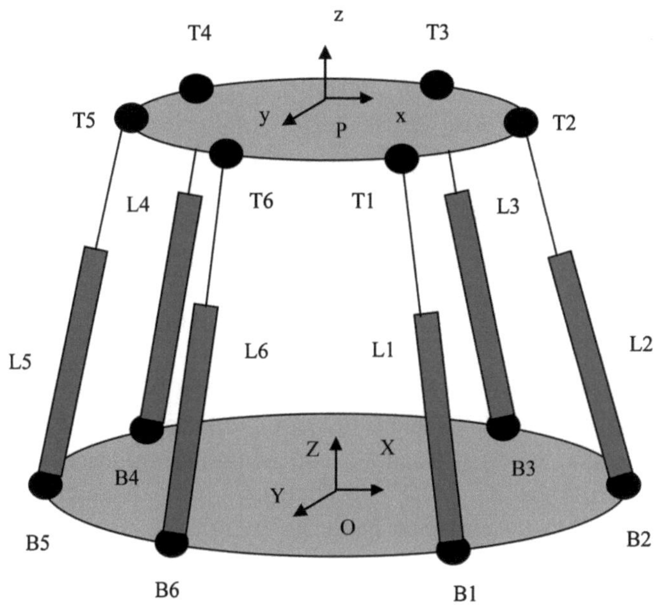

Fig. 1.5 Structure of 6 DOF Stewart platform

force matrix, $G(q) = [G_1(q), \cdots, G_n(q)] \in R^{n\times 1}$ is the gravity force vector, $\tau = [\tau_1, \cdots, \tau_n]^T \in R^{n\times 1}$ is the actuating force vector.

1.4.2 Fully Adaptive Synchronized Tracking Control

The position synchronization errors of Stewart platform can be defined as follows [34]:

$$\begin{cases} \varepsilon_1 = c_1\Delta q_1 - c_2\Delta q_2 \\ \varepsilon_2 = c_2\Delta q_2 - c_3\Delta q_3 \\ \quad \vdots \\ \varepsilon_n = c_n\Delta q_n - c_1\Delta q_1 \end{cases}, \tag{1.21}$$

where $\Delta q_i = q_i^d - q_i$ is the position error, ε_i represents the synchronization error of ith joint, and c_i is the coupling coefficient which can be chosen as $c_i = 1/q_i^d$. The cross-coupling error can use the definition (1.11).

The fully adaptive feedforward feedback synchronized tracking controller is designed as

$$\tau_i = K_i^M c_i^{-1}(\dot{u}_i - \dot{c}_i\dot{q}_i) + K_i^C c_i^{-1} u_i + K_i^G + sign\left(c_i^{-1} r_i\right)\hat{K}_i^N(t) + \hat{K}_{ri}(t)\, c_i^{-1} r_i + c_i^T K_{\varepsilon i}(\varepsilon_i - \varepsilon_{i-1}). \tag{1.22}$$

$$\begin{cases} \dot{\tilde{K}}_{ri}(t) = \dot{\hat{K}}_{ri}(t) = k_{ri}\left\|c_i^{-1} r_i\right\|^2 \\ \dot{\tilde{D}}_i^M(t) = \dot{\hat{D}}_i^M(t) = d_i^M\left\|c_i^{-1} r_i\right\|\left\|c_i^{-1}(\dot{u}_i - \dot{c}_i\dot{q}_i)\right\| \\ \dot{\tilde{D}}_i^C(t) = \dot{\hat{D}}_i^C(t) = d_i^C\left\|c_i^{-1} r_i\right\|\left\|c_i^{-1} u_i\right\| \\ \dot{\tilde{D}}_i^G(t) = \dot{\hat{D}}_i^G(t) = d_i^G\left\|c_i^{-1} r_i\right\| \end{cases} \tag{1.23}$$

for $i = 1, \cdots, n$, where $K_i^M, K_i^C > 0$ are the feedforward gains to compensate the effects of $M(q)_i\, c_i^{-1}(\dot{u}_i - \dot{c}_i\dot{q}_i)$ and $C(q, \dot{q})_i\, s_i^{-1} u_i$, respectively, $K_i^G > 0$ is the gravity compensator. $\hat{K}_{ri}(t) > 0$ and $K_{\varepsilon i} > 0$ are positive feedback control gains. For $i = 1, \cdots, n$, where $\hat{K}_{ri}(0), \hat{D}_i^M(0), \hat{D}_i^C(0), \hat{D}_i^G(0) > 0$ are the initial value and $k_{ri}, d_i^M, d_i^C, d_i^G > 0$ are adaptive gains.

$$\hat{K}_i^N(t) = \hat{D}_i^M(t)\left\|c_i^{-1}(\dot{u}_i - \dot{c}_i\dot{q}_i)\right\| + \hat{D}_i^C(t)\left\|c_i^{-1} u_i\right\| + \hat{D}_i^G(t), \tag{1.24}$$

where $\hat{D}_i^M(t), \hat{D}_i^C(t), \hat{D}_i^G(t) \geq 0$.

The approach can be implemented without any prior knowledge of Stewart platform manipulator, such as initial conditions and parameters in model (1.20).

1.4.3 Summary

A fully adaptive feedforward feedback synchronized tracking control approach is developed for precision control of 6DOF Stewart Platform. By incorporating cross-coupling error technique, the proposed approach can guarantee that both position error and synchronization error converge to zero asymptotically. The tracking performances are improved for the actuators working in coordinating manner. The gains of the proposed controller can be updated online without requiring any prior knowledge of Stewart platform manipulator. It can be implemented in practice easily.

1.5 Finite-Time Formation Control of Multiple Mobile Robots

Multiple mobile robot system is a typical mechanical system. In this section, the synchronized control is adopted to address the formation issues. The finite-time control has some super advantages such as fast converging speed and strong robustness. The terminal sliding mode is introduced to achieve the finite-time formation control. The interested reader can find more information in paper [12].

1.5.1 Multiple Mobile Robot Systems

In recent years, there has been an increasing research interest in formation control for swarms of mobile robots due to its strong practical background [34]. These practical applications include cooperative robot reconnaissance [35], exploration [36], manipulation [37], formation flight of spacecraft [38], constellation of satellite [39], and unmanned vehicles' control [40–42], to name a few. There are mainly three types of approaches to solve formation control for multiple mobile robots, such as behavior-based control [43–46], virtual structure approach [47–50], and leader-following algorithm [51–54].

Note that most of the formation control methods mentioned above can only achieve asymptotic stability of both consensus and position tracking target, that is, it needs infinite time to achieve their control objective. Finite-time stability control approach has faster convergence rate and stronger robustness to system uncertainty and external disturbance [55–58]. Terminal sliding mode control (TSMC) is a finite-time stability control method. It predescribes a terminal sliding mode (TSM) which has finite-time convergence property. By designing some appropriate finite-time feedback control law, one can make system states reach TSM in finite-time. Then, the system states can converge to equilibrium point in finite-time along the predescribed TSM.

Synchronized control can be used to solve the formation issues, which is shown in Fig. 1.6. The formation of a group of mobile robots changes to a rectangle from an ellipse. The synchronized control may guarantee that the formation is an ellipse at each time before the final position.

Consider a group of n planar and fully actuated mobile robots along the boundary (curve) of a two-dimensional compact set. Dynamic equation of robot i can be given as follows [59, 60]:

$$\ddot{q}_i = u_i + \xi_i (i = 1, \cdots, n), \tag{1.25}$$

where $q_i = \begin{bmatrix} x_i & y_i \end{bmatrix}^T \in R^2$ is the coordinates in $x - -y$ plane, $u_i = \begin{bmatrix} u_{ix}, u_{iy} \end{bmatrix} \in R^2$ is the control input of robot i,$\xi_i = \begin{bmatrix} \xi_{ix} & \xi_{iy} \end{bmatrix}^T \in R^2$ is the external disturbance, which is bounded, i.e., $|\xi_{ix}| \leq D_{ix}$,$|\xi_{iy}| \leq D_{iy}$, and $D_{ix}, D_{iy} > 0$ is positive real number.

Define position error for robot i as

$$e_i = q_i - q_i^d, \tag{1.26}$$

where $e_i = \begin{bmatrix} e_{ix}, e_{iy} \end{bmatrix} \in R^2$ is the position error.

In terms of formulation (1.26), synchronization error can be defined as

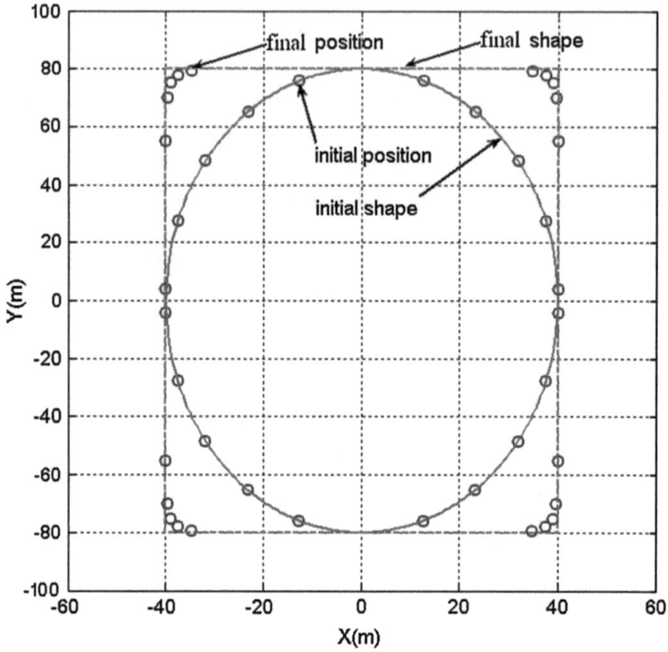

Fig. 1.6 Formation switch by using synchronized control

$$\begin{cases} \varepsilon_1 = 2c_1e_1 - c_2e_2 - c_ne_n \\ \varepsilon_2 = 2c_2e_2 - c_1e_1 - c_3e_3 \\ \vdots \\ \varepsilon_i = 2c_ie_i - c_{i-1}e_{i-1} - c_{i+1}e_{i+1} \\ \vdots \\ \varepsilon_n = 2c_ne_n - c_{n-1}e_{n-1} - c_1e_1 \end{cases}, \tag{1.27}$$

where $\varepsilon_i = \left[\varepsilon_{ix}, \varepsilon_{iy}\right] \in R^2$ is the synchronization error of the ith robot, $i = 1, \cdots, n$.

Equation (1.27) can be written in a matrix form:

$$\begin{bmatrix} \varepsilon_1 \\ \varepsilon_2 \\ \vdots \\ \varepsilon_{n-1} \\ \varepsilon_n \end{bmatrix} = \begin{bmatrix} 2c_1 & -c_2 & 0 & \cdots & -c_n \\ -c_1 & 2c_2 & -c_3 & \cdots & 0 \\ \vdots & \ddots & \ddots & \ddots & \vdots \\ 0 & \cdots & -c_{n-2} & -2c_{n-1} & -c_n \\ -c_1 & 0 & \cdots & -c_{n-1} & 2c_n \end{bmatrix} \begin{bmatrix} e_1 \\ e_2 \\ \vdots \\ e_{n-1} \\ e_n \end{bmatrix}, \tag{1.28}$$

$$\varepsilon = Te,$$

where $\varepsilon = \left[\varepsilon_1^T, \cdots, \varepsilon_n^T\right]^T \in R^{2n}, e = \left[e_1^T, \cdots, e_n^T\right]^T \in R^{2n}, T \in R^{2n\times 2n}$ is the synchronization transformation matrix.

By using cross-coupling error technique [22], define

$$E = e + \gamma\varepsilon, \tag{1.29}$$

where $E \in R^{2n}$ is coupling error, $E_i = \left[E_{ix}, E_{iy}\right]^T, i = 1, \cdots, n, \gamma \in R^{2n\times 2n}$ is positive definite diagonal matrix representing coupling parameter.

By Eq. (1.28), Eq. (1.29) can be rewritten as

$$E = \left(\bar{I} + \gamma T\right) e, \tag{1.30}$$

where $\bar{I} \in R^{2n\times 2n}$ is the identity matrix.

The philosophy of the finite-time synchronized control is to regulate motions of the mobile robots while tracking the desired positions q_i^d to ensure that the robots maintain the desired formation. Note that all the operations are completed in finite-time.

1.5.2 Finite-Time Formation Control

For simplicity of expression, the following notions [61] are introduced with simple expression:

$$\begin{aligned}
&sig\,(x)^{\omega} = \left[|x_1|^{\omega} sign\,(x_1), \cdots, |x_n|^{\omega} sign\,(x_n)\right]^T \\
&x^{\omega} = \left[x_1^{\omega}, \cdots, x_n^{\omega}\right]^T \\
&|x|^{\omega} = \left[|x_1|^{\omega}, \cdots, |x_n|^{\omega}\right]^T \\
&|x| = \left[|x_1|, \cdots, |x_n|\right]^T \\
&sign\,(x) = \left[sign\,(x_1), \cdots, sign\,(x_n)\right]^T,
\end{aligned} \tag{1.31}$$

where $x \in R^n$, $\omega \in R$, $sign\,(\cdot)$ is the sign function.

To design TSMC, the following new nonsingular TSM is defined:

$$s_i = E_i + \Lambda_i sig\left(\dot{E}_i\right)^{\alpha}, \tag{1.32}$$

where $s_i = \left[s_{ix}, s_{iy}\right]^T \in R^2$, $\Lambda_i = diag\left(\lambda_{ix}, \lambda_{iy}\right)$, $1 < \alpha < 2, i = 1, \cdots, n$.

Finite-time formation control law can be designed as

$$u = u_0 + u_1 + u_2. \tag{1.33}$$

$$u_0 = -\gamma \left(\bar{I} + \gamma T\right)^{-1} \left(2\dot{T}\dot{e} + \ddot{T}e\right) - \alpha^{-1}\Lambda^{-1} \left(\bar{I} + \gamma T\right)^{-1} diag(\left|\dot{E}\right|^{2-\alpha}) sign\left(\dot{E}\right) + \ddot{q}^d. \tag{1.33a}$$

$$u_1 = -K_1 s - K_2 sig\,(s)^{\beta}. \tag{1.33b}$$

$$u_2 = -\eta sign\,(s), \tag{1.33c}$$

where

$$q^d = \left[\left(q_1^d\right)^T, \cdots, \left(q_n^d\right)^T\right]^T \in R^{2n}, \quad K_1 = diag\,(K_{1i}) \in R^{2n\times 2n},$$
$$K_2 = diag\,(K_{2i}) \in R^{2n\times 2n}, \Lambda = diag\,(\Lambda_i) \in R^{2n\times 2n}, \eta = diag\,(\eta_i) \in R^{2n\times 2n}.$$

The main purpose of this section is to design a finite-time formation control for multiple mobile robots systems. All mobile robots are considered as a group of particles whose orientations are controlled by the resultant force of u_{ix} and u_{iy}. The proposed approach can be used in control of many types of multiple mobile robots.

1.5.3 Summary

A new finite-time synchronized control approach is systematically studied to accommodate the demand for multiple mobile robots formation control. With the design, the cross-coupling error, position error, and synchronization error converge to zero in finite-time simultaneously. In realization, the formation control problem is successfully converted into a finite-time synchronized control problem. This is a good example to view a problem from a different angle to lead to new ideas and solutions.

1.6 Adaptive Backstepping Sliding Mode Synchronized Control for Multi-Agent Systems

Multi-agent is an important class of mechanical systems. There are many results on the consensus control of multi-agent. The system consensus and synchronization have many similarities. Here a leader–follower multi-agent is selected to illustrate the consensus control. The proposed approach is a good illustration for the operation of such mechanical systems. For more details, the reader can refer to paper [15].

1.6.1 Leader–Follower Multi-Agent Systems

It has been observed that, among the many studies of multi-agent systems, leader–follower multi-agent system has been a predominantly attempted domain due to its explicit physical meaning and engineering background [62–64]. In this class of systems, the leader is usually independent of its followers, but can affect the followers' behaviors [65]. By controlling the leader's behavior, one can achieve the control objective comparatively easily. This consensus protocol can simplify system design, implementation, and reduce operation energy and cost [66]. Some significant interest in the applications of the field in various areas involve teams of manned or unmanned aerial, ground, space or underwater vehicles, robots, and mobile sensors [67].

Based on the brief review above and justification, this chapter proposes a systematical controller design procedure, by means of a novel adaptive backstepping sliding mode control (ABSMC) approach, for leader–follower multi-agent systems. It is the author's belief that the integration of the backstepping and sliding mode principles will provide a comprehensive package of merits to deal with the above analyzed issues. It should be mentioned that these cannot be achieved by each individual independently. Unlike the traditional controller design method which may suffer difficulties to find an appropriate Lyapunov function, backstepping technique is a systematic and recursive design methodology, and can provide a simple procedure to determine Lyapunov function and synthesize controller [68–71]. Sliding mode control is a robust control scheme, which has accumulated a large cohort of theoretical and practical achievements due to its insensitivity/robustness to system uncertainty and external disturbance [72, 73]. Further, an adaptive law is presented to estimate the upper bounds of control input of active leader and lumped uncertainty online to facilitate the controller design. Compared with the existing results [60, 62, 74] the proposed approach has two striking features. One is that the upper bounds of control input of active leader and lumped system uncertainty are not required during controller design. The other is the simple and systematical design procedure. It should be mentioned that the main differences between the proposed approach with the existing approaches [60, 62, 74] lie on the aspects in control input of active leader, system uncertainty, and controller design procedure. The proposed approach uses an adaptive law to deal with the control input of an active leader and lumped system uncertainty, while it designs ABSMC by backstepping method. The existing literature needs to estimate control input of active leader and lumped system uncertainty before controller designed, alternatively an observer is needed to estimate them. Most of the existing literature requires strong skills to find an appropriate Lyapunov function [62, 74]. Although there are coordinate controllers developed for multi-agent systems with sliding mode control [67], the effectively integrated studies of these two principles have not been developed for multi-agent systems. In summary, the motivation of this study is to offer an alternative, but more effective, solution for designing controller for multi-agent system in an integrated procedure rooted in both backstepping and sliding mode principles.

In a leader–follower system, its active leader dynamic equation is expressed by [60, 62, 74] as

$$\begin{cases} \dot{x}_0 = v_0 \\ M_0 \dot{v}_0 = u_0 \end{cases}, \tag{1.34}$$

where $x_0 \in R^m$ and $v_0 \in R^m$ are position and velocity vectors, respectively, $u_0 \in R^m$ is a time-varying control input vector and satisfies $\|u_0\| \le \bar{u}_0$, $\bar{u}_0 > 0$ is a positive real number, $M_0 \in R^{m\times m}$ is an inertia matrix. In general, the leader's state keeps changing throughout the entire motion process and its behavior is independent of its followers.

The ith follower's dynamic equation is described as follows:

$$\begin{cases} \dot{x}_i = v_i \\ m_i \dot{v}_i = u_i + f_i \end{cases}, \tag{1.35}$$

where $x_i \in R^m$ and $v_i \in R^m$ are position and velocity vectors, respectively, $m_i \in R^{m\times m}$ is an inertia matrix, $u_i \in R^m$ is a control input vector of the ith agent. $f_i \in R^m$ denotes the lumped system uncertainty induced by modeling errors and external disturbance, and $\|f_i\| \le F$,$F > 0$ is a positive real number.

1.6.2 The Foundation on Graph Theory

Consider a class of leader–follower multi-agent systems. Graph theory is an effective mathematical tool to describe coordination and communication among agents. Let $\mathcal{G} = \{\mathcal{V}, \mathcal{E}\}$ be a directed graph, where $\mathcal{V} = \{0, 1, 2, \cdots, n\}$ denotes the set of nodes, node i for the$i$th agent, $\mathcal{E}$ represents the set of edges. An edge in $\mathcal{G}$ is an ordered pair (i, j), it means that the ith agent can send information to the$j$th agent directly, but not necessarily vice versa. Compared to directed graph, the pairs of nodes in an undirected graph are unordered, that is, the edge (i, j) denotes that agent i and agent j can mutually send information. Hence, an undirected graph can be viewed as a special case of a directed graph. A rooted tree is a directed graph, in which every node has a sole parent except the root. The root has a directed path to every other node. A rooted spanning tree of $\mathcal{G}$ is a rooted tree that contains all nodes of $\mathcal{G}$ [75].

$A = \left(a_{ij}\right) \in R^{(n+1)\times(n+1)}$ is defined as weighted adjacency matrix of $\mathcal{G}$ with nonnegative elements, only if there is an edge between agent i and agent j $a_{ij} > 0$, otherwise $a_{ij} = 0$. Let $D = diag\,(d_0, d_1, \cdots, d_n) \in R^{(n+1)\times(n+1)}$ be a diagonal matrix, where $d_i = \sum_{j=0}^{n} a_{ij}$ for $i = 0, 1, \cdots, n$. Laplacian of the weighted graph can be defined as [60]

$$L = D - A \in R^{(n+1)\times(n+1)}. \tag{1.36}$$

The connection weight between agent i and the leader is represented by b_i. If there is an edge between agent i and the leader, then $b_i > 0$. Two lemmas about Laplacian matrix and graph theory are listed as follows:

Lemma 1: [76] The directed graph $\mathcal{G} = \{\mathcal{V}, \mathcal{E}\}$ has a rooted spanning tree if $\{\mathcal{V}, \mathcal{E}\}$ has at least one node with a directed path to all other nodes.

Lemma 2: [77] The Laplacian matrix L of a directed graph $\mathcal{G} = \{\mathcal{V}, \mathcal{E}\}$ has at least one zero eigenvalue and all of the nonzero eigenvalues are in the open right-half plane. In addition, L has exactly one zero eigenvalue if and only if $\mathcal{G}$ has a rooted spanning tree. Furthermore, $Rank\,(L) = n$ if and only if L has a simple zero eigenvalue.

Let the $\bar{\mathcal{G}} = \left\{\bar{\mathcal{V}}, \bar{\mathcal{E}}\right\}$ be the sub graph, which is formed by n followers.

$$\bar{A} = \begin{bmatrix} a_{11} & a_{12} & \cdots & a_{1n} \\ \vdots & \vdots & \ddots & \vdots \\ a_{n1} & a_{n2} & \cdots & a_{nn} \end{bmatrix}_{n\times n}. \tag{1.37}$$

Let $\bar{D} = diag\left(\bar{d}_1, \bar{d}_2, \cdots, \bar{d}_n\right)$ be a diagonal matrix, $\bar{d}_i = \sum\limits_{j=1}^{n} a_{ij}, i = 1, 2, \cdots, n$.

The Laplacian of graph $\bar{\mathcal{G}}$ is defined as

$$\bar{L} = \bar{D} - \bar{A}. \tag{1.38}$$

And a_{ij} is defined as

$$a_{ij} = \begin{cases} 1, & \text{if}\,(j, i) \in \mathcal{E} \\ 0, & \text{otherwise} \end{cases} \tag{1.39}$$

The connection weight between agent i and leader is represented by:

$$\bar{B} = diag\,(b_1, b_2, \cdots, b_n)\,, \tag{1.40}$$

where

$$b_i = \begin{cases} 1, & \text{if agent } i \text{ is connected to the leader} \\ 0, & \text{otherwise} \end{cases} \tag{1.41}$$

1.6.3 Adaptive Backstepping Sliding Mode Synchronized Control

Define synchronization error as follows [60]:

$$\begin{cases} e_i^x = \sum\limits_{j=1}^{n} a_{ij}\left(x_i - x_j\right) + b_i\left(x_i - x_0\right) \\ e_i^v = \sum\limits_{j=1}^{n} a_{ij}\left(v_i - v_j\right) + b_i\left(v_i - v_0\right) \end{cases}, \tag{1.42}$$

where e_i^x, e_i^v are the position and velocity consensus errors, respectively. The physical meaning of this definition is if all of the followers converge to the trajectory of the active leader the consensus objective will be achieved. This is called position tracking control in this study.

The error dynamic equation can be written in terms of matrix and vector:

$$\begin{cases} \dot{E}_x = E_v \\ \dot{E}_v = \left(\bar{L} + \bar{B}\right) M^{-1} U + \left(\bar{L} + \bar{B}\right) f - \bar{B}\vec{1} M_0^{-1} u_0 \end{cases}, \tag{1.43}$$

where

$$E_x = \left[\left(e_1^x\right)^T, \cdots, \left(e_n^x\right)^T\right]^T \in R^{mn}, \quad E_v = \left[\left(e_1^v\right)^T, \cdots, \left(e_n^v\right)^T\right]^T \in R^{mn},$$
$$U = \left[u_1^T, \cdots, u_n^T\right]^T \in R^{mn}, \quad M = diag\left(m_1, \cdots, m_n\right) \in R^{mn \times mn}$$
$$f = \left[f_1^T, \cdots, f_n^T\right]^T \in R^{mn}, \vec{1} = [1, \cdots, 1]^T \in R^{mn}.$$

Let $Z_1 = E_x$ and define the following stabilizing function:

$$\alpha = C Z_1, \tag{1.44}$$

where $C \in R^{n \times n}$ is a diagonal positive definite matrix. Then, Z_2 is defined as

$$Z_2 = \dot{Z}_1 + \alpha. \tag{1.45}$$

According to Eqs. (1.43) and (1.44), the error dynamic (1.42) is rewritten as

$$\begin{cases} \dot{Z}_1 = Z_2 - \alpha \\ \dot{Z}_2 = \left(\bar{L} + \bar{B}\right) M^{-1} U + \left(\bar{L} + \bar{B}\right) M^{-1} f - \bar{B}\vec{1} M_0^{-1} u_0 + C\dot{Z}_1 \end{cases} \tag{1.46}$$

Sliding mode is defined as

$$S = K Z_1 + Z_2, \tag{1.47}$$

where $K \in R^{n \times n}$ is a diagonal positive definite matrix.

For dynamic Eq. (1.45), the following ABSMC can be designed as:

$$\begin{aligned} U = \; & M\left(\bar{L} + \bar{B}\right)^{-1}\left(-K\left(Z_2 - C Z_1\right) - C\dot{Z}_1 - \eta\left(S + \beta sign\left(S\right)\right)\right. \\ & \left. - diag\left(\bar{B}\vec{1} M_0^{-1} \hat{u}_0\right) sign\left(S\right)\right) + \tilde{U} \end{aligned} \tag{1.48}$$

$$\tilde{U} = \begin{cases} -\frac{\left(S^T(\bar{L}+\bar{B})M^{-1}\right)^T}{\left\|S^T(\bar{L}+\bar{B})M^{-1}\right\|}\sqrt{n}\hat{F} & \left\|S^T\left(\bar{L}+\bar{B}\right)M^{-1}\right\| \neq 0 \\ 0 & \left\|S^T\left(\bar{L}+\bar{B}\right)M^{-1}\right\| = 0 \end{cases} \tag{1.49}$$

$$\begin{cases} \dot{\hat{u}}_0 = -\gamma_1 S^T diag\left(\bar{B}\vec{1}M_0^{-1}\right) sign\left(s\right) \\ \dot{\hat{F}} = \gamma_2 n \left\|S^T\left(\bar{L}+\bar{B}\right)M^{-1}\right\| \end{cases}, \tag{1.50}$$

where $\eta \in R^{n\times n}$ and $\beta \in R^{n\times n}$ are diagonal positive definite matrices, $sign(\cdot)$ is a sign function, for a vector $y \in R^n$, $sign(y) = [sign(y_1), \cdots, sign(y_n)]^T$, $\gamma_1 > 0$ and $\gamma_2 > 0$ are positive real numbers, $\hat{u}_0(0)$ and $\hat{F}(0)$ are the initial values of $\bar{u}_0$ and F, respectively.

Due to employing adaptive law (1.23), the upper bounds of control input of active leader and lumped system uncertainty can be updated online. Previously they were required before controller design [60], which was a very difficult task to estimate in practice. Obviously the proposed approach relaxes the request on these constraints.

The proposed procedure reduces the major burden in determining Lyapunov function to design a consensus controller through backstepping routine. It should be mentioned that the adaptive law is inherent to the specific consensus algorithm proposed by this study. In other words, it cannot be used by other consensus algorithms directly. Note that the design philosophy is mainly based on Lyapunov method. If the readers want to design an adaptive law to cope with control input of active leader and lumped system uncertainty, they should consider the specific multi-agent system and the employed control algorithm. Then they can use Lyapunov method to design.

1.6.4 Summary

In regard to the consensus control issues associated in leader–follower multi-agent systems, this study has developed a new ABSMC consensus control algorithm by effectively integrating the merits of backstepping and sliding mode control principles. The challenging issues on online estimation of the control input of active leader and lumped system uncertainty have been resolved by a proposed adaptive law, which has to be estimated before controller designed in conventional consensus control strategies (it may be difficult or unrealistic to estimate them before controller design in applications). Furthermore, the corresponding stability analysis has established an assurance for systems in safe operation. The future work is the application of the consensus control to the general mechanical systems.

1.7 Conclusions

The synchronized control can be efficiently and effectively used for many complex tasks of mechanical systems. This chapter uses several design examples to illustrate the synchronized control design philosophy and formulation. In spite of the different kinematics, the controller design has three key points in common, that is, synchronization error, cross-coupling error, and the control principles. It should be noted that the synchronization is very important in the transition process. In conventional nonsynchronized control, the synchronization errors will be vanished in the steady state; however, their synchronization errors are much larger than those of the synchronized control. Larger synchronization error may degrade control performance or even damage mechanical equipment. The synchronized control algorithms summarized in this chapter have many potential applications and also are general references for many other mechanical systems (not described in the study) such as the CNC machine tools, surface mounting technology (SMT) machines, etc. The future work on the synchronized control may be based on three aspects:

(1) Find the new synchronization issue in the mechanical systems, which may lead to the new synchronized control approach.
(2) It is very interesting to find the quantitative relationship between the synchronized performance and computational work. It may lead to an optimal design of the synchronization error.
(3) Bench test these existing synchronization approaches in industrial applications and therefore improve to make them feasible to real problems.
(4) Find the efficient computing algorithm of the realization for the synchronized controller.

Acknowledgments This work is partially supported by the National Nature Science Foundation of China under Grant 61004080, 61273188, Shandong Provincial Natural Science Foundation under Grant ZR2011FM003, China and the Fundamental Research Funds for the Central Universities of China, Development of key technologies project of Qingdao Economic and Technological Development Zone under Grant 2011-2-52, Taishan Scholar Construction Engineering Special funding.

References

1. Gueaieb W, Al-Sharhan S, Miodrag B (2007) Robust computationally efficient control of cooperative closed-chain manipulators with uncertain dynamics. Automatica 43(5):842–851
2. Gueaieb W, Karray F (2007) A robust hybrid intelligent position/force control scheme for cooperative manipulators. IEEE Trans Mechatron 12(2):109–125
3. Nijmeijer H, Rodriguez-Angeles A (2003) Synchronization of mechanical systems. World Scientific, Singapore
4. Martinez-Rosas JC, Arteaga MA, Castillo-Sanchez AM (2006) Decentralized control of cooperative robots without velocity-force measurements. Automatica 42(2):329–336
5. Gudino-Lau J, Artegag MA (2005) Dynamic model and simulation of cooperative robots: a case study. Robotica 23(5):615–624

6. Sun D, Mills JK (2002) Adaptive synchronized control for coordination of multirobot assembly tasks. IEEE Trans Robot Autom 18(4):498–510
7. Sun D (2003) Position synchronization of multiple motion axes with adaptive coupling control. Automatica 39(6):997–1005
8. Rodriguez-Angeles A, Nijmeijer H (2004) Mutual synchronization of robots via estimated state feedback: a cooperative approach. IEEE Trans Control Syst Technol 12(4):542–554
9. Chung S-J, Slotine EJ-J (2009) Cooperative robot control and concurrent synchronization of Lagrangian systems. IEEE Trans Rob 25(3):686–700
10. Zhao D, Li S, Gao F (2008) Fully adaptive feedforward feedback synchronized tracking control for stewart platform systems. Int J Control Autom Syst 6(5):689–701
11. Zhao D, Li S, Gao F (2009) Finite time position synchronised control for parallel manipulators using fast terminal sliding mode. Int J Syst Sci 40(8):829–843
12. Zhao D, Zou T (2012) A finite-time approach to formation control of multiple mobile robots with terminal sliding mode. Int J Syst Sci 34(11):1998–2014 (2012)
13. Zhao D, Li S, Gao F, Zhu Q (2009) Robust adaptive terminal sliding mode-based synchronised position control for multiple motion axes systems. IET Control Theory Appl 3(1):136–150
14. Zhao D, Li D, Zhu Q (2011) Low pass filter based position synchronization sliding mode control for multiple robotic manipulators systems. Inst Mech Eng Part I J Syst Control Eng 225(18):1136–1148
15. Zhao D, Zou T, Li S, Zhu Q (2012) Adaptive backstepping sliding mode control for leader-follower multi-agent systems. IET Control Theory Appl 6(8):1109–1117
16. Liu Y-C, Chopra N (2012) Controlled synchronization of heterogeneous robotic manipulators in the task space. IEEE Trans Rob 28(1):268–275
17. Wang H (2013) Passivity based synchronization for networked robotic systems with uncertain kinematics and dynamics. Automatica 49(3):755–761
18. Nuno E, Basanez L, Ortega R (2011) Passivity-based control for bilateral teleoperation: a tutorial. Automatica 47(3):485–495
19. Cui R, Yan W (2012) Mutual synchronization of multiple robot manipulators with unknown dynamics. J Intell Rob Syst 68(2):105–119
20. Panwara V, Kumarb N, Sukavanamc N, Bormb J-H (2012) Adaptive neural controller for cooperative multiple robot manipulator system manipulating a single rigid object. Appl Soft Comput 12(1):216–227
21. Sun D (2010) Synchronization and Control of Multiagent Systems. CRC Press, Taylor and Francis Group, Boca Raton
22. Koren Y (1980) Cross-coupled biaxial computer control for manufacturing systems. ASME J Dyn Syst Meas Contr 102(4):265–272
23. Spong MW, Hutchinson S, Vidyasagar M (2006) Robot modeling and control. Wiley, New York
24. Ge SS, Hang CC, Woon LC (1997) Adaptive neural network control of robot manipulators in task space. IEEE Trans Industr Electron 44(6):746–752
25. Sun D, Shao X, Feng G (2007) A model-free cross-coupled control for position synchronization of multi-axis motions: theory and experiments. IEEE Trans Control Syst Technol 15(2): 306–314
26. Slotine J-JE, Li W (1990) Applied nonlinear control. Prentice Hall, Englewood Cliffs
27. Merlet JP (2000) Parallel robots. Kluwer Academic Publishers, Dordrecht
28. Dasgupta B, Mruthyunjaya TS (2000) The Stewart Platform manipulator: a review. Mech Mach Theory 35(1):15–40
29. Ting Y, Chen Y-S, Jar H-C (2004) Modeling and control for a Gough-Stewart platform CNC machine. J Rob Syst 21(11):609–623
30. Kim HS, Cho YM (2005) Kyo-II: robust nonlinear task space control for 6DOF parallel manipulator. Automatica 41(9):1591–1600
31. Tsai LW (1999) Robot analysis the mechanics of serial and parallel manipulators. Wiley, New York

32. Angeles J (2003) Fundamentals of robotic mechanical systems, theory, methods, and algorithms. Springer, New York
33. Sun D, Ren L, Mills JK, Wang C (2006) Synchronous tracking control of parallel manipulators using cross-coupling approach. Int J Rob Res 25(11):1137–1147
34. Xie G, Wang L (2009) Moving formation convergence of a group of mobile robots via decentralised information feedback. Int J Syst Sci 40(10):1019–1027
35. Balch T, Arkin R (1998) Behavior-based formation control for multi-robot teams. IEEE Trans Rob Autom 14(6):926–939
36. Fox D, Burgard W, Kruppa H, Thrun S (2000) A probabilistic approach to collaborative multi-robot localization. Auton Rob 8(3):325–344
37. Tanner HG, Loizou SG, Kyriakopoulos KJ (2003) Nonholonomic navigation and control multiple mobile manipulators. IEEE Trans Rob Autom 19(1):53–64
38. Yang G, Yang Q, Kapila V, Palmer D, Vaidyanathan R (2002) Fuel optimal manoeuvres for multiple spacecraft formation reconfiguration using multi-agent optimization. Int J Robust Nonlinear Control 12(2–3):243–283
39. McInnes CR (1995) Autonomous ring formation for a planar constellation of satellites. AIAA J Guidance Control Dyn 18(5):1215–1217
40. Giulietti F, Pollini L, Innocent M (2000) Autonomous formation flight. IEEE Control Syst Mag 20(6):34–44
41. Stilwell DJ, Bishop BE (2000) Platoons of underwater vehicles. IEEE Control Syst Mag 20(6):45–52
42. Ding BC, Xie LH, Cai WJ (2010) Distributed model predictive control for constrained linear systems. Int J Robust Nonlinear Control 20(11):1285–1298
43. Berman S, Edan Y, Hamshidi M (2003) Navigation of decentralized autonomous automatic guided vehicles in material handling. IEEE Trans Rob Autom 19(4):743–749
44. Lawton JRT, Beard RW, Young BJ (2003) A decentralized approach to formation maneuvers. IEEE Trans Rob Autom 19(6):933–941
45. Long M, Gage A, Murphy R, Valavanis K (2005) Application of the distributed field robot architecture to a simulated deming task. In: Proceeding of IEEE International Conference on Robotics and Automation, Barcelona, Spain, pp 3204–3211
46. Parker LE (1998) ALLIANCE: an architecture for fault tolerant multirobot cooperation. IEEE Trans Rob Autom 14(2):220–240
47. Beard RW, Lawton H, Hadaegh FY (2001) A coordination architecture for spacecraft formation control. IEEE Trans Control Syst Technol 9(6):777–790
48. Egerstedt M, Hu X (2001) Formation constrained multi-agent control. IEEE Trans Rob Autom 17(6):947–951
49. Kang W, Xi N, Sparks A (2000) Formation control of autonomous agents in 3D workspace. In: Proceeding of IEEE International Conference on Robotics and Automation, San francisco, pp 1755–1760
50. Lewis MA, Tan KH (1997) High precision formation of mobile robots using virtual structures. Auton Rob 4(4):387–403
51. Das AK, Fierro R, Kumar V, Ostrowski JP, Spletzer J, Taylor CJ (2002) A vision-based formation control framework. IEEE Trans Rob Autom 18(5):813–825
52. Deasi JP, Kumar V, Ostrowski P (2001) Modeling and control of formations of nonholonomic mobile robots. IEEE Trans Rob Autom 17(6):905–908
53. Huang J, Farritor SM, Qadi A, Goddard S (2006) Localization and follow-the-leader control of a heterogeneous group of mobile robots. IEEE/ASME Trans Mechatron 11(2):205–215
54. Takahashi H, Nishi H, Ohnishi K (2004) Autonomous decentralized control for formation of multiple mobile robots considering ability of robot. IEEE Trans Industr Electron 51(6): 1272–1279
55. Amato F, Ambrosinl R, Ariola, M Cosentino C (2009) Finite-time stability of linear time-varying systems with jumps. Automatica 45(5):1354–1358
56. Bhat S, Bernstein D (2000) Finite-time stability of continuous autonomous systems. SIAM J Control Optim 38(3):751–766

57. Hong Y, Jiang ZP (2006) Finite time stabilization of nonlinear systems with parametric and dynamic uncertainties. IEEE Trans Autom Control 51(12):1950–1956
58. Zhao D, Li S, Zhu Q, Gao F (2010) Robust finite-time control approach for robotic manipulators. IET Control Theory Appl 4(1):1–15
59. Sun D, Wang C, Shang W, Feng G (2009) A synchronization approach to trajectory tracking of multiple mobile robots while maintaining time-varying formations. IEEE Trans Rob 25(5):1074–1084
60. Khoo S, Xie L, Man Z (2009) Robust finite-time consensus tracking algorithm for multirobot system. IEEE/ASME Trans Mechatron 14(2):219–228
61. Haimo VT (1986) Finite time controllers. SIAM J Control Optim 24(4):760–770
62. Hong Y, Gao L, Cheng D, Hu J (2007) Lyapunov-based approach to multiagent systems with switching jointly connected interconnection. IEEE Trans Autom Control 52(5):943–948
63. Dimarogonas D, Kyriakopoulos K (2008) A connection between formation infeasibility and velocity alignment in kinematic multi-agent systems. Automatica 44(10):2648–2654
64. Lin P, Jia Y (2009) Consensus of second-order discrete-time multi-agent systems with nonuniform time-delays and dynamically changing topologies. Automatica 45(9):2154–2158
65. Consolini L, Morbidi F, Prattichizzo D, Tosques M (2008) Leader-follower formation control of nonholonomic mobile robots with input constraints. Automatica 44(5):1343–1349
66. Ni W, Cheng D (2010) Leader-following consensus of multi-agent systems under fixed and switching topologies. Syst Control Lett 59(3–4):209–217
67. Gazi V, Fidan B (2007) Coordination and control of multi-agent dynamic systems: models and approaches. Lecture Notes in Computer Science (including subseries Lecture Notes in Artificial Intelligence and Lecture Notes in Bioinformatics) 4433 LNCS. Springer, Berlin, pp 71–102
68. Wu Z-J, Xie X-J, Zhang S-Y (2006) Stochastic adaptive backstepping controller design by introducing dynamic signal and changing supply function. Int J Control 79(12):1635–1646
69. Ozbay U, Zergeroglu E, Sivrioglu S (2008) Adaptive backsetpping control of variable speed wind turbines. Int J Control 81(6):910–919
70. Xia Y, Fu M, Shi P, Wu Z, Zhang J (2009) Adaptive backstepping controller design for stochastic jump systems. IEEE Trans Autom Control 54(12):2853–2859
71. Xia Y, Zhu Z, Fu M (2011) Back-stepping sliding mode control for missile systems based on extended state observer. IET Control Theory Appl 5(1):93–102
72. Spurgeon SK (2008) Sliding mode observers: a Survey. Int J Control 39(8):751–764
73. Lee H, Utkin VI, Malinin A (2009) Chattering reduction using multiphase sliding mode control. Int J Control 82(9):1720–1737
74. Hong Y, Chen G, Bushnell L (2008) Distributed observers design for leader–follower control of multi-agent networks. Automatica 44(3):846–850
75. Ren W (2007) Multi-vehicle consensus with a time-varying reference state. Syst Controller Lett 56(7–8):474–483
76. Ren W, Beard RW (2007) Distributed consensus in multi-vehicle cooperative control. Springer, New York
77. Ren W, Beard RW (2005) Consensus seeking in multiagent systems under dynamically changing interaction topologies. IEEE Trans Autom Control 50(5):655–661

Chapter 2
Control Reconfiguration on Deadlocked Gimballed Thrust of Launch Vehicle

Lei Liu and Yongji Wang

Abstract An unprecedented challenge of the new generation launch vehicle control is its four parallel strap-on engines are oscillateable by the servomechanism. It increases the manoeuvrability of the vehicle and the control complexity simultaneously. The chapter investigates actuator failure compensation for new generation launch vehicle control. A control reconfiguration scheme of fault-tolerant control is developed to enhance the reliability of launch vehicle attitude control systems and prevent the control invalidation caused by the deadlock of the oscillating actuator, based on the congruity of the composite moment before and after the failure happens. This reconfiguration scheme is capable of utilizing the remaining control authority to achieve the desired performance in the presence of deadlock at certain detectable angle occurring in one or several actuators at unknown time instants.

Keywords Launch vehicle · Fault tolerance · Control reconfiguration · Attitude tracking

2.1 Introduction

Heavy lift launch vehicle (HLLV) is capable of lifting more than 14 tons into low earth orbit (LEO) and at least 5 tons to geostationary transfer orbit (GTO) than medium lift launch vehicles. The Chinese Government announced in its November 2000 white paper titled "China's Space Activities" that "the next generation of launch vehicles

L. Liu (✉) · Y. Wang
Key Laboratory of Ministry of Education for Image Processing and Intelligent Control,
School of Automation, Huazhong University of Science and Technology,
Wuhan, 430074, Hubei, People's Republic of China
e-mail: lei.liu.chn@gmail.com

Y. Wang
e-mail: wangyjch@hust.edu.cn

L. Liu et al. (eds.), *Applied Methods and Techniques for Mechatronic Systems*,
Lecture Notes in Control and Information Sciences 452,
DOI: 10.1007/978-3-642-36385-6_2, © Springer-Verlag Berlin Heidelberg 2014

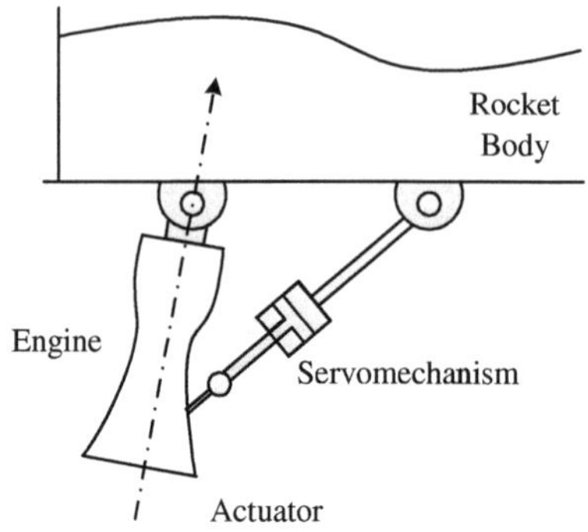

Fig. 2.1 Gimballed thrust vector control system

with non-toxic, non-polluting, high-performance and low-cost qualities" [1] is able to deliver up to 25 tons payload into the LEO or 14 tons payload in the GTO. The new generation launch vehicle will be equipped with much powerful thrust to support heavier actual load to the target orbit.

The symbolical structure of the Gimballed Thrust Vector Control (GTVC) system in launch vehicle is shown in Fig. 2.1. The GTVC consists of power and servomechanism components, which manipulate the direction of the thrust to control the attitude and angular velocity of the vehicle. However, along with the additional gimballed thrust used on the new generation launch vehicle to increase the efficiency of the attitude control, the probability of engines failure is raised. Thus, to enhance the dependability of the launch vehicle, the fault-tolerance control method is a crucial part in vehicle attitude control.

The fault-tolerant control (FTC) of dynamic system is developed in accompany with the analytical redundancy-based fault diagnostics technology, and connects with the fault detection and diagnosis (FDD) module closely and ordinarily. The FTC is a certain kind of close-loop control system, which keeps the system stable with acceptable performance while the actuators or some parts of the system somehow fail [2].

There are two common manners for FTC, passive FTC and active FTC. Passive FTC is to design the proper controller with certain fixed structure and choose the parameters to satisfy both normal situation and failure conditions [3]. On the other hand, the controller should be adjusted after the failure occurred. Most of the active FTC requires FDD subsystem to provide the failure information of the system, to guarantee the priority of the failure. From the principle of the controller, we have following two kinds of the Active FTC ways. One is the Controller Gain Scheduling method, which designs the control law by choosing variant gain parameters based on the fault conditions. The other one is called Online Controller Design, which will make up a new controller structure with the proper parameters, which is called control reconfiguration method [4–6]. The common strategies for control reconfiguration method focusing on the controller parameters regulation are Pseudo-inverse method, Control mixer method, MIMO model self-adaptive method, and Quantitative feedback theory (QFT), etc. [7–11]. Some other widely used control reconfiguration methods to regulate the structure of the controller while fault is occurred, are model-based control [12], model reference adaptive control [13], Conjugate Gradient Method [14], artificial neutral networks [15], T-S Fuzzy Models [16] etc. Comparing with the strategy proposed in this chapter, the above parameters based and structure-based fault tolerant control methods are based on model of failure launch vehicle. And the controller is designed both for the normal flight conditions and fault occurred conditions [17, 18]. However, the model of the failure vehicle is often hard to be obtained, because of the randomness of the occurrence. Moreover, the model-based control method needs huge precomputation to determine the structure and parameters of the controller, and then load them to the vehicle computer. The regulation will be always hysteretic and stiff. Therefore, in this chapter, we tend to provide a different FTC strategy to reconfigure the control action immediately when the fault is occurred.

The Passive FTC method is essential to design a robust close-loop controller to guarantee the stability of the failure system, but it sacrifices the performance of the system. However, the Active FTC needs to deal with the FDD and FTC problem at the same time. Therefore, in our chapter, the FDD module and fault information is assumed available from the servomechanism directly before the control reconfiguration FTC design.

The structure of the chapter is organized as follows: First, a nonlinear launch vehicle model that incorporates independently adjustable oscillating angles of both core engines and strap-on engines is presented and linearized to describe the vehicle's pitching, yawing, and rolling motion. The dynamic analysis and linearizing process are discussed in detail in Appendix. Then, the proposed reconfiguration scheme is provided to the linear vehicle model in the presence of servomechanism malfunction during operation. Furthermore, the inherent stability of the system is analyzed based on the distribution of the roots of its eigen-equation. The Nyquist curves and Bode diagrams are used to study the system's stability after the proper frequency correction. Based on the proper attitude controller, the reconfigurability analysis is derived. The ability of the control reconfiguration is analyzed under some practical assumptions, to study the tolerable range of the servomechanism to be deadlocked. At last, simulation results are presented to assess the effectiveness of this reconfiguration failure compensation design.

2.2 Problem Statement and Preliminaries

As described in Fig. 2.2, the attitude of the strap-on boosted Launch Vehicle has three basic orientations in the rocket coordinate system, which are called pitching, yawing, and rolling channels in the attitude control principles, separately.

For the strap-on boosted Launch Vehicle, the first stage composes with the strap-on thrusts around the body and the core thrusts in the center, of which the gimballed thrusts are the essential actuators of each channel and they affect the attitude of the launch vehicle. Therefore, any malfunction on actuators will cause serious disasters.

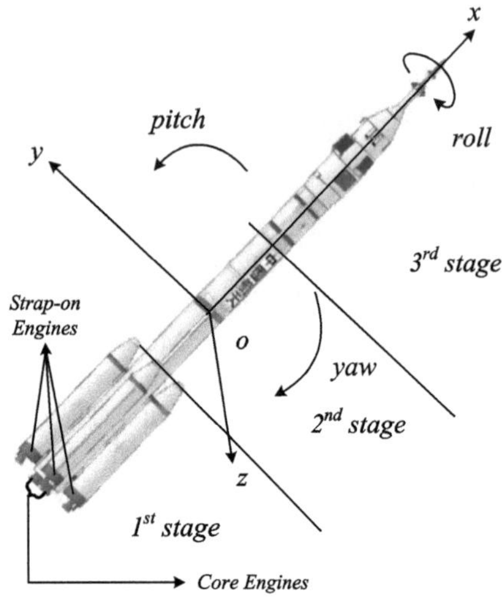

Fig. 2.2 Basic attitude angle and stages of launch vehicle

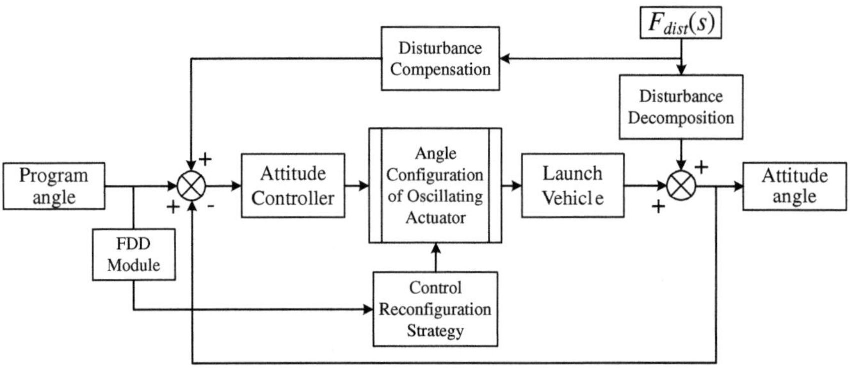

Fig. 2.3 Brief control and reconfiguration structure diagram

Generally, the common actuator failures happen on the servomechanism and the thrust engines. However, the thrust failure would be fatal and irreversible, because the loss of thrust force will affect the initial track by the guidance process. Thus, in this chapter, the servomechanism failure problem is concerned, particularly the gimballed mechanism is deadlocked [19]. In this situation, the vehicle will lose its control stability immediately. However, the propulsion of the vehicle is still maintained. The presented control reconfiguration method is provided to maintain the control action by the remaining control authority to achieve the desired performance.

The brief control and reconfiguration structure is designed to achieve the attitude tracking control and fault-tolerance functions at the same time (see Fig. 2.3). In this structure, the system input is the program angles of the attitude, which is generated from the guiding system. After that is the attitude controller module. The frequency-domain method is widely used in the stable attitude controller design, specially for the inherent instable launch vehicle model, because of its proven reliability. Practically, the disturbance always existed because of the wind and air, and assumed detectable or observable. The direct compensation method will be convenient and effective, which is not the essential problem of this chapter.

On the other hand, the fault-tolerant system is designed to provide service in the presence of faults. Providing redundant manipulation is a common method of recovering from faults, but it is not feasible in cases where space or weight restrictions limit the amount of room available for spares, for instance, the deep-space probe and the launch vehicle. A possible fault recovery method under these circumstances is to reconfigure the control actions by the existing equipments [20, 21].

At first, the failed actuator and its state will be detected by the servo feedback module to provide the prior information for control reconfiguration of the gimballed actuators. Furthermore, the system dynamics and the architectural feature of the actuators are the essential factors as well. Therefore, the further study is constructed by the upper problems.

2.3 Dynamic Model

The dynamic model of the launch vehicle is considered as a variable mass rigid body. Its dynamic motion can be obtained from Newton's second law which states that the summation of all external forces acting on a body is equal to the time rate of change of the momentum of the body, and the summation of the external moments acting on the body is equal to the time rate of change of the moment of angular momentum. For the flight dynamics, normally, an axis system fixed to the Earth can be used as an inertial reference frame. Newton's second law can be expressed in the following equations [22]:

$$\begin{cases} \sum \mathbf{F} = \frac{\mathrm{d}}{\mathrm{d}t}(m\mathbf{V}) = \mathbf{G} + \mathbf{P} + \mathbf{A} + \mathbf{C} + \mathbf{S} + \mathbf{E} + \mathbf{R} \\ \sum \mathbf{M} = \frac{\mathrm{d}}{\mathrm{d}t}\mathbf{H} = \frac{\mathrm{d}}{\mathrm{d}t}(J\omega) + \omega \times (J\omega) \end{cases} \tag{2.1}$$

If we let δm be an element of mass of the vehicle, $\mathbf{V}$ be the velocity of the elemental mass relative to an inertial frame, and $\delta\mathbf{F}$ be the composite force acting on the elemental mass; then Newton's second law yields:

$$\delta\mathbf{F} = \delta m \frac{\mathrm{d}\mathbf{V}}{\mathrm{d}t} \tag{2.2}$$

Similarly, we consider about the composite moment on the elemental mass of the vehicle $\delta\mathbf{M}$ and the moment of inertia δJ.

$$\frac{\mathrm{d}}{\mathrm{d}t}(J\omega) = \delta J\frac{\mathrm{d}\omega}{\mathrm{d}t} \tag{2.3}$$

Therefore, the dynamics including the motion of the mass center and the rotation around the mass center in the three axis directions (see Fig. 2.2) can be presented as:

$$\begin{cases} \delta m\frac{\mathrm{d}\mathbf{V}}{\mathrm{d}t} = \sum \mathbf{F}_x = \mathbf{F}_{Gx} + \mathbf{F}_{Px} + \mathbf{F}_{Ax} + \mathbf{F}_{Cx} + \mathbf{F}_{Ex} + \mathbf{F}_x \\ \delta m\mathbf{V}\frac{\mathrm{d}\theta}{\mathrm{d}t} = \sum \mathbf{F}_y = \mathbf{F}_{Gy} + \mathbf{F}_{Py} + \mathbf{F}_{Ay} + \mathbf{F}_{Cy} + \mathbf{F}_{Ey} + \mathbf{F}_y \\ -\delta m\mathbf{V}\frac{\mathrm{d}\sigma}{\mathrm{d}t} = \sum \mathbf{F}_z = \mathbf{F}_{Gz} + \mathbf{F}_{Pz} + \mathbf{F}_{Az} + \mathbf{F}_{Cz} + \mathbf{F}_{Ez} + \mathbf{F}_z \\ \delta J_x\frac{\mathrm{d}\omega_x}{\mathrm{d}t} + (\delta J_z - \delta J_y)\omega_z\omega_y = \sum \mathbf{M}_x = \mathbf{M}_{Ax} + \mathbf{M}_{Cx} + \mathbf{M}_{Ex} + \mathbf{M}_x \\ \delta J_y\frac{\mathrm{d}\omega_y}{\mathrm{d}t} + (\delta J_x - \delta J_z)\omega_x\omega_z = \sum \mathbf{M}_y = \mathbf{M}_{Ay} + \mathbf{M}_{Cy} + \mathbf{M}_{Ey} + \mathbf{M}_y \\ \delta J_z\frac{\mathrm{d}\omega_z}{\mathrm{d}t} + (\delta J_y - \delta J_x)\omega_y\omega_x = \sum \mathbf{M}_z = \mathbf{M}_{Az} + \mathbf{M}_{Cz} + \mathbf{M}_{Ez} + \mathbf{M}_z \end{cases} \tag{2.4}$$

Based on the equations above, the derivation of the formula for nonlinear modeling of the vehicle and its linearizing process is studied in the Appendix. Then, the linear dynamic model of the vehicle can be formulated in the following three channels, corresponding to the three axis directions before.
In pitching channel:

$$\begin{cases} \Delta\dot{\theta} = c_1^{\varphi}\Delta\alpha + c_2^{\varphi}\Delta\theta + c_{3\text{core}}^{\varphi}\delta_{\varphi}^{\text{core}} + c'^{\varphi}_{3\text{core}}\ddot{\delta}_{\varphi}^{\text{core}} + c_{3\text{strap-on}}^{\varphi}\delta_{\varphi}^{\text{strap-on}} \\ +c'^{\varphi}_{3\text{strap-on}}\ddot{\delta}_{\varphi}^{\text{strap-on}} + c'^{\varphi}_1\alpha_w + \overline{F}_{yc} \\ \Delta\ddot{\varphi} + b_1^{\varphi}\Delta\dot{\varphi} + b_2^{\varphi}\Delta\alpha + b_{3\text{core}}^{\varphi}\delta_{\varphi}^{\text{core}} + b'^{\varphi}_{3\text{core}}\ddot{\delta}_{\varphi}^{\text{core}} + b_{3\text{strap-on}}^{\varphi}\delta_{\varphi}^{\text{strap-on}} \\ +b'^{\varphi}_{3\text{strap-on}}\ddot{\delta}_{\varphi}^{\text{strap-on}} + b_2^{\varphi}\alpha_w = \overline{M}_z \\ \Delta\varphi = \Delta\alpha + \Delta\theta \end{cases} \tag{2.5}$$

in yawing channel:

$$\begin{cases} \dot{\sigma} = c_1^{\psi}\beta + c_2^{\psi}\sigma + c_{3\text{core}}^{\psi}\delta_{\psi}^{\text{core}} + c'^{\psi}_{3\text{core}}\ddot{\delta}_{\psi}^{\text{core}} + c_{3\text{strap-on}}^{\psi}\delta_{\psi}^{\text{strap-on}} + c'^{\psi}_{3\text{strap-on}}\ddot{\delta}_{\psi}^{\text{strap-on}} \\ +c'^{\psi}_1\beta_w - \overline{F}_{zc} \\ \ddot{\psi} + b_1^{\psi}\dot{\psi} + b_2^{\psi}\beta + b_{3\text{core}}^{\psi}\delta_{\psi}^{\text{core}} + b'^{\psi}_{3\text{core}}\ddot{\delta}_{\psi}^{\text{core}} + b_{3\text{strap-on}}^{\psi}\delta_{\psi}^{\text{strap-on}} + b'^{\psi}_{3\text{strap-on}}\ddot{\delta}_{\psi}^{\text{strap-on}} \\ +b_2^{\psi}\beta_w = \overline{M}_y \\ \psi = \beta + \sigma \end{cases} \tag{2.6}$$

and in rolling channel:

$$\ddot{\gamma} + d_1\dot{\gamma} + d_{3\text{core}}\delta_{\gamma}^{\text{core}} + d'_{3\text{core}}\ddot{\delta}_{\gamma}^{\text{core}} + d_{3\text{strap-on}}\delta_{\gamma}^{\text{strap-on}} + d'_{3\text{strap-on}}\ddot{\delta}_{\gamma}^{\text{strap-on}} = \overline{M}_x \tag{2.7}$$

In the equations above, the coefficients of the linear system is decided by the certain physics parameters as in the Appendix. Here, we assume that the variation of the vehicle motion is small in a short-time period. Therefore, by the small disturbances linearizing method, the complicated nonlinear system can be indicated by the variate coefficients linear system approximately.

Notice that, the oscillation angles in the system differential equations, such as $\delta_{\varphi}^{\text{strap-on}}$, $\delta_{\varphi}^{\text{core}}$, $\delta_{\psi}^{\text{strap-on}}$, $\delta_{\psi}^{\text{core}}$, etc., are the equivalent angles of the vehicle and cannot

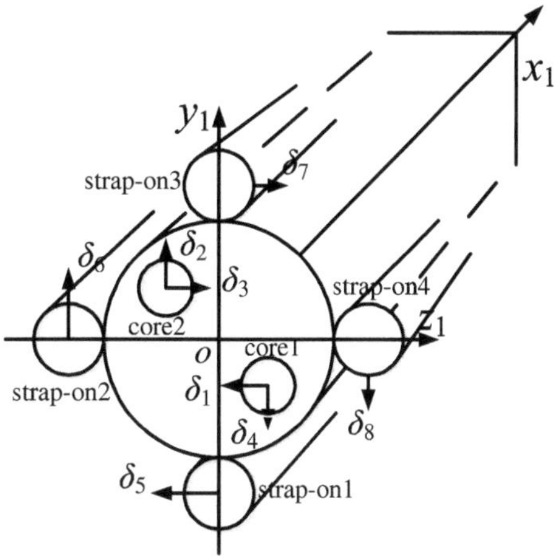

Fig. 2.4 Schematic diagram of propellant rocket engines

reflect the physics property or internal relationship between the oscillation angles of the actuators. Dr. Wu, who was the President of China Academy of Launch Vehicle Technology (CALT), has revealed the outline of Chinese new generation launch vehicle in [23] and discussed the structure of the first stage with four strap-on engines and two moveable core engines, which is different from the previous design of the "Long March" launch vehicles.

As shown in Fig. 2.4, $ox_1y_1z_1$ is the rocket coordinate system from the bottom of the launch vehicle. The "core1" and "core2" stand for two core thrusts, and "strap-on1" to "strap-on4" are symbols of four strap-on thrusts. The "strap-on" thrusts are also called booster engines in some literatures. The core engines are in the first and third quartiles in oy_1z_1 plane, while four strap-on engines are on y, z axis of the plane. The movements of the gimballed thrusts in different directions are labeled by δ_i with the arrows denoting their positive orientations.

Generally, to simplify the kinematic and dynamic models of the vehicle, we use the equivalent oscillating angle $\delta_{\varphi}^{\text{core}}$ and $\delta_{\varphi}^{\text{strap-on}}$ as the equivalent angles to rewrite the dynamic model of the launch vehicle with core stages and the strap-on stages, which denote the combined oscillating angle of the physical helms of the engines. Obviously, the corresponding relationship between the equivalent angles: $\delta_{\varphi}^{\text{core}}$, $\delta_{\varphi}^{\text{strap-on}}$ and the oscillating angles of the engines: $\delta_i, i = 1, 2, \ldots, 8$ will satisfy the following relationships:

For core stages:

$$\begin{cases} \delta_{\varphi}^{\text{core}} = \delta_4 - \delta_2 \\ \delta_{\psi}^{\text{core}} = \delta_3 - \delta_1 \\ \delta_{\gamma}^{\text{core}} = (\delta_1 + \delta_2 + \delta_3 + \delta_4)/2 \end{cases} \tag{2.8}$$

For strap-on stages:

$$\begin{cases} \delta_{\varphi}^{\text{strap-on}} = (\delta_8 - \delta_6)/2 \\ \delta_{\psi}^{\text{strap-on}} = (\delta_7 - \delta_5)/2 \\ \delta_{\gamma}^{\text{strap-on}} = (\delta_5 + \delta_6 + \delta_7 + \delta_8)/4 \end{cases} \tag{2.9}$$

2.4 Control Reconfiguration Strategies

In this section, the actuator deadlock compensation strategies based on the linearized dynamic model are researched. First, we study the correlation of the actuators to look for the necessary redundancy for the failure actuator compensation. Then, the control reconfiguration strategies of different faulty conditions are derived and formulated, which are the essential results of the chapter.

2.4.1 *Association Analysis*

To study the association of the actuators of the strap-on and core thrusts, the state space representation of the system can be established as follows, from the linearized dynamic model is:

The state equation in the pitching channel is:

$$\begin{bmatrix} \dot{\theta} \\ \dot{\varphi} \\ \ddot{\varphi} \end{bmatrix} = \begin{bmatrix} c_2^{\varphi} - c_1^{\varphi} & c_1^{\varphi} & 0 \\ 0 & 0 & 1 \\ b_2^{\varphi} & -b_2^{\varphi} & -b_1^{\varphi} \end{bmatrix} \begin{bmatrix} \theta \\ \varphi \\ \dot{\varphi} \end{bmatrix} + \begin{bmatrix} \frac{c_{3\text{strap-on}}^{\varphi}}{2} & -\frac{c_{3\text{strap-on}}^{\varphi}}{2} & c_{3\text{core}}^{\varphi} & -c_{3\text{core}}^{\varphi} \\ 0 & 0 & 0 & 0 \\ -\frac{b_{3\text{strap-on}}^{\varphi}}{2} & \frac{b_{3\text{strap-on}}^{\varphi}}{2} & -b_{3\text{core}}^{\varphi} & b_{3\text{core}}^{\varphi} \end{bmatrix} \cdot \begin{bmatrix} \delta_8 \\ \delta_6 \\ \delta_4 \\ \delta_2 \end{bmatrix} + \begin{bmatrix} c_1^{\prime\varphi} \\ 0 \\ -b_2^{\varphi} \end{bmatrix} \alpha_w + \begin{bmatrix} \bar{F}_{yc} \\ 0 \\ \bar{M}_z \end{bmatrix} \tag{2.10}$$

The one in the yawing channel is:

$$\begin{bmatrix} \dot{\sigma} \\ \dot{\psi} \\ \ddot{\psi} \end{bmatrix} = \begin{bmatrix} c_2^{\psi} - c_1^{\psi} & c_1^{\psi} & 0 \\ 0 & 0 & 1 \\ b_2^{\psi} & -b_2^{\psi} & -b_1^{\psi} \end{bmatrix} \begin{bmatrix} \sigma \\ \psi \\ \dot{\psi} \end{bmatrix} + \begin{bmatrix} \frac{c_{3\text{strap-on}}^{\psi}}{2} & -\frac{c_{3\text{strap-on}}^{\psi}}{2} & c_{3\text{core}}^{\psi} & -c_{3\text{core}}^{\psi} \\ 0 & 0 & 0 & 0 \\ -\frac{b_{3\text{strap-on}}^{\psi}}{2} & \frac{b_{3\text{strap-on}}^{\psi}}{2} & -b_{3\text{core}}^{\psi} & b_{3\text{core}}^{\psi} \end{bmatrix} \cdot \begin{bmatrix} \delta_7 \\ \delta_5 \\ \delta_3 \\ \delta_1 \end{bmatrix} + \begin{bmatrix} c_1^{\prime\psi} \\ 0 \\ -b_2^{\psi} \end{bmatrix} \beta_w + \begin{bmatrix} -\bar{F}_{zc} \\ 0 \\ \bar{M}_y \end{bmatrix} \tag{2.11}$$

And in the rolling channel is:

$$\begin{bmatrix} \dot{\gamma} \\ \ddot{\gamma} \end{bmatrix} = \begin{bmatrix} 0 & 1 \\ 0 & -d_1 \end{bmatrix} \begin{bmatrix} \gamma \\ \dot{\gamma} \end{bmatrix} + \begin{bmatrix} 0 & 0 \\ -d_{3\text{strap-on}} & -d_{3\text{core}} \end{bmatrix} \cdot \begin{bmatrix} \frac{1}{4}_{1\times 4} & 0 \\ 0 & \frac{1}{2}_{1\times 4} \end{bmatrix} \begin{bmatrix} \delta_5 \\ \vdots \\ \delta_8 \\ \delta_1 \\ \vdots \\ \delta_4 \end{bmatrix} + \begin{bmatrix} 0 \\ \bar{M}_x \end{bmatrix} \tag{2.12}$$

From the state equation representation of the launch vehicle, the oscillating angles of the core thrusts are δ_2 and δ_4, and the ones of strap-on thrusts are δ_6 and δ_8, which will give the thrust in pitching and rolling channels. On the other hand, the oscillating angles δ_1, δ_3, δ_5, and δ_7 of the core and strap-on thrusts support for the motion in yawing and rolling channels simultaneously.

Usually, the rolling motion is always very little and control ignored. Considering the situation that there exists the rolling motion of the launch vehicle, the motion related actuators will be interchanged during the flight process. The system will become coupled and complicated. Therefore, the condition of the following results of control reconfiguration strategies is to eliminate the rolling motion of the launch vehicle by the attitude controller. Then the states coupling barely existed between the rolling channel and other two channels.

2.4.2 Control Reconfiguration of One Engine Deadlock

As we discussed before, the gimballed thrusts consist of the power and actuation components that control the main engine's direction to steer the vehicle in the right direction.

Without loss of generality, we choose the pitching channel as an example for our analysis. Considering a fault scenario that the oscillating angle of the strap-on thrust No.2 (strap-on2 in Fig. 2.4) δ_6 is deadlocked in a constant value $\delta_6' = \tilde{\delta}_6$, the distinction between the deadlocked angle and normal angle is: $\Delta\delta_6 = \delta_6' - \delta_6 = \tilde{\delta}_6 - \delta_6$. From the state space representation of the dynamic model in Eqs. (2.10), (2.11), and (2.12), the oscillation angle δ_6 of the failed strap-on engine No. 2 will affect the motion in pitching and rolling channels simultaneously, except the yawing channel. To guarantee the vehicle's trajectory tracking precision, the adequate manipulation for ce and moment from the normal operated thrust is required to maintain the control authority and compensate the appended moments from the deadlocked engine, and the composite control moment remaining on the vehicle.

First of all, we propose the following parameters: the thrust of each core engine is N_c, with the relative oscillation angles: δ_1, δ_2, δ_3, and δ_4, and the thrust of each strap-on engine is N_s with the relative oscillation angles: δ_5, δ_6, δ_7, and δ_8. The distance from the oscillating axis of the engine to the center of mass is L, and the vertical distances from direct-axis to the strap-on and core thrusts are R and r, correspondingly. Since the oscillating angle is always less than eight degrees, the following equations can be obtained approximately:

$$\sin\delta \approx \delta \tag{2.13}$$

Then, the approximate control moments of gimballed thrusts to each axis in space are:

$$\begin{cases} M_{xc} = N_s(\delta_5 + \delta_6 + \delta_7 + \delta_8)R + N_c(\delta_1 + \delta_2 + \delta_3 + \delta_4)r \\ M_{yc} = [N_s(\delta_5 - \delta_7) + N_c(\delta_1 - \delta_3)]L \\ M_{zc} = [N_s(\delta_8 - \delta_6) + N_c(\delta_4 - \delta_2)]L \end{cases} \tag{2.14}$$

After that, assuming the reconfigured angles of the gimballed thrusts except failed strap-on2 are δ'_1 to δ'_8, which are the expected angles to keep the control moment. Then, balance equations of the control moment are:

$$\begin{cases} N_s(\delta_5 + \delta_6 + \delta_7 + \delta_8)R + N_c(\delta_1 + \delta_2 + \delta_3 + \delta_4)r = N_s(\delta'_5 + \tilde{\delta}_6 + \delta'_7 + \tilde{\delta}_8)R \\ +N_c(\delta'_1 + \delta'_2 + \delta'_3 + \delta'_4)r \\ [N_s(\delta_5 - \delta_7) + N_c(\delta_1 - \delta_3)]\,L = [N_s(\delta'_5 - \delta'_7) + N_c(\delta'_1 - \delta'_3)]L \\ [N_s(\delta_8 - \delta_6) + N_c(\delta_4 - \delta_2)]\,L = [N_s(\tilde{\delta}_8 - \tilde{\delta}_6) + N_c(\delta'_4 - \delta'_2)]L \end{cases} \tag{2.15}$$

Since the oscillating angle δ_6 would not influence the state of the yawing channel, the oscillating angles in the yawing channel satisfy: $\delta'_1 = \delta_1$, $\delta'_3 = \delta_3$, $\delta'_5 = \delta_5$, and $\delta'_7 = \delta_7$.

Thus, the Eq. (2.15) can be reduced as follows:

$$\begin{cases} N_s(\delta_8 + \delta_6 - \delta'_8 - \tilde{\delta}_6)R = N_c(\delta'_4 + \delta'_2 - \delta_4 - \delta_2)r \\ N_s(\delta_8 - \delta_6 - \delta'_8 + \tilde{\delta}_6) = N_c(\delta'_4 - \delta'_2 - \delta_4 + \delta_2) \end{cases} \tag{2.16}$$

Moreover, there is a proportional relation between the corresponded core thrusts and strap-on thrusts, which is denoted by a parameter k: $\delta_8 = k \cdot \delta_4$ and $\delta'_8 = k \cdot \delta'_4$.

$$\begin{cases} \dfrac{N_s}{N_c}\dfrac{R}{r}(\delta_8 + \delta_6 - \tilde{\delta}_6) + \delta_4 + \delta_2 = (1 + \dfrac{N_s}{N_c}\dfrac{R}{r}k)\delta'_4 + \delta'_2 \\ \dfrac{N_s}{N_c}(\delta_8 - \delta_6 + \tilde{\delta}_6) + \delta_4 - \delta_2 = (1 + \dfrac{N_s}{N_c}k)\delta'_4 - \delta'_2 \end{cases} \tag{2.17}$$

From the simultaneous equations above, the reconfigured oscillating angles should be regulated as follow, with the failed strap-on thrust No.2 deadlocked at $\tilde{\delta}_6$.

$$\begin{cases} \delta'_6 = \tilde{\delta}_6 = \delta_6 + \Delta\delta_6 \\ \delta'_8 = \delta_8 - \dfrac{N_s(R-r)k}{[2N_c r + N_s(R+r)k]}\Delta\delta_6 \\ \delta'_2 = \delta_2 - \dfrac{N_s}{N_c}\dfrac{N_c(R+r)+2N_s kR}{[2N_c r + N_s(R+r)k]}\Delta\delta_6 \\ \delta'_4 = \delta_4 - \dfrac{N_s(R-r)}{[2N_c r + N_s(R+r)k]}\Delta\delta_6 \end{cases} \tag{2.18}$$

In which, N_s, N_c, R, r, and k are the fundamental parameters and the constants given before. $\Delta\delta_6$ is a determined quantity to describe the deviation of the deadlocked engine.

The Eq. (2.18) denotes that when the strap-on thrust No. 2 is deadlocked at a certain angle, other normal working engines, such us the strap-on thrust No.4, core thrust No.2, and core thrust No.4 should regulate to a relative angles to maintain the control moment of the pitching and rolling channels. The regulative angles are relative to the scale of the launch vehicle R and r, the thrust power of strap-on thrusts N_s, and core thrusts N_c and the proportional relationship between the oscillating angles k. On the other hand, the oscillating angles of other four engines: strap-on1 and strap-on3, core1 and core3, should retain their state before.

The similar results of other engines will be achieved in yawing and rolling channels because of the symmetry of the vehicle.

2.4.3 Control Reconfiguration of Two Engines Deadlocked

From analysis of one thrust deadlocked situation, we find that the control reconfiguration is the independent and symmetric process between the pitching and yawing channels based on the no-rolling assumption. The symmetry of the launch vehicle makes the yawing channel identical to the pitching channel. Thus, we can separate the two thrusts deadlocked analysis into two situations:

(I) Two failed thrusts are in different channels
(II) Two failed thrusts are in the same channel.

In situation (I), it is obviously to obtain the result that we can establish two groups of simultaneous equations in pitching and yawing channels relatively to guarantee that the control moment of the launch vehicle is kept invariable. Then the solutions of the equations are the reconfigured oscillating angles we need. So, we emphasize on the second situation that two deadlocked thrusts are in the same channel.

Considering the pitching channel without loss of generality, we choose strap-on2 and strap-on4 corresponding to the oscillating angles δ_6 and δ_8 for example. They are assumed to be deadlocked at the certain angle $\tilde{\delta}_6$ and $\tilde{\delta}_8$. Thus,

$$\begin{aligned} \delta_6' &= \tilde{\delta}_6 = \delta_6 + \Delta\delta_6 \\ \delta_8' &= \tilde{\delta}_8 = \delta_8 + \Delta\delta_8 \end{aligned} \tag{2.19}$$

Similarly, based on the assumption of Eq. (2.13), the corresponding balance relation of the control moments satisfy the Eq. (2.14) as well. Then, substitute the failure occurred angle $\tilde{\delta}_6$ and $\tilde{\delta}_8$ in Eq. (2.14) by Eq. (2.19).

$$\begin{cases} N_s(\delta_5+\delta_6+\delta_7+\delta_8)R + N_c(\delta_1+\delta_2+\delta_3+\delta_4)r = N_s(\delta'_5+\tilde{\delta}_6+\delta'_7+\tilde{\delta}_8)R \\ +N_c(\delta'_1+\delta'_2+\delta'_3+\delta'_4)r \\ \left[N_s(\delta_5-\delta_7)+N_c(\delta_1-\delta_3)\right]L = [N_s(\delta'_5-\delta'_7)+N_c(\delta'_1-\delta'_3)]L \\ [N_s(\delta_8-\delta_6)L+N_c(\delta_4-\delta_2)]\,L = [N_s(\tilde{\delta}_8-\tilde{\delta}_6)+N_c(\delta'_4-\delta'_2)]L \end{cases} \tag{2.20}$$

The oscillating angles performing in the yawing channel: δ_1, δ_3, δ_5, and δ_7 can be removed.

$$\begin{cases} N_s(\delta_8+\delta_6-\tilde{\delta}_8-\tilde{\delta}_6)R = N_c(\delta'_4+\delta'_2-\delta_4-\delta_2)r \\ N_s(\delta_8-\delta_6-\tilde{\delta}_8+\tilde{\delta}_6) = N_c(\delta'_4-\delta'_2-\delta_4+\delta_2) \end{cases} \tag{2.21}$$

However, because the strap-on thrusts are both deadlocked, the proportion relation k between strap-on and core thrust is noneffective.

Solving the Eq. (2.21), we can obtain the reconfigured oscillating angles of the core engines in pitching channel:

$$\begin{cases} \delta'_2 = \delta_2 - \frac{N_s}{2N_c}\left(\frac{R}{r}+1\right)\Delta\delta_6 - \frac{N_s}{2N_c}\left(\frac{R}{r}-1\right)\Delta\delta_8 \\ \delta'_4 = \delta_4 - \frac{N_s}{2N_c}\left(\frac{R}{r}-1\right)\Delta\delta_6 - \frac{N_s}{2N_c}\left(\frac{R}{r}+1\right)\Delta\delta_8 \end{cases} \tag{2.22}$$

Therefore, the normal operative thrusts need to be regulated according to Eq. (2.22) to compensate the control force by the failed thrusts to maintain the competent operation for attitude manipulation.

Moreover, consider another situation that the deadlocked thrusts belongs to strap-on booster and core booster, separately. Assume they are strap-on2 and core2, for instance. So, we have the fault condition description in Eq. (2.23)

$$\begin{cases} \delta_2' = \tilde{\delta}_2 = \delta_2 + \Delta\delta_2 \\ \delta_6' = \tilde{\delta}_6 = \delta_6 + \Delta\delta_6 \end{cases} \tag{2.23}$$

And the regulated equations are:

$$\begin{cases} N_s(\delta_8+\delta_6-\delta'_8-\tilde{\delta}_6)R = N_c(\delta'_4+\tilde{\delta}_2-\delta_4-\delta_2)r \\ N_s(\delta_8-\delta_6-\delta'_8+\tilde{\delta}_6) = N_c(\delta'_4-\tilde{\delta}_2-\delta_4+\delta_2) \end{cases} \tag{2.24}$$

Then solving the equations above, we have:

$$\begin{cases} \delta'_4 = \delta_4 + \frac{2N_s R}{N_c(R-r)}\Delta\delta_6 + \frac{R+r}{R-r}\Delta\delta_2 \\ \delta'_8 = \delta_8 - \frac{R+r}{R-r}\Delta\delta_6 - \frac{2N_c r}{N_s(R-r)}\Delta\delta_2 \end{cases} \tag{2.25}$$

The control reconfiguration strategy is obtained.

Finally, consider about the situation that the deadlocked thrusts are strap-on2 and core4.

$$\begin{cases} \delta'_4 = \tilde{\delta}_4 = \delta_4 + \Delta\delta_4 \\ \delta'_6 = \tilde{\delta}_6 = \delta_6 + \Delta\delta_6 \end{cases} \tag{2.26}$$

The regulated equations are:

$$\begin{cases} N_s(\delta_8 + \delta_6 - \delta'_8 - \tilde{\delta}_6)R = N_c(\tilde{\delta}_4 + \delta'_2 - \delta_4 - \delta_2)r \\ N_s(\delta_8 - \delta_6 - \delta'_8 + \tilde{\delta}_6) = N_c(\tilde{\delta}_4 - \delta'_2 - \delta_4 + \delta_2) \end{cases} \tag{2.27}$$

The solution of the equations is:

$$\begin{cases} \delta'_2 = \delta_2 + \frac{R-r}{R+r}\Delta\delta_4 - \frac{2N_s R}{N_c(R+r)}\Delta\delta_6 \\ \delta'_8 = \delta_8 - \frac{2N_c r}{N_s(R+r)}\Delta\delta_4 - \frac{R-r}{R+r}\Delta\delta_6 \end{cases} \tag{2.28}$$

2.4.4 Control Reconfiguration of More Than Two Engines Deadlocked

From the state space equations of the launch vehicle in Eqs. (2.10) and (2.11), four engines are implemented to provide the thrust power and control its attitude for the pitching and yawing channel, respectively. First, if there are three thrusts failed in the same channel, the vehicle will be a single-input but multi-output system. In this situation, the lack of control process will break the equilibrium of the moment in the corresponding channel and produce the trend of rotational motion. Then, the attitude of the vehicle will be unable to regulate. Therefore, it is forbidden that there are three engines deadlocked at the same time in the same channel. Obviously, it is meaningless if more than three thrusts are failed in the six engines driving the launch vehicle. The critical situation that the vehicle is recoverable is two thrusts failed in the same channel.

After that, considering the situation that more than two thrusts deadlocked in the different channels, the control reconfiguration strategy could be composed by the results of one engine or two engines deadlocked that we discussed before. The concrete analysis in detail is omitted temporarily here.

2.5 Stability Analysis

In the pitching channel for instance, from the analysis in the Appendix, the parameter $b_2^{\varphi} = m_z^{\alpha} q S_m l_k / J_z$ and its physics derivation is obtained obviously. Here, $m_z^{\alpha} = C_{y1}^{\alpha}(x_{\text{press}} - x_{\text{mass}})/l_k$ is called the Aerodynamic Moment Coefficient (or Static Stability Moment Coefficient), which presents the static stability of the vehicle by the distribution of the press center and mass center. We put m_z^{α} into the expression of b_2^{φ} here, to find its implication of the stability of the vehicle.

$$b_2^{\varphi} = \frac{m_z^{\alpha} q S_m l_k}{J_z} = \frac{C_{y1}^{\alpha}(x_{\text{press}} - x_{\text{mass}}) q S_m}{J_z} \tag{2.29}$$

The following analysis is available.

Lemma 1 *Assume the lift coefficient C_{y1}^{α}, the character area S_m and the moment of inertia J_z are the positive constants. The velocity head $q = \frac{1}{2}\rho V_w^2 \geq 0$ with the atmospheric density $\rho > 0$.*

Using a positive constant K to indicate the product of the coefficients above:

$$K = \frac{C_{y1}^{\alpha} q S_m}{J_z} \geq 0 \tag{2.30}$$

Then, we have:

$$b_2^{\varphi} = K(x_{\text{press}} - x_{\text{mass}}) \tag{2.31}$$

While $b_2^{\varphi} > 0$, $x_{\text{press}} > x_{\text{mass}}$. In this way, the press center is at the lower place of the vehicle than the mass center. Then, the attack angle will be decreased autonomously by the aerodynamic moment without the control and the vehicle is asymptotic stable.

On the other hand, if $b_2^{\varphi} < 0$, $x_{\text{press}} < x_{\text{mass}}$. The press center is at the upper place of the vehicle than the mass center. Then, the vehicle is static unstable without control under the aerodynamic moment of force.

Based on the Lemma 1, the sign of the system coefficient b_2^{φ} denotes the inherent aerodynamic stability of the launch vehicle. The dynamic stability of the launch vehicle is studied from the comprehensive influences of aerodynamic, air dam, gravitational force, etc.

Considering the pitching channel for instance, the system has two input signal as the oscillating angle from the strap-on thrusts $\delta_{\text{strap-on}}^{\varphi}$ and core thrusts $\delta_{\text{core}}^{\varphi}$. Considering the relation of the input and output is $\Delta\varphi = G_1 \Delta\delta_{\text{strap-on}}^{\varphi} + G_2 \Delta\delta_{\text{core}}^{\varphi}$. The transfer functions G_1 and G_2 of the system are:

$$G_1 = \frac{\Delta\varphi}{\Delta\delta_{\text{strap-on}}^{\varphi}} = \frac{q_3 s^3 + q_2 s^2 + q_1 s + q_0}{s^3 + p_2 s^2 + p_1 s + p_0} \tag{2.32}$$

with the parameters:

$$
\begin{aligned}
q_0 &= b_2^{\varphi} c_{3\text{strap-on}}^{\varphi} + (c_2^{\varphi} - c_1^{\varphi}) b_{3\text{strap-on}}^{\varphi} \\
q_1 &= -b_{3\text{strap-on}}^{\varphi} \\
q_2 &= b_{3\text{strap-on}}^{\prime\varphi}(c_2^{\varphi} - c_1^{\varphi}) + b_2^{\varphi} c_{3\text{strap-on}}^{\prime\varphi} \\
q_3 &= -b_{3\text{strap-on}}^{\prime\varphi} \\
p_0 &= -b_2^{\varphi} c_2^{\varphi} \\
p_1 &= b_2^{\varphi} + c_1^{\varphi} b_1^{\varphi} - b_1^{\varphi} c_2^{\varphi} \\
p_2 &= b_1^{\varphi} + c_1^{\varphi} - c_2^{\varphi}
\end{aligned}
$$

and

$$
G_2 = \frac{\Delta\varphi}{\Delta\delta_{\text{core}}^{\varphi}} = \frac{q_3 s^3 + q_2 s^2 + q_1 s + q_0}{s^3 + p_2 s^2 + p_1 s + p_0} \tag{2.33}
$$

with the parameters:

$$
\begin{aligned}
q_0 &= b_2^{\varphi} c_{3\text{core}}^{\varphi} + (c_2^{\varphi} - c_1^{\varphi}) b_{3\text{core}}^{\varphi} \\
q_1 &= -b_{3\text{core}}^{\varphi} \\
q_2 &= b_{3\text{core}}^{\prime\varphi}(c_2^{\varphi} - c_1^{\varphi}) + b_2^{\varphi} c_{3\text{core}}^{\prime\varphi} \\
q_3 &= -b_{3\text{core}}^{\prime\varphi} \\
p_0 &= -b_2^{\varphi} c_2^{\varphi} \\
p_1 &= b_2^{\varphi} + c_1^{\varphi} b_1^{\varphi} - b_1^{\varphi} c_2^{\varphi} \\
p_2 &= b_1^{\varphi} + c_1^{\varphi} - c_2^{\varphi}
\end{aligned}
$$

Study the eigen-equation of the related transfer functions:

$$
D(s) = s^3 + (b_1^{\varphi} + c_1^{\varphi} - c_2^{\varphi}) s^2 + (b_2^{\varphi} + c_1^{\varphi} b_1^{\varphi} - b_1^{\varphi} c_2^{\varphi}) s - b_2^{\varphi} c_2^{\varphi}
$$

and the attitude stability of the system in the typical phase of flight.

2.5.1 Rocket Taking-Off

At this moment, the air damping coefficient $b_1^{\varphi} \approx 0$. The eigen-equation of the system can be simplified approximately as:

$$
D(s) \approx s^3 + (c_1^{\varphi} - c_2^{\varphi}) s^2 + b_2^{\varphi} s - b_2^{\varphi} c_2^{\varphi}
$$

Study the coefficients of the eigen-equation. The thrust and aerodynamic resistance related coefficient $c_1^{\varphi} = \frac{(2P_{\text{core}} + 4P_{\text{strap-on}})\cos\alpha_0 + C_y^{\alpha} q S_m}{mV}$ and the gravity-related coefficient $c_2^{\varphi} = \frac{g \sin\theta_0}{V}$.

Consider the discrepancy:

$$c_1^{\varphi} - c_2^{\varphi} = \frac{(2P_{\text{core}}+4P_{\text{strap-on}})\cos\alpha_0 + C_y^{\alpha} q S_m - mg\sin\theta_0}{mV} > 0$$

and the coefficient $c_2^{\varphi} > 0$, however, the other two coefficients b_2^{φ} and $-b_2^{\varphi} c_2^{\varphi}$ cannot be positive at the same time. The eigen-equation will have positive root, and therefore, the system has unstable components, described as follow.

If $b_2^{\varphi} < 0$, the vehicle is static unstable by aerodynamic moment. Assume the eigen-equation can be transformed to following polynomial multiplication form.

$$D(s) = (s + c_1^{\varphi} - c_2^{\varphi})(s^2 + 2\xi\omega s + \omega^2)$$

in which, $\omega = \sqrt{-\frac{b_2^{\varphi} c_2^{\varphi}}{c_1^{\varphi} - c_2^{\varphi}}}$, $\xi\omega = \frac{b_2^{\varphi} c_1^{\varphi}}{2(c_1^{\varphi} - c_2^{\varphi})^2} < 0$.

Therefore, there is a pair of complex roots in the right half s plane. The attitude motion is oscillating divergent.

If $b_2^{\varphi} > 0$, the vehicle is static stable. The eigen-equation can be decomposed as:

$$D(s) = (s + c_1^{\varphi} - c_2^{\varphi})\left[s - \frac{b_2^{\varphi} c_1^{\varphi}}{2(c_1^{\varphi} - c_2^{\varphi})^2} + \sqrt{\frac{b_2^{\varphi} c_2^{\varphi}}{c_1^{\varphi} - c_2^{\varphi}}}\right] \cdot \left[s + \frac{b_2^{\varphi} c_1^{\varphi}}{2(c_1^{\varphi} - c_2^{\varphi})^2} - \sqrt{\frac{b_2^{\varphi} c_2^{\varphi}}{c_1^{\varphi} - c_2^{\varphi}}}\right]$$

in which, $\frac{b_2^{\varphi} c_1^{\varphi}}{2(c_1^{\varphi} - c_2^{\varphi})^2} - \sqrt{\frac{b_2^{\varphi} c_2^{\varphi}}{c_1^{\varphi} - c_2^{\varphi}}} < 0$

Therefore, there will be a positive real root in the right half s plane, and the attitude motion will be monotonic divergent.

2.5.2 With Maximum Aerodynamic Moment Coefficient $|m_z^{\alpha}|$

Since we have: $b_2^{\varphi} = \frac{m_z^{\alpha} q S_m l_k}{J_z}$, the coefficient $|b_2^{\varphi}|$ is maximum as well.

The eigen-equation is:

$$\begin{aligned} D(s) &= s^3 + (b_1^{\varphi} + c_1^{\varphi} - c_2^{\varphi})s^2 + (b_2^{\varphi} + c_1^{\varphi} - b_1^{\varphi} c_2^{\varphi})s - b_2^{\varphi} c_2^{\varphi} \\ &\approx s^3 + (b_1^{\varphi} - c_2^{\varphi})s^2 + (b_2^{\varphi} - b_1^{\varphi} c_2^{\varphi})s - b_2^{\varphi} c_2^{\varphi} \\ &\approx (s - c_2^{\varphi})(s^2 + b_1^{\varphi} s + b_2^{\varphi}) \end{aligned}$$

If $b_2^{\varphi} > 0$, vehicle is static stable.

$$s^2 + b_1^{\varphi} s + b_2^{\varphi} = s^2 + 2\xi\omega + \omega^2$$

in which $\omega = \sqrt{b_2^{\varphi}}$, $\xi\omega = \frac{b_1^{\varphi}}{2} > 0$.

Thus, the eigen-equation has a real root $s = c_2^{\varphi}$ in the right half s plane. The attitude motion is monotonic divergent slowly because of the gravity.

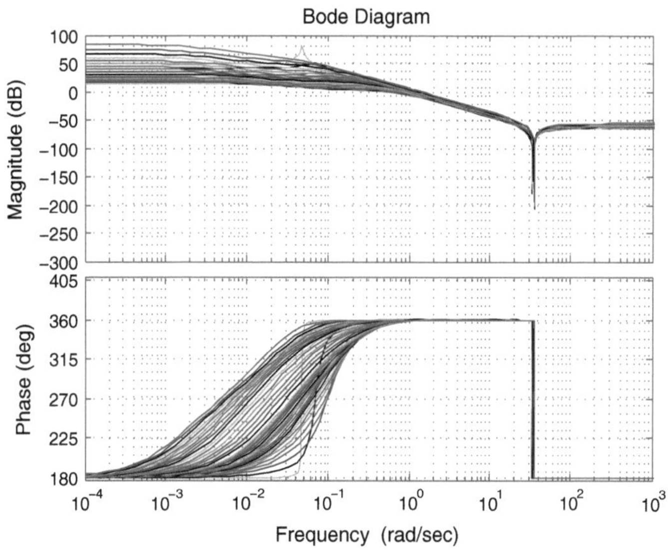

Fig. 2.5 Bode diagram before correction

If $b_2^\varphi < 0$, vehicle is static unstable.

$$s^2 + b_1^\varphi s + b_2^\varphi = s^2 + 2\xi\omega + \omega^2$$
$$\approx \left(s - \sqrt{|b_2^\varphi|}\right)\left(s + \sqrt{|b_2^\varphi|}\right)$$

There are two real roots $s = c_2^\varphi$ and $s = \sqrt{|b_2^\varphi|}$ in the right half s plane. The system will be divergent quickly by the effect of aerodynamic force.

2.5.3 First-Stage Engines Power-Off

Since the air density is very slight at the hight that the first stage thrusts are power-off, the aerodynamic moment can be ignored and the coefficients b_1^φ and b_2^φ are negligible. The simplified eigen-equation is:

$$D(s) \approx s^3 + (c_1^\varphi - c_2^\varphi)s^2 = s^2(s + c_1^\varphi - c_2^\varphi)$$

There are a couple of poles $s = 0$ in the original point, and another single pole $s = -(c_1^\varphi - c_2^\varphi) < 0$ in the left half plane. The system is marginally stable, which will be divergent with the external disturbance.

We can draw the Bode diagram and the Nyquist curve to indicate the stability of the certain type of the vehicle in Figs. 2.5 and 2.6.

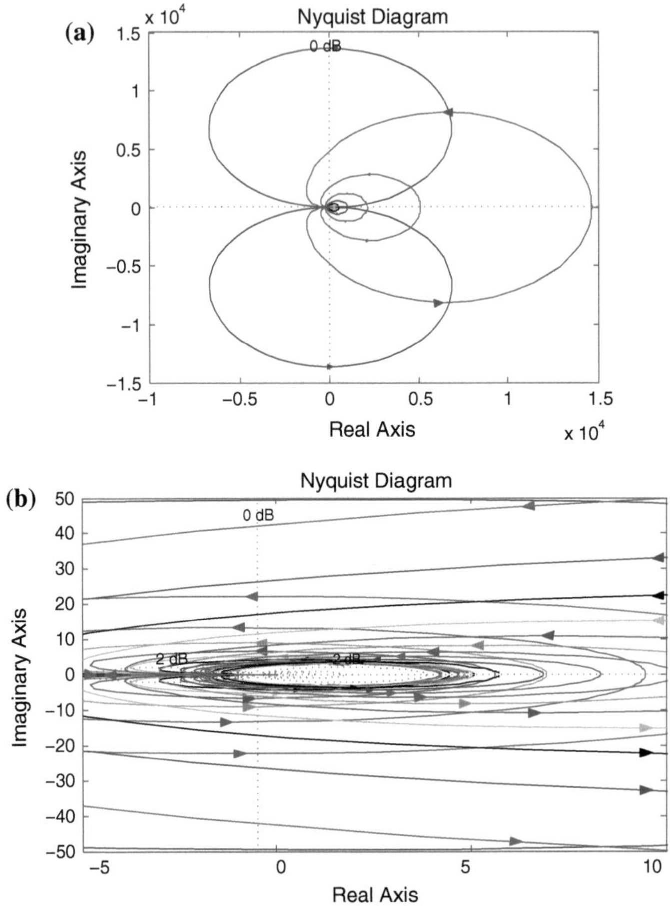

Fig. 2.6 Nyquist curves before correction and partial enlargement at $(-1, 0)$

Obviously, the system is inherent unstable. Thus, an error feedback based attitude controller by the classic frequency correction method should be designed to stabilize the system. Then, we can plot the bode diagram to indicate the magnitude-phase characteristics of the corrected system (see Fig. 2.7), and the Nyquist curve after frequency-domain correction (see Fig. 2.8) as well.

Moreover, the step-response is proposed to investigate the static and dynamic property of the controller. Obviously, the system can be stabilized in 1.5 s and the overshoot is less than 20 % (see Fig. 2.9).

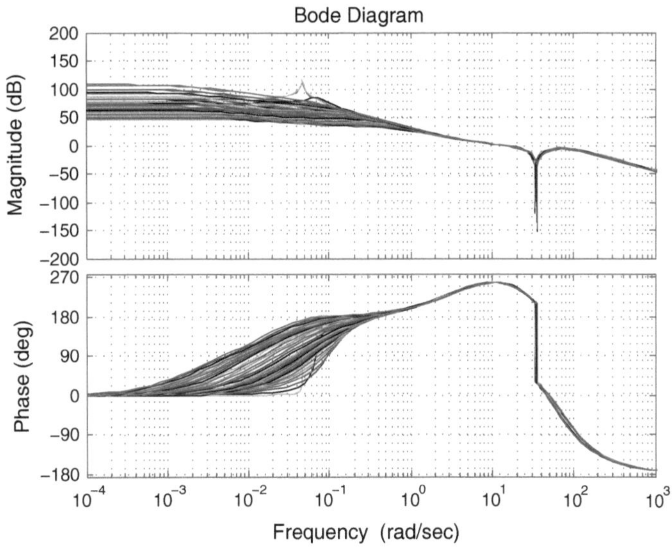

Fig. 2.7 Bode diagram after correction

2.6 Reconfigurability Analysis

The above research is to find the proper compensation strategies to regulate the oscillating angles for control reconfiguration of the launch vehicle, when the thrusts are deadlocked at determined angles. However, for a physical system, the oscillation angle of the servomechanism is always bounded. Thus, in some limit cases, the above reconfiguration strategy cannot achieve the respected compensation because of its physics restriction. Therefore, in this section, rationally, the reconfigurability of the compensation strategy is studied under the certain numeral assumptions.

To simplify the discussion, the following assumption is proposed before the reconfigurability analysis:

Assumption 6.1 *Consider a certain kind of the launch vehicle with:*

(I) thrust power of the strap-on engines and core engines: $N_s = N_c$*;*
(II) proportion relation between the oscillating angle of strap-on and core engines: k=*1;*
(III) radius of the strap-on engine cluster circle and core engines cluster circle: R=*3*r.

Based on the Assumption 6.1 above, we consider the situation that deadlock occurs on the strap-on thrust No. 2 for instance, so $\delta_6' = \tilde{\delta}_6$.

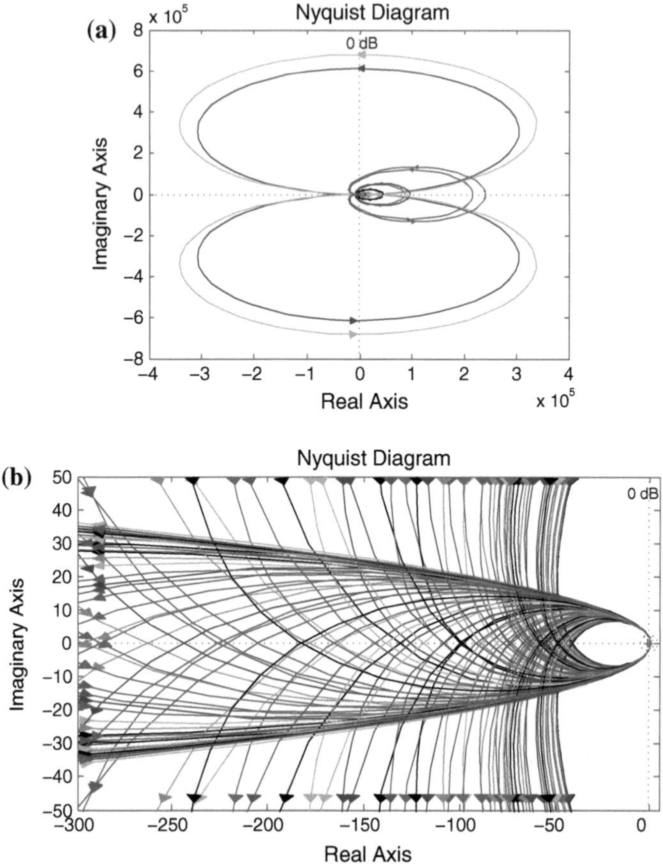

Fig. 2.8 Nyquist curves after correction (**a**) and partial enlargement (**a**) at $(-1, 0)$

From the Eq. (2.18) and the assumptions, we have:

$$\begin{cases} \delta'_2 = \delta_2 - \frac{5}{3}\Delta\delta_6 \\ \delta'_4 = \delta_4 - \frac{1}{3}\Delta\delta_6 \\ \delta'_6 = \tilde{\delta}_6 = \delta_6 + \Delta\delta_6 \\ \delta'_8 = \delta_8 - \frac{1}{3}\Delta\delta_6 \end{cases} \tag{2.34}$$

Give the bounds of the oscillating angle of both strap-on and core engines are:

$$-\overline{\delta} \leq \delta_i \leq \overline{\delta} \quad \text{(with } i = 1, 2, 3, \ldots, 8). \tag{2.35}$$

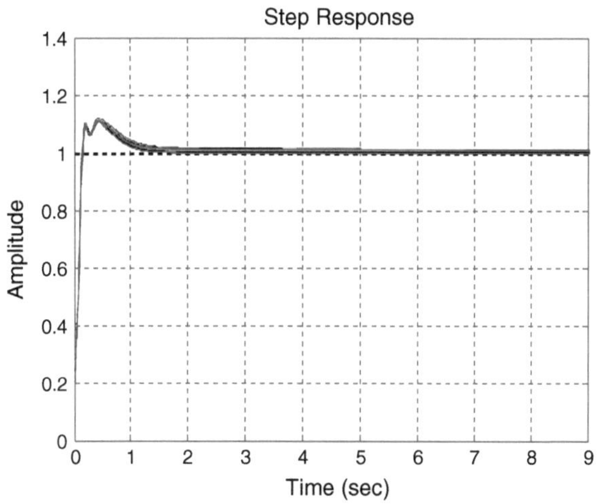

Fig. 2.9 Step-response curves of the system

Theoretically, the maximum disparity of the angle $\Delta\delta_i = \delta_i - \delta_i'$ satisfies:

$$-2\bar{\delta} \leq \Delta\delta_i \leq 2\bar{\delta} \tag{2.36}$$

Substitute the disparity between δ_2 and δ_2' in Eq. (2.34) by Eq. (2.36), with $i = 2, 4, 8$.

$$\begin{cases} -2\bar{\delta} \leq \frac{1}{3}\Delta\delta_6 \leq 2\bar{\delta} \\ -2\bar{\delta} \leq \frac{5}{3}\Delta\delta_6 \leq 2\bar{\delta} \\ -2\bar{\delta} \leq \frac{1}{3}\Delta\delta_6 \leq 2\bar{\delta} \end{cases} \tag{2.37}$$

and then we have:

$$-1.2\bar{\delta} \leq \Delta\delta_6 \leq 1.2\bar{\delta} \tag{2.38}$$

Thus, if the disparity between the deadlocked angle of the failed thrust and its required angle is bigger than $|1.2\bar{\delta}|$, the reconfiguration strategies are unavailable for the control action compensation.

On the other hand, if the disparity angle of the failed actuator is within the interval of $\left[-1.2\bar{\delta}, 1.2\bar{\delta}\right]$, because the oscillation angle satisfies the inequality (2.35), the reconfigurable interval of the deadlocked angle should be within the range of $\left[-0.2\bar{\delta}, 0.2\bar{\delta}\right]$ to guarantee the reliability of the compensation.

So, the reconfigurability condition of the oscillation angle δ_6' deadlock is:

$$-0.2\bar{\delta} \leq \delta_6' \leq 0.2\bar{\delta} \tag{2.39}$$

Because of the symmetry, the same result will be obtained when one of other oscillation engines is deadlocked.

Second, the situation of two thrusts in the certain channel deadlocked at the same time is considered. Supposing that the booster thrusts strap-on2 and strap-on4 are deadlocked corresponding to the oscillating angles δ_6 and δ_8, the Eq. (2.22) can be transformed based on the Assumption 6.1:

$$\begin{cases} \delta'_2 = \delta_2 - 2\Delta\delta_6 - \Delta\delta_8 \\ \delta'_4 = \delta_4 - \Delta\delta_6 - 2\Delta\delta_8 \end{cases} \tag{2.40}$$

Then, a set of inequalities is available from Eq. (2.36):

$$\begin{cases} -2\bar{\delta} \le 2\Delta\delta_6 + \Delta\delta_8 \le 2\bar{\delta} \\ -2\bar{\delta} \le \Delta\delta_6 + 2\Delta\delta_8 \le 2\bar{\delta} \end{cases} \tag{2.41}$$

Therefore, from the inequalities (2.41) above, the reconfigurability conditions for the composed oscillating angle δ_6 and δ_8 when two thrusts strap-on2 and strap-on4 are failed simultaneously.

$$-\tfrac{4}{3}\bar{\delta} \le \Delta\delta_6 + \Delta\delta_8 \le \tfrac{4}{3}\bar{\delta} \tag{2.42}$$

Obviously, the control operation can only be reconfigured when the error of the composed oscillating angle is not bigger than $\frac{4}{3}\bar{\delta}$. Thus, to guarantee the control reconfigurability of the vehicle, the failed thrusts should not be deadlocked at the fringe place. For example, if $\Delta\delta_6 = \Delta\delta_8 = \bar{\delta}$, the inequalities (2.41) are unsubstantiated.

Similarly, we can derive the same results when thrusts strap-on2 and core2 are deadlocked, and thrusts strap-on2 and core4 are deadlocked.

2.7 Simulation Result

We have studied a launch vehicle system which can be described by the dynamic model in Eqs. (2.5), (2.6) and (2.7) to demonstrate the performance of the control reconfiguration strategies while a helm of an engine is deadlocked.

First of all, an attitude controller is designed to stabilize the program angle tracking control. In this structure, the control commands produced from the controller are the equivalent oscillating angles: δ_φ, δ_ψ and δ_γ. However, the physical actuators are the oscillating angles of the helms of four strap-on thrusts and two core thrusts: δ_1 to δ_8 in Fig. 2.3. Thus, an angle composing and decomposing block of the oscillating engines is required between the controller and launch vehicle to transform the equivalent angles to the oscillating angles of the physical actuators. In this way, when the deadlocking fault is occurred, the above control reconfiguration strategy will be

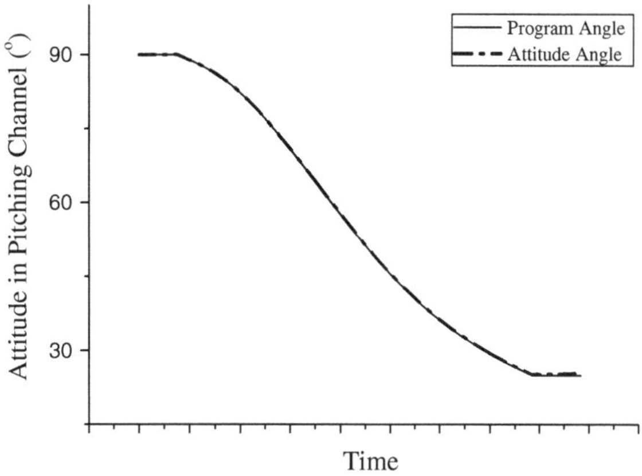

Fig. 2.10 Attitude control in pitching channel

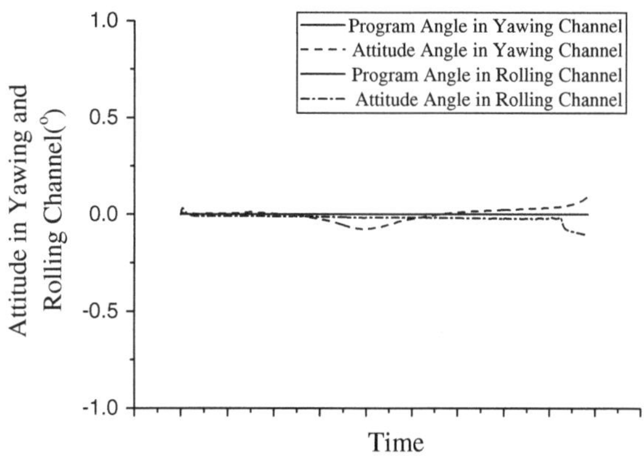

Fig. 2.11 Attitude control in yawing and rolling channel

operated to reconfigure the control efforts of the gimballed thrust that still works. Furthermore, a disturbance compensation part is developed to decrease the effect of the general disturbance (such as wind and air damping) and perturbation (such as model uncertainties). The attitude control and fault-tolerance structure in our simulation is shown in Fig. 2.4.

Then, assuming the motion of the launch vehicle satisfies the following assumptions: the vehicle is launched vertically and then turns to about 30° in pitching channel in the certain program turning time period, and meanwhile, other two channels of Yawing and Rolling should maintain 0° attitude angle under the influence of inside and outside disturbances.

The curves in Figs. 2.10 and 2.11 compare the program angles and the flight attitude angles with disturbance by using the designed attitude controller, in which the solid curves denote the program angles in three channels given by the guidance process and the dash curves and dot dash curves are the flight attitude angles from simulation. Thereby, although the disturbance can cause some turbulence to the

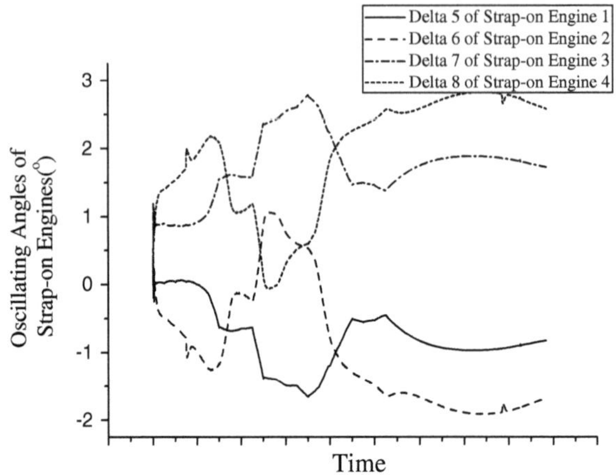

Fig. 2.12 Curve of the oscillating angles of strap-on engines without deadlocked engine

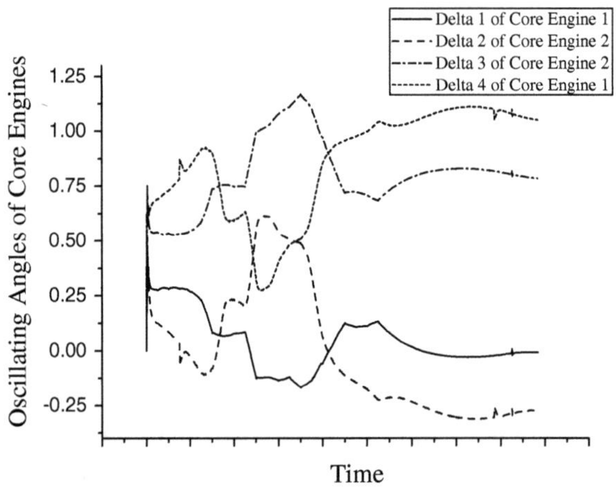

Fig. 2.13 Curves of the oscillating angles of core thrusts without deadlocked thrust

rocket flight, the attitudes of the vehicle can follow the program angles precisely with the errors of about $\pm 0.1°$.

Based on the attitude controller, we add the control reconfiguration strategies into the system to study the fault-tolerance ability of the system. Here we assume that the fault occurs randomly, and then stays at a certain angle. In order to test the command angle of each engine, we give the dynamic tracking curves of the oscillating angles of both strap-on and core engines with and without the deadlocked fault.

The command angles of oscillation without the deadlocked engine are shown in Figs. 2.12 and 2.13.

It is obvious that the command angles of oscillating angles of each engine are in couple in order to counteract the effect to the rolling channel. Moreover, the command

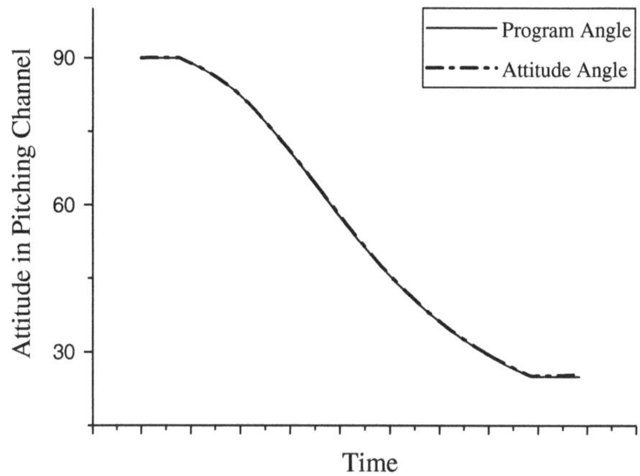

Fig. 2.14 The pitching control results with one deadlocked thrust

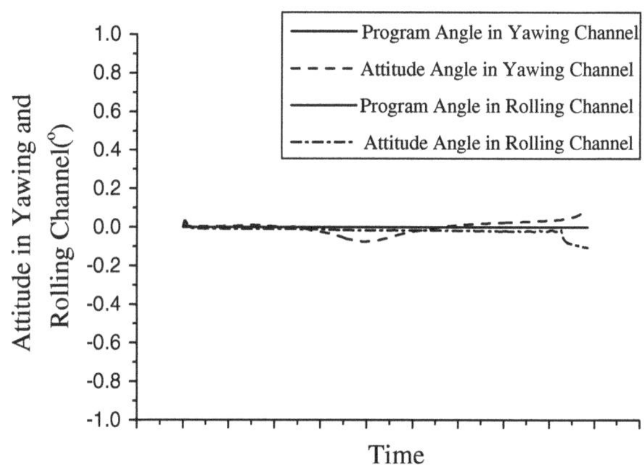

Fig. 2.15 The yawing and rolling control result with one deadlocked thrust

angles of the thrusts are all less than 3°, which satisfies the physical limitation 10° of the helm mechanism.

After that, considering the faulty situation, assuming that the strap-on thrust II is deadlocked randomly, we choose a random deadlocked time point and simulate the situation that one strap-on2 is deadlocked. First, the attitude control results with one damaged thrust is shown in Figs. 2.14 and 2.15, comparing with the original control performance in Figs. 2.10 and 2.11. We can see that the attitude angles in pitching, yawing, and rolling channel can track the program angles accurately with one thrust deadlocked.

Then, we displace the command angles of both strap-on and core thrusts in Figs. 2.16 and 2.17. In Fig. 2.16, the position feedback of the servomechanism shows

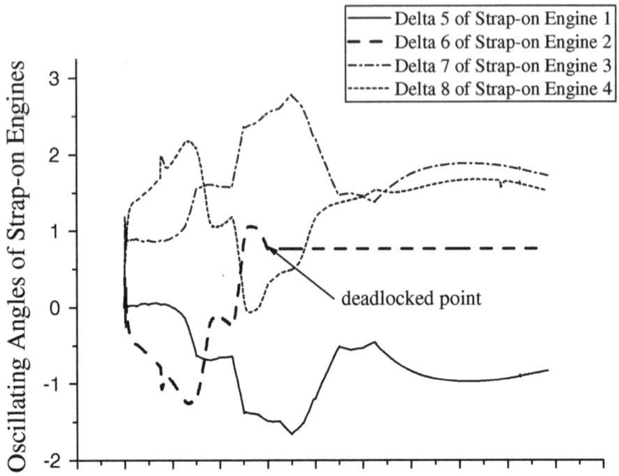

Fig. 2.16 Curve of the oscillating angles of strap-on thrusts with one deadlocked thrust

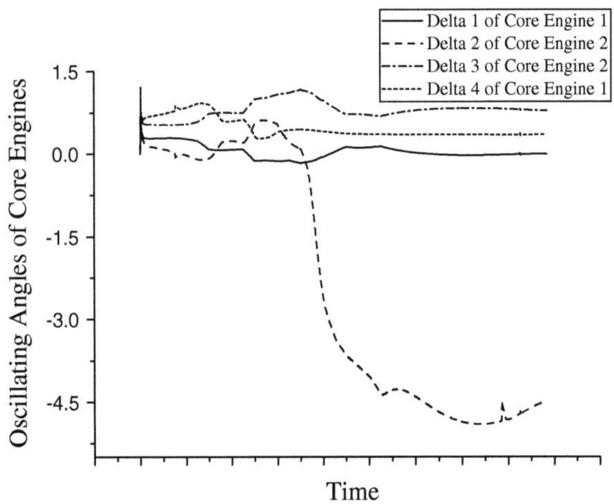

Fig. 2.17 Curves of the oscillating angles of core thrusts with one deadlocked thrust

that the strap-on thrust II is deadlocked at about $\delta_6 = 0.8°$. Then its corresponding angle δ_8 (Delta 8) affords more to control the attitude to counteract the deadlocked angle.

Similarly, the oscillating angles of the core thrusts contribute in the fault compensation as well. The oscillating angle δ_2 (Delta 2) of core thrust 2 and its corresponding angle δ_4 (Delta 4) in the pitching channel, are changing simultaneously while δ_6 (Delta 6) is deadlocked. Because of the symmetry between δ_2 and δ_6, δ_2 will change more than the other thrusts.

2.8 Conclusion

A control reconfiguration FTC strategy is introduced in this chapter for the new generation "CZ" launch vehicle in China under the no rolling movement assumption. The attitude controller is designed by the frequency method based on its dynamic model to achieve the stable movement of the vehicle without fault. From the analysis of the core and strap-on thrusts, which involve in the control in our new generation vehicle, the redundancy in control of the thrusts provides the additional manipulation capability to the vehicle with damaged thrusts. After that, the reconfiguration strategies are proposed for several conventional scenarios of thrusts deadlocked, which is real-time operated but without increasing the robustness of the system. It can be improved in the attitude controller design in our further work. However, we find that the vehicle motion is unable to be recovered in some extreme situations. So, the reconfigurability analysis is given to obtain the recoverable conditions of the vehicle, which is very harsh when more than one thrust is deadlocked. Finally, the chapter gives a set of simulation results, including the real-time attitude tracking control in three channels and the oscillating angles of all thrusts, with and without the fault occurred. It is illustrated that, the reconfiguration strategy has achieved the regulation of the control operation between the failed thrusts and normal thrusts to maintain the adequate manipulation in attitude control.

Acknowledgments This work is supported in part by the National Nature Science Foundation of China No 61203081 and 61174079, the Doctoral Fund of Ministry of Education of China NO.20120142120091, and the Precision manufacturing technology and equipment for metal parts NO.2012DFG70640.

Appendix: Dynamic Modeling Analysis

From the dynamics of the motion of the mass center and the rotation around the mass center in three axis directions x, y, and z.

$$\begin{cases} \delta m \frac{\mathrm{d}\mathbf{V}}{\mathrm{d}t} = \sum \mathbf{F}_x = \mathbf{F}_{Gx} + \mathbf{F}_{Px} + \mathbf{F}_{Ax} + \mathbf{F}_{Cx} + \mathbf{F}_{Ex} + \mathbf{F}_x \\ \delta m \mathbf{V} \frac{\mathrm{d}\theta}{\mathrm{d}t} = \sum \mathbf{F}_y = \mathbf{F}_{Gy} + \mathbf{F}_{Py} + \mathbf{F}_{Ay} + \mathbf{F}_{Cy} + \mathbf{F}_{Ey} + \mathbf{F}_y \\ -\delta m \mathbf{V} \frac{\mathrm{d}\sigma}{\mathrm{d}t} = \sum \mathbf{F}_z = \mathbf{F}_{Gz} + \mathbf{F}_{Pz} + \mathbf{F}_{Az} + \mathbf{F}_{Cz} + \mathbf{F}_{Ez} + \mathbf{F}_z \\ \delta J_x \frac{\mathrm{d}\omega_x}{\mathrm{d}t} + (\delta J_z - \delta J_y)\omega_z\omega_y = \sum \mathbf{M}_x = \mathbf{M}_{Ax} + \mathbf{M}_{Cx} + \mathbf{M}_{Ex} + \mathbf{M}_x \\ \delta J_y \frac{\mathrm{d}\omega_y}{\mathrm{d}t} + (\delta J_x - \delta J_z)\omega_x\omega_z = \sum \mathbf{M}_y = \mathbf{M}_{Ay} + \mathbf{M}_{Cy} + \mathbf{M}_{Ey} + \mathbf{M}_y \\ \delta J_z \frac{\mathrm{d}\omega_z}{\mathrm{d}t} + (\delta J_y - \delta J_x)\omega_y\omega_x = \sum \mathbf{M}_z = \mathbf{M}_{Az} + \mathbf{M}_{Cz} + \mathbf{M}_{Ez} + \mathbf{M}_z \end{cases}$$

We consider the motion in the pitching channel of flight period for instance.

$$\begin{cases} \delta m \mathbf{V} \frac{\mathrm{d}\theta}{\mathrm{d}t} = \mathbf{F}_{Gy} + \mathbf{F}_{Py} + \mathbf{F}_{Ay} + \mathbf{F}_{Cy} + \mathbf{F}_{Ey} + \mathbf{F}_y \\ \delta J_z \frac{\mathrm{d}\omega_z}{\mathrm{d}t} + (\delta J_y - \delta J_x)\omega_y\omega_x = \mathbf{M}_{Az} + \mathbf{M}_{Cz} + \mathbf{M}_{Ez} + \mathbf{M}_z \end{cases}$$

The controlling forces and moments are $\mathbf{F}_C = \mathbf{F}_C^{\text{core}} + \mathbf{F}_C^{\text{strap-on}}$, $\mathbf{M}_C = \mathbf{M}_C^{\text{core}} + \mathbf{M}_C^{\text{strap-on}}$, and the inertial forces and moments of the gimballed thrusts are: $\mathbf{F}_E = \mathbf{F}_E^{\text{core}} + \mathbf{F}_E^{\text{strap-on}}$, $\mathbf{M}_E = \mathbf{M}_E^{\text{core}} + \mathbf{M}_E^{\text{strap-on}}$.

Consider the dynamics of the general launch vehicle without vibration and shaking elements can be symbolized separately by the simultaneous equations as follow in three attitudes [24]:

From the dynamic equation of the flight.

$$\begin{aligned}
\mathbf{F}_{Gy} &= -mg\cos\theta_0 \\
\mathbf{F}_{Py} &= \sum_{\text{core\&strap-on}} P_i \cos\beta \sin\alpha \\
&\overset{\beta\to 0}{\approx} (2P_{\text{core}} + 4P_{\text{strap-on}})\sin\alpha \\
\mathbf{F}_{Ay} &= C_y^{\alpha} q S_m(\alpha + \alpha_w) \\
\mathbf{F}_{Cy}^{\text{core}} &= 2P_{\text{core}}\delta_{\varphi}^{\text{core}} \\
\mathbf{F}_{Cy}^{\text{strap-on}} &= 2P_{\text{strap-on}}\delta_{\varphi}^{\text{strap-on}} \\
\mathbf{F}_{Cy} &= \mathbf{F}_{Cy}^{\text{core}} + \mathbf{F}_{Cy}^{\text{strap-on}} \\
&= 2P_{\text{core}}\delta_{\varphi}^{\text{core}} + 2P_{\text{strap-on}}\delta_{\varphi}^{\text{strap-on}}
\end{aligned}$$

$$\begin{aligned}
\mathbf{F}_{Ey}^{\text{core}} &= m_R l_R \ddot{\delta}_{\varphi}^{\text{core}} - m_R \dot{W}_x \delta_{\varphi}^{\text{core}} \\
\mathbf{F}_{Ey}^{\text{strap-on}} &= m_R l_R \ddot{\delta}_{\varphi}^{\text{strap-on}} - m_R \dot{W}_x \delta_{\varphi}^{\text{strap-on}} \\
\mathbf{F}_{Ey} &= \mathbf{F}_{Ey}^{\text{core}} + \mathbf{F}_{Ey}^{\text{strap-on}} \\
&= m_R^{\text{core}} l_R^{\text{core}} \ddot{\delta}_{\varphi}^{\text{core}} - m_R^{\text{core}} \dot{W}_x \delta_{\varphi}^{\text{core}} + m_R^{\text{strap-on}} l_R^{\text{strap-on}} \ddot{\delta}_{\varphi}^{\text{strap-on}} \\
&\quad - m_R^{\text{strap-on}} \dot{W}_x \delta_{\varphi}^{\text{strap-on}}
\end{aligned}$$

About the moment:

$$\begin{aligned}
\mathbf{M}_{Asz} &= -F_{Ay}(x_{\text{press}} - x_{\text{mass}}) \\
&= -C_y^{\alpha} q S_m(\alpha + \alpha_w)(x_{\text{press}} - x_{\text{mass}}) \\
&= -m_z^{\alpha} q S_m l_k(\alpha + \alpha_w)
\end{aligned}$$

with the dimensionless Steady Stability Moment Coefficient: $m_z^{\alpha} = C_y^{\alpha}(x_{\text{press}} - x_{\text{mass}})/l_k$, which is called Aerodynamic Moment Coefficient as well. Here n, x_{press} and x_{mass} are the distance from the theory top of the vehicle to the aerodynamic press center and to the mass center,
autoedited1, respectively.

$$\mathbf{M}_{Adz} = -\frac{m_{dz}^{\omega} q S_m l_k^2 \omega_z}{V} = -\frac{m_{dz}^{\omega} q S_m l_k^2}{V}\dot{\varphi}$$

is the damping moment from atmospheric.

Thus,

$$\begin{aligned}
\mathbf{M}_{Az} &= \mathbf{M}_{Asz} + \mathbf{M}_{Adz} \\
&= -m_z^{\alpha} q S_m l_k (\alpha + \alpha_w) - \frac{m_{dz}^{\omega} q S_m l_k^2}{V} \dot{\varphi} \\
\mathbf{M}_{Cz}^{\text{core}} &= -\mathbf{F}_{Cy}^{\text{core}} \rho_C \\
&= -2 P_{\text{core}} \delta_{\varphi}^{\text{core}} \rho_C \\
\mathbf{M}_{Cz}^{\text{strap-on}} &= -\mathbf{F}_{Cz}^{\text{strap-on}} \rho_C \\
&= -2 P_{\text{strap-on}} \delta_{\varphi}^{\text{strap-on}} \rho_C \\
\mathbf{M}_{Cz} &= \mathbf{M}_{Cz}^{\text{core}} + \mathbf{M}_{Cz}^{\text{strap-on}} \\
&= -2 P_{\text{core}} \delta_{\varphi}^{\text{core}} \rho_C - 2 P_{\text{strap-on}} \delta_{\varphi}^{\text{strap-on}} \rho_C
\end{aligned}$$

$$\begin{aligned}
\mathbf{M}_{Ez}^{\text{core}} &= -J_R^{\text{core}} \ddot{\delta}_{\varphi}^{\text{core}} - m_R^{\text{core}} l_R^{\text{core}} \rho_C \ddot{\delta}_{\varphi}^{\text{core}} - m_R^{\text{core}} \dot{W}_x l_R^{\text{core}} \delta_{\varphi}^{\text{core}} \\
\mathbf{M}_{Ez}^{\text{strap-on}} &= -J_R^{\text{strap-on}} \ddot{\delta}_{\varphi}^{\text{strap-on}} - m_R^{\text{strap-on}} l_R^{\text{strap-on}} \rho_C \ddot{\delta}_{\varphi}^{\text{strap-on}} \\
&\quad - m_R^{\text{strap-on}} \dot{W}_x l_R^{\text{strap-on}} \delta_{\varphi}^{\text{strap-on}}
\end{aligned}$$

$$\begin{aligned}
\mathbf{M}_{Ez} &= \mathbf{M}_{Ez}^{\text{core}} + \mathbf{M}_{Ez}^{\text{strap-on}} \\
&= -J_R^{\text{core}} \ddot{\delta}_{\varphi}^{\text{core}} - m_R^{\text{core}} l_R^{\text{core}} \rho_C \ddot{\delta}_{\varphi}^{\text{core}} - m_R^{\text{core}} \dot{W}_x l_R^{\text{core}} \delta_{\varphi}^{\text{core}} - J_R^{\text{strap-on}} \ddot{\delta}_{\varphi}^{\text{strap-on}} \\
&\quad - m_R^{\text{strap-on}} l_R^{\text{strap-on}} \rho_C \ddot{\delta}_{\varphi}^{\text{strap-on}} - m_R^{\text{strap-on}} \dot{W}_x l_R^{\text{strap-on}} \delta_{\varphi}^{\text{strap-on}}
\end{aligned}$$

First, the ideal situation without disturbances is considered to form its equilibrium equation based on Newton's second law in the pitching channel.

$$\begin{aligned}
m V \dot{\theta}_0 &= -mg \cos\theta_0 + (2 P_{\text{core}} + 4 P_{\text{strap-on}}) \sin\alpha_0 + C_y^{\alpha} q S_m \alpha_0 \\
&\quad + (2 P_{\text{core}} \delta_{\varphi 0}^{\text{core}} + 2 P_{\text{strap-on}} \delta_{\varphi 0}^{\text{strap-on}}) + (m_R^{\text{core}} l_R^{\text{core}} \ddot{\delta}_{\varphi 0}^{\text{core}} - m_R^{\text{core}} \dot{W}_x \delta_{\varphi 0}^{\text{core}} \\
&\quad + m_R^{\text{strap-on}} l_R^{\text{strap-on}} \ddot{\delta}_{\varphi 0}^{\text{strap-on}} - m_R^{\text{strap-on}} \dot{W}_x \delta_{\varphi 0}^{\text{strap-on}}) \\
&= -mg \cos\theta_0 + (2 P_{\text{core}} + 4 P_{\text{strap-on}}) \sin\alpha_0 + C_y^{\alpha} q S_m \alpha_0 + (2 P_{\text{core}} - m_R^{\text{core}} \dot{W}_x) \delta_{\varphi 0}^{\text{core}} \\
&\quad + (2 P_{\text{strap-on}} - m_R^{\text{strap-on}} \dot{W}_{x1}) \delta_{\varphi 0}^{\text{strap-on}} + m_R^{\text{core}} l_R^{\text{core}} \ddot{\delta}_{\varphi 0}^{\text{core}} + m_R^{\text{strap-on}} l_R^{\text{strap-on}} \ddot{\delta}_{\varphi 0}^{\text{strap-on}}
\end{aligned}$$

Then, study the practical system with deviation of the Euler angles of the vehicle: $\Delta\theta$ and $\Delta\alpha$.

$$
\begin{aligned}
mV(\dot{\theta}_0 + \Delta\dot{\theta}) = & -mg\cos(\theta_0 + \Delta\theta) + (2P_{\text{core}} + 4P_{\text{strap-on}})\sin(\alpha_0 + \Delta\alpha) \\
& + C_y^{\alpha} q S_m(\alpha_0 + \Delta\alpha + \alpha_w) + (2P_{\text{core}} - m_R^{\text{core}}\dot{W}_x)(\delta_{\varphi 0}^{\text{core}} + \delta_{\varphi}^{\text{core}}) \\
& + (2P_{\text{strap-on}} - m_R^{\text{strap-on}}\dot{W}_x)\cdot(\delta_{\varphi 0}^{\text{strap-on}} + \delta_{\varphi}^{\text{strap-on}}) + m_R^{\text{core}} l_R^{\text{core}} \\
& \cdot(\ddot{\delta}_{\varphi 0}^{\text{core}} + \ddot{\delta}_{\varphi}^{\text{core}}) + m_R^{\text{strap-on}} l_R^{\text{strap-on}} \cdot(\ddot{\delta}_{\varphi 0}^{\text{strap-on}} + \ddot{\delta}_{\varphi}^{\text{strap-on}}) + F_y \\
& \overset{\substack{\Delta\theta\to 0\\ \Delta\alpha\to 0}}{\approx} -mg(\cos\theta_0 - \sin\theta_0\Delta\theta) + (2P_{\text{core}} + 4P_{\text{strap-on}})(\sin\alpha_0 + \cos\alpha_0\Delta\alpha) \\
& + C_y^{\alpha} q S_m(\alpha_0 + \Delta\alpha) + (2P_{\text{core}} - m_R^{\text{core}}\dot{W}_x)(\delta_{\varphi 0}^{\text{core}} + \delta_{\varphi}^{\text{core}}) \\
& + (2P_{\text{strap-on}} - m_R^{\text{strap-on}}\dot{W}_x)(\delta_{\varphi 0}^{\text{strap-on}} + \delta_{\varphi}^{\text{strap-on}}) + m_R^{\text{core}} l_R^{\text{core}}(\ddot{\delta}_{\varphi 0}^{\text{core}} \\
& + \ddot{\delta}_{\varphi}^{\text{core}}) + m_R^{\text{strap-on}} l_R^{\text{strap-on}}(\ddot{\delta}_{\varphi 0}^{\text{strap-on}} + \ddot{\delta}_{\varphi}^{\text{strap-on}}) + C_y^{\alpha} q S_m \alpha_w + F_y
\end{aligned}
$$

$$
\begin{aligned}
mV\Delta\dot{\theta} = & \, mg\sin\theta_0\Delta\theta + (2P_{\text{core}} + 4P_{\text{strap-on}})\cdot\cos\alpha_0\Delta\alpha + C_y^{\alpha} q S_m \Delta\alpha + (2P_{\text{core}} \\
& - m_R^{\text{core}}\dot{W}_x)\delta_{\varphi}^{\text{core}} + (2P_{\text{strap-on}} - m_R^{\text{strap-on}}\dot{W}_x)\delta_{\varphi}^{\text{strap-on}} \\
& + m_R^{\text{core}} l_R^{\text{core}}\ddot{\delta}_{\varphi}^{\text{core}} + m_R^{\text{strap-on}} l_R^{\text{strap-on}}\cdot\ddot{\delta}_{\varphi}^{\text{strap-on}} + C_y^{\alpha} q S_m \alpha_w + F_y \\
= & \left[(2P_{\text{core}} + 4P_{\text{strap-on}})\cos\alpha_0 + C_y^{\alpha} q S_m\right]\cdot\Delta\alpha + mg\sin\theta_0\Delta\theta + (2P_{\text{core}} \\
& - m_R^{\text{core}}\dot{W}_x)\cdot\delta_{\varphi}^{\text{core}} + m_R^{\text{core}} l_R^{\text{core}}\ddot{\delta}_{\varphi}^{\text{core}} + (2P_{\text{strap-on}} - m_R^{\text{strap-on}}\dot{W}_x)\delta_{\varphi}^{\text{strap-on}} \\
& + m_R^{\text{strap-on}} l_R^{\text{strap-on}}\ddot{\delta}_{\varphi}^{\text{strap-on}} + C_y^{\alpha} q S_m \alpha_w + F_y
\end{aligned}
$$

$$
\begin{aligned}
\Delta\dot{\theta} = & \frac{(2P_{\text{core}} + 4P_{\text{strap-on}})\cos\alpha_0 + C_y^{\alpha} q S_m}{mV}\Delta\alpha + \frac{g\sin\theta_0}{V}\Delta\theta + \frac{2P_{\text{core}} - m_R^{\text{core}}\dot{W}_x}{mV}\delta_{\varphi}^{\text{core}} \\
& + \frac{m_R^{\text{core}} l_R^{\text{core}}}{mV}\ddot{\delta}_{\varphi}^{\text{core}} + \frac{2P_{\text{strap-on}} - m_R^{\text{strap-on}}\dot{W}_x}{mV}\delta_{\varphi}^{\text{strap-on}} + \frac{m_R^{\text{strap-on}} l_R^{\text{strap-on}}}{mV}\ddot{\delta}_{\varphi}^{\text{strap-on}} \\
& + \frac{C_y^{\alpha} q S_m}{mV}\alpha_w + \frac{F_y}{mV} \\
= & \, c_1^{\varphi}\Delta\alpha + c_2^{\varphi}\Delta\theta + c_{3\text{core}}^{\varphi}\delta_{\varphi}^{\text{core}} + c_{3\text{core}}^{\prime\varphi}\ddot{\delta}_{\varphi}^{\text{core}} + c_{3\text{strap-on}}^{\varphi}\delta_{\varphi}^{\text{strap-on}} \\
& + c_{3\text{strap-on}}^{\prime\varphi}\ddot{\delta}_{\varphi}^{\text{strap-on}} + c_1^{\prime\varphi}\alpha_w + \overline{F}_{yc}
\end{aligned}
$$

In which, the coefficients are:

$$
\begin{aligned}
c_1^{\varphi} &= \frac{(2P_{\text{core}} + 4P_{\text{strap-on}})\cos\alpha_0 + C_y^{\alpha} q S_m}{mV} \\
c_2^{\varphi} &= \frac{g\sin\theta_0}{V} \\
c_{3\text{core}}^{\varphi} &= \frac{2P_{\text{core}} - m_R^{\text{core}}\dot{W}_{x1}}{mV} \\
c_{3\text{core}}^{\prime\varphi} &= \frac{m_R^{\text{core}} l_R^{\text{core}}}{mV} \\
c_{3\text{strap-on}}^{\varphi} &= \frac{2P_{\text{strap-on}} - m_R^{\text{strap-on}}\dot{W}_{x1}}{mV}
\end{aligned}
$$

$$\begin{aligned}
c'^{\varphi}_{3\text{strap-on}} &= \frac{m_R^{\text{strap-on}} l_R^{\text{strap-on}}}{mV} \\
c'^{\varphi}_{1} &= \frac{C_y^{\alpha} q S_m}{mV} \\
\overline{F}_{yc} &= \frac{F_y}{mV}
\end{aligned}$$

Similarly, from the

$$\begin{aligned}
&J_z(\ddot{\varphi}_0 + \Delta\ddot{\varphi}) + (J_y - J_x)(\dot{\gamma} - \dot{\varphi}_0\psi)(\dot{\psi} + \dot{\varphi}_0\gamma) \\
&= -m_z^{\alpha} q S_m l_k(\alpha_0 + \Delta\alpha + \alpha_w) - \tfrac{m_{dz}^{\omega} q S_m l_k^2}{V}(\dot{\varphi}_0 + \Delta\dot{\varphi}) - 2P_{\text{core}}\rho_C(\delta_{\varphi 0}^{\text{core}} + \delta_{\varphi}^{\text{core}}) \\
&-2P_{\text{strap-on}}\rho_C(\delta_{\varphi 0}^{\text{strap-on}} + \delta_{\varphi}^{\text{strap-on}}) - J_R^{\text{core}}(\ddot{\delta}_{\varphi 0}^{\text{core}} + \ddot{\delta}_{\varphi}^{\text{core}}) - m_R^{\text{core}} l_R^{\text{core}}\rho_C(\ddot{\delta}_{\varphi 0}^{\text{core}} + \ddot{\delta}_{\varphi}^{\text{core}}) \\
&-m_R^{\text{core}}\dot{W}_x l_R^{\text{core}}(\delta_{\varphi 0}^{\text{core}} + \delta_{\varphi}^{\text{core}}) - J_R^{\text{strap-on}}(\ddot{\delta}_{\varphi 0}^{\text{strap-on}} + \ddot{\delta}_{\varphi}^{\text{strap-on}}) - m_R^{\text{strap-on}} l_R^{\text{strap-on}} \\
&\cdot\rho_C(\ddot{\delta}_{\varphi 0}^{\text{strap-on}} + \ddot{\delta}_{\varphi}^{\text{strap-on}}) - m_R^{\text{strap-on}}\dot{W}_x l_R^{\text{strap-on}}(\delta_{\varphi 0}^{\text{strap-on}} + \delta_{\varphi}^{\text{strap-on}}) + M_z
\end{aligned}$$

$$\begin{aligned}
&= -m_z^{\alpha} q S_m l_k(\alpha_0 + \Delta\alpha) - \tfrac{m_{dz}^{\omega} q S_m l_k^2}{V}(\dot{\varphi}_0 + \Delta\dot{\varphi}) - (2P_{\text{core}}\rho_C + m_R^{\text{core}}\dot{W}_x l_R^{\text{core}})(\delta_{\varphi 0}^{\text{core}} + \delta_{\varphi}^{\text{core}}) \\
&-(2P_{\text{strap-on}}\rho_C + m_R^{\text{strap-on}}\dot{W}_x l_R^{\text{strap-on}})(\delta_{\varphi 0}^{\text{strap-on}} + \delta_{\varphi}^{\text{strap-on}}) - (J_R^{\text{core}} + m_R^{\text{core}} l_R^{\text{core}}\rho_C) \\
&\cdot(\ddot{\delta}_{\varphi 0}^{\text{core}} + \ddot{\delta}_{\varphi}^{\text{core}}) - (J_R^{\text{strap-on}} + m_R^{\text{strap-on}} l_R^{\text{strap-on}}\rho_C)(\ddot{\delta}_{\varphi 0}^{\text{strap-on}} + \ddot{\delta}_{\varphi}^{\text{strap-on}}) - m_z^{\alpha} q S_m l_k \alpha_w + M_z
\end{aligned}$$

Thus we have:

$$\begin{aligned}
&(\ddot{\varphi}_0 + \Delta\ddot{\varphi}) + \tfrac{J_y - J_x}{J_z}(\dot{\gamma} - \dot{\varphi}_0\psi)(\dot{\psi} + \dot{\varphi}_0\gamma) + \tfrac{m_z^{\alpha} q S_m l_k}{J_z}(\alpha_0 + \Delta\alpha) + \tfrac{m_{dz}^{\omega} q S_m l_k^2}{V J_z}(\dot{\varphi}_0 + \Delta\dot{\varphi}) \\
&+\tfrac{2P_{\text{core}}\rho_C + m_R^{\text{core}}\dot{W}_x l_R^{\text{core}}}{J_z}(\delta_{\varphi 0}^{\text{core}} + \delta_{\varphi}^{\text{core}}) + \tfrac{2P_{\text{strap-on}}\rho_C + m_R^{\text{strap-on}}\dot{W}_x l_R^{\text{strap-on}}}{J_z}(\delta_{\varphi 0}^{\text{strap-on}} + \delta_{\varphi}^{\text{strap-on}}) \\
&+\tfrac{J_R^{\text{core}} + m_R^{\text{core}} l_R^{\text{core}}\rho_C}{J_z}(\ddot{\delta}_{\varphi 0}^{\text{core}} + \ddot{\delta}_{\varphi}^{\text{core}}) + \tfrac{J_R^{\text{strap-on}} + m_R^{\text{strap-on}} l_R^{\text{strap-on}}\rho_C}{J_z}(\ddot{\delta}_{\varphi 0}^{\text{strap-on}} + \ddot{\delta}_{\varphi}^{\text{strap-on}}) \\
&+\tfrac{m_z^{\alpha} q S_m l_k}{J_z}\alpha_w = \tfrac{M_z}{J_z}
\end{aligned}$$

Considering the ideal situation:

$$\begin{aligned}
&\ddot{\varphi}_0 + \tfrac{J_y - J_x}{J_z}(\dot{\gamma} - \dot{\varphi}_0\psi)(\dot{\psi} + \dot{\varphi}_0\gamma) + \tfrac{m_z^{\alpha} q S_m l_k}{J_{z1}}\alpha_0 + \tfrac{m_{dz}^{\omega} q S_m l_k^2}{V J_{z1}}\dot{\varphi}_0 + \tfrac{2P_{\text{core}}\rho_C + m_R^{\text{core}}\dot{W}_x l_R^{\text{core}}}{J_z}\delta_{\varphi 0}^{\text{core}} \\
&+\tfrac{2P_{\text{strap-on}}\rho_C + m_R^{\text{strap-on}}\dot{W}_x l_R^{\text{strap-on}}}{J_z}\delta_{\varphi 0}^{\text{strap-on}} + \tfrac{J_R^{\text{core}} + m_R^{\text{core}} l_R^{\text{core}}\rho_C}{J_z}\ddot{\delta}_{\varphi 0}^{\text{core}} \\
&+\tfrac{J_R^{\text{strap-on}} + m_R^{\text{strap-on}} l_R^{\text{strap-on}}\rho_C}{J_z}\ddot{\delta}_{\varphi 0}^{\text{strap-on}} = 0
\end{aligned}$$

$$\begin{aligned}
&\Delta\ddot{\varphi} + \tfrac{m_{dz}^{\omega} q S_m l_k^2}{V J_z}\Delta\dot{\varphi} + \tfrac{m_z^{\alpha} q S_m l_k}{J_z}\Delta\alpha + \tfrac{2P_{\text{core}}\rho_C + m_R^{\text{core}}\dot{W}_x l_R^{\text{core}}}{J_z}\delta_{\varphi}^{\text{core}} \\
&+\tfrac{2P_{\text{strap-on}}\rho_C + m_R^{\text{strap-on}}\dot{W}_x l_R^{\text{strap-on}}}{J_z}\delta_{\varphi}^{\text{strap-on}} + \tfrac{J_R^{\text{core}} + m_R^{\text{core}} l_R^{\text{core}}\rho_C}{J_z}\ddot{\delta}_{\varphi}^{\text{core}} \\
&+\tfrac{J_R^{\text{strap-on}} + m_R^{\text{strap-on}} l_R^{\text{strap-on}}\rho_C}{J_z}\ddot{\delta}_{\varphi}^{\text{strap-on}} + \tfrac{m_z^{\alpha} q S_m l_k}{J_z}\alpha_w = \tfrac{M_z}{J_z}
\end{aligned}$$

$$\begin{aligned}
&\Delta\ddot{\varphi} + b_1^{\varphi}\Delta\dot{\varphi} + b_2^{\varphi}\Delta\alpha + b_{3\text{core}}^{\varphi}\delta_{\varphi}^{\text{core}} + b'^{\varphi}_{3\text{core}}\ddot{\delta}_{\varphi}^{\text{core}} + b_{3\text{strap-on}}^{\varphi}\delta_{\varphi}^{\text{strap-on}} \\
&+b'^{\varphi}_{3\text{strap-on}}\ddot{\delta}_{\varphi}^{\text{strap-on}} + b_2^{\varphi}\alpha_w = \overline{M}_z
\end{aligned}$$

The related coefficients are:

$$b_1^{\varphi} = \frac{m_{dz}^{\omega} q S_m l_k^2}{J_z V}$$
$$b_2^{\varphi} = \frac{m_z^{\alpha} q S_m l_k}{J_z}$$
$$b_{3\text{core}}^{\varphi} = \frac{2P_{\text{core}}\rho_C + m_R^{\text{core}} \dot{W}_x l_R^{\text{core}}}{J_z}$$
$$b_{3\text{core}}^{\prime\varphi} = \frac{J_R^{\text{core}} + m_R^{\text{core}} l_R^{\text{core}} \rho_C}{J_z}$$
$$b_{3\text{strap-on}}^{\prime\varphi} = \frac{2P_{\text{strap-on}}\rho_C + m_R^{\text{strap-on}} \dot{W}_x l_R^{\text{strap-on}}}{J_z}$$
$$b_{3\text{strap-on}}^{\prime\varphi} = \frac{J_R^{\text{strap-on}} + m_R^{\text{strap-on}} l_R^{\text{strap-on}} \rho_C}{J_z}$$
$$\overline{M}_z = \frac{M_z}{J_z}$$

Therefore, the formulation expression of the launch vehicle in its Pitching channel can be denoted as follows:

$$\begin{cases} \Delta\dot{\theta} = c_1^{\varphi}\Delta\alpha + c_2^{\varphi}\Delta\theta + c_{3\text{core}}^{\varphi}\delta_{\varphi}^{\text{core}} + c_{3\text{core}}^{\prime\varphi}\ddot{\delta}_{\varphi}^{\text{core}} + c_{3\text{strap-on}}^{\varphi}\delta_{\varphi}^{\text{strap-on}} + c_{3\text{strap-on}}^{\prime\varphi}\ddot{\delta}_{\varphi}^{\text{strap-on}} \\ +c_1^{\prime\varphi}\alpha_w + \overline{F}_{yc} \\ \Delta\ddot{\varphi} + b_1^{\varphi}\Delta\dot{\varphi} + b_2^{\varphi}\Delta\alpha + b_{3\text{core}}^{\varphi}\delta_{\varphi}^{\text{core}} + b_{3\text{core}}^{\prime\varphi}\ddot{\delta}_{\varphi}^{\text{core}} + b_{3\text{strap-on}}^{\varphi}\delta_{\varphi}^{\text{strap-on}} + b_{3\text{strap-on}}^{\prime\varphi}\ddot{\delta}_{\varphi}^{\text{strap-on}} \\ +b_2^{\varphi}\alpha_w = \overline{M}_z \\ \Delta\varphi = \Delta\alpha + \Delta\theta \end{cases}$$

Similarly, we have the linearized model in other channels.

In the Yawing channel:

$$\begin{cases} \dot{\sigma} = c_1^{\psi}\beta + c_2^{\psi}\sigma + c_{3\text{core}}^{\psi}\delta_{\psi}^{\text{core}} + c_{3\text{core}}^{\prime\psi}\ddot{\delta}_{\psi}^{\text{core}} + c_{3\text{strap-on}}^{\psi}\delta_{\psi}^{\text{strap-on}} + c_{3\text{strap-on}}^{\prime\psi}\ddot{\delta}_{\psi}^{\text{strap-on}} \\ +c_1^{\prime\psi}\beta_w - \bar{F}_{zc} \\ \ddot{\psi} + b_1^{\psi}\dot{\psi} + b_2^{\psi}\beta + b_{3\text{core}}^{\psi}\delta_{\psi}^{\text{core}} + b_{3\text{core}}^{\prime\psi}\ddot{\delta}_{\psi}^{\text{core}} + b_{3\text{strap-on}}^{\psi}\delta_{\psi}^{\text{strap-on}} + b_{3\text{strap-on}}^{\prime\psi}\ddot{\delta}_{\psi}^{\text{strap-on}} \\ +b_2^{\psi}\beta_w = \bar{M}_y \\ \psi = \beta + \sigma \end{cases}$$

and in Rolling channel:

$$\ddot{\gamma} + d_1\dot{\gamma} + d_{3\text{core}}\delta_{\gamma}^{\text{core}} + d_{3\text{core}}^{\prime}\ddot{\delta}_{\gamma}^{\text{core}} + d_{3\text{strap-on}}\delta_{\gamma}^{\text{strap-on}} + d_{3\text{strap-on}}^{\prime}\ddot{\delta}_{\gamma}^{\text{strap-on}} = \bar{M}_x.$$

References

1. Cheng T, Wang X, Li D (2003) The new generation launch vehicles of long march family. In: The 54th international astronautical congress of the International Astronautical Federation, Bremen, Germany
2. Patton RJ (1993) Robustness issues in fault tolerant control. In: The processing of international conference on fault diagnosis, Toulouse, France, pp 1081–1117
3. Stengel RR (1991) Intelligent failure tolerant control. IEEE Control Syst Mag 11(3):14–23
4. Vidyasagar M (1987) Some results on simultaneous stabilization with multiple domains of stability. Automatica 23:535–540
5. Wang Z, Zhao DJ, Wang YJ, Liu DB (2012) Reconfiguration of shipboard power system using discrete particle swarm optimisation. Int J Model Identif Control 15(4):277–283
6. Wu DN, Gao WB, Chen M (1990) Algorithm for simultaneous stabilization of single-input systems via dynamic feedback. Int J Control 51(3):631–642
7. Vishwakarma R, Turner D, Lewis A, Chen YK, Xu YG (2012) The use of pseudo-inverse methods in reconstructing loads on a missile structure. Int J Model Identif Control 17(3): 242–250
8. Wang ZJ, Guo SJ (2012) Rolling flight control using pseudo inverse control allocation for UAVs with multiple seamless warping control surfaces. Int J Model Identif Control 16(1):15–23
9. Pei Y, Li PQ, Yao YP (1998) The application of function reconfiguration for digital fly-by-wire flight control system. In: The proceedings of the 17th Digital Avionics Systems conference, pp 76–81
10. Gopinathan M, Boskovic JD, Mehra RK, Rago C (1998) A multiple model predictive scheme for fault-tolerant flight control design. In: The proceedings of the 37th IEEE conference on decision and control, pp 1376–1381
11. Chandler P, Mears M, Pachter M (1994) On-line optimizing networks for reconfiguratable control. AIAA-94-3643-CP, pp 263–268
12. Chen Z, Hao LN, Xue DY, Xu XH, Liu YM (2011) Real-time compensation control for hysteresis and creep in IPMC actuators. Int J Model Identif Control 12(1/2):182–189
13. Jiang Y, Hu Q, Ma G (2008) Design of robust adaptive integral variable structure attitude controller with application to flexible spacecraft. Int J Innov Comput Inf Control 4(9): 2431–2440
14. Kemih K, Merabtine N, Benslama M, Filali S (2007) Generalized predictive control using conjugate gradient method applied to micro satellites attitude control. ICIC Express Lett 1(2): 99–104
15. Marcello R, Nitin D, Gao Z (1994) On-line learning non-linear direct neural controllers for reconstrucible control systems. AIAA-94-3643-CP, pp 141–147
16. Gao Z, Jiang B, Shi P, Xu Y (2010) Fault accommodation for near space vehicle attitude dynamics via T-S fuzzy models. Int J Innov Comput Inf Control 6(11):4843–4856
17. Sadeghi T, Lai K (1995) Integrated fault tolerant control law. In: The proceedings of the IEEE 1995 national aerospace and electronics conference, pp 491–512
18. Robert C, Eslinger A, Chandler PR (1993) Self-repairing flight control system overview. IEEE Trans Autom Control 4:143–151
19. Du F, Wang X, Sun H (2008) Fault-tolerant control on missile rudder deadlock. J Proj Rockets Missiles Guidance 28(5):67–70
20. Das RK, Sen S, Dasgupta S (2007) Robust and fault tolerant controller for attitude control of a satellite launch vehicle. IET Control Theory Appl 1(1):304–312
21. Blakelock JH (1991) Automatic control aircraft and missile. A Wiley-Interscience, New York
22. Nelson RC (1989) Flight stability and automatic control. McGraw-Hill Book Company, New York
23. Wu Y, He L (2009) Attitude control technology of new generation launch vehicles. J Beijing Univ Aeronaut Astronaut 35(11):1294–1297
24. Xu Y (1999) Ballistic Missile, Launch vehicle control system design and analysis. China Astronautic Publishing House, Beijing

Chapter 3
Synthesis of an Advanced State Feedback Control for Continuous Nonlinear Polynomial Systems

Houssem Jerbi and Wiem Jebri Jemai

Abstract In this chapter, the problem of approximate linearization of affine nonlinear control systems by a static state feedback is considered. First of all, we propose an analytical method, based on the development into generalized Taylor series expansions and the Kronecker product tools, in order to simplify the complex implementation of the input-state feedback linearization formalism. Next, to improve the synthesized polynomial feedback control, the genetic algorithm, as an optimization method, is used. Finally, the new approach presented in this work is applied to investigate the control problem of a chemical reactor. Moreover, we prove that the controlled process is locally asymptotically stable in a wide region around the operating point, in the Lyapunov sense.

Keywords Feedback linearization · Polynomial nonlinear control · Optimization routine · Genetic algorithm · Lyapunov stability

3.1 Introduction

For nonlinear systems, considerable heed has been paid in the literature, to design the controllers which are based on input-state feedback linearization technique [18, 24, 30, 32, 35]. This technique is used to alter the nonlinear system to a linear one which is characterized by a new state representation [17, 18, 38]. Unfortunately, this technique is mainly related to the solution of a set of PDEs (Partial Differential

H. Jerbi (✉)
College of Engineering of Hail, University of Hail, Hail, P.O. Box 2440, Kingdom of Saudi Arabia
e-mail: h.jerbi@uoh.edu.sa; houssem.jerbi@gmail.com

W. J. Jemai
Research Unit MACS High School of Engineering, University of Gabès, Zrig, Gabès, 6029, Tunisia
e-mail: wiem_jeb@yahoo.fr

L. Liu et al. (eds.), *Applied Methods and Techniques for Mechatronic Systems*, Lecture Notes in Control and Information Sciences 452,
DOI: 10.1007/978-3-642-36385-6_3,

Equations) which can be hardly solved for high-order systems [8, 18, 24, 27, 29]. To overcome this difficulty, many researches are focused to reduce the requirements of the considered technique [2, 4, 7, 9, 11, 15, 23, 26, 36, 41]. In a recent work [21, 22], we have been studying the control problem of input-affine SISO nonlinear systems based on the input-state feedback linearization methodology. This study has focused on the development of a novel approach which relaxes the complex implementation of the input-state feedback linearization technique in order to design a polynomial control via a nonlinear analytical coordinates transformation. This new approach, which is essentially based on the use of the Taylor series expansions and the Kronecker product tool, leads to a recurrent algorithm: solution of the considered problem. Some parameters characterizing this algorithm are chosen arbitrarily. Therefore, an optimization problem arises from the best choice of these parameters. We addressed this problem in [19, 20]. In this work, we will furthermore improve the previous developments using, as optimization approach, the most extended group of methods representing the application of Evolutionary Algorithms (EAs): Genetic Algorithms (GAs). The recourse to GA is justified by the fact that this optimization algorithm raises important features which make it a powerful optimization tool [34]. First, it is an iterative and stochastic algorithm and the randomness is an essential specificity of GA. A second very important point is that the GA considers a set of solutions. Thus, at each iteration, the algorithm keeps in memory more than one solution which will be recombined to provide the better ones. Finally, for illustration purposes, this study is applied to design a controller of a chemical reactor [12]. The effectiveness of the synthesized controller is checked via a stability analysis in the Lyapunov sense. The outline of this chapter is as follows: In Sect. 3.2, we give some preliminaries about the input-state feedback linearization and we present, briefly, the new analytical method presented in [21]. This facilitates the implementation of the feedback linearization theory. Section 3.3 is devoted to the optimization problem using the GAs. The problem of a chemical reactor control is studied in Sect. 3.4, to evaluate the applicability and the effectiveness of the proposed work. Besides in the same section, a stability analysis via the determination of a polynomial Lyapunov function of the considered process is conducted. Finally, the chapter is concluded with a summary in Sect. 3.5.

3.2 Preliminaries and Motivations

Consider the class of continuous, single-input affine nonlinear control systems given by

$$\dot{x} = f(x) + g(x)u, \tag{3.1}$$

where $f(0) = 0$, $x \in \Re^n$, and $u \in \Re$.

The vector field $f(x)$is assumed to be analytic (C^∞) to ensure the convergence of the Taylor series expansion of the vector field. A general linearization problem is defined as follows:

Definition 2.1 The system (3.1) is said to be (locally) static state feedback linearizable at the origin x_0, if it can be transformed into a controllable linear system [13, 18, 37]

$$\dot{z} = Az + bv \tag{3.2}$$

via a locally defined state diffeomorphism.

$$z = \phi(x), \quad \phi(0) = 0 \tag{3.3}$$

and a locally defined static state feedback

$$u = \alpha(x) + \beta(x)v. \tag{3.4}$$

Theorem 2.1 *It has been shown* [18, 24, 30, 35] *that the aforementioned State-Space Exact Linearization problem is solvable if and only if:*

(i) The distribution G_{n-1} has dimension n near x_0,
(ii) The distribution G_{n-2} is involutive,

where

$$G_i = \text{span}\left\{ad_f^k g(x) : 0 \le k \le i\right\} \tag{3.5}$$

defined as distributions with $ad_f^k g(x)$ is the k-th order Lie bracket of the vector field of $g(x)$ with respect to the vector field $f(x)$.

Provided that conditions (i) and (ii) are satisfied by (3.1), the first step is to transform (3.1)–(3.2) by using the exact feedback linearization method outlined above. Since the transformed system (3.2) is a linear controllable one, the next step is to employ linear pole-placement techniques in order to arbitrarily assign the poles of the closed-loop system [6]. In the particular case, one can calculate a constant gain vector k, given that the static state feedback law:

$$v = -kz \tag{3.6}$$

induces the closed-loop dynamics :

$$\dot{z} = (A - bk)z \tag{3.7}$$

when applied to the linear system (3.2).

The present work is based on the effective use of mathematical tools: the Kronecker product, the non-redundant state formulation, and the polynomial representations. Using these tools, the system (3.1) becomes

$$\begin{aligned}\dot{x} &= \sum_{i\geq 1} F_i X^{[i]} + g_0 u \\ &= \sum_{i\geq 1} f_i x^{[i]} + g_0 u.\end{aligned} \tag{3.8}$$

$X^{[i]}$ is the i-th redundant Kronecker power of a n-dimensional vector x [5], and $x^{[i]} \in \Re^n$ is the i-th nun-redundant Kronecker power of vector X [3, 32]. Knowing that $X^{[i]} = R_n^i x^{[i]}$ or $x^{[i]} = \left(R_n^i\right)^{\dagger} X^{[i]}$, where $\left(R_n^i\right)^{\dagger}$ is the Moore-Penrose pseudo-inverse of R_n^i.

Also, the control law and the diffeomorphism, which will be found in the sequel, can be written as

$$\begin{cases} u = \sum_{i\geq 1} \alpha_i x^{[i]} + \sum_{j\geq 0} \beta_j x^{[j]} v \\ z = \phi(x) = \sum_{k\geq 1} \phi_k x^{[k]} \end{cases} \tag{3.9}$$

Note that the constructive solution developed in the literature leads to a linear system (3.2) in Brunosky canonical form [18, 37]. In the present work, we will transform the original system (3.1) into a linear one assuming that it is stable (not necessary in Brunowsky form).

To determine the diffeomorphism, we will determine the nonlinear transformation $S(z)$, under the canonical form, defined as

$$x = \phi^{-1}(z) = S(z) = \sum_{i\geq 1} S_i^1 z^{[i]}. \tag{3.10}$$

On the one hand, and using the Kronecker power, the expression (3.8) will be written as

$$\begin{aligned} x =& \sum_{i\geq 1} f_1 S_i^1 z^{[z]} + \sum_{i\geq 2} f_2 S_i^2 z^{[z]} + \cdots \\ &+ \sum_{i\geq p} f_p S_i^p z^{[i]} + g_o u, \end{aligned} \tag{3.11}$$

where

$$S_p^n = \sum_{j=1}^{p-n+1} \left(S_{p-j}^{n-1} \otimes S_j^1\right). \tag{3.12}$$

On the other hand, the derivative $\dot{x}$ can be obtained using the derivative of i-th Kronecker power of vector x [5], as

$$\dot{x} = \sum_{i\geq 1} S_i^1 V^{[i]} \left((A - bk) \otimes I_{n^{(i-1)}}\right) z^{[i]}, \tag{3.13}$$

where $V^{[i]} = \sum_{j=0}^{i-1} \left(U_{n^j \times n} \otimes I_{n^{(i-j-1)}}\right)$ and $U_{p\times q}$ designates the Kronecker permutation matrix [5].

Then, the control law expressed in terms of z is given as follows:

$$u = \sum_{i\geq 1} \alpha_1 S_i^1 z^{[i]} + \cdots + \sum_{i\geq 1} \alpha_p S_i^p z^i \\ - \left(\beta_0 kz + \cdots + \sum_{i\geq p} \left(\beta_p S_i^p \otimes k\right) z^{[i+1]}\right. \tag{3.14}$$

Replacing u by its new expression (3.14) in (3.11) yields the expression of the derivative $\dot{x}$ in terms of z. The identification of elements of this equation with those of Eq. (3.13) leads to the following recurrent algorithm which exposes the solution of the studied problem:

(1) For $p = 1$

$$\begin{cases} S_1^1 = I_n \\ f_1 + g_0\alpha_1 = A - bk \\ A - bk \quad \text{is stable} \\ \beta_0 \text{ arbitrary} \end{cases} \tag{3.15}$$

(2) For $p \geq 2$

$$\begin{cases} \operatorname{vec}\left(S_p^1\right) = -\operatorname{pinv}\left(A_p\right) \operatorname{vec}\left(\sum_{i=2}^{p} f_i S_p^i R_n^p\right) \\ \alpha_p = \operatorname{pinv}(g_0) g_0 \begin{bmatrix} \left[-\sum_{i=2}^{p-1} \alpha_i S_p^i R_n^p\right] + \\ \left[\sum_{i=1}^{p-1} (\beta_i S_{p-1}^i \otimes k)\right] \end{bmatrix} \operatorname{pinv}(S_p^p R_n^p) \\ A_p = \left[\left(R_n^p\right)^T \otimes (f_1 + g_0\alpha_1)\right] \\ - \left(\left[V^{(p)}[(A - bk) \otimes (I_n p - 1)] R_n^p\right]^T \otimes I_n\right) \\ \beta_{p-1} \text{ arbitrary} \end{cases} \tag{3.16}$$

where vec(·) designates the vectorization operator and pinv(·) designates the Moore-Penrose inverse.

Using this algorithm, we can deduce the components of the nonlinear transformation defined by (3.9) as

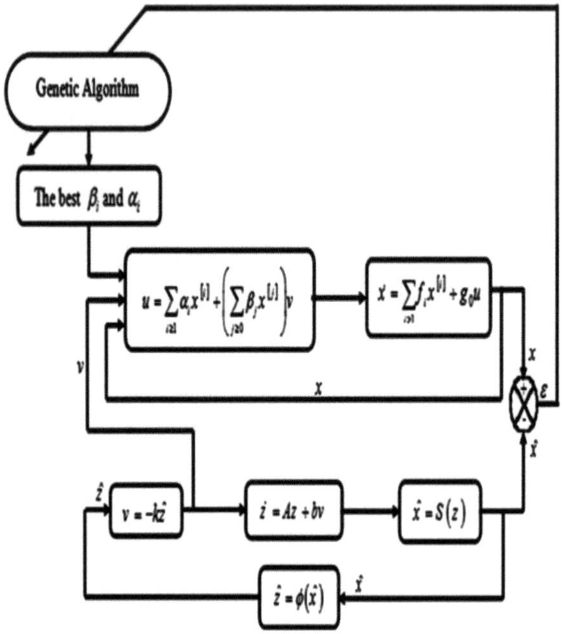

Fig. 3.1 Block diagram of the polynomial feedback control law, using GA, applied to a nonlinear system

$$\begin{cases} \phi_1^1 = \left(S_1^1\right)^{-1} \\ \phi_k^1 = -\left(S_1^1\right)^{-1}\left(\sum\limits_{i=2}^{k} S_i^1 \phi_k^i\right) \\ \phi_k^i = \sum\limits_{j=1}^{k-i+1}\left(\phi_{k-j}^{i-1} \otimes S_j^1\right). \end{cases} \tag{3.17}$$

A problem which will be studied, now, is the influence of parameters β_p on the effectiveness of the solution.

Since the choice of parameters β_p is arbitrary, we have resorted to genetic algorithm as an optimization method to determine the optimal values of these parameters that yield the best value of the performance criterion.

The block diagram shown in Fig. 3.1 summarizes the proposed analytical method to find a polynomial controller, using GA, for a nonlinear system.

3.3 Genetic Algorithm and Optimization Method

The GAs are an optimization procedure that is developed based on the principles of natural selection and are thus able to evolve solutions to real-world problems [14, 16, 33]. A GA aims at evolving an optimal solution by letting a set of random solutions undergo a sequence of probabilistic transformations governed by a selection scheme biased toward high-quality solutions.

The GA starts with a random population containing a predefined number of individuals (or solutions), which are represented by a genetic string (incorporating the variable information). Each individual has an associated fitness measure, typically representing an objective value [39]. The concept that the fittest (or best) individuals in a population will produce a fitter offspring is then implemented in order to reproduce the next population. Selected individuals are chosen for reproduction (or crossover) at each generation, with an appropriate mutation factor to randomly modify the genes of an individual, in order to develop the new population.

The result is another set of individuals based on the original subjects leading to subsequent populations with better (min. or max.) individual fitness. Therefore, the algorithm identifies the individuals with the optimizing fitness values, and those with lower fitness will naturally get discarded from the population.

For the present work, in order to achieve an effective nonlinear transformation leading to a satisfactory control law, we optimize the random parameters of the control using the GA optimization algorithm. The fitness function is defined as

$$\min_{\beta_p} \frac{\left\| x - \hat{x} \right\|_2}{\left\| x \right\|_2}, \tag{3.18}$$

where x and $\hat{x}$ are the solutions given by

$$\begin{cases} \dot{x} = f_1 x + f_2 x^{[2]} + g_0 u \\ u = \alpha\left(x\right) + \beta\left(x\right) v \\ \hat{x} = S\left(z\right) \\ x\left(0\right) = x_0 \neq 0 \\ v = -kz \end{cases} \tag{3.19}$$

$\|.\|_2$ denotes norm 2.

Figure 3.2 shows a flowchart of the GA optimization problem for the studied system.

3.4 Illustrative Example and Simulations Results

3.4.1 Mathematical Model

In this section, we illustrate the applicability and effectiveness of the proposed approach in a chemical reactor control problem: an isothermal nonlinear CSTR [12].

A dimensionless model for such a reactor is given by [31]

$$\begin{cases} \dot{x}_1 = -x_1 - Da_1 x_1 + u \\ \dot{x}_2 = Da_1 x_1 - x_2 - Da_2 x_2^2 \end{cases} \tag{3.20}$$

where x_1 and x_2 are the dimensionless concentrations of A and B, respectively. The control variable u corresponds to the inlet concentration of A and the Damkholer numbers, Da_1 and Da_2, relate the effect of the reaction terms to the residence time (see [31] for further details on the reacting system model and the definition of variables and parameters).

The polynomial form of (3.20) can be written as follows:

$$\dot{X} = F_1 X + F_2 X^{[2]} + g_0 U \tag{3.21}$$

with

$$F_1 = \begin{bmatrix} -1 - Da_1 & 0 \\ Da_1 & -1 \end{bmatrix}, \quad F_2 = \begin{bmatrix} 0 & 0 & 0 & 0 \\ 0 & 0 & 0 & -Da_2 \end{bmatrix},$$

and $g_0 = \begin{bmatrix} 1 \\ 0 \end{bmatrix}$.

Let us consider the following variable change:

$$\begin{cases} x = X - X_n \\ u = U - U_n \end{cases} \tag{3.22}$$

where X_n is the state vector in the nominal operating point and U_n is the control.

So we can rewrite (3.21) in the following form:

$$\dot{x} = f_1 x + f_2 x^{[2]} + g_0 u, \tag{3.23}$$

where

$$f_1 = \begin{bmatrix} -1 - Da_1 & 0 \\ Da_1 & -(1 + 2Da_2 X_2 n) \end{bmatrix},$$

$$f_2 = \begin{bmatrix} 0 & 0 & 0 & 0 \\ 0 & 0 & 0 & -Da_2 \end{bmatrix}, \quad \text{and} \quad g_0 = \begin{bmatrix} 1 \\ 0 \end{bmatrix}.$$

3.4.2 Simulation Results

The nonlinear system (3.23) will be changed into a linear one given by (3.7) with:

$$A = \begin{bmatrix} -0.5 & -3 \\ 4 & -2 \end{bmatrix}; \ b = \begin{bmatrix} 0 \\ 1 \end{bmatrix}$$

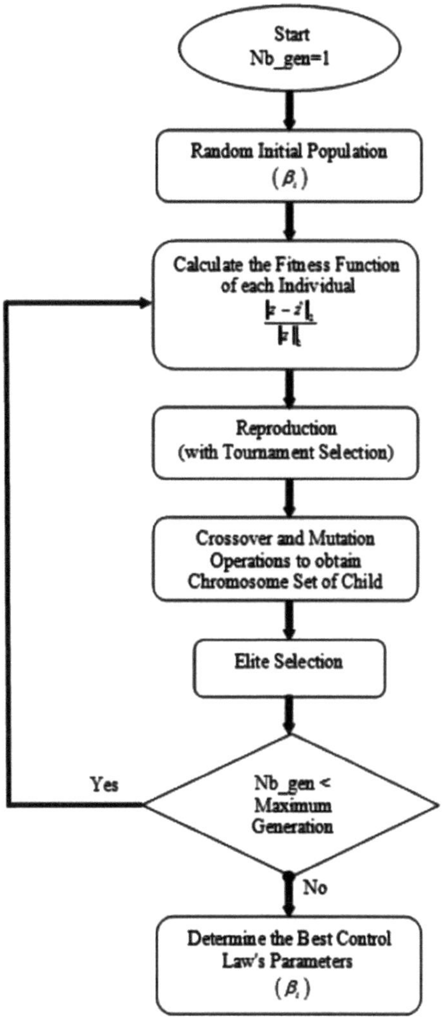

Fig. 3.2 Flowchart of the Genetic Algorithm optimization

and the new control law $v = -k\hat{z}$ is defined such that the linear system poles are -1 and -2 which correspond to a gain matrix $k = \begin{bmatrix} 2 & 3 \end{bmatrix}$.

The GA parameters are set as follows:

- Chromosome $= [\beta_0, \beta_1, \beta_2, \beta_3]$
- Binary coding (the size of chromosome is equal to 4 bits)
- Population size $= 20$
- Crossover rate $= 0.65$
- Mutation rate $= 0.02$
- Maximum generation $= 50$

The polynomial feedback control truncated to third order is given by

$$\begin{aligned} u &= \alpha(x) + \beta(x)v \\ &= \alpha_1 x + \alpha_2 x^{[2]} + \alpha_3 x^{[3]} + \alpha_4 x^{[4]} \\ &\quad + \beta_0 + \beta_1 x + \beta_2 x^{[2]} + \beta_3 x^{[3]} v \end{aligned} \tag{3.24}$$

The implementation of the GA leads to the optimal values of the control law parameters:
$\beta_0 = 0$, $\beta_1 = \begin{bmatrix} 0.4 & 0.8 \end{bmatrix}$,
$\beta_2 = \begin{bmatrix} 0.4 & 0.6 & 0.8 & 0.6 \end{bmatrix}$, and $\beta_3 = \begin{bmatrix} \beta_{13} & \beta_{23} \end{bmatrix}$
with $\beta_{13} = \begin{bmatrix} 0.2 & 0.8 & 0 & 1.2 \end{bmatrix}$ and
$\beta_{23} = \begin{bmatrix} 1.5 & 1.4 & 1.5 & 1.1 \end{bmatrix}$.
$\alpha_1 = \begin{bmatrix} 2.5 & -3 \end{bmatrix}$, $\alpha_2 = \begin{bmatrix} 0.8 & 1.4 & 1.4 & 2.4 \end{bmatrix}$,
$\alpha_3 = \begin{bmatrix} \alpha_{31} & \alpha_{32} \end{bmatrix}$
$\alpha_{31} = \begin{bmatrix} -.2941 & 1.0275 & 1.0275 & 2.3882 \end{bmatrix}$

$\alpha_{32} = \begin{bmatrix} 1.0275 & 2.3882 & 2.382 & 0.2471 \end{bmatrix}$
and $\alpha_4 = \begin{bmatrix} \alpha_{14} & \alpha_{24} \end{bmatrix}$ with
$\alpha_{14} = \begin{bmatrix} -0.70 & 0.65 & 0.65 & 2.84 & 0.65 & 2.84 & 2.84 & 3.44 \end{bmatrix}$
and $\alpha_{24} = \begin{bmatrix} 0.65 & 2.84 & 0.65 & 2.84 & 3.44 & 2.84 & 3.44 & 3.44 \end{bmatrix}$

In order to illustrate the performance of the control law under which the parameters are optimized via the GA, we simulate the behaviors of the state variables $(x_1; \hat{x}_1)$ and $(x_2; \hat{x}_2)$.

The initial condition have been chosen as
$x(0) = \begin{bmatrix} 0.5 & -0.5 \end{bmatrix}^T$.

The simulation results presented in Figs. 3.3 and 3.4 illustrate the correlation between the two solutions of the system (3.19). This good result confirms that the satisfactory polynomial feedback obtained, using the GA, allows the linearization of the nonlinear system (3.20) via the diffeomorphism.

A satisfactory dynamical evolution of the control law signal appears in Fig. 3.5.

Figure 3.6 shows the evolution of the quadratic error. Initially, the error is rather significant since the choice of the parameter is not adequate. After that, this error decreases gradually during generations.

3.4.3 Stability Analysis

In this section, we will analyze the stability of the synthesized polynomial feedback control law based on the determination of a polynomial Lyapunov function [1, 10] Via this Lyapunov function, we will determine an initial asymptotic stability region for the isothermal nonlinear CSTR controlled model using (3.24).

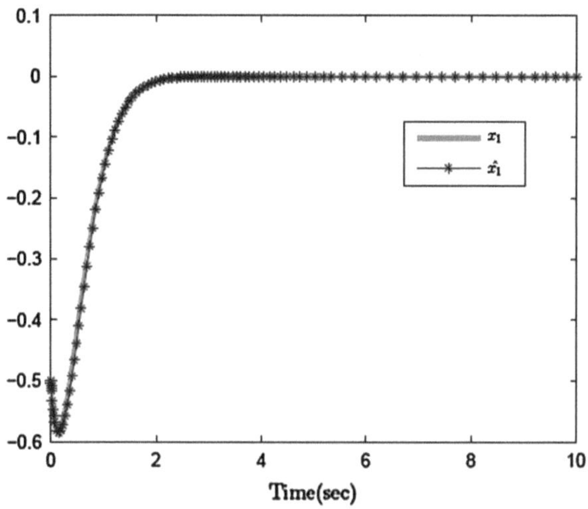

Fig. 3.3 Evolution of the state variables x_1 and $\hat{x}_1$

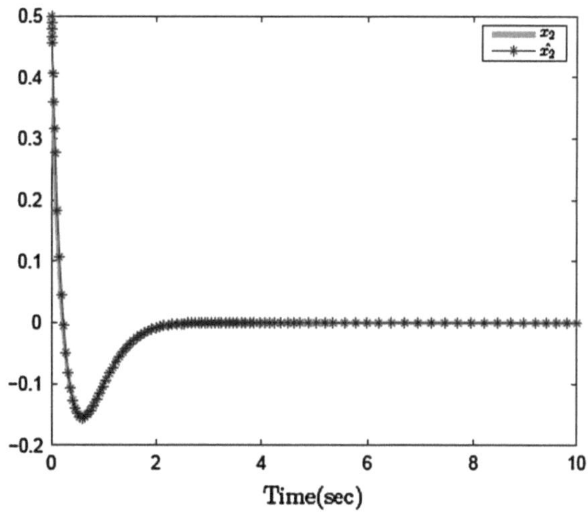

Fig. 3.4 Evolution of the state variables x_2 and $\hat{x}_2$

Definition 4.3.1 Carleman Linearization [25]
Consider the following nonlinear system:

$$\dot{x} = f(x). \tag{3.25}$$

The Taylor series can be written as

$$f(x) = \sum_{l} A^l x^{[l]} \tag{3.26}$$

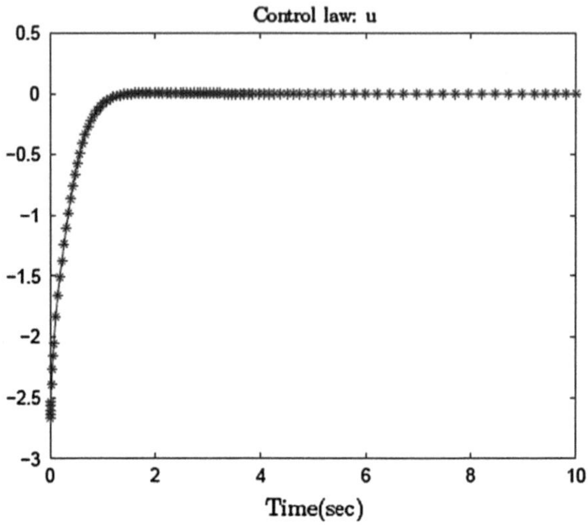

Fig. 3.5 Evolution of control signal

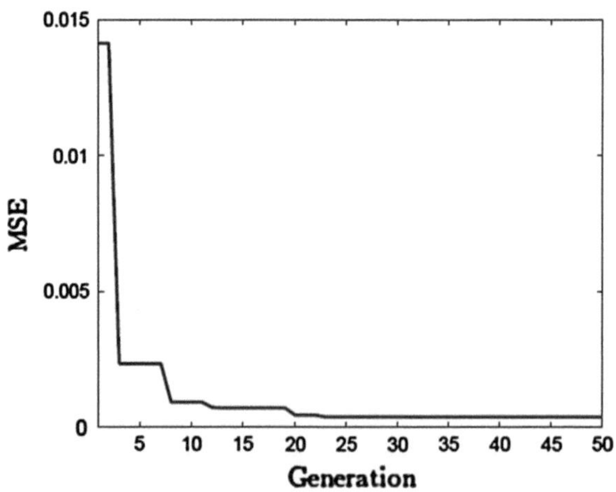

Fig. 3.6 Evolution of the quadratic error

with some algebraic manipulations, this leads to

$$\dot{x}^{[k]} = \sum_{i=1}^{k} x \otimes \cdots \otimes \underbrace{\dot{x}}_{i} \otimes \cdots \otimes x. \tag{3.27}$$

Substituting (3.26) in (3.27), we obtain

$$\dot{x}^{[k]} = \sum_{i=1}^{k} A_k^l x^{[k+l-1]}, \tag{3.28}$$

where

$$A_k^l = \sum_{i=1}^{k} I \otimes \cdots \otimes I \otimes A^l \otimes I \otimes \cdots \otimes I \tag{3.29}$$

or:

$$A_k^l = A_1^l \otimes I^{[k-l]} + I \otimes A_{k-1}^l. \tag{3.30}$$

Let us set

$$w = \langle x, x \otimes x, x \otimes x \otimes x, \ldots \rangle^T. \tag{3.31}$$

So we obtain the linear system as

$$\dot{w} = Aw \tag{3.32}$$

with

$$A = \begin{pmatrix} A_1^1 & A_1^2 & A_1^3 & \cdots & \cdots & \cdots \\ 0 & A_2^1 & A_2^2 & A_2^3 & \cdots & \cdots \\ 0 & 0 & A_3^1 & A_3^2 & A_3^3 & \cdots \\ \vdots & \vdots & \vdots & \vdots & \vdots & \vdots \end{pmatrix}.$$

.

Theorem 4.3.1 Stability in sense of Lyapunov
Consider the system:

$$\dot{x} = f(x), \quad f(0) = 0. \tag{3.33}$$

Assume that within a neighborhood Ω of the origin,

(i) $V(x)$ is locally positive definite
(ii) $\dot{V}(x)$ is negative semi-definite
(iii) The set $R = \{x \in \Omega / \dot{V}(x) = 0\}$ contains no solutions to (3.32) apart for the origin.

Then, the origin is asymptotically stable.

To determine a Lyapunov function for stable nonlinear systems, we use a defined Lyapunov equation.

Corollary 4.3.1 Lyapunov equation [6, 40]
A linear time invariant system given by:

$$\dot{x} = Ax \tag{3.34}$$

is asymptotically stable if

$$A^T P + PA = -Q. \tag{3.35}$$

The Lyapunov function is

$$V = x^T P x. \tag{3.36}$$

Theorem 4.3.2 McCann's theorem [28]

Consider the linear system defined by (3.32) *and let P satisfy the following Lyapunov function:*

$$A^T P + PA = -I. \tag{3.37}$$

Then,

$$V = \langle (x, x \otimes x, \ldots), P(x, x \otimes x, \ldots) \rangle^T, \tag{3.38}$$

$$\dot{V} = -\left\| (x, x \otimes x, x \otimes x \otimes x, \ldots) \right\|^2. \tag{3.39}$$

The nonlinear system (3.32) *is asymptotically stable if* (3.36) *is soluble for P (positive definite).*

Setting in autonomous form of the system (3.23)

The problem of guaranteed stability region around an operating point requires the setting into the autonomous form of the state Eq. (3.1) as follows:

$$\dot{x} = \sum_{i=1}^{q} P_i x^{[i]}, \tag{3.40}$$

where P_1 must be a Hurwitz matrix.

Using the polynomial forms of u and z given by (3.9), the feedback can expressed as

$$u = \sum_{i \geq 1} \alpha_i x^{[i]} + \sum_{k \geq 1} \mu_k x^{[k]} \tag{3.41}$$

with

$$\mu_k = \sum_{l=0}^{k-1} \beta_l \otimes (-k\phi_{k-l}). \tag{3.42}$$

Replacing u by its polynomial expression (3.41) in Eq. (3.8), we easily obtain the autonomous form of this model expressed under the form (3.40), where:

$$P_i = f_i + g_0 (\alpha_i + \mu_i). \tag{3.43}$$

In our case, using (3.42) and (3.43), the system (3.1) is written into autonomous form (3.40) truncated in order 7, with:

$$P_1 = \begin{bmatrix} -0.5 & -2 \\ 2 & -5 \end{bmatrix}, P_2 = \begin{bmatrix} 0 & 0.2 & -0.2 & 0 \\ 0 & 0 & 0 & -1 \end{bmatrix}$$

$$P_3 = \begin{bmatrix} P_{13} & P_{12} \end{bmatrix}$$

$$P_{13} = \begin{bmatrix} -1.09 & -0.17 & -0.17 & 058 & -0.57 \end{bmatrix}$$

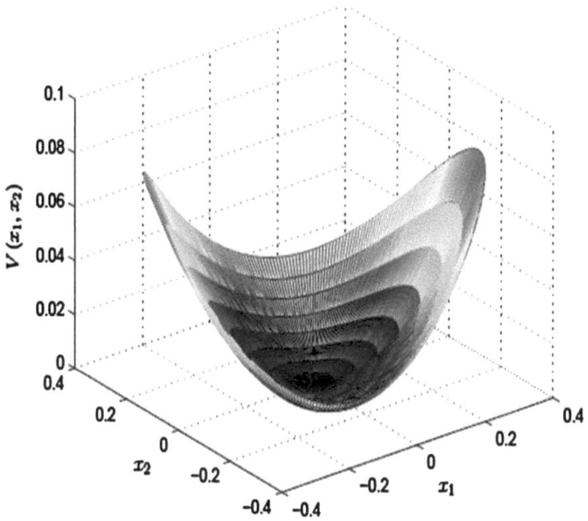

Fig. 3.7 Illustrative Diagram $V(x_1, x_2)$ of Eq. (3.44)

$P_{23} = \begin{bmatrix} -0.011\ 1.18\ -1.55\ \text{zeros}(1,8) \end{bmatrix}$
$P_4 = \begin{bmatrix} P_{14}\ P_{24}\ P_{34} \end{bmatrix}$
$P_{14} = \begin{bmatrix} -1.1\ 0.05\ -0.94\ 0.44\ 0.65\ 2.84\ 0 \end{bmatrix}$
$P_{24} = \begin{bmatrix} 0.44\ -0.15\ \text{zeros}(1,8)\ -2.34\ -0.65\ 0.04 \end{bmatrix}$
$P_{34} = \begin{bmatrix} 0.75\ -0.15\ -1.05\ 1.2\ -0.88\ \text{zeros}(1,8) \end{bmatrix}$
$for\ i = 5\ to\ \ 7,\ P_i = \text{zeros}(2, 2^i)$

Determination of Lyapunov function for (3.23)

Truncating A, defined by (3.32), to seventh order and applying McCann's theorem, we obtain the following Lyapunov function:

$$V(x_1, x_2) = V_2x^{[2]} + V_3x^{[3]} + V_4x^{[4]} + V_5x^{[5]}, \tag{3.44}$$

where:
$V_2 = \begin{bmatrix} 0.4\ -0.02\ 0\ 0.18 \end{bmatrix}$
$V_3 = \begin{bmatrix} 0\ 0\ 0\ 0.1\ 0\ 0\ 0\ -0.02 \end{bmatrix}$
$V_4 = \begin{bmatrix} -0.01\ -0.07\ 0\ 0.3\ \text{zeros}(1,3)\ -0.09\ \text{zeros}(1,7)\ 0.06 \end{bmatrix}$
$V_5 = \begin{bmatrix} 0\ 0\ 0\ -0.015\ \text{zeros}(1,28) \end{bmatrix}$
and

$$\dot{V}(x_1, x_2) = -{x_1}^2 - {x_2}^2 - {x_1}^4 - 1.6{x_1}^3x_2 - 1.8{x_1}^2x_2^2 - {x_2}^4. \tag{3.45}$$

The two Eqs. (3.44) and (3.45) are represented in Figs. 3.7 and 3.8. It is obvious that $V\ (x_1, x_2)$ is positive and $\dot{V}\ (x_1, x_2)$ is negative, then and according to the theorem's statement 4.3.1 above, we deduce the asymptotic stability of the studied point.

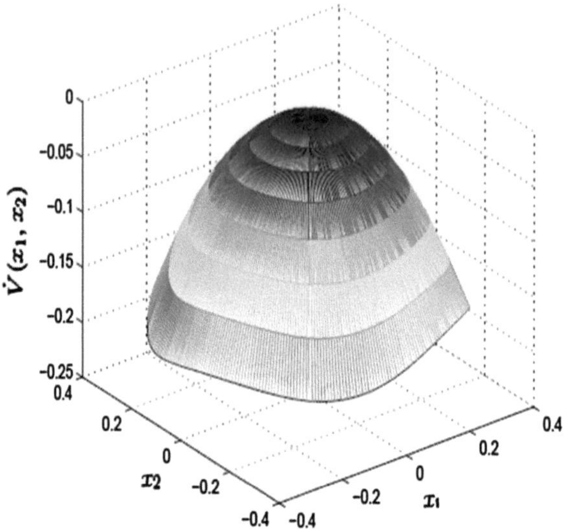

Fig. 3.8 Illustrative Diagram $\dot{V}(x_1, x_2)$ of Eq. (3.45)

3.5 Conclusion

To sum up, this chapter dealt, first, with the problem of the design of a controller for SISO affine nonlinear systems. The idea of the proposed approach is the use of the customary input-state feedback linearization technique without going through its stringent requirements. The solution of the problem is given using a recurrent algorithm characterized by arbitrary parameters which are optimized via the GA method. The new idea is applied successfully to design the controller of a chemical reactor. The stability of designed controller is ensured and checked via an analysis in the Lyapunov sense.

References

1. Bacha A, Jerbi H (2008) Backward iteration approaches for the stability domain of estimation of discrete nonlinear polynomial systems. Int J Model Ident Control 5(4):313–319
2. Baili S, Guoyuan Q, Barend JVW (2012) Output feedback predictive control for uncertain non-linear switched systems. Int J Model Ident Control 17(3):195–205
3. Benhadj Braiek E, Rotella F (1995) State observer design for a class of nonlinear systems. SAMS 17:265–277
4. Bortoff S (1997) Approximate state-feedback linearization using spline functions. Automatica 33(8):1449–1458
5. Brewer J (1978) Kronecker products and matrix calculus in system theory. IEEE Trans Circuits Syst 25(9):772–781

6. Camilli F, Grüne L, Wirth F (2000) Zubov's method for perturbed differential equations. In: Proceedings of the mathematical theory of networks and systems
7. Chanchareon R, Pongsin N, Reeseewat P, Jiamluksanapaisal S (2006) Controller design based on computed feedback linearization. In: IEEE conference on robotics, automation and mechatronics, pp 1–6
8. Chiasson J, Bodson M (1993) Technical notes and correspondence. IEEE Trans Autom Control 38(11):1662–1666
9. Choi H, Lim J (2005) Stabilizing a class of nonlinear systems based on approximate feedback linearization. IEICE Trans Fundam E88-A(6):1626–1630
10. Dongkai Z, Chaoli W, Jiqing Q, Hua C (2012) State-feedback stabilisation for stochastic nonholonomic systems with Markovian switching. Int J Model Ident Control 16(3):212–228
11. Fuh C (2009) Optimal control of chaotic systems with input saturation using an input-state linearization scheme. Commun Nonlinear Sci Numer Simul 14(8):3424–3431
12. Garc'ıa-Sandoval J, Gonzalez-Alvarez V, Castillo-Toledo B, Pelayo-Ortiz C (2008) Robust discrete control of nonlinear processes: application to chemical reactors. Compu Chem Eng 32(12):3246–3253
13. Ge S, Sun Z, Lee T (2001) Nonregular feedback linearization for a class of second-order nonlinear systems. Automatica 37(11):1819–1824
14. Goldberg D (1989) Genetic algorithms in search, optimization, and machine learning. Addison-wesley, Boston
15. Guardabassi G, Savaresi S (2001) Approximate linearization via feedback-an overview. Automatica 37(1):1–15
16. Holland J (1992) Adaptation in natural and artifcial systems: an introductory analysis with applications to biology, control, and artifcial intelligence. The MIT Press, Cambridge
17. Hunt L, Su R, Meyer G (1983) Global transformations of nonlinear systems. IEEE Trans Autom Control 28(1):24–31
18. Isidori A (1995) Nonlinear control systems. Springer Verlag, New York
19. Jemai W, Jerbi H, Abdelkrim M (2010a) A method based on the Monte Carlo optimization schemes for the control of nonlinear polynomial systems. In: Proceedings of the 6th WSEAS international conference on Dynamical systems and control, pp 184–189. World Scientific and Engineering Academy and Society (WSEAS)
20. Jemai W, Jerbi H, Abdelkrim M (2010b) Synthesis of an approximate feedback nonlinear control based on optimization methods. WSEAS Trans Syst Control 5(8):646–655
21. Jemai W, Jerbi H, Abdelkrim M (2011) Nonlinear state feedback design for continuous polynomial systems. Int J Control Autom Syst (IJCAS) 9(3):566-573
22. Jianfeng L, Youmei L, Zhiwu H, Weihua G, Huosheng H (2012), Modelling and control design for an electro-pneumatic braking system in trains with multiple locomotives. 17(2):99–108
23. Johansen T, Hunt K (2000) A computational approach to approximate input/state feedback linearization. In: Proceedings of the 39th IEEE conference on decision and Control, vol 5, pp 4467–4472
24. Khalil H, Grizzle J (2002) Nonlinear systems. Prentice hall, New Jersey
25. Kowalski K, Steeb W (1991) Nonlinear dynamical systems and Carleman linearization. World Scientific Publishing Co. Pte, Ltd, Singapore
26. Lei C, Xiu-Juan Z, Huai-Yu W, Quan-Min Z, Yong-Ji W, Hassan N (2012) A new procedure for multi-mode sequential flocking with application to multiple non-holonomic mobile robot motion control: implementation and analysis. Int J Model Ident Control 16(1):50–59
27. Marino R, Tomei P (1995) Nonlinear control design: geometric, adaptive and robust. Prentice Hall, New Jersey
28. McCann R (1979) Asymptotically stable dynamical systems are linear. Pac J Math 81(2): 475–479
29. Mehta S, Chiasson J (1998) Nonlinear control of a series DC motor: theory and experiment. IEEE Trans Industr Electron 45(1):134–141
30. Nijmeijer H, Van der Schaft A (1990) Nonlinear dynamical control systems. Springer, New York

31. Ray W (1981) Advanced process control. McGraw-Hill Companies, New York
32. Rotella F, Dauphin-Tanguy G (1988) Nonlinear systems: identification and optimal control. Int J Control 48(2):525–544
33. Qi-ming F, Quan L, Xio-yan W, Le Z (2011) Relevance feedback techniques and genetic algorithm for image retrieval based on multiple features. Int J Model Ident Control 14(4): 279–285
34. Sivanandam S, Deepa S (2008) Introduction to genetic algorithms. Springer, Berlin, New York
35. Slotine J, Li W et al (1991) Applied nonlinear control. Prentice hall, Englewood Cliffs
36. Son J, Lim J (2008) Stabilization of approximately feedback linearizable systems using singular perturbation. IEEE Trans Autom Control 53(6):1499–1503
37. Sontag E (1998) Mathematical control theory: deterministic finite dimensional systems. Springer, Berlin
38. Su R (1982) On the linear equivalents of nonlinear systems. Syst Control Lett 2(1):48–52
39. Tiaoping F, Shang YL (2011) An improved multi-objective particle swarm optimization algorithm. 12(1–2):66–71
40. Vannelli A, Vidyasagar M (1985) Maximal Lyapunov functions and domains of attraction for autonomous nonlinear systems. Automatica 21(1):69–80
41. Vinodha R, Abraham Lincoln S (2011) Design and implementation of simple adaptive control schemes on simulated model of CSTR process. Int J Model Ident Control 14(3):159–169

Biography notes: Houssem JERBI was born in Sfax, Tunisia, in 1971. He received a Master's degree in Electrical Engineering from the National School of Engineering of Tunis, Tunisia, in 1995, and the Ph.D. degree in Automatic Control from the National School of Engineering of Tunis, Tunisia, in 2001. Currently, he is an Associate Professor in Automatic Control at the University of Hail—Kingdom of Saudi Arabia. His research interests include Nonlinear Systems, Power Systems, Control Systems, and Optimization Theory.
E-mail: h.jerbi@uoh.edu.sa, houssem.jerbi@gmail.com
Wiem Jebri Jemai received the Diploma in Electrical Engineering (Control) from the National School of Engineers of Gabès (ENIG), Tunisia in 2002, and the Master's degree in Automatic Control from the National School of Engineers of Sfax (ENIS), Tunisia in 2004. She is a member of Research Unit: Modeling, Analysis, and Control of Systems (MACS), at the ENIG, University of Gabès, from 2008.

E-mail: wiem_jeb@yahoo.fr

Chapter 4
Recent Advances in Nonsingular Terminal Sliding Mode Control Method

Shengbo Eben Li and Kun Deng

Abstract The terminal sliding mode (TSM) control method has become a hot topic in recent years due to its special merit on finite-time convergence and good robustness. One critical issue is how to balance the singularity of control law and the fast convergence of closed-loop system. The chapter reviews the research history of the singularity and introduces the recent advance on nonsingular and fast terminal sliding mode (NFTSM) control method. The synthesis of NFTSM controller synthesis is based on a newly proposed nonsingular fast terminal function and a terminal attractor with nonnegative exponential coefficient. Both theoretical analyses and computer simulations have proved its effectiveness under the condition that plant uncertainties are bounded.

4.1 Introduction

The Sliding Mode Control (SMC) method has been widely recognized for its good robustness to certain parameter variations and external disturbances [1, 2]. The SMC family has many variants, e.g., discrete-time SMC [3], adaptive SMC [4], dynamic SMC [5], backstepping SMC [6, 7], et al., due to its easy-to-use property and relatively simple structure. The terminal sliding mode (TSM) control method is also one of the most famous variants in recent years. The closed-loop system of SMC often has two modes, i.e., reaching mode and sliding mode. The former is to enable the

S. Eben Li
State Key Lab of Automotive Safety and Energy, Department of Automotive Engineering, Tsinghua University, Beijing,100084, China
e-mail: lisb04@gmail.com

K. Deng (✉)
Coordinated Science Laboratory, University of Illinois at Urbana-Champaign, Urbana, IL, 61801, USA
e-mail: kundeng2@illinois.edu

L. Liu et al. (eds.), *Applied Methods and Techniques for Mechatronic Systems*, Lecture Notes in Control and Information Sciences 452,
DOI: 10.1007/978-3-642-36385-6_4,

state trajectory to converge to the sliding hyperplane, and the latter is to force the state to converge to the equilibrium point. The design of sliding mode controller accordingly contains two tasks: one is to design a suitable reaching law for the reaching mode, and the other is to choose an appropriate sliding function for the sliding mode. In TSM, a terminal attractor is employed as its reaching law and a terminal function is selected as its sliding function. This novel design generates special merits on system performance, such as finite-time convergence, good robustness to uncertainties, and small steady-state errors [8–11].

The terminal function plays a central role in designing TSM controllers. Its finite-time convergence property is the major merit of the TSM control method. This type of function may have other names, e.g., terminal attractor or terminal sliders, depending on their application, but actually possesses similar mathematical forms. To our knowledge, Venkataraman and Gulati [8] from California Institute of Technology were the first researchers to introduce terminal function to sliding mode control field [8]. Considering a second-order robotic system, an earlier terminal function (TF) is selected as

$$s = \dot{x} + \alpha \cdot x^{p/q}. \tag{4.1}$$

In Eq. (4.1), $x \in \mathbb{R}$ is the system state, $\alpha \in \mathbb{R}^+$ and $p, q \in N$, satisfying $p < q$. One of its critical features is that the exponent p/q is on state x instead of on the derivative. It is well known that a linear sliding function generates exponential stability, and the state will infinitely approach, but never exactly equal the equilibrium point. On the contrary, the terminal sliding function enables a finite-time convergence property and the state can reach the equilibrium point in a discontinuous behavior. For multi-input multi-output (MIMO) systems, Man and Yu [12] extended this type of terminal function to high-dimensional situations [12]. The following research also proved its finite-time convergence property for some cascading systems [13]. This control method was applied to n-link rigid robotic manipulators and its good robustness to large uncertain dynamics was also observed [14].

Compared with linear sliding function, earlier terminal function might have slow convergence speed although it can still converge in finite time. The reason is as follows: since $p < q$ in Eq. (4.1), the exponent of state x is smaller than 1. Thus, when far away from equilibrium point, i.e., $x \gg 1$, the derivative $\dot{x}$ in sliding mode will be much smaller than that in a similar linear sliding function, which thus largely slows down the convergence process. To address this issue, Yu and Man et al. [15] proposed a terminal function with fast convergence characteristics [15], called fast terminal function (FTF)

$$s = \dot{x} + \alpha \cdot x^{\gamma} + \beta \cdot x^{\rho}. \tag{4.2}$$

In Eq. (4.2), $0 < \gamma < 1, \rho > 1, \gamma, \rho \in \mathbb{R}^+$. Compared with Eq. (4.1), Eq. (4.2) contains a polynomial term of order higher than 1. When $x \ll 1$, the high-order term is negligible, and hence Eq. (4.2) can be approximated by Eq. (4.1). When $x \gg 1$, the high-order term is able to strengthen control inputs, forcing Eq. (4.2) to converge faster than any linear sliding function. Yu and Guo et al. [16]

obtained its global description and applied it to design both reaching law and sliding function [16]. This method can be easily extended by introducing new types of mathematical functions, e.g., Kang and Xi et al [17]:

$$s = \dot{x} + \frac{\alpha}{k}\left(e^{kx} - 1\right)^{q/p} x^{-kx}. \tag{4.3}$$

In Eq. (4.3), $0 < k < 1$ and all other parameters are similar to those in Eq. (4.1). The theoretical analysis shows that on the sliding hyperplane the system is asymptotically stable and system state converges faster than Eq. (4.1).

One major drawback of TSM control is the singularity issue. When the system state is close to zero, the negative exponent of state in the control law will generate infinite values, not able to be implemented by actual actuators. This phenomenon is named the singularity of controllers. To resolve this issue, Yu and Man et al. [15] first modified the high-order fast terminal function and pointed out that the control input will be bounded in mathematics if some parameters satisfy the following inequality [15]:

$$\rho_{k+1} > \gamma_{k+1} > \frac{n-k-1}{n-k}, 0 \le k \le n-1, \tag{4.4}$$

where ρ, γ are the same as Eq. (4.2), n is the dimension of state, k represents the k-th state. With this inequality condition, [18] and [19] designed a terminal sliding mode control law for a nonlinear dynamic system. Simulation showed that the accompanying coefficient of state approaches zero at a higher speed than state itself, thus eliminating the singular phenomenon.

However, for actual plants, the coefficient of state may not be exactly equal to zero when close to equilibrium points. The instantaneous singularity still exists because of parameter perturbations, external disturbances, and measurement errors. To obtain a truly nonsingular control law, Feng et al. [20] developed a nonsingular terminal function (NTF), realizing that the state converges in finite time during the sliding mode and the control law has no negative exponential term [20], shown as follows:

$$s = \frac{1}{\beta}\dot{x}^{p/q} + x. \tag{4.5}$$

In Eq. (4.5), $p > 0, q > 0$, q is integer, and $1 < p/q < 2$. Combined with the global reaching conditions, a nonsingular terminal sliding mode (NTSM) control law can be designed [21, 22]. But this kind of terminal function has slow convergence speed in the region far away from equilibrium point. Moreover, a switching term inevitably exists in the control law, causing the commonly known chattering problem (another famous issue in sliding mode control). This largely limits its application in engineering practice. Zhang et al. [23] used the hysteresis layer to weaken chattering to some extent [23]. Hu et al. [24] developed an adaptive TSM control method. The method estimates the boundary of plant uncertainties and adjusts the control gain in a real-time manner [24]. The method can somewhat reduce the chattering issue in the condition of small uncertainties by avoiding using unnecessary large gain.

The designing method of terminal attractor is similar to the terminal function. By replacing state x with sliding mode variable s, all of the abovementioned terminal functions can be transferred to terminal attractors. In general, for the sake of simplicity, there are two commonly used terminal attractors, derived from Eqs. (4.1) and (4.2):

$$\dot{s} + \alpha \cdot s^{p/q} = 0$$
$$\dot{s} + \alpha \cdot s^{\gamma} + \beta \cdot s^{\rho} = 0 \tag{4.6}$$

In summary, the rapidity and singularity are the two key concerns in terminal sliding mode control. The fast terminal function or attractor can be used to improve convergence. The singularity is more difficult to handle in reality. The solutions up to now include the inequality constraint method and nonsingular terminal function. In real-world applications, the former still has singularity due to unexpected plant uncertainties; the latter avoid this difficulty, but might cause other issues such as slow convergence and chattering phenomenon. These shortcomings limit their application in practice.

To comprehensively deal with singularity, chattering, and slow convergence, this chapter introduces a nonsingular and fast terminal sliding mode control method. The remainder of this chapter is structured as follows. Section 4.2 reviews the conventional TSM control method. Section 4.3 introduces a newly proposed nonsingular fast terminal function and its corresponding terminal attractor, followed by how to synthesize the control law. Section 4.4 proves the convergence characteristics in both sliding and reaching modes, and the global existence of the sliding hyperplane. Section 4.5 demonstrates the success of this method. Section 4.6 concludes this chapter.

4.2 Problem Description

Consider a second-order SISO nonlinear system

$$\begin{cases} \dot{x}_1 = x_2 \\ \dot{x}_2 = f(x,t) + u, \end{cases} \tag{4.7}$$

where $x = [x_1, x_2]^T \in \mathbb{R}^2$, $u \in \mathbb{R}$, $f : (\mathbb{R}^2, \mathbb{R}) \to \mathbb{R}$. The conventional nonsingular terminal function (NTF) is defined for the sliding hyperplane [20]

$$s = x_1 + \frac{1}{\beta} x_2^{p/q}, \tag{4.8}$$

where $\bar{\beta} \in \mathbb{R}^+, \eta \in \mathbb{R}^+$ and p, q are odds, satisfying $1 < p/q < 2$. The corresponding nonsingular terminal sliding mode (NTSM) control law is [20]

$$u = -f(x,t) - \frac{\bar{\beta} q}{p} x_2^{2-p/q} - \eta \mathrm{sgn}(s). \tag{4.9}$$

In Eq.(4.9), the exponent of x_2 is always larger than zero, and therefore the control law completely avoids singularity. When the system is in sliding mode, we have $s = 0$ and

$$\dot{x}_1 = -\bar{\beta}^{q/p} \cdot x_1^{q/p} \tag{4.10}$$

However, the exponent of x_1 is smaller than 1 in Eq. (4.10). This is the reason causing slow convergence in large region. Another inherent shortcoming accompanying the NTSM controller is that Eq. (4.9) has a switching structure, resulting in unavoidable chattering issue. It is undesirable for precise tracking control. Even worse, it might be able to excite high-order unmodeled dynamics and cause instability.

4.3 Nonsingular and Fast TSM Control

As discussed before, the main reason for slow convergence is due to the exponent of state. Intuitively, if Eq. (4.10) has a polynomial term of degree higher than 1, the derivative of state will be increased in large region and consequently conquer this problem. Hence, we proposed a nonsingular fast terminal function (NFTF) with non-integer exponents for the sliding mode [25]

$$s = x_1 + \frac{1}{\alpha} x_1^{g/h} + \frac{1}{\beta} x_2^{p/q}, \tag{4.11}$$

where $\alpha \in \mathbb{R}^+$, $\beta \in \mathbb{R}^+$, $p, q, g, h \in N^+$ are odds, satisfying

$$\begin{aligned} &1 < p/q < 2, \\ &g/h > p/q. \end{aligned} \tag{4.12}$$

The reaching law is then defined as a terminal attractor with nonnegative exponential coefficient [25]

$$\dot{s} = \left(-\phi s - \gamma s^{m/n}\right) x_2^{p/q-1}, \tag{4.13}$$

where $\phi \in \mathbb{R}^+$, $\gamma \in \mathbb{R}^+$, $m, n \in N^+$ are odds, satisfying

$$1 < m/n < 2. \tag{4.14}$$

Substituting Eq. (4.11) and Eq. (4.13) into Eq. (4.7), we have a nonsingular fast terminal sliding mode (NFTSM) control law [25]

$$u = -f(x,t) - \frac{\beta q}{p}\left(\phi s + \gamma s^{m/n}\right) - \frac{\beta q}{p} x_2^{2-p/q}\left(1 + \frac{g}{\alpha h} x_1^{g/h-1}\right). \qquad (4.15)$$

The merit of NFTSM control method lies in the special structure of functions used in defining sliding mode (4.11) and reaching law (4.13). In sliding status, the sliding mode function dominates the closed-loop performance. Compared with other sliding functions, e.g., NTF, the NFTF contains a high-order exponential term of x_1. It has the ability to increase control gain when far away from equilibrium points, thus increasing the convergence speed in large region. Outside of sliding status, the terminal attractor in reaching law is used to match the specially designed NFTF. The synthesized control law eliminates any potential singularity in control inputs while not using any switching functions, thus no chattering issue.

4.4 Performance Analysis

4.4.1 Convergence Analysis for Sliding Mode

In the sliding mode, i.e., $s = 0$, the system dynamic is characterized by the sliding hyperplane. The qualitative analysis in Sect. 4.3 points out that NFTF has finite convergence time and converges faster than conventional NTF. In this section, we rigorously prove the convergence property of NFTF and NTF using the Lyapunov method. Then, we use computer simulations to compare the convergence speeds of NFTF and NTF. Before presenting the main proof we introduce a useful lemma below.

Lemma 1: Consider a Gauss hyper-geometric function

$$F(A, B, C, z) = \sum_{k=0}^{+\infty} \frac{(A)_k (B)_k}{(C)_k\, k!} z^k.$$

If $A, B, C \in \mathbb{R}^+$ and $C - A - B > 0$, then the function $F(A, B, C, z)$ converges for any $z < 0$. (The proof is given in [25])

Theorem 1: For system (4.7), any state, lying on the NFTF hyperplane (4.11), converges to zero in finite time T. The convergence time T is given by

$$T = 2\tau_1^{-q/p} \frac{p}{p-q} V(0)^{\frac{p-q}{2p}} F\left(A, B, C, -\frac{\tau_2}{\tau_1} V(0)^{\frac{g-h}{2h}}\right), \qquad (4.16)$$

where $\tau_1 = 2^{\frac{p+q}{2q}}\beta$, $\tau_2 = ^{\frac{p}{2q}+\frac{g}{2h}} \frac{\beta}{\alpha'}$ $A = \frac{q}{p'}$ $B = \frac{(p-q)h}{p(g-h)'}$ $C = \frac{pg-qh}{p(g-h)'}$ $V(t) = \frac{1}{2}x_1^2(t)$ and $F(\cdot)$ is the Gauss hyper-geometric function.

Proof: When the system state lies on the sliding hyperplane, we have $s = 0$. Substituting $s = 0$ into Eq. (4.11), we obtain

$$\dot{x}_1^{p/q} = -\beta x_1 - \frac{\beta}{\alpha} x_1^{g/h}. \tag{4.17}$$

Define a Lyapunov function

$$V(t) = 0.5 x_1^2(t), \tag{4.18}$$

and obtain its derivative as

$$\dot{V}^{\frac{p}{q}} = -\beta x_1^{\frac{p+q}{q}} - \frac{\beta}{\alpha} x_1^{\frac{p}{q}+\frac{g}{h}}. \tag{4.19}$$

Note that p, q, g, and h are all odds. Thus $\dot{V} < 0$ for any $x \neq 0$. Using the Lyapunov theory of stability, we know that the closed-loop system is stable in the sliding mode, and the state at least asymptotically converges to the equilibrium point. Using Eq. (4.17), we find that the equilibrium point is unique, i.e., $x = 0$. To solve for the convergence time, we substitute $x_1^2 = 2V$ into Eq. (4.19) and obtain

$$\dot{V}^{\frac{p}{q}} = -2^{\frac{p+q}{2q}} \beta V^{\frac{p+q}{2q}} - 2^{\frac{p}{2q}+\frac{g}{2h}} \frac{\beta}{\alpha} V^{\frac{p}{2q}+\frac{g}{2h}}. \tag{4.20}$$

Using $\dot{V} = dV/dt$ to Eq. (4.20), we have

$$\mathrm{d}t = -\left(2^{\frac{p+q}{2q}} \beta V^{\frac{p+q}{2q}} + 2^{\frac{p}{2q}+\frac{g}{2h}} \frac{\beta}{\alpha} V^{\frac{p}{2q}+\frac{g}{2h}}\right)^{-q/p} \mathrm{d}V. \tag{4.21}$$

Let us define $\tau_1 = 2^{\frac{p+q}{2q}} \beta$, $\tau_2 = 2^{\frac{p}{2q}+\frac{g}{2h}} \frac{\beta}{\alpha}$, $A = \frac{q}{p}$, $B = \frac{(p-q)h}{p(g-h)}$, $C = \frac{pg-qh}{p(g-h)}$. Integrating both sides of Eq. (4.21), we obtain the convergence time T as

$$T = -\left. 2\tau_1^{-q/p} \frac{p}{p-q} V^{\frac{p-q}{2p}} F\left(A, B, C, -\frac{\tau_2}{\tau_1} V^{\frac{g-h}{2h}}\right)\right|_{V(0)}^{V(T)}. \tag{4.22}$$

It is clear that the system in sliding mode converges to the equilibrium point, i.e.,$x = 0$, and thus we have $V(T) = 0$. According to the property of Gauss hypergeometric function, we know that $F(\cdot,\cdot,\cdot,0) = 1$. Substituting $V(T) = 0$ and $F(\cdot,\cdot,\cdot,0) = 1$ into Eq. (4.22), we obtain Eq. (4.16).

Moreover, considering $q < p$ and $h < g$, we have $A > 0$, $B >, C > 0$, and $C - A - B > 0$. Using Lemma 1 and the fact that $V(0) > 0$, we know that $F(A, B, C - \frac{\tau_2}{\tau_1} V(0)^{\frac{g-h}{2h}})$ is bounded. Thus, the convergence time T is finite, which implies that on the sliding mode hyperplane the system state converges to the equilibrium point in finite time. (End of Proof)

Table 4.1 Selected parameters of NFTF and NTF

	p	q	g	h	α	β	$\bar{\beta}$
NFTF	5	3	7	3	7	2	–
NTF	5	3	–	–	–	–	2.3

As discussed in Sect. 4.3, NFTF has faster convergence speed than NTF due to the high-order terms of x_1. To verify this observation, we use simulations to compare their convergence speeds in the sliding stage. When system lies on the NFTF sliding hyperplane, we have $s = 0$. By combing Eq. (4.7) and Eq. (4.11), we cancel x_2 and obtain sliding equation (4.17). Similarly, we can obtain the sliding equation (4.10) for the system lying on the NTF sliding hyperplane. To quantitatively verify the convergence speed of NFTF and NTF, we assume that the two controllers have comparable coefficients except the factor g/h. Note that by setting $g/h = 1$, Eq. (4.17) reduces to the form of Eq. (4.10)

$$\dot{x}_1 = -\left(\beta + \frac{\beta}{\alpha}\right)^{q/p} x_1^{q/p}.$$

To ensure that Eqs. (4.17) and (4.10) have comparable coefficients, it is required that

$$\bar{\beta} = \beta + \frac{\beta}{\alpha}.$$

Hence, we select the following parameters for NFTF and NTF, as shown in Table 4.1.

We simulate the sliding Eqs. (4.10) and (4.17) using identical initial condition $x_1 = 9$. The state response curve of x_1 is depicted in Fig. 4.1. As shown in Fig. 4.1 (a), NFTF and NTF have the same initial states in the sliding stage. Note that NFTF is delayed by $\delta = 0.44\,s$ to start compared to NTF, but both converge to zeros around $t = 3.7\ s$. The state response curve of x_2, i.e., the derivative of x_1, is depicted in Fig. 4.1 (b). We observe that NFTF has larger state derivatives in magnitude than NTF when $x_1 \gg 1$ and almost same derivatives when $x_1 < 1$. This means that with the same parameters NFTF has faster convergence speed in sliding stage, which is consistent with the qualitative analysis presented before.

4.4.2 Convergence Analysis for Reaching Law

When the system is in the reaching stage, the convergence speed of the sliding mode s depends on the terminal attractor. It is proved in the following that terminal attractor (4.13) does not affect the existence of NFTF hyperplane and guarantees at least asymptotical convergence for the reaching stage. The proof is given below.

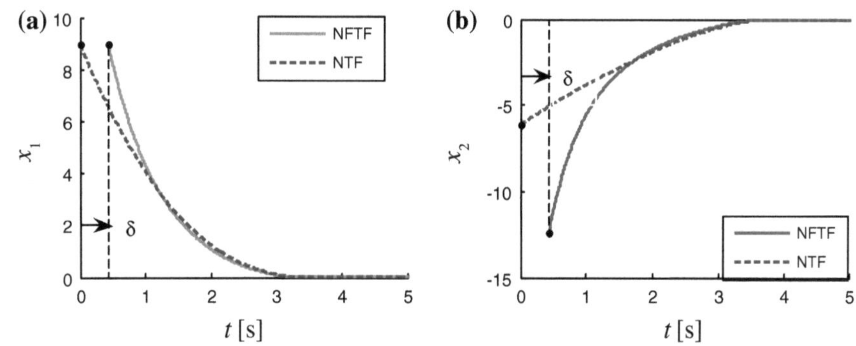

Fig. 4.1 Convergent speed comparison between NFTF and NTF **a** Response curve of state x_1 **b** Response curve of state x_2.

Theorem 2: For the system (4.7) with NFTSM control law (4.15), the sliding mode hyperplane (4.11) globally exists, and the system state asymptotically reaches the sliding mode hyperplane from any initial point.

Proof: Define a Lyapunov function

$$V = 0.5 \cdot s^2. \tag{4.23}$$

Then,

$$\dot{V} = s \cdot \dot{s} = -\left(\phi s^2 + \gamma s^{m/n+1}\right) x_2^{p/q-1}. \tag{4.24}$$

Divide the $x_1 - x_2$ phase plane into two regions

$$\begin{array}{l} D = \{\boldsymbol{x} | x_1 \neq 0, x_2 \neq 0\}, \\ \overline{D} = \{\boldsymbol{x} | x_1 \neq 0, x_2 = 0\}. \end{array} \tag{4.25}$$

For any $\boldsymbol{x} \in$ D, since p, q, m, and n are all odds, we have $x_2^{p/q-1} > 0$ and $s^{m/n+1} > 0$, which implies that $\dot{V} < 0$. According to the Lyapunov theory of stability, the sliding mode s asymptotically converges in region D, and the sliding mode hyperplane exists for the NFTSM control law. To prove that s still converges for $x \in \overline{\text{D}}$, we substitute the control law (4.13) into the system equation (4.7) and obtain

$$\dot{x}_1 = x_2, \tag{4.26}$$

$$\dot{x}_2 = -\frac{\beta q}{p}\left[\left(\phi s + \gamma s^{m/n}\right) + x_2^{2-p/q}\left(1 + \frac{g}{\alpha h} x_1^{g/h-1}\right)\right]. \tag{4.27}$$

On the $x_1 - x_2$ phase plane, we obtain the gradient equation of the state trajectory using Eq. (4.26) and Eq. (4.27)

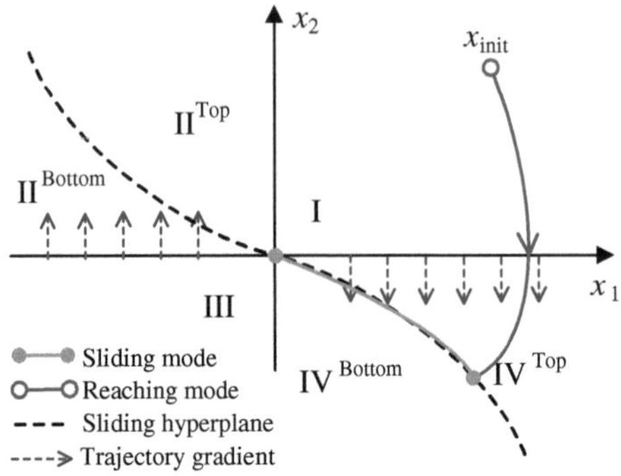

Fig. 4.2 State trajectory in reaching mode

$$\frac{dx_2}{dx_1} = \frac{\dot{x}_2}{\dot{x}_1} = \frac{-\frac{\beta q}{p}\left[\left(\phi s + \gamma s^{m/n}\right) + x_2^{2-p/q}\left(1 + \frac{g}{\alpha h}x_1^{g/h-1}\right)\right]}{x_2}. \tag{4.28}$$

Note that $p/q < 2$, and p, q, g, h, m, n are all odds. For $x_1 > 0$ and $x_2 = 0$, we have

$$\frac{dx_2}{dx_1} = -\infty, \quad \dot{x}_2 = -\frac{\beta q}{p}\left[\phi\left(x_1 + \frac{1}{\alpha}x_1^{g/h}\right) + \gamma\left(x_1 + \frac{1}{\alpha}x_1^{g/h}\right)^{m/n}\right] < 0. \tag{4.29}$$

It is observed from Eq. (4.29) that, on the x_1-positive-axis, the state trajectory is perpendicular to the x_1-axis and the state trajectory points to negative. Moreover, the velocity of the state x is not zero. Similarly, on the x_1-negative-axis, the state trajectory is perpendicular to the x_1-axis and the state trajectory points to positive. The velocity of the state x is not zero. Figure 4.2 is the illustration of the state trajectory in the reaching stage. The thick dashed line indicates the NFTF hyperplane, the thin dashed lines with arrows indicate the direction of state trajectory on $\overline{D}$, x_{init} denotes the initial state, and the solid line denotes state trajectory starting from x_{init}.

It is observed from Fig. 4.2 that the state trajectory is symmetric for NFTSM control system. Without loss of generality, we only consider the situations when the initial state lies on the upper side of the sliding hyperplane. We discuss the different situations based on the location of the initial state:

(1) Suppose $x_{\text{init}} \in \text{IV}^{\text{Top}}$. Since the state trajectory points to the negative direction on the x_1-positive-axis, the system state will not enter region I from IV^{Top} by passing through the x_1-positive-axis. Thus, we have $x \in D$ and the sliding mode s asymptotically converges in the reaching stage;
(2) Suppose $x_{\text{init}} \in \text{II}^{\text{Top}} \cup \text{I}$ and the state trajectory passes through the x_1-positive-axis. Since the state trajectory points to the negative direction on the x_1-positive-axis and the velocity is not zero, the state trajectory will not

stay on the x_1-positive-axis and must enter region $\mathrm{IV}^{\mathrm{Top}}$ by passing through the axis. Thus, the sliding mode s converges to zero finally;

(3) Suppose $x_{\mathrm{init}} \in \Pi^{\mathrm{Top}} \cup \mathrm{I}$ and the state trajectory does not pass through the x_1-positive-axis. Thus, we have $x \in D$ and the sliding mode s asymptotically converges in the reaching stage.

In summary, the state trajectory starting from arbitrary starting point either directly reaches the sliding hyperplane, or first passes through $\overline{D}$ and then reaches the sliding hyperplane. In both cases, the state trajectory does not stay in the region $\overline{D}$, and thus NFTF hyperplane globally exists. In conclusion, the system state at least asymptotically converges to the sliding hyperplane from arbitrary initial point. (End of Proof)

4.4.3 *Effect of Stagnation*

The nonnegative exponential efficient of state x_2 in the terminal attractor (4.13) is necessary to match the newly proposed NFTF for comprehensive performance enhancement. However, it also has a potential drawback if considering the whole closed-loop system, called "effect of stagnation. " This is because state x_2 may approach zero before the sliding mode s converges to zero. In such cases, even though the state can still converge in finite time, yet its convergence speed could be very slow. The following phenomenon is observed when the effect of stagnation occurs: (4.1) state x_1 converges slowly, and x_2 is quite small compared to x_1; (4.2) state x_2 is close to a constant and its derivative is almost zero. In the following, we will analyze this effect and try to find corresponding solutions.

When the effect of stagnation occurs, the derivative of x_2 is approximately equal to zero. Substituting $\dot{x}_2 \approx 0$ into Eq. (4.27), we have

$$\left(\phi s + \gamma s^{m/n}\right) + x_2^{2-p/q}\left(1 + \frac{g}{\alpha h} x_1^{g/h-1}\right) \approx 0. \tag{4.30}$$

We consider two situations here: One is for the situation when the system state is close to the equilibrium point, and the other is for the situation when the system state is far away from the equilibrium point. We obtain two approximations of Eq. (4.30):

(1) When the system state is close to the equilibrium point, we have $x_1, x_2 \ll 1$. Then $(\phi s + \gamma s^{m/n}) \approx \gamma x_1^{m/n}$ and $1 + \frac{g}{ah} x_1^{g/h-1} \approx 1$. The Eq. (4.30) is reduced to

$$\dot{x}_1 \approx -\gamma^{\frac{q}{2q-p}} x_1^{\frac{qm}{(2q-p)n}}. \tag{4.31}$$

(2) When the system state is far away from the equilibrium point, we have x_1, $x_2 \gg 1$. Then, $(\phi s + \gamma s^{m/n}) \approx \phi x_1^{g/h}$ and $1 + \frac{g}{ah} x_1^{g/h-1} \approx \frac{g}{ah} x_1^{g/h-1}$. Eq. (4.30) is reduced to

$$\dot{x}_1 \approx -\left(\frac{\phi\alpha h}{g}\right)^{\frac{q}{2q-p}} x_1^{\frac{g}{g-h}\cdot\frac{q}{2q-p}}. \tag{4.32}$$

According to Eq. (4.31) and (4.32), even if the condition $q < p < 2q$ is satisfied, the term $2q - p$ may be still smaller than q for some choices of q and p. In such cases, if in addition $\gamma < 1$ and $\phi ah/g < 1$, then the coefficients on the right-hand side of Eqs. (4.31) and (4.32) will be very small. This will largely decrease the convergence speed of the state x_1, which is the major cause of the effect of stagnation. On the contrary, to avoid the effect of stagnation, we need to increase the coefficients on the right-hand side of Eq. (4.31) and Eq. (4.32). This means that we need to choose control parameters to satisfy one of the following conditions:

(1) One can choose suitable ϕ and γ to make $\gamma < 1$ and $\phi ah/g > 1$;
(2) If the condition (4.1) is not satisfied, one can choose suitable p and q to make $q/(2q-p)$ very close to 1.

4.4.4 System Robustness

In practice, due to the modeling errors and external disturbances, the system convergence speed might be reduced [26–28]. In some cases, the system can even not converge to the equilibrium point, and the steady-state tracking trajectory can only reach a neighborhood region of the equilibrium point. Thus, the robustness of the control algorithm largely affects its effectiveness in practice. For control law (4.15), to analyze its robustness to model uncertainty and ability to external disturbance, we consider the following second-order nonlinear and uncertain SISO system:

$$\begin{cases} \dot{x}_1 = x_2 \\ \dot{x}_2 = f(x) + \Delta f(x) + u + d(t) \end{cases}, \tag{4.33}$$

where $f(\boldsymbol{x})$ and $\Delta f(\boldsymbol{x})$ are C^1 vector field, $f(\boldsymbol{x})$ and $d(t)$ are modeling error and external disturbance, satisfying $|\Delta f(\boldsymbol{x}) + d(t)| < L$.

Theorem 3: For the uncertain system (4.33), if the control law is (4.15), then system state reaches the region Π in a finite time and the convergence speed is not smaller than the dynamics defined by Eq. (4.34):

$$\dot{s} \cdot sgn(s) = -\Psi x_2^{p/q-1}, \tag{4.34}$$

where

$$\begin{aligned} \Psi &= \phi\,|s| + \gamma\,|s|^{m/n} - \tfrac{pL}{\beta q}, \\ \Pi &= \left\{x : \phi\,|s| + \gamma\,|s|^{m/n} \le \tfrac{pL}{\beta q}\right\}. \end{aligned} \tag{4.35}$$

Proof: Note that the system (4.33) contains the terms with uncertainty. We substitute the control law (4.15) into Eq. (4.33), yielding

$$\begin{aligned}\dot{x}_2 = & -\frac{\beta q}{p}\left[\left(\phi s+\gamma s^{m/n}\right)+x_2^{2-p/q}\left(1+\frac{g}{\alpha h}x_1^{g/h-1}\right)\right] \\ & +\Delta f(x)+\mathrm{d}(t).\end{aligned} \tag{4.36}$$

Differentiate Eq. (4.11) and arrange the results as

$$\dot{x}_2 = \frac{\beta q}{p}x_2^{1-p/q}\dot{s}-\frac{\beta q}{p}x_2^{2-p/q}\left(1+\frac{g}{\alpha h}x_1^{g/h-1}\right). \tag{4.37}$$

Combining Eq. (4.36) and Eq. (4.37), we obtain

$$\dot{s} = -\left(\phi s+\gamma s^{m/n}-\frac{p}{\beta q}\left[\Delta f(x)+d(t)\right]\right)x_2^{p/q-1}. \tag{4.38}$$

Since $|\Delta f(\boldsymbol{x})+\mathrm{d}(t)| < L$, we have

$$\dot{s}\cdot sgn(s) < -\left(\phi\,|s|+\gamma\,|s|^{m/n}-\frac{pL}{\beta q}\right)x_2^{p/q-1} = -\Psi x_2^{p/q-1}. \tag{4.39}$$

If $\Psi > 0$, using Theorem 2, we know that sliding mode s asymptotically converges and convergence speed is not smaller than the dynamics defined by Eq. (4.34). If $\Psi \le 0$, sliding mode s might not continuously converge to equilibrium point, but at least reach the region confined by $\Psi \le 0$, i.e., region Π here. (End of Proof)

Generally, we can assume that both modeling errors and external disturbances are bounded. According to Theorem 3 and the fact that the factor $m/n > 1$ in the definition of Π, we know that the steady-state error of sliding mode s is less than the steady-state error of control system designed based on linear attractor. By choosing large enough ϕ, γ and $n/m > 1$, we can make the convergence region of sliding mode small enough. Then, the control system has good robustness to the modeling errors and external disturbances.

4.5 Simulation Verification

Consider a nonlinear and uncertain SISO system

$$\begin{cases}\dot{x}_1 = x_2 \\ \dot{x}_2 = 2x_1^2+x_2^2+\Delta f(x)+u+\mathrm{d}(t)\end{cases},$$

Table 4.2 Settings of simulation conditions

Simulation condition	Uncertainty	Sliding mode	Reaching mode	Initial Condition
A	No			$x = [8, 8]^T$
B	No	$g = 1, h = 1$		$x = [8, 8]^T$
C	No			$x = [4, 4]^T$
D	Yes			$x = [4, 4]^T$
E	No		$\phi = 2, \gamma = 2$	$x = [4, 4]^T$
F	No		$\phi = 0.1, \gamma = 0.5$	$x = [4, 4]^T$
G	Yes		$\phi = 2, \gamma = 4$	$x = [0, 0]^T$
H	Yes		$\phi = 2, \gamma = 8$	$x = [0, 0]^T$
I	Yes		$\phi = 2, \gamma = 12$	$x = [0, 0]^T$

where $\Delta f(\boldsymbol{x}) = 2\cos(x_1 + x_2)$ and $d(t) = 2\sin(3t)$. We design the control law using the method described in this chapter. We choose the parameters for terminal sliding hyperplane as shown in Table 4.1. Other parameters are taken as $m = 1, n = 3, \phi = 1.2$, and $\gamma = 1.2$. In the following, we will use simulation to demonstrate some critical properties of NFTSM, e.g., finite-time convergence, nonsingularity, non-chattering, effect of stagnation and closed-loop robustness. By tuning some control parameters, we choose the following settings of simulation conditions as shown in Table 4.2.

Conditions A and B are used to demonstrate the finite-time convergence characteristics of the NFTSM control method and NTSM control method, respectively. The state response curves are shown in Fig. 4.3. The solid line is for simulation condition A and the dashed line is for simulation condition B. We observed from Fig. 4.3 that simulation condition A has faster convergence speed than B, which is because NFTSM has fast convergence characteristics. This implies that the NFTSM control method has better convergence performance than the NTSM method in a large region of state space.

Conditions C and D are used to demonstrate the nonsingularity and non-chattering characteristics of the NFTSM control method. The simulation results are shown in Figs. 4.4 and 4.5. Figure 4.4 shows the state response curves of the closed-loop system. The solid line is for the simulation condition C and the dashed line is for the simulation condition D. If there is no modeling error and no external disturbance, system state converges in finite time to the equilibrium point, as shown in solid lines. If modeling errors and external disturbances exist, the system convergence speed decreases but the system can still converge to a neighborhood region of the equilibrium point, as shown by dashed lines. The control inputs are shown in Fig. 4.4. We observe that the control inputs are both bounded and continuous in time around the equilibrium points, thus completely avoiding the singularity or chattering problems.

Conditions E and F are used to analyze how the selection of terminator parameters ϕ and γ affect the effect of stagnation. The state response curves are shown in Fig. 4.6. The solid line is for the simulation condition E and the dashed line is for the simulation

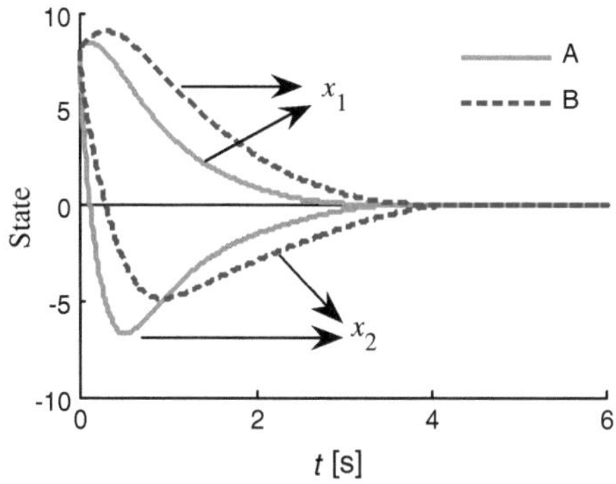

Fig. 4.3 State response in A and B

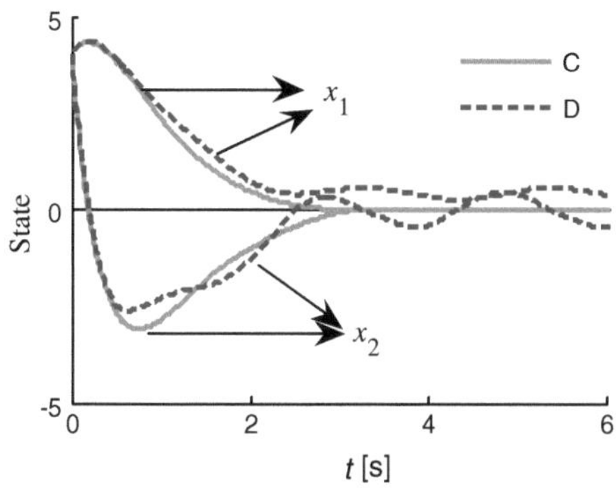

Fig. 4.4 State response in C and D

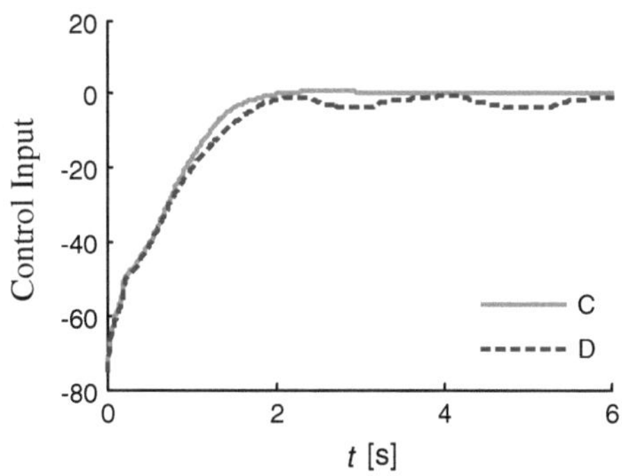

Fig. 4.5 Control input in C and D

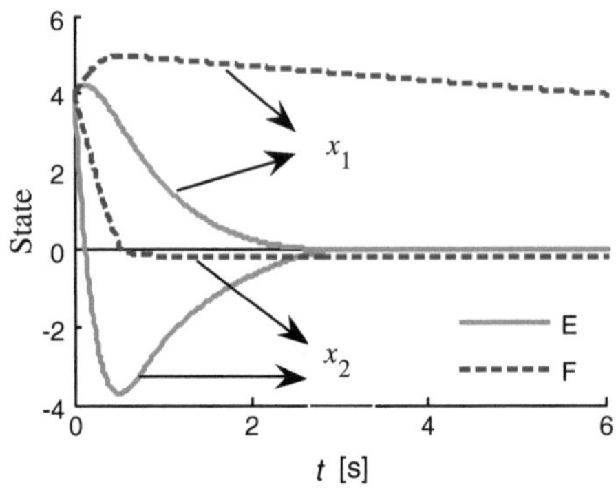

Fig. 4.6 State response in E and F

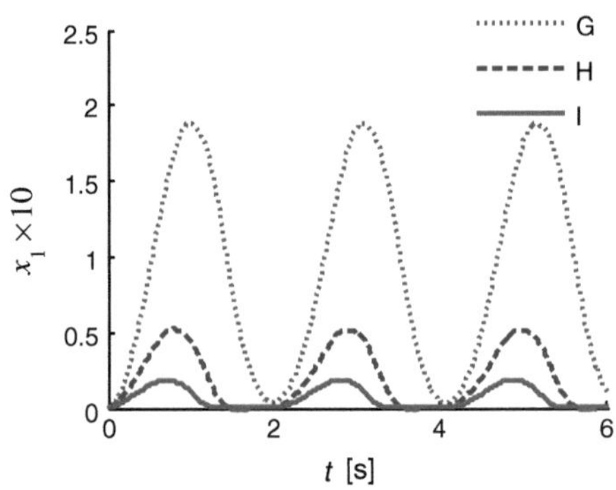

Fig. 4.7 State x_1 response in G, H, and I

condition F. As shown by the dashed lines, if $\phi, \gamma < 1$, the state x_2 is much smaller than x_1 and x_2 is almost a constant. This causes the slow convergence of x_1 and the effect of stagnation occurs in the system. As shown by the solid lines, if $\phi, \gamma < 1$, the effect of stagnation disappears and the system state converges to the equilibrium point in finite time, which is consistent with the analysis results in Sect. 4.4.3.

Conditions G, H, and I are used to demonstrate the steady-state convergence characteristics of the NFTSM method. The simulation results are shown in Figs. 4.7, 4.8, and 4.9. Simulation conditions G, H, and I correspond to dot lines, dashed lines, and solid lines. According to Figs. 4.7 and 4.8, we find that the steady-state errors of state x_1 and x_2 decrease as the parameter γ increases. According to Fig. 4.9, the steady-state errors of the sliding mode s are less than 0.20, 0.053, and 0.018, respectively. Using Theorem 3, we can calculate the steady-state errors as 0.31, 0.064, and 0.021, which are quite close to the simulated error values. This implies that the estimation method presented in Theorem 3 works very well to estimate the steady-state errors of the closed-loop system.

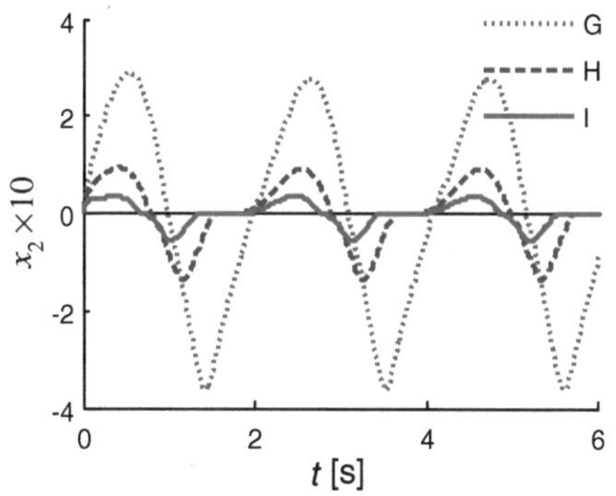

Fig. 4.8 State x_2 response in G, H, and I

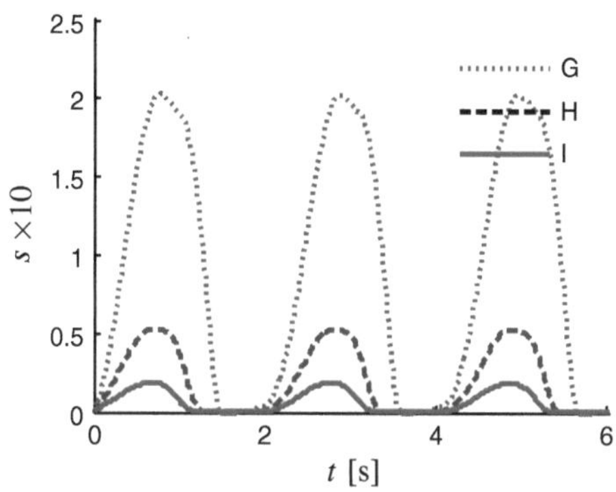

Fig. 4.9 Sliding mode response in G, H, and I

Based on the above simulation results, we know that the NFTSM control proposed in this chapter has no singularity or chattering problems. By properly choosing parameters for the control system, we can guarantee the fast convergence of the closed-loop system, and provide good robustness to modeling errors and external disturbances.

4.6 Conclusions

To resolve the singularity and chattering problems associated with the existing TSM control method, this chapter proposed an NFTSM control method for the second-order nonlinear and uncertain system. The NFTSM control is nonsingular, time-continuous, and has faster convergence speed. The theoretical analysis and simulation experiments show that:

(1) The proposed NFTF has finite-time convergence property, and has faster convergence speed than the NTF in the region far away from the equilibrium point. The terminal attractor with negative exponential coefficient guarantees the global existence of the sliding hyperplane, and enables the asymptotical convergence of system state to the sliding hyperplane from arbitrary starting point.
(2) The designed NSTSM controller has no state negative exponents, thus completely eliminating the singularity phenomenon from the TSM control. Moreover, the NFTSM control is time continuous, thus eliminating the chattering phenomenon.
(3) By properly choosing control parameters, the closed-loop system can avoid the effect of stagnation. If the modeling errors and external disturbances are bounded, then system state asymptotically converges to the neighborhood region of equilibrium point, still having good convergence speed and tracking performance.

Acknowledgments The authors greatly appreciate the NSF of China with grant number 51205228 for the support to this research.

References

1. Young KD, Utkin VI, Ozguner U (1996) A control engineer's guide to sliding mode control. In: Proceedings of 1996 IEEE international workshop on variable structure systems, VSS'96, pp 1–14
2. Utkin VI (1993) Sliding mode control design principles and applications to electric drives. IEEE Trans Industr Electron 40(1):23–36
3. Furuta K (1990) Sliding mode control of a discrete system. Syst Control Lett 14(2):145–152
4. Xu H, Mirmirani MD, Ioannou PA (2004) Adaptive sliding mode control design for a hypersonic flight vehicle. J Guidance Control Dyn 27(5):829–838
5. Koshkouei AJ, Burnham KJ, Zinober AS (2005) Dynamic sliding mode control design. IEE Proc Control Theory Appl 152(4):392–396
6. Bouabdallah S, Siegwart R (2005) Backstepping and sliding-mode techniques applied to an indoor micro quadrotor. In: Proceedings of the 2005 IEEE international conference on robotics and automation, ICRA 2005, Apr 2005, pp 2247–2252
7. Shang A, Wang Z (2013) Adaptive backstepping second order sliding mode control of nonlinear systems. Int J Model Ident Control 19(2):195–201
8. Venkataraman ST, Gulati S (1991) Terminal sliding modes: a new approach to nonlinear control synthesis. In: Fifth International Conference on advanced robotics, 1991. 'Robots in Unstructured Environments', 91 ICAR., Jun 1991, pp 443–448
9. Zhuang KY, Zhang KQ (2002) Terminal sliding mode control for high-order nonlinear dynamic systems. J Zhejiang Univ 36(5):482–539 (Engineering Science)
10. Bhave M, Janardhanan S, Dewan L (2013) An efficient control of rigid robotic manipulator with uncertainties using higher order sliding mode control. Int J Model Ident Control 19(2):179–185
11. Du H, Li S (2012) Finite-time cooperative attitude control of multiple spacecraft using terminal sliding mode control technique. Int J Model Ident Control 16(4):327–333
12. Man Z, Yu X H (1996). Terminal sliding mode control of MIMO linear systems. In: Proceedings of the 35th IEEE on decision and control, 1996, vol 4, pp 4619–4624
13. Wu Y, Yu X, Man Z (1998) Terminal sliding mode control design for uncertain dynamic systems. Syst Control Lett 34(5):281–287
14. Man Z, Paplinski AP, Wu HR (1994) A robust MIMO terminal sliding mode control scheme for rigid robotic manipulators. IEEE Trans Autom Control 39(12):2464–2469

15. Yu X, Man Z, Wu Y (1997) Terminal sliding modes with fast transient performance. In: Proceedings of the 36th IEEE conference on decision and control, 1997, vol 2, pp 962–963
16. Yu S, Guo G, Man Z, Du J (2006) Global fast terminal sliding mode control for robotic manipulators. Int J Model Ident Control 1(1):72–79
17. Kang Y, Xi HS, Ji HB, Wang J (2003) Fast terminal sliding mode control of uncertain multivariable linear systems. J Univ Sci Technol China 33(6):718–725
18. Yu S, Yu X, Man Z (2000) Robust global terminal sliding mode control of SISO nonlinear uncertain systems. In: Proceedings of the 39th IEEE conference on decision and control, vol 3, pp 2198–2203
19. Yu X, Man Z (2002) Fast terminal sliding-mode control design for nonlinear dynamical systems. IEEE Trans Circ Syst I Fundam Theory Appl 49(2):261–264
20. Feng Y, Bao S, Yu XH (2002) Design method of non-singular terminal sliding mode control systems. Control Decis 17(2):194–198
21. Feng Y, Yu X, Man Z (2002) Non-singular terminal sliding mode control of rigid manipulators. Automatica 38(12):2159–2167
22. Feng Y, Yu X, Man Z (2001). Non-singular terminal sliding mode control and its application for robot manipulators. In: IEEE international symposium on circuits and systems, ISCAS, May 2001, vol 3, pp 545–548
23. Zhang KD, Hu YM, Hu ZH (2007) Sliding mode control of low chattering non-singular terminal. J Guangdong Univ Technol 24(3):32–36
24. Hu JB, Shi MH, Zhang KY (2005) Terminal sliding mode control for a class of nonlinear systems. Control Theory Appl 22(3):495–502
25. Li SB, Li KQ, Wang JQ, Gao F (2009) Nonsingular and fast terminal sliding model control method. Inf Control 38(1):1–8
26. Yoshimura T (2012) Adaptive sliding mode control for uncertain discrete-time systems using an improved reaching law. Int J Model Ident Control 16(4):380–391
27. Parra-Vega V, Hirzinger G (2000). Finite-time tracking for robot manipulators with singularity-free continuous control: a passivity-based approach. In: Proceedings of the 39th IEEE conference on decision and control, 2000, vol 5, pp 5085–5090
28. Soliman HM, Bayoumi EH, Soliman M (2012) Robust guaranteed-cost sliding control for brushless DC motors by LMI. Int J Model Ident Control 17(3):251–260

Chapter 5
Flocking Behavior via Leader's Backstepping on Nonholonomic Robot Group

Lei Cheng, Jun Wang, Huaiyu Wu, Wenxia Xu, Wenhao Zhang, Pian Jin and Quanmin Zhu

Abstract This chapter aims to improve flocking control for a group of nonholonomic robots. It introduces a new flocking control algorithm with potential-based flocking as its foundation. By incorporating Leader's Backstepping algorithm into the flocking strategy, an improved flocking performance is obtained, which leads the flock to the target point swiftly in a smoothed trajectory. Simulations in this chapter test and verify the effectiveness of the algorithm, in which key parameters' influences on system performance are discussed.

Keywords Flocking · Backstepping · Nonholonomic robot

5.1 Introduction

Multi-robot system is a complex self-organized decentralized dynamic group which arises from the applications [1–3], including advanced transportation systems, search-and-rescue operations, distributed sensors networks, formation flight, and military surveillance. Flocking is a bionic methodology with inspiration drawn from nature, such as the schooling of fish. Flocking particularly means that a group of

L. Cheng (✉) · J. Wang · H. Wu · W. Zhang
Engineering Research Center of Metallurgical Automation and Measurement Technology, Ministry of Education, Wuhan University of Science and Technology, Wuhan, 430081, China
e-mail: chenglei@wust.edu.cn

L. Cheng
Henan Provincial Open Laboratory for Control Engineering Key Disciplines, Henan Polytechnic University, Jiaozuo, 454000, China

W. Xu · P. Jin
School of Automation, Huazhong University of Science and Technology, Wuhan , 430074, China

W. Zhang · Q. Zhu
Faculty of Environment and Technology, University of the West of England, Bristol, BS161QY, UK

L. Liu et al. (eds.), *Applied Methods and Techniques for Mechatronic Systems*, Lecture Notes in Control and Information Sciences 452,
DOI: 10.1007/978-3-642-36385-6_5,

individuals are able to maintain the organized collective motion without centralized coordination, which seems very suitable to ground cooperative movement scenario of multiple mobile robots.

Reynolds first introduced the principles of flocking [4], where later scholars placed major attention and formed main methods for flock behavioral modeling and analysis built on the foundation of "potential field" [5]. On this basis, multiple issues regarding flocking have been discussed and researched under environments free of obstacles [6–14]. By employing artificial forces and local control laws, avoiding inter-member collisions and consistency of multi-agent flocking system were explored. Leaderless strategies were usually adopted in flocking topologies, as in [6–8]. To mimic nature, the style of 'leader-following' [15] can be introduced to flock, namely, the other members following the leader in a planned route, which can enhance the efficiency of multi-robot system in the real environment.

The majority of researches [6, 9, 11, 13, 16, 17] tend to replace general mobile robot with simplified particle model in flocking-related research. Instead of being a simple agent, the real mobile robot is nonholonomic for the wheel constraint, to keep the consistency of moving direction and orientation other than agent's possible discontinuous direction of motion. When the nonholonomic mobile robot is approaching the goal in flocking motion, it may possess a spiral trajectory driven by the potential force which is generated by the goal attraction [14]. Therefore, an approach that ensures a smoother trajectory of flocking is demanded, to compensate for the lack of rapidness and efficiency in arriving at the goal.

Conversely, the Backstepping approach has the merit of producing a smoother trajectory [18]. It is a systematic method for controller design targeting uncertainty system, and a regression design that combines Lyapunov function selection and controller design. Starting with minimum order of differential equations, it achieves the desired virtual control performance with step-by-step design built on the basis of the employment of virtual control thereby eventually realizing a genuine control law [19]. Backstepping was first introduced by Kokotovic [20] and applied to the design of trajectory tracking controller for the enhanced model of mobile robot [21, 22]. Then it was extended to the area of multiple robots. With regression design and virtual control method, Backstepping approaches were engaged in stable coordination [23–25], desired geometric pattern formation [26–30], and collision avoidance issues [30–33] related with multi-robot control. The above researches were mainly focused on group members' mutual and inner virtual position tracking. Different from these domains, to perform flocking strategy, special Leader robot's external Backstepping design was utilized in our task to guide the whole multi-robot flock to avoid spiral trajectory due to the goal potential attraction.

The research conducted in this chapter is built on the approaches stated above, which have respective merits in advancing the flocking performance. With potential-based flocking being the primary method [34, 35], Backstepping mechanism is introduced for Leader robot which can lead the flock to the goal swiftly, while attaining a trajectory with a smooth transition.

The rest of the chapter is organized as follows. In Sect. 5.2, we briefly introduce multi-robot flocking based on potential, constructing a foundation where additional methodologies take effect. In Sect. 5.3, integrated Backstepping algorithm, constitutes the mechanism implemented on the flock. In Sect. 5.4, a series of experiments simulate the flocking process and verify the performance of the methodology mentioned in this chapter. Finally Sect. 5.5 draws the conclusions.

5.2 Nonholonomic Robot's Flocking Behavior Based on Potential

Flocking mechanism is the foundation of our investigation in this chapter. To be convenient for investigation of the latter part, its strategy is concisely presented in this section [34].

The algorithm of multi-robot flocking consists of two parts:

(1) Agent-based flocking algorithm;
(2) Control law transformation from agent to mobile robot.

The motion of group should be divided into two parts: Leader's motion and followers' motion. Generally, Leader's movement can follow the common potential-based goal attractive principle [36]. Besides, N followers should obey the following rules. Considering follower agents (r_i, v_i, u_i) dynamics:

$$\begin{aligned} \dot{r}_i &= v_i \\ \dot{v}_i &= u_i \end{aligned} \quad i = 1, 2, \ldots N \tag{5.1}$$

where $r_i = (x_i, y_i)^T$ is the position vector of agent i, $v_i = (\dot{x}_i, \dot{y}_i)$ is its velocity vector, and $u_i = (u_{x_i}, u_{y_i})^T$ is control (acceleration) input.

The position vector of the ith agent relative to the Leader (r_L, v_L, u_L) is denoted by $r_{iL} = r_i - r_L$, and the velocity of the ith agent relative to the Leader is $\dot{r}_{iL} = v_i - v_L$.

Artificial potential function V_{ij} is a differentiable, nonnegative, radially unbounded function of the distance $\|r_{ij}\|$ between agents i and j [37]. V_{ij} can be selected as

$$V_{ij} = \frac{L^2}{\|r_{ij}\|^2} + \log \|r_{ij}\|^2 \tag{5.2}$$

where, L is configuration coefficient, which can determine the structure of the flock.

The control law u_i can then be defined as

$$u_i = -A \sum_{j \in N_i} \nabla_{r_{ij}} V_{ij} - \dot{r}_{iL} + \dot{v}_L \tag{5.3}$$

where $A > 0$, A is gradient adjustment coefficient and N_i is neighboring set of agent i.

Consider a system of N agents with dynamics (5.1), where each agent is steered by control law (5.3) with the assumption that the neighboring graph is connected.

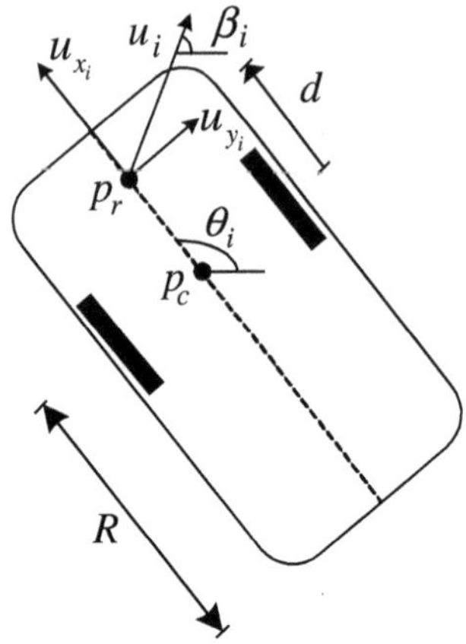

Fig. 5.1 Mobile robot controlled by virtual force u_i

Then all agent velocity vectors become asymptotically the same, relative distances between agents maintain constant, and collisions between interconnected agents are avoided [38].

From expressions of u_i in (5.3), u_i can be considered as the virtual force operated on the mobile robot. Figure 5.1 illustrates a mobile robot controlled by u_i.

In Fig. 5.1, R is the radius of the mobile robot, the displacement between rotation center p_c and virtual force beating point p_r is d. The attitude of the robot is represented by $(x_{p_c}, y_{p_c}, \theta_i)$. The direction of u_i is represented by β_i.

In Fig. 5.1,

$$u_{x_i} = u_i \cdot \cos(\beta_i - \theta_i) \tag{5.4a}$$

$$u_{y_i} = u_i \cdot \cos(\beta_i - \theta_i - \frac{\pi}{2}). \tag{5.4b}$$

From principles of mechanics, Eq. (5.5) represents the control transformation from agent-based system (u_i) to mobile robots' control input in flocking motion $q_i = (v_i, \omega_i)^T$. m_i is defined as the virtual mass, and I_i is defined as the virtual moment of inertia.

$$\dot{v}_i = \frac{u_i \cdot \cos(\beta_i - \theta_i)}{m_i}. \tag{5.5a}$$

$$\dot{\omega}_i = \frac{u_i \cdot \sin(\beta_i - \theta_i) \cdot d}{I_i}. \tag{5.5b}$$

5.3 Motion Control of Leader Robot by Backstepping

Flocking control is a potential-based methodology. For the above flocking, potential exerts impact on it. On account of mobile robot nonholonomic constraint, it may cause spiral trajectory driven by the general goal potential attraction. More specifically,

Fig. 5.2 Leader robot movement schematic diagram

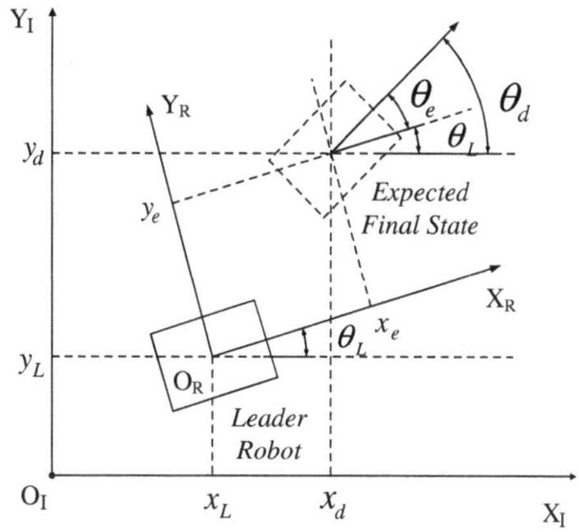

virtual potential force u_i on the agent, shown in Sect. 5.2, functions in the X and Y directions on robot axis u_{x_i} and u_{y_i}, when it is applied to nonholonomic mobile robots, and hence constitutes an acceleration on both axes. When the goal is not remotely located from the robot, the difference between the angle linking robot and target $\angle l$ (namely β in Sect. 5.2), and the current direction of the robot θ is relatively large, and then a winding path will be produced.

In this section, Backstepping is implemented in motion control for nonholonomic mobile robots to improve the undesired trajectory of the leader robot, whose basic idea is referred from [18].

5.3.1 Principles of Leader's Backstepping

Backstepping is a regression design methodology that combines Lyapunov function selection and controller design. Its principles can be described as: with constant comparison of the current status and desired status of the controlled plant, synchronously performing iterative deviation control, and eventually fitting the plant to desired value.

First of all, through appropriate coordinate transformation, we define a set of pose error state vectors, to relate it to the research in this chapter. We virtualize $p_d(x_d, y_d)$ to the Expected Final State (*EFS*) with position and orientation information $p_d(x_d, y_d, \theta_d, v_d, \omega_d)$, and θ_d can be designed to approach $\angle l$ (the angle of line linking the leader and the target point). Unlike general dynamic Backstepping-based tracking problem [19], we take into consideration the stationary target point whose virtualized desired final state has the proportion (v_d, ω_d) with 0 value, and convert the problem of tracking the goal's trajectory into a problem of the stabilization of the pose error.

Principle of Leader robot's movement with Backstepping is as shown in Fig. 5.2:

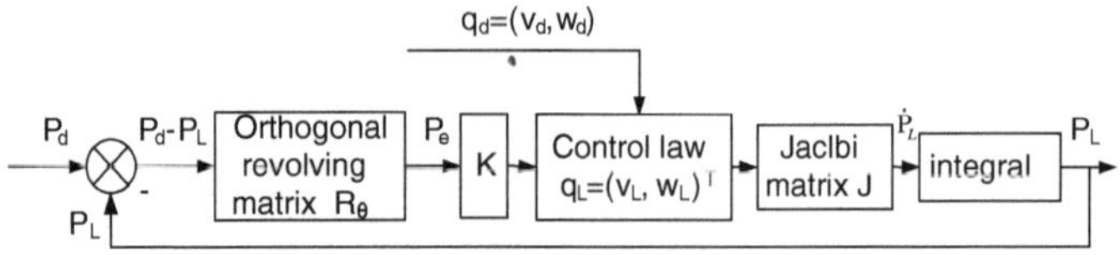

Fig. 5.3 Mobile robot motion control system structure diagram

$O : \{X_I, Y_I\}$ is the reference frame of overall situation, $O : \{X_R, Y_R\}$ is the robot reference frame, θ_L is the leader robot's heading direction. $p_L = (x_L, y_L, \theta_L)^T$ is leader robot's current pose, $p_d(x_d, y_d, \theta_d)$ is the *EFS* when the leader robot reaches the target. $q_L = (v_L, \omega_L)^T$ is the state of leader robot's motion. In Backstepping scheme, the leader robot's kinematics equation is

$$\dot{p}_L = \begin{bmatrix} \dot{x}_L \\ \dot{y}_L \\ \dot{\theta}_L \end{bmatrix} = Jq_L = \begin{bmatrix} \cos\theta_L & 0 \\ \sin\theta_L & 0 \\ 0 & 1 \end{bmatrix} L \begin{bmatrix} v_L \\ \omega_L \end{bmatrix}. \tag{5.6}$$

J is Yakebi Matrix.

Then the error between the current and desired position and orientation of leader robot may be represented as

$$p_e = \begin{bmatrix} x_e \\ y_e \\ \theta_e \end{bmatrix} = R_\theta \begin{bmatrix} x_d - x_L \\ y_d - y_L \\ \theta_d - \theta_L \end{bmatrix} = \begin{bmatrix} \cos\theta & \sin\theta & 0 \\ -\sin\theta & \cos\theta & 0 \\ 0 & 0 & 1 \end{bmatrix} (p_d - p_L). \tag{5.7}$$

R_θ is orthogonal revolving matrix [39], we have differential equation:

$$\dot{p}_e = \begin{bmatrix} \dot{x}_e \\ \dot{y}_e \\ \dot{\theta}_e \end{bmatrix} = \begin{bmatrix} \omega_L y_e - v_L + v_d \cos\theta_e \\ -\omega_L x_e + v_d \sin\theta_e \\ \omega_d - \omega_L \end{bmatrix}. \tag{5.8}$$

Moreover, Leader's control input pair is

$$\begin{bmatrix} v_L(t) \\ \omega_L(t) \end{bmatrix} = K \cdot p_e = K \begin{bmatrix} x_e \\ y_e \\ \theta_e \end{bmatrix}. \tag{5.9}$$

Make $p_e \to 0$, namely $(x_e, y_e, \theta_e)^T$ has constraints and $\lim\limits_{t\to\infty} \left\|(x_e, y_e, \theta_e)^T\right\| = 0$, and make the error between the current and desired state to be 0, which means the leader reaches the target stably. It may finally obtain the nonholonomic mobile robot motion control system, and its structure diagram is as shown in Fig. 5.3.

5.3.2 Controller Design

Supposing $(v_d, \omega_d) \equiv 0$, according to nonholonomic Leader robot kinematics model, construct the new hypothesized error feedback variable:

$$\bar{x}_e = x_e - k_1 \sin(k_2\omega_L) y_e. \tag{5.10}$$

Constant $k_1 > 0$. $\dot{y}_e = -\omega_L x_e + v_d \sin\theta_e$, if the control action causes $\lim_{t\to\infty} x_e = k_1 \sin(\omega_L) y_e$, and $\lim_{t\to\infty} \theta_e = 0$. Thus, $\lim_{t\to\infty} \dot{y}_e = -k_1\omega_L \sin(k_2\omega_L) y_e$. At the same time, calculate the time derivative $V_y = \frac{1}{2}y_e^2$ of the Lyapunov scalar function which includes the part of y_e, obtaining $\dot{V}_y = y_e\dot{y}_e$. And because $\omega_L \sin(k_2\omega_L) \geq 0$, $\lim_{t\to\infty} \dot{y}_e = -k_1\omega_L \sin(k_2\omega_L) y_e^2 \leq 0$, according to the Barbalat lemma [40], obtains $t \to 0$, y_e converges to 0. Thus it can show that the designed hypothesized feedback variable may indirectly control y_e.

Above all, by seeking the input control $q_L = (v_L, \omega_L)^T$ to get $\lim_{t\to\infty} x_e = k_1 \sin(k_2\omega_L) y_e$, and $\lim_{t\to\infty} \theta_e = 0$, then the overall system achieves the state of global asymptotic stability.

According to the above analysis, get the Lyapunov function:

$$V = \frac{1}{2}\bar{x}_e^2 + \frac{1}{2}y_e^2 + k_3(1 - \cos\frac{\theta_e}{2}) \tag{5.11}$$

$k_3 > 0$, which is constant. This Lyapunov scalar function has two characteristics: (1) $V \geq 0$; (2) V contains all parameters that can reflect the stability of control system, when $V = 0$, $(\bar{x}_e, y_e, \theta_e)^T = 0$.

Therefore,

$$\dot{\bar{x}}_e = \dot{x}_e - k_1k_2 \cos(k_2\omega_L)\dot{\omega}_L y_e - k_1 \sin(k_2\omega_L)\dot{y}_e \tag{5.12}$$

$$\dot{y}_e = -\omega_L x_e + v_d \sin\theta_e \tag{5.13}$$

$$\dot{\theta}_e = \omega_d - \omega_L. \tag{5.14}$$

Calculating the time derivative of the Lyapunov function (5.11) then obtains:

$$\begin{aligned}
\dot{V} &= \bar{x}_e\dot{\bar{x}}_e + y_e\dot{y}_e + \frac{k_3}{2}\sin\frac{\theta_e}{2}\theta_e \\
&= \bar{x}_e[-v_L + v_d\cos\theta_e - k_1k_2\cos(k_2\omega_L)\dot{\omega}_L y_e - k_1\sin(k_2\omega_L)(-\omega_L x_e + v_{dr}\sin\theta_e)] \\
&\quad - k_1\omega_L\sin(k_2\omega_L)y_e^2 + \frac{k_3}{2}\sin\frac{\theta_e}{2}(\omega_d - \omega_L + \frac{1}{k_3}y_e v_d\cos\frac{\theta_e}{2}).
\end{aligned} \tag{5.15}$$

According to Lyapunov principle, set the system control law as:

$$\begin{cases} v_L = v_d \cos\theta_e - k_1 k_2 \cos(k_2\omega_L)\dot{\omega}_L y_e - k_1 \sin(k_2\omega_L)(-\omega_L x_e + v_d \sin\theta_e) \\ +a[x_e - k_1 \sin(k_2\omega_L) y_e] \\ \omega_L = \omega_d + \frac{1}{k_3} y_e v_d \cos\frac{\theta_e}{2} + b \sin\frac{\theta_e}{2} \end{cases} \tag{5.16}$$

$$\dot{\omega}_L = \dot{\omega}_d + \frac{1}{k_3}(\dot{y}_e v_d + y_e \dot{v}_d)\cos(\frac{\theta_e}{2}) - \frac{1}{k_3} v_d y_e \sin(\frac{\theta_e}{2})\dot{\theta}_e + \frac{1}{2} b \cos(\frac{\theta_e}{2})\dot{\theta}_e \tag{5.17}$$

obtain that:

$$\dot{V} = -a\,[x_e - k_1 \sin(k_2\omega_L) y_e]^2 - k_1 \omega_L \sin(k_2\dot{\omega}_L) y_e^2 - \frac{k_3}{2} b \sin^2\frac{\theta_e}{2}. \tag{5.18}$$

As k_1, k_2, k_3, a, b are the constants which are bigger than 0, and $\omega_L \sin(k_2\omega_L) \geq 0$, so $\dot{V} \leq 0$. V is positive differentiable function and has bound, $\dot{V}$ is half-negative uniform continuous function, so according to the Barbalat Lemma [40], when $t \to \infty$, $\dot{V} \to 0$. Thus $\lim_{t\to\infty} \theta_e = 0$, $\lim_{t\to\infty} \bar{x}_e = 0$, $\lim_{t\to\infty} \omega_L \sin(k_2\omega_L) y_e^2 = 0$, all converges to 0. Further, by ω_L is not zero constantly, we get $\lim_{t\to\infty} y_e = 0$, then $x_e \to 0$.

On the background of the research conducted in this chapter $(v_d, \omega_d) \equiv 0$, Eq. (5.16) can be simplified as the following format:

$$\begin{cases} v_L = -k_1 k_2 \cos(k_2\omega_L)\dot{\omega}_L y_e - k_1 \sin(k_2\omega_L)(-\omega_L x_e) + a[x_e - k_1 \sin(k_2\omega_L) y_e] \\ \omega_L = b \sin\frac{\theta_e}{2} \end{cases} \tag{5.19}$$

Under the function of control law (5.19), closed-loop system pose error has overall bound, and $\lim_{t\to\infty} [|x_e(t)| + |y_e(t)| + |\theta_e(t)|] = 0$, then the leader robot reaches the target progressively and iteratively.

5.4 Simulation Design and Results

A series of simulation experiments are conducted in this section to verify the strategies introduced in this chapter. A detailed illustration is presented regarding the design of simulation platform, experiment results and analysis, etc.

5.4.1 Platform

The simulation platform for multi-robot flocking system in this chapter is improved from MuRoS—Multi-Robot Simulator [41]. Virtual nonholonomic mobile robot is usually presented as circles with directions.

Given that there is a limit to maximal translational velocity $v_{\max}$ and maximal rotational velocity $\omega_{\max}$. If $|v_i(t)| \geq v_{\max}$, $v_i(t) = \pm v_{\max}$. Then when $|\omega_i(t)| \geq \omega_{\max}$,

$\omega_i(t) = \pm\omega_{\max}$. Furthermore, if Δt is small enough:

$$\dot{v}(t) = \frac{v(t) - v(t - \Delta t)}{\Delta t} \tag{5.20a}$$

$$\dot{\omega}(t) = \frac{\omega(t) - \omega(t - \Delta t)}{\Delta t}. \tag{5.20b}$$

Basic parameter setting: the maximal translational velocity of followers: $v_{\max} = 70\,\text{mm/s}$, and the maximal translational velocity of Leader is half of that. $\omega_{\max} = 18°/\text{s}$, $d = 2\,\text{mm}$, $L = 100$, $m_i = 0.1$, $I_i = 0.2$, $A = 5000$, in the simulation, $\Delta t = 0.01\,\text{s}$, and the radius of mobile robot in simulation: $R = 10\,\text{mm}$.

5.4.2 Demonstrations

The subsequent parts from D1-D2 simulate the theory put forward in this part. To begin with, the simulation verifying Backstepping-based multi-robot flocking is conducted which is compared to pure potential-based one; then the current strategy performance is evaluated and compared under different settings of parameters.

D1: *Comparison with two strategies*

Traditional potential-based flocking strategy strictly depends on Lagrangian representations in the mathematical sense. In the following experiments, assume the Leader robot moves toward the given goal, the attraction of goal is denoted by constant $F_a = 1000$. Then, derived from Sect. 5.2, the dynamics of Leader with potential is represented as

$$\dot{v}_L(t) = F_a \cdot \cos(\beta_L(t) - \theta_L(t))/m_L \tag{5.21a}$$

$$\dot{\omega}_L(t) = F_a d \cdot \sin(\beta_L(t) - \theta_L(t))/I_L \tag{5.21b}$$

Follower's dynamics are both flocking algorithm shown in Sect. 5.2, whose principles originated from Tanner's work [7].

The verification was conducted by comparing two algorithms in the same environment, along with a flock comprising six robot individuals who had the same initial state. See Figs. 5.4 and 5.5, where "×" denotes the goal, connecting lines denote the connective relation of the robots, and dashed lines denote the trajectory of the robots.

Figure 5.4 is the simulation result for flocking motion with artificial potential field. Figure 5.4a shows the initial status for the multi-robot system while the target is beside the group. In Fig. 5.4b, c, the goal is already near the leader, but the leader does not directly move toward it, instead of heading toward the goal directly, the leader possesses a "spiral trajectory" around the target point. In Fig. 5.4d, when the leader arrives at the target point ($t = 84.06\,\text{s}$), the follower reaches the stable status, while the leader possesses a spiral trajectory which is in much disorder for the followers.

Figure 5.5 is the simulation result for multi-robot flocking via Leader's Backstepping purely. In Fig. 5.5a, the scenario for the system is similar to that in Fig. 5.4. In Fig. 5.5b–d, following the leader, the followers proceed toward the target point. When the leader arrives at the target point, the flock reaches stable as well without the "spiral trajectory." And compared with potential-based multi-robot flocking motion, Backstepping-based multi-robot flocking motion only uses 46.45 s, which is almost half duration of potential-based multi-robot flocking motion.

Discussion: This set of experiments demonstrates the process of multi-robot flocking under various strategies, and proves that the Backstepping-based motion control algorithm for multi-robot flock can effectively enable the robots to tackle the "spiral trajectory" problem. That means the consistency of rapidly convergent motion, safe movement to the goal, could both be achieved.

D2: *Analysis of parameters adjustment*

In the simulation experiments, the Backstepping-based motion control law for Leader robot is expressed as Eq. (5.19). In Eq. (5.19), k_1, k_2, a, and b need to be adjusted respectively. To test its performance with different parameters, the environment is simplified where obstacles are removed, Figs. 5.6 and 5.7, Tables 5.1, and 5.2 are simulation results on leader's Backstepping in multi-robot flocking according to different parameters. All the experiments had the same initial state, with a flock comprising four robot individuals.

(1) Adjustment for parameter k_1

In Fig. 5.6b, the multi-robot group reaches the goal precisely with $k_1 = 0.01$, while the overshoot of the Leader robot in Fig. 5.6c where $k_1 = 10$ exceeds that in Fig. 5.6b. The overshoot keeps increasing as $k_1 = 100$ in Fig. 5.6d, respectively. It is concluded that when $k_1 < 0.5$ the overshoot reaches minimum, and the overshoot becomes larger with the increase of k_1.

(2) Adjustment for parameter k_2

Figure 5.7b demonstrates the simulation results for multi-robot motion based on Backstepping when $k_2 = 10$, in which case the group is not moving toward the goal. Figure 5.7c shows the successful arrival of the multi-robot group when $k_2 = 20$. In Fig. 5.7d, the group reaches the goal as swiftly as the situation in Fig. 5.7c. Therefore it is concluded that by setting $k_2 > 20$, the multi-robot group can be guaranteed to move toward the goal and reach it rapidly.

(3) Adjustment for parameter a

Table 5.1 illustrates the error for different values of a. It is evident that the error is extremely small when $a > 100$.

(4) Adjustment for parameter b

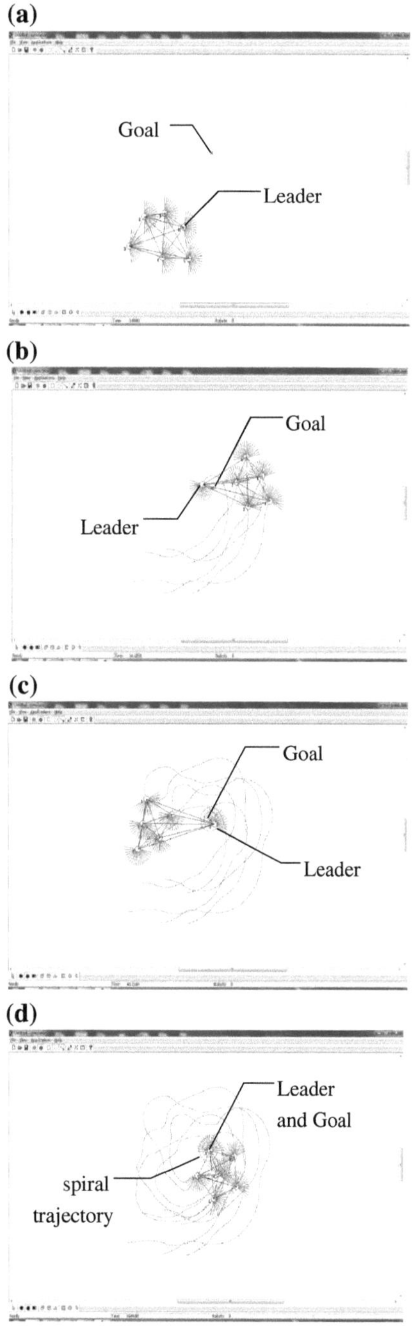

Fig. 5.4 Potential-based multi-robot flocking motion. **a** Initial state, **b** $t = 14.19\,\mathrm{s}$, **c** $t = 43.32\,\mathrm{s}$, **d** $t = 84.06\,\mathrm{s}$

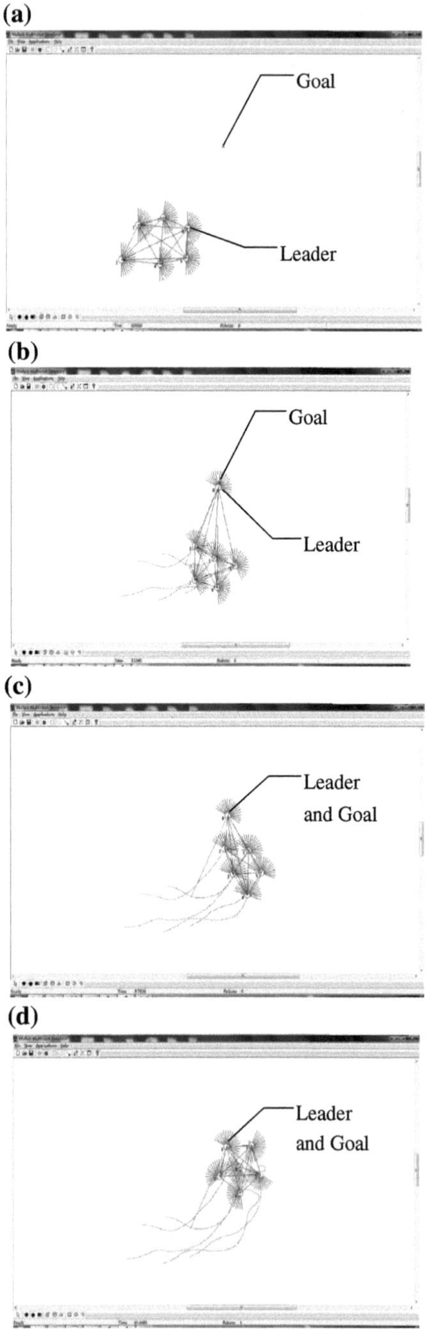

Fig. 5.5 Backstepping-based multi-robot flocking motion. **a** Initial state, **b** $t = 5.13\,\mathrm{s}$, **c** $t = 8.79\,\mathrm{s}$, **d** $t = 46.45\,\mathrm{s}$

Fig. 5.6 Adjustment for parameter k_1. **a** Initial state, **b** $k_1 = 0.01$, **c** $k_1 = 10$, **d** $k_1 = 100$

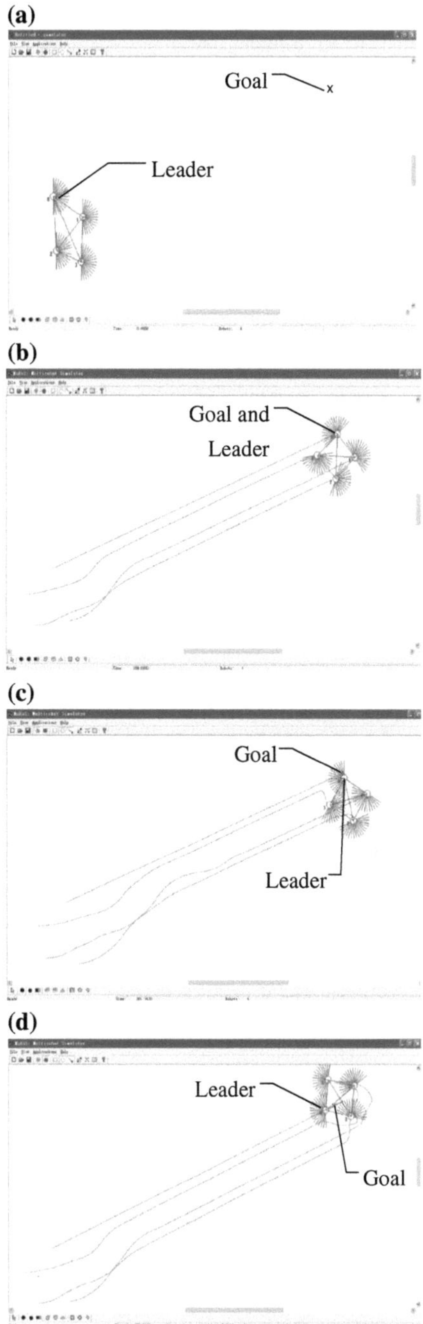

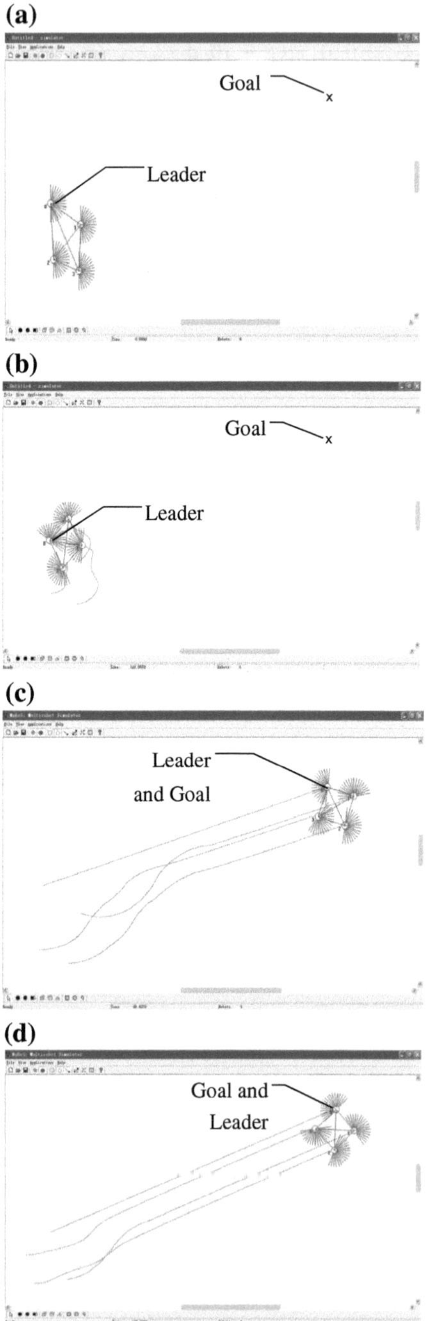

Fig. 5.7 Adjustment for parameter k_2. **a** Initial state, **b** $k_2 = 10$, **c** $k_2 = 20$, **d** $k_2 = 200$

Table 5.1 Errors of positions and orientations for values of a

a	x_e	y_e	θ_e
100	0.2519	−0.0885	−0.4382
10	0.2300	0.0217	−1.0261
1	2.0036	−0.3610	2.6985
0.5	2.9598	−0.8559	2.6897
0.1	238.53	−4.8209	−3.1184

Table 5.2 Errors of positions and orientations for values of b

b	x_e	y_e	θ_e
100	21.3348	3.9851	−0.4147
50	12.2446	1.5055	−1.0261
20	4.6249	0.6739	−0.3952
10	0.2582	−0.0930	−0.4448
2	0.7476	−0.0566	−0.4575
0.2	0.2576	−0.0867	−0.4187
0.01	0.2101	−0.0934	−0.4241

Table 5.2 shows the errors for different values of b. It is therefore concluded that the errors approach 0 when $b < 0.2$.

Discussion: From the above analysis, the role of each parameter is described as follows: The proper adjustment of k_1 will reduce the overshoot of the error between positions and orientations of Leader robot and goal; parameter k_2 affects largely the stability of Leader's motion control based on Backstepping; as well as a, parameter b will affect the error between the positions and orientations of Leader robot and go al. A suitable parameter may be selected with $k_1 < 0.5, k_2 > 20, a > 100$, and $b < 0.2$, which could support the running of flocking strategy with Leader's Backstepping.

5.5 Conclusion

In the chapter, we presented a unique and practical flocking mechanism for a group of nonholonomic wheeled robots, combined with the Leader's motion control based on Backstepping, in order to realize the trajectory of flock with a smooth transition, and move to the goal more swiftly. Simulation results and analysis on modified MuRoS platform demonstrated that the proposed mechanism allowed the robots to perform group tasks adaptively, flexibly, and efficiently.

Acknowledgments This work was supported by the National Natural Science Foundation in China (Grant No. 60705035, 61075087, 61203331, 61005065), Key Program of Hubei Province Natural Science Foundation of China (Grant No. 2010CDA005), Key Program of Open Foundation of Hubei Province Key Laboratory of Systems Science in Metallurgical Process of China (Grant No. Z201102), Open Foundation of Henan Provincial Open Laboratory for Control Engineering Key

Disciplines (Grant No. KG2011-01), Scientific Research Plan Key Project of Hubei Provincial Department of Education (Grant No. D20131105). This is greatly acknowledged.

References

1. Xiao BJ, Su HM, Zhao YL, Chen X (2013) Ant colony optimisation algorithm-based multi-robot exploration. Int J Model Ident Control 18(1):41–46
2. Rigatos G (2008) Multi-robot motion planning using swarm intelligence. Int J Adv Rob Syst 5(2):139–144
3. Zhang XY, Peng J, Hu HS, Lin K, Wang J (2012) Target attraction-based ant colony algorithm for mobile robots in rescue missions. Int J Model Ident Control 17(2):133–142
4. Reynolds C (1987) Flocks, birds, and schools: a distributed behaviour model. Comput Graphics 21(4):25–34
5. Chen SM (2007) Review of the modeling and control of swarm behaviours. Comput Eng Sci 29(7):102–105
6. Jadbabaie A, Lin J, Morse AS (2002) Coordination of groups of mobile autonomous agents using nearest neighbor rules. IEEE Trans Autom Control 48(6):988–1001
7. Tanner HG, Jadbabaie A, Pappas GJ (2007) Flocking in fixed and switching networks. IEEE Trans Autom Control 52(5):863–868
8. Olfati-Saber R, Murray RM (2004) Consensus problems in networks of agents with switching topology and time-delays. IEEE Trans Autom Control 49(9):1520–1533
9. Ji M, Egerstedt M (2007) Distributed coordination control of multiagent systems while preserving connectedness. IEEE Trans Rob 23(4):693–703
10. Li SQ, Shuai L, Cheng XY, Tang ZM, Yang JY (2005) A descriptive model of robot team and the dynamic evolution of robot team cooperation. Int J Adv Rob Syst 2(2):139–143
11. Zavlanos MM, Tanner HG, Jadbabaie A, Pappas GJ (2009) Hybrid control for connectivity preserving flocking. IEEE Trans Autom Control 54(12):2869–2875
12. Ekanayake SW, Pathirana PN (2010) Formations of robotic swarm: an artificial force based approach. Int J Adv Rob Syst 7(3):173–190
13. Dimarogonas D, Kyriakopoulos KJ (2007) On the rendezvous problem for multiple nonholonomic agents. IEEE Trans Autom Control 52(5):916–922
14. Cheng L, Yu H, Wu HY, Wang YJ (2008) A sequential flocking control system for multiple mobile robots. Control Theory Appl/Kongzhi Lilun yu Yingyong 25(6):1117–1120
15. Tran VH, Lee SG (2011) A stable formation control using approximation of translational and angular accelerations. Int J Adv Rob Syst 8(1):65–75
16. Saber RO (2006) Flocking for multi-agent dynamic systems: algorithms and theory. IEEE Trans Autom Control 51(3):401–420
17. Li W, Chen Z, Liu Z (2009) Formation control of multi-agent system based on potential function in complex environment. Int J Syst Control Commun 1(4):525–539
18. Cheng L, Cao L, Wu HY (2011) Trajectory tracking control of nonholonomic mobile robots by Backstepping. In: Proceedings of 2011 international conference on modelling, identification and control, pp 134–139
19. Yang JH, Wu J, Hu YM (2002) Backstepping method and its applications to nonlinear robust control. Control Decis 17:641–647
20. Saberi A, Kokotovic PV, Sussmam HJ (1990) Global stabilization of partially linear composite systems. SIAM J Control Optim 28:1491–1503
21. Zohar I, Ailon A, Rabinovici R (2011) Mobile robot characterized by dynamic and kinematic equations and actuator dynamics: trajectory tracking and related application. Rob Auton Syst 59:343–353
22. Bouteraa Y, Ghommam J, Derbel N (2011) Coordinated Backstepping control of multiple robot system of the leader-follower structure. In: International multi-conference on systems, signals and devices, pp 1–5

23. Aguiar AP, Hespanha JP (2007) Trajectory-tracking and path-following of underactuated autonomous vehicles with parametric modeling uncertainty. IEEE Trans Autom Control 52(8):1362–1379
24. Ghommam J, Saad M, Mnif, F (2010) Robust adaptive formation control of fully actuated marine vessels using local potential functions. In: 2010 IEEE international conference on robotics and automation anchorage convention district, Anchorage, Alaska, USA, pp 3001–3007
25. Chiu CH, Peng YF, Lin YW (2011) Intelligent Backstepping control for wheeled inverted pendulum. Expert Syst Appl 38:3364–3371
26. Li XH, Xiao J, Cai ZJ (2005) Backstepping based multiple mobile robots formation control. In: IEEE/RSJ international conference on intelligent robots and systems, City University of New York, NY, USA, pp 887–892
27. Castro R, Álvarez J, Martínez J (2009) Robot formation control using Backstepping and sliding mode techniques. In: International conference on 6th electrical engineering, computing science and automatic control, Mexico City, Mexico, pp 1–6
28. Dong WJ (2010) Formation control of multiple wheeled mobile robots with uncertainty. In: 49th IEEE conference on decision and control, Hilton Atlanta Hotel, Atlanta, GA, USA, pp 4492–4497
29. Dong WJ (2011) Flocking of multiple mobile robots based on Backstepping. IEEE Trans Syst Man Cybern Part B Cybern 41(2):414–424
30. Chen YY, Tian YP (2008) A Backstepping design for directed formation control of three-coleader agents in the plane. Int J Robust Nonlinear Control 19:729–745
31. Do KD (2008) Formation tracking control of unicycle-type mobile robots With limited sensing ranges. IEEE Trans Control Syst Technol 16(3):527–538
32. Li Q, Jiang ZP (2008) Formation tracking control of unicycle teams with collision avoidance. In: Proceedings of the 47th IEEE conference on decision and control, Cancun, Mexico, pp 496–501
33. Han TT, Ge SS (2011) Cooperative control design for circular flocking of underactuated hovercrafts. In: 50th IEEE conference on decision and control and European control conference (CDC-ECC), Orlando, FL, USA, pp 4891–4896
34. Cheng L, Xu WX, Wu HY, Zhu QM, Wang YJ, Nouri H (2012) A new procedure for multi-mode sequential flocking with application to multiple non-holonomic mobile robot motion control: mode description and integration principle. Int J Model Ident Control 15(1):39–47
35. Cheng L, Zheng XJ, Wu HY, Zhu QM, Wang YJ, Nouri H (2012) A new procedure for multi-mode sequential flocking with application to multiple non-holonomic mobile robot motion control: implementation and analysis. Int J Model Ident Control 16(1):50–59
36. Chaimowicz L (2002) Dynamic coodination of cooperative robots: a hybrid system approach. Doctoral Thesis, University of Minas Gerais, Collaborated with GRASP Laboratory at University of Pennsylvania
37. Tanner HG, Jadbabaie A, Pappas GJ (2003) Stable flocking of mobile agents, part I: fixed topology. In: Proceedings of IEEE conference on decision and control conference proceedings, Maui, HI, USA, pp 2010–2015
38. Yu H, Wang YJ, Cheng L (2005) Control of stable flocking motion of multiply-agent with a leader. J Huazhong Univ Sci Technol. Nat Sci Ed 33(8):56–58
39. Siegwart R, Nourbakhsh IR (2006) Introduction to autonomous mobile robots. LI Ren-hou Translation. Xi'an Jiaotong University Press, Xi'an
40. Min YY, Liu YG (2007) Barbalat Lemma and its applicatioin in analysis of system stability. J Shandong Univ 37(1):51–55
41. http://www.grasp.upenn.edu/chaimo/muros_download.html

Chapter 6
Performance Comparison Between NCTF and PV Techniques for the Control of Linear Motion Servo System

Jiwa Abdullah, Noor Hisham Jalani and Jamaludin Jalani

Abstract Motion control system is a challenging problem in the area of control systems. It is very useful to demonstrate concepts in linear control such as the stabilization of unstable systems. Motion control system plays important roles in industrial equipment such as machine tools, semiconductor manufacturing systems, and robot systems. One type of motion control system is the point-to-point (PTP) positioning system, which is used to move an object from one point to another point. Linear motion servo system is a machine that moves the cart from one point to another point. This system consists of a cart driven by a DC motor, via a rack and pinion mechanism to ensure consistent and continuous traction. Till date, many types of controllers have been proposed and evaluated for positioning systems. Two types of controllers discussed in this thesis are Proportional-Velocity (PV) and Nominal Characteristic Trajectory Following (NCTF). The experimental results showed that the PV was successfully implemented which controlled the settling time, rise time, and steady-state error of the desired position. However, the overshoot performances show its disadvantages. Additionally, the PV controller design is a time-consuming process, since model and parameters of the linear motion servo system are needed. Therefore, the needs for higher performance controller become important for the simplicity of the controller design. Hence, the investigation proceeds with the non-model-based NCTF controller to control the cart position of the linear motion servo system. The NCTF controller consists of a Nominal Characteristic Trajectory (NCT) and PI compensator. The NCTF controller was designed based on a simple open-loop experiment of the object. The experimental results showed that the NCTF controller

J. Abdullah (✉)
Faculty of Electrical and Electronic Engineering, Universiti Tun Hussein Onn Malaysia, Batu Pahat, Malaysia
e-mail: jiwa@uthm.edu.my

N. H. Jalani
Advanced Technology Training Center, Batu Pahat, Malaysia

J. Jalani
Faculty of Engineering Technology, Universiti Tun Hussein Onn Malaysia, Batu Pahat, Malaysia

L. Liu et al. (eds.), *Applied Methods and Techniques for Mechatronic Systems*, Lecture Notes in Control and Information Sciences 452,
DOI: 10.1007/978-3-642-36385-6_6,

is more effective for controlling the position of linear motion servo system than the PV controller.

Keywords Proportional-velocity · Nominal characteristic trajectory following · Performance comparison · Motion control system · Mathematical modeling

6.1 Introduction

Motion control system is a challenging problem in the area of control systems. It is very useful to demonstrate concepts in linear control such as the stabilization process. Motion control system plays important roles in machine tools, semiconductor manufacturing systems, and robotic systems. One type of motion control systems is point-to-point (PTP) positioning system, which is used to move an object from one point to another. Till date, many types of controllers have been proposed and evaluated for positioning systems. The objective of this chapter is to propose an appropriate control method for linear motion control system. It is expected to have a simple structure and is easy to design. The design of the proposed controller is based solely on the simple open-loop experiment. The chapter proposed a linear motion controller based on Nominal Characteristic Trajectory Following (NCTF) [1, 2] which is used to control the position of a cart horizontally. Parametric estimation is the most essential step to prove the concept [3].

The rest of the chapter is organized as follows. In Sect. 6.2, a critical review on the existing controller methodology, characteristics, and features is done. Mathematical modeling of motion servo system is illustrated in Sect. 6.3. It involves the necessary efforts of model derivations and parameter identification. In Sect. 6.4, the design, implementation, and performance of a benchmark PV controller is discussed. Since there are limitations in the performance of a PV controller, a new controller, NCTF, is proposed and deliberated in Sect. 6.5. The NCTF controller was designed, implemented, and its performance was studied rigorously. Lastly, the chapter concludes with a summary.

6.2 Related Work

In recent years, many types of positioning system controllers were proposed and evaluated. Omar et al. [4] proposed PD controller for positioning system. However, it was a well-known fact that controlling the position by using a PD controller was not very effective in eliminating the steady-state error due to the error in the friction values. Wahyudi et al. [5] used classical PID controller for position control in which the parameters were designed and optimized using Nonlinear Control Design (NCD) Blockset of Matlab/Simulink. Simulink is the most effective to show the working principle of most control mechanisms [6]. The effectiveness of PID and

PD controllers were experimented utilizing a lab-scale gantry crane system. It was concluded that PID and PD were successfully implemented to control the position and swing of the system especially for small load displacement. However, the positioning performance degraded due to the actuator's saturation. On the other hand, it was noted that PD controller was rather sensitive to the length variation. The classical PID and PD controllers were well known as a model-based design control systems and thus can be implemented easily. However, for the gantry crane system, proper tuning procedures were still needed. Furthermore, the proposed PID controller shows poor performance. Hence, there is the need for a higher performance controller based on the non-model method.

An adaptive control strategy was proposed by Yang et al. [7]. However, the proposed controller design requires nonlinear control theory approaches which demand rigorous mathematical analysis. Fuzzy logic controllers for location positioning have also been proposed by several researchers. Dongmei Yu et al. [8] proposed a tracking control strategy that meets the position tracking performance and robustness of the closed loop design to external disturbance. One important characteristic was the ability to model parameters variation without the need for parameter identification. In this study, a controller based on intelligent control, embedded with fuzzy logic is proposed to improve the tracking and robust performances without the parameters identification.

Akin Delibasi et al. [9] introduced the position control of a DC motor using Fuzzy Logic and PID Control algorithms. Fuzzy Logic and PID controllers are designed based on labview programming module. The real-time position control of the DC motor was realized by using DAQ device. The authors claimed that the conventional PID controllers generally do not work well for nonlinear systems, and particularly complex and vague systems that have no precise mathematical models. To overcome these difficulties, various types of modified conventional PID controllers such as autotuning and adaptive PID controllers were developed lately. When compared to the conventional controller, the main advantage of the fuzzy logic is that there is no stringent requirement on mathematical modeling is required [10].

Wahyudi et al. [11] have proposed an intelligent gantry system that contains two fuzzy logic controllers for controlling both the position and anti-swing motion of the payload. The results gave smaller overshoot, shorter settling time, and smaller steady-state errors than the PID controller. Therefore it can be concluded that the proposed fuzzy position control system is better than the classical PID controller. However, the fuzzy logic controller uses mapping method which needs delayed feedback controller before fuzzy logic can be designed and implemented. This method is not practically applicable and is time-consuming because the fuzzy logic controller cannot be implemented directly as it relies on other controllers to fulfill the design specifications.

Beside fuzzy logic, Wahyudi et al. [1] have also proposed the NCTF controller as a practical controller for PTP positioning systems. The author claimed that, NCTF control system has a good positioning performance. The NCTF controller is also effective to compensate for the effect of friction which is the main source of positioning inaccuracy. In general, advanced controllers tend to be complicated and require

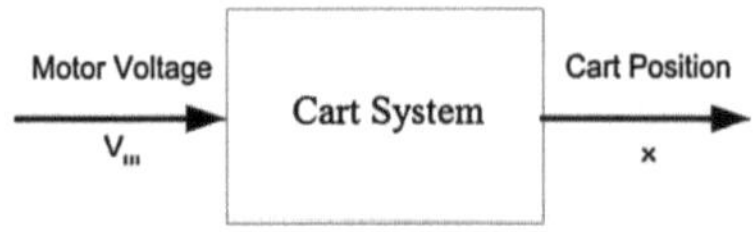

Fig. 6.1 The cart system

deep knowledge of controller theory and design. However, in a practical working environment, engineers who are not experts in control systems are often required to design controllers. Additionally, exact modeling and parameter identifications are generally troublesome and time-consuming tasks. Hence, the ease of design and simple controller structure are very important in practical application. Thus, the aim of this chapter is to investigate the new non-model-based controller for controlling PTP positioning system that was proposed by Wahyudi et al. [12].

Kaiji Sato et al. [13, 14] study the performances of motion control using NCTF controller. It was evaluated from experimental motion control results, including tracking, contouring, and positioning control. Its performances were compared with those using the conventional NCTF and the PID controllers. The results prove that, the Continuous Motion NCTF controller is more suitable than the conventional NCTF controller in all the motion control including positioning.

6.3 Modeling the Cart System

Modeling is a process of capturing the relevant information about the system. A model is developed to solve a problem in a short period; as well as economical, to ease the manipulation of variables of the plant system and able to test the model if it is the best solution for the particular problem. In this chapter, the model of cart systems is developed and explained mathematically. As illustrated in Fig. 6.1, the plant input is the voltage to the DC motor. The plant output is selected to be the cart linear position on the rack.

The open-loop transfer function, *G(s)*, can be written as $G(s) = \frac{x(s)}{V_m(s)}$, where x is cart position and V_m is motor voltage. Applying Newton's second law of motion to the cart system and by assuming cart's Coulomb friction is neglected, the equation of motion can be represented as

$$M\ddot{x}(t) = F_c(t) - D_{eq}\dot{x}(t) - F_{ai}(t), \tag{6.1}$$

where F_c is the cart driving force which is generated by the motor and F_{ai} is the inertial force due to the motor's armature in rotation and the other parameters are given in Table 6.1. The driving force, F_c, generated by the DC motor act on the cart through the motor pinion can be expressed as

$$F_c = \frac{\eta_g K_g T_m}{r_{mp}}. \tag{6.2}$$

Table 6.1 Parameter of the cart system

Parameter	Value/Unit (mm)
Motor nominal input voltage, V_m	6 V
Motor armature resistance, R_m	2.6 Ω
Motor armature inductance, L_m	0.18 mH
Motor torque constant, K_i	0.00767 Nm/A
Motor efficiency, η_m	100 %
EMF constant, K_m	0.00767 Vs/rad
Rotor moment inertia, J_m	3.9×10^{-7} kgm^2
Planetary gearbox gear ratio, K_g	3.71
Planetary gearbox efficiency, η_g	100 %
Motor pinion radius, r_{mp}	6.35×10^{-3} m
Position pinion radius, r_{pp}	1.48×10^{-2} m
Equivalent viscous damping coefficient at the motor pinion, B_{eq}	5.4 Nms/rad
Viscous damping coefficient at the pendulum pivot, B_p	0.0024 Nms/rad
Pendulum length from pivot to center of gravity, l_p	0.3302 m
Pendulum moment of inertia, I_p	7.88×10^{-3} kgm^2
Pendulum mass, M_p	0.230 kg
Cart mass, M	0.94 kg

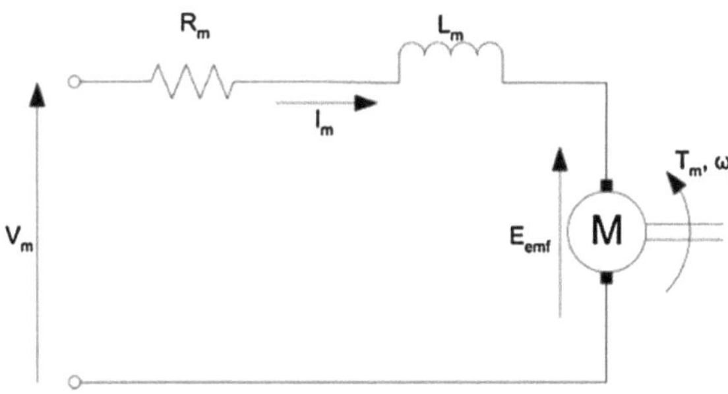

Fig. 6.2 DC motor electric circuit

Next, consider the electrical components of the DC motor. Figure 6.2 represents the classic electrical schematic drawing of the armature circuit for a standard DC motor. Using Kirchhoff's voltage law, the following equation is obtained:

$$V_m - R_m I_m - L_m\left(\frac{\partial}{\partial t} I_m\right) - E_{\text{emf}} = 0. \tag{6.3}$$

However, since $L_m \ll R_m$, the motor inductance can be neglected. Hence Eq. (6.3) is reduced into the following forms:

$$I_m = \frac{V_m - E_{\text{emf}}}{R_m}. \tag{6.4}$$

Since the back-emf voltage created by the motor, E_{emf}, is proportional to the motor shaft velocity ω_m, which is given by $E_{\text{emf}} = R_m \omega_m$, then Eq. (6.4) now becomes

$$I_m = \frac{V_m - K_m \omega_m}{R_m}. \tag{6.5}$$

In order to account for the DC motor electrical losses, the motor efficiency is introduced to calculate the torque generated by the DC motor, $T_m = \eta_m K_t I_m$. Substituting this and Eq. (6.5) into Eq. (6.8) leads to

$$F_c = \frac{\eta_g K_g \eta_m K_t (V_m - K_m \omega_m)}{R_m r_{mp}}. \tag{6.6}$$

By considering the rack and pinion and the gearbox mechanisms, the motor angular velocity can be written as a function of the cart linear velocity, such as

$$\omega_m = \frac{K_g \left(\frac{\mathrm{d}}{\mathrm{d}t} x(t)\right)}{r_{mp}}. \tag{6.7}$$

Substituting Eq. (6.7) into Eq. (6.6) and rearranging leads to

$$F_c = \frac{\eta_g K_g \eta_m K_t \left(V_m r_{mp} - K_g K_m \left(\frac{\mathrm{d}}{\mathrm{d}t} x(t)\right)\right)}{R_m r_{mp}^2}. \tag{6.8}$$

As seen at the motor pinion, the armature inertial force due to the motor rotation and acting on the cart can be expressed as a function of the armature inertial torque

$$F_{ai} = \frac{\eta_g K_g T_{ai}}{r_{mp}}. \tag{6.9}$$

Applying Newton's second law of motion to the motor shaft,

$$J_m \left(\frac{\mathrm{d}^2}{\mathrm{d}t^2} \theta_m(t)\right) = T_{ai}(t). \tag{6.10}$$

The mechanical configuration of the cart's rack and pinion system gives the following relationship:

$$\theta_m = \frac{K_g x}{r_{mp}}. \tag{6.11}$$

Substituting Eqs. (6.9) and (6.10) into Eq. (6.8) produces the following expression for the armature inertial force:

$$F_{ai} = \frac{\eta_g K_g^2 J_m \left(\frac{d^2}{dt^2} x(t)\right)}{r_{mp}^2}. \tag{6.12}$$

Finally, substituting Eqs. (6.8) and (6.12) into Eq. (6.1), applying the Laplace transform and rearranging yields the desired open-loop transfer function for the cart system, such that,

$$G(s) = \frac{r_{mp}\eta_g K_g \eta_m K_t}{((R_m M r_{mp}^2 + R_m \eta_g K_g^2 J_m)s + \eta_g K_g^2 \eta_m K_t K_m + B_{eq} R_m r_{mp}^2)s}. \tag{6.13}$$

Mathematical model of cart system is derived by using Newton's second law of motion, Kirchhoff's voltage law, and Laplace transform theorem. The obtained model of the cart system will be applied for PV and NCTF controllers design as well as simulation purposes.

6.4 Design and Implementation of PV Control System

6.4.1 Characteristics of PV Controller

Generally, characteristic equations of the PV and PD controller closed-loop transfer function are equal. The classical PD controller for position control system was proposed by Omar et al. [4]. The author claimed that PD controller is already well known. However, its negative point was that it produced higher steady-state error. Moreover, complex mathematical formulation is needed to design the PD controller. In this section, the design of PV controller for linear motion servo system is elaborated. The simulations, experiments, and evaluation are explained. The performances of PV controller are analyzed and its conclusions deduced. Consider a standard closed-loop position control system with a unity feedback loop as shown in Fig. 6.3. Its transfer function, *T(s)*, is given as follows:

$$\frac{x(s)}{x_d(s)} = \frac{G_c(s)G(s)}{1 + G_c(s)G(s)H(s)}. \tag{6.14}$$

Equation (6.14) expresses a plant model that has no zero and two poles. In order to design controllers satisfying the given requirements, control theory techniques generally provide approximate design formulae, which are based, for quadratic lag system with no zero, in the following standard equation:

$$T(s) = \frac{K_{dc}\omega_n^2}{s^2 + 2\xi\omega_n s + \omega_n^2}, \tag{6.15}$$

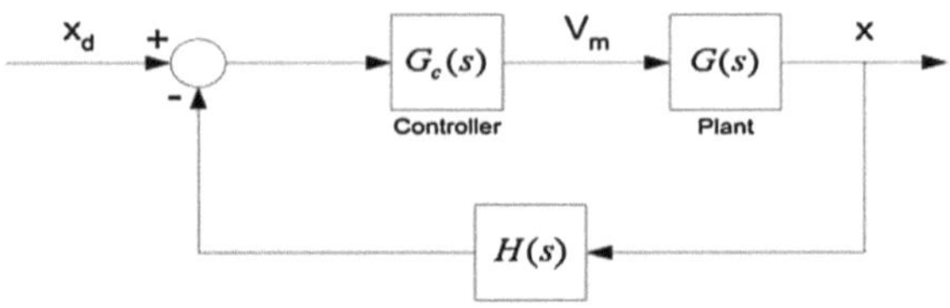

Fig. 6.3 Standard closed-loop position control system

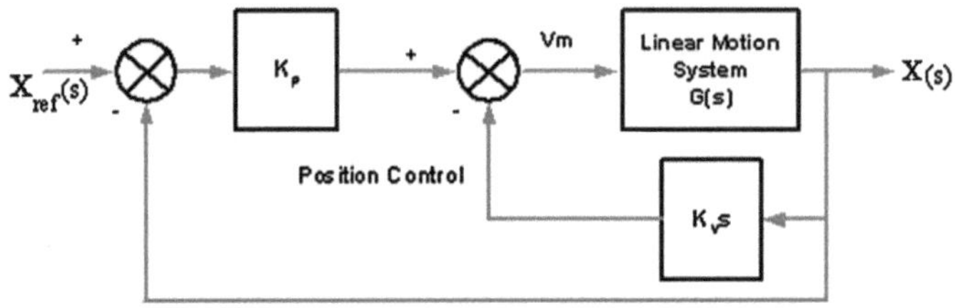

Fig. 6.4 Schematic diagram of linear motion servo system using PV controller

where $K_{dc}$is the system's DC gain. The characteristic equation of the closed-loop transfer function expressed in its standard form by Eq. (6.15) is as follows:

$$s^2 + 2\xi\omega_n s + \omega_n^2. \tag{6.16}$$

In the classical sense, Proportional-Derivative (PD) controller has the following transfer function:

$$G_c(s) = K_p + K_d s. \tag{6.17}$$

As expressed by Eq. (6.17), placing a controller in the forward path will introduce a zero in the closed-loop transfer function; hence there exists a mismatch between the closed-loop transfer function and the standard form of Eq. (6.15). Therefore, the design formulae derived from Eq. (6.15) would no longer be valid. In this case, adding an integral gain (I) to the forward path does not have to be considered since the open-loop transfer function, as seen in Eq. (6.9), is already of type 1, with a pole located at the origin of the s-plane.

6.4.2 Proportional-Velocity Control Scheme

The PV controller introduces two corrective terms: one is proportional (by K_p) to the position error and the other is proportional (by K_v) to the velocity (or the derivative of the actual position) of the plant. Coincidentally, the characteristic equations of the PV and PD controller closed-loop transfer function are equal. The structure of the existing linear motion servo system is shown in Fig. 6.4. The function of the controller is to control the movement of the cart toward the desired position x_{ref} as fast as possible with no or small overshoot. The variable x is the distance which varies between 0 and x_{ref}. Equation (6.18) expresses the PV control law, where x_{ref} is the reference signal

$$V_m(t) = K_p(x_{\text{ref}}(t) - x(t)) - K_v\left(\frac{\mathrm{d}}{\mathrm{d}t}x(t)\right). \tag{6.18}$$

6.4.3 PV Controller Design

These solutions of PV controller design are as follows: one possible block diagram reduction is to move the second summing point ahead of the K_p block and combine the two parallel feedback loops. This leads to a general closed-loop block diagram similar to the one presented in Fig. 6.4. The analogy is characterized by the following transfer functions:

$$H(s) = \frac{K_v}{K_p} + 1 \qquad \text{and} \qquad G_c(s) = K_p. \tag{6.19}$$

Substituting $G_c(s)$and $H(s)$ of Eq. (6.19) into Eq. (6.14) leads to the following closed-loop transfer function:

$$\frac{x(s)}{x_{\text{ref}}(s)} = \frac{K_pG(s)}{1 + K_pG(s) + G(s)K_vs}. \tag{6.20}$$

Replacing $G_c(s)$with the cart system open-loop transfer function as determined in Eq. (6.13) leads to the following closed-loop transfer function for the system:

$$\begin{aligned}\frac{x(s)}{x_{\text{ref}}(s)} &= K_t\eta_mK_g\eta_gr_{mp}K_p/(s^2R_mMr_{mp}^2 + s^2R_m\eta_gK_g^2J_m \\ &+ s\eta_gK_g^2\eta_mK_tK_m + sB_{eq}R_mr_{mp}^2 + r_{mp}\eta_gK_g\eta_mK_tK_vs + r_{mp}\eta_gK_g\eta_mK_tK_p).\end{aligned} \tag{6.21}$$

The following equation expresses the normalized characteristic equation of the closed-loop system:

$$\begin{aligned}s^2 &+ \frac{(r_{mp}\eta_gK_g\eta_mK_tK_v + \eta_gK_g^2\eta_mK_tK_m + B_{eq}R_mr_{mp}^2)s}{R_mMr_{mp}^2 + R_m\eta_gK_g^2J_m} \\ &+ \frac{r_{mp}\eta_gK_g\eta_mK_tK_p}{R_mMr_{mp}^2 + R_m\eta_gK_g^2J_m} = 0.\end{aligned} \tag{6.22}$$

By identification with the standard characteristic equation, it becomes the following expression for ω_n:

$$\omega_n = \frac{\sqrt{R_m(Mr_{mp}^2 + \eta_g K_g^2 J_m) r_{mp} \eta_g K_g \eta_m K_t K_p}}{R_m M r_{mp}^2 + R_m \eta_g K_g^2 J_m}. \tag{6.23}$$

By identification with the standard characteristic equation, the expression for ξ as a function of ω_n is given as

$$\xi = \frac{1}{2} \frac{\eta_g K_g^2 \eta_m K_t K_m + B_{eq} R_m r_{mp}^2 + r_{mp} \eta_g K_g \eta_m K_t K_v}{R_m M r_{mp}^2 \omega_n}. \tag{6.24}$$

Substituting Eq. (6.23) into Eq. (6.24) yields

$$\xi = \frac{1}{2} \frac{\eta_g K_g^2 \eta_m K_t K_m + B_{eq} R_m r_{mp}^2 + r_{mp} \eta_g K_g \eta_m K_t K_v}{\sqrt{R_m(Mr_{mp}^2 + \eta_g K_g^2 J_m) r_{mp} \eta_g K_g \eta_m K_t K_p}}. \tag{6.25}$$

(i) According to Eqs. (6.23) and (6.25), above, the following variations can be inferred:

(a) Assuming $K_v$constant, as K_p is increased, $\omega_n$will increase and ξ will decrease, resulting in a faster but also less damped closed-loop response.
(b) Assuming $K_p$constant, as K_v is increased, ω_n will not change and ξ will increase. This would translate into a similar t_p but with smaller *PO* thus adding damping into the closed-loop system.
(c) The needed K_p and K_v are calculated to meet the required specifications. Specifically, the steps to achieve this are as follows:

i. Solving Eq. (6.23) for K_p leads to the following expression for K_p, as a function of ω_n:

$$K_p = \frac{\omega_n^2 R_m(Mr_{mp}^2 + \eta_g K_g^2 J_m)}{r_{mp} \eta_g K_g \eta_m K_t}. \tag{6.26}$$

ii. Substituting Eq. (6.26) into Eq. (6.25) and solving for $K_v$yields:

$$K_v = \frac{2\xi\omega_n R_m M r_{mp}^2 + 2\xi\omega_n R_m \eta_g K_g^2 J_m - \eta_g K_g^2 \eta_m K_t K_m - B_{eq} R_m r_{mp}^2}{r_{mp} \eta_g K_g \eta_m K_t}. \tag{6.27}$$

iii. x_{ref} The minimum damping factor to meet the maximum *PO* requirement can be obtained by,

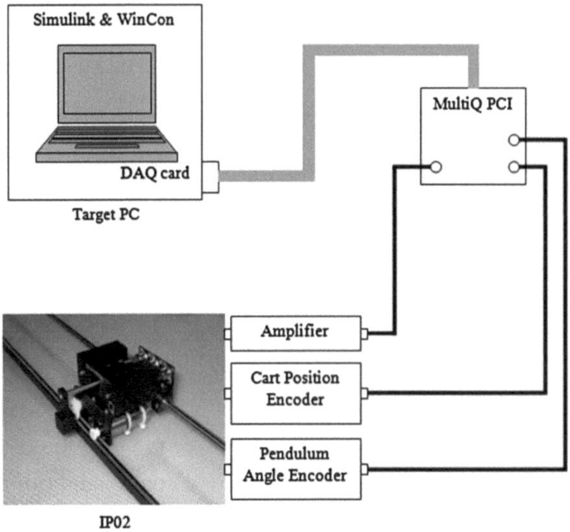

Fig. 6.5 Serial connection between host and target computer to the linear motion servo system

$$\xi = \frac{\left|\ln\left(\frac{PO}{100}\right)\right|}{\sqrt{\ln\left(\frac{PO}{100}\right)^2 + \pi^2}}. \tag{6.28}$$

$$\omega_n = \frac{\pi}{t_p\sqrt{1-\xi^2}}. \tag{6.29}$$

Substituting ξ, from Eq. (6.28) into Eq. (6.29) resulted in ω_n as a function of the two time specifications, t_p and *PO,*

$$\omega_n = \frac{\sqrt{\ln\left(\frac{PO}{100}\right)^2 + \pi^2}}{t_p}. \tag{6.30}$$

iv. According to performance requirements set as PO=10% and t_p = 0.15s, natural frequency ω_n= 26.0 rad/s and damping ratio ξ= 0.59 are obtained.

Finally, using the initial controller parameters of K_p and K_v with values of 419.1239 and 11.4697, respectively, and then applying the numerical computation of Eqs. (6.26) and (6.27) with the value of ω_n and ξ corresponding to the given time requirements gives the PV controller gains.

Table 6.2 Positioning performances for PV controller

Controllers	Positions (mm)	Overshoot (%)	Settling time (s)	Rise time (s)	Steady-state error (mm)
PV simulated	50	7.323	0.235	0.080	0
	100	5.872	0.268	0.102	0
	200	5.267	0.331	0.144	0
PV experimental	50	10.194	0.242	0.076	0.044
	100	10.671	0.291	0.106	0.043
	200	11.627	0.372	0.149	0.086

6.4.4 Experimental Setup

The experimental setup is shown in Fig. 6.5. The PV control algorithms were implemented on a personal computer and is operated with a 1 ms sampling time. The MathWork's MATLAB/Simulink and Quanser Toolbox is used for real-time controller implementation through Ardence as a real-time extension (RTX) kernel and WinCon as real-time control software. The PV controllers are developed in Simulink using its block, and then its M-code is generated, compiled, and finally a real-time executable code is produced and downloaded into the computer. The Quanser DAQ card is used to acquire analog signal from the cart's encoders and send back the output signal to the linear motion servo system. The amplifier is used to amplify the output voltage signal from DAQ card and limit between −10 and 10 V.

6.4.5 Results and Discussions

The obtained parameters of PV controller were tested in the simulation and experiment. The different input references were used to show the performances of the controller. Figure 6.6 shows the responses of the PV controller when the 50, 100, and 200 mm step input references were being used, respectively. The square input references are also used in order to show continuously the responses of the PV controller. Figure 6.7 shows the responses of the PV controller when the 50, 100, and 200 mm square input references were being used, respectively. The detailed performances are shown in Table 6.2. Here, the performances of position control system were evaluated based on overshoot, settling time, rise time, and steady-state error.

The results for the position control showed that PV controller gave a bigger overshoot. By simulation, the bigger overshoot is when 50 mm step input reference was used. However, in experiments, the PV controller gave bigger overshoot when the 100 and 200 mm step input references were used. However, the simulation and experiment results gave slightly different settling time and rise time. The steady-state error by experiment was higher as compared with the simulation. The results confirmed that

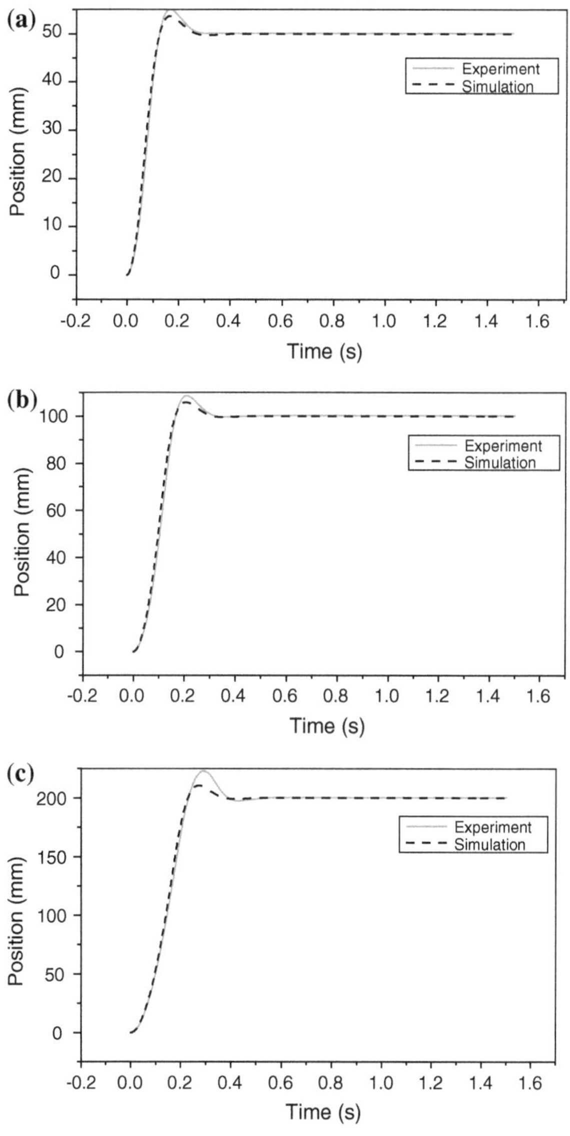

Fig. 6.6 Results for position control when system is applied step input. **a** Response to 50 mm step input; **b** Response to 100 mm step input; **c** Response to 200 mm step input

the PV controller successfully controlled the position of linear motion servo system but with the bigger overshoot.

The effectiveness of the PV controller was examined through simulations and experiments of the linear motion servo system. The results showed that the PV controller successfully controlled resulting in the settling time, rise time, and steady-state errors at the desired values. However, significant overshoot is recorded showing

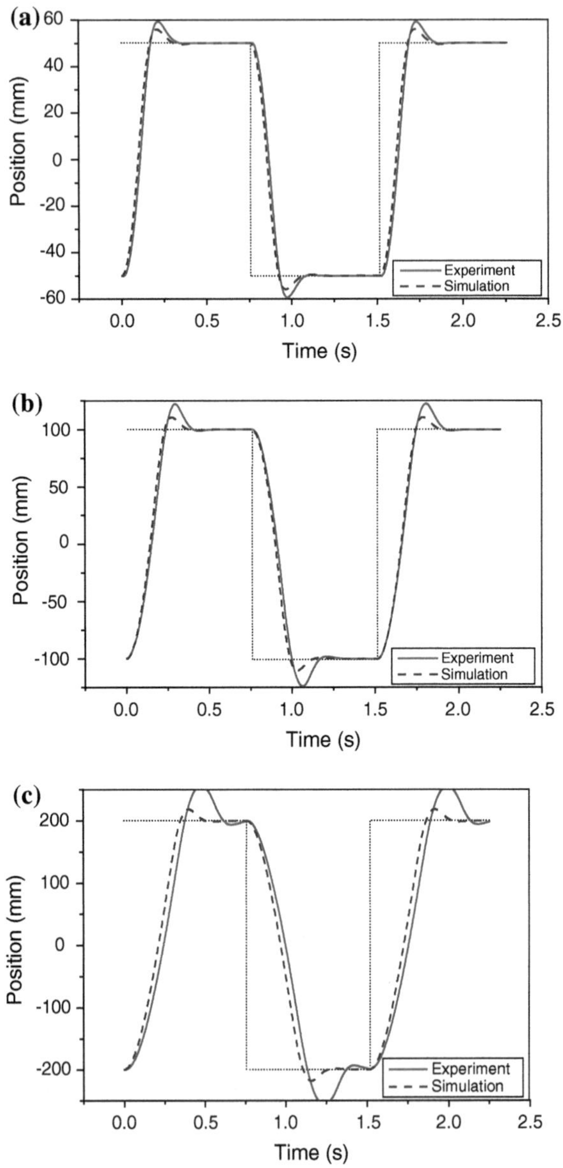

Fig. 6.7 Results for position control when the system is applied with square input. **a** Response to 50 mm square input; **b** Response to 100 mm square input; **c** Response to 200 mm square input

that its performance is relatively poor. The need for higher performance and simple controller based on non-model becomes necessity. Hence, the investigation of non-model-based on NCTF controller was introduced for further study of the linear motion servo system.

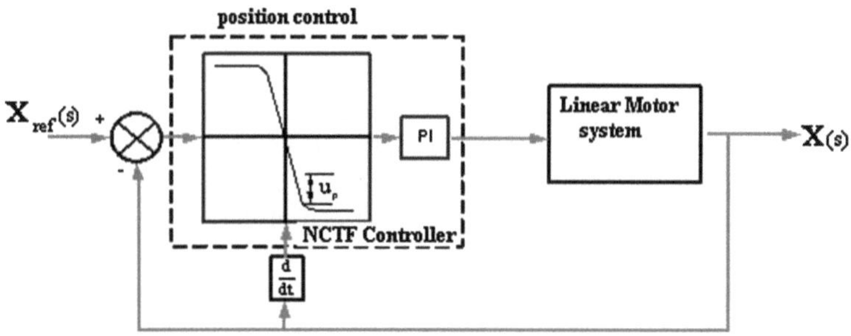

Fig. 6.8 Schematic diagram of using NCTF controller

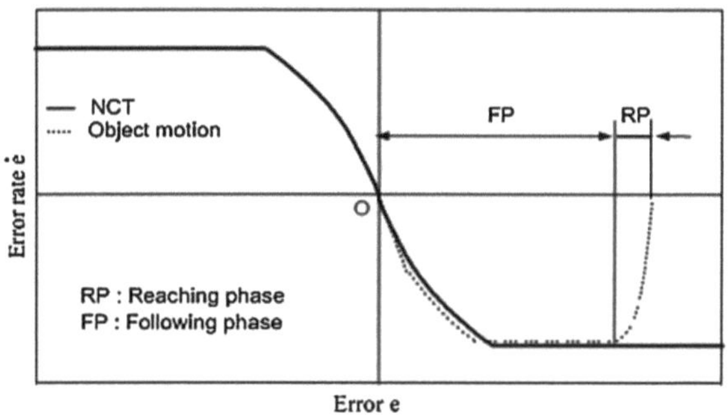

Fig. 6.9 NCT and object motion

6.5 Design and Implementation of NCTF Controller

6.5.1 Basic Concepts of NCTF Controller

The new non-model-based controller, known as NCTF controller, was introduced by Wahyudi [1]. The author claims that the NCTF controller is good for PTP positioning system. The structure of the proposed controller for linear motion servo system is shown in Fig. 6.8. The proposed NCTF controller consists of NCT and PI compensator. The aim of NCTF is to make the object move following the NCT and terminate at the origin of the phase plane. The parameter $\mathbf{u}_p$ represents the difference between the actual error rate and NCT. If the parameter $\mathbf{u}_p$ is zero, the object moves perfectly following the NCT. In order for the parameter $\mathbf{u}_p$ to reach zero, PI compensator is used to control the object.

The structure of the NCTF control system is shown in Fig. 6.8. The NCTF controller consists of a nominal characteristic trajectory (NCT) and a compensator. The NCTF controller operates with the following assumptions; (i) A DC or an AC servomotor is used as an actuator of the object; (ii) PTP positioning systems are used with x_{ref} is set to constant, and $\dot{x}_{ref} \equiv 0$. The objective of the NCTF controller is to make the object movement follow NCT and terminate it at the origin of the phase plane

$(e, \dot{e})$. Figure 6.9 shows an example of object movement controlled by the NCTF controller. The motion comprises of two phases, the first being the reaching phase and next, the following phase. In the reaching phase, the compensator forces the object motion to reach the NCT as fast as possible. Next, in the following phase, the compensator controls the object movement so that it follows the NCT. The object motion stops at the origin, which represents the end of the positioning movement. Hence, in NCTF system, the NCT governs the positioning response performances.

The NCTF controller is designed based on a simple open-loop experiment of the object as stated below:

(i) The open-loop-drive the object with stepwise inputs and measure the displacement and velocity responses of the object. Figure 6.6 shows the velocity and displacement responses due to the stepwise inputs. The rated input to the actuator u_r is used as height of the stepwise inputs.
(ii) The NCT is constructed using object responses. The velocity and the displacement responses are used to determine the NCT. Since the main objective of PTP system is to stop an object at certain position, a deceleration process (curve in area A of Fig. 6.10 is used. Variable h is the maximum motion velocity. From the curve in area A and h in Fig. 6.10a, the value of NCT can be derive from Fig. 6.10b is determined [15]. Since the NCT is constructed based on the actual responses of the object, it includes the effect of nonlinearity such as friction and saturation.
(iii) Design the compensator based on the NCT information

$$u = K_p u_p + K_i \int u_p dt, \tag{6.31}$$

where K_p and K_i are proportional and integral gains, respectively. Using the PI compensator parameters K_p and K_i, and the NCT characteristic near the origin (see Fig. 6.6b), the transfer function of the closed-loop positioning system utilizing NCTF controller can be approximated as follows:

$$\frac{X(s)}{X_{\text{ref}}(s)} = G(s) = G_1(s)G_2(s), \tag{6.32}$$

where $G_1(s) = \frac{\alpha}{s+\alpha}$; $G_2(s) = \frac{2\xi\omega_n+\omega_n^2}{s^2+2\xi\omega_n+\omega_n^2}$; $K_p = \frac{2\xi\omega_n}{K\alpha}$; and $K_i = \frac{\omega_n^2}{K\alpha}$ where α is the simplified object parameter, ξ is the damping ratio, and ω_n is the natural frequency. When ξ and ω_n are large enough, G(s) becomes nearly equal to G_1(s), which signify the condition when the object motion follows the NCT as the objective of the NCTF control system [16]. Moreover, large ξ and ω_n also make the closed-loop system robust to friction or inertia variation of the object in continuous systems [12]. Finally, by using ξ and ω_n the PI compensator parameters are designed with the following values of K, $K_p = \frac{2\varsigma\omega_n u_r}{mh}$, and $K_i = \frac{\omega_n^2 u_r}{mh}$.

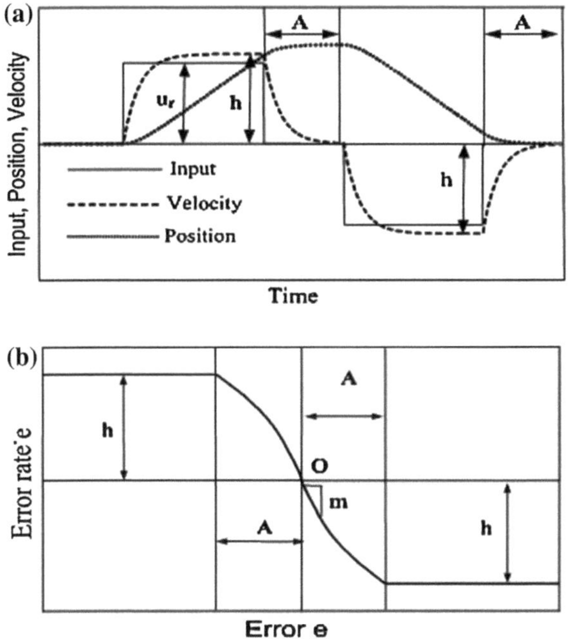

Fig. 6.10 NCT determination. **a** Stepwise inputs and responces; **b** Nominal characteristics trajectory (NCT) [15]

6.5.2 Controller and Compensator Design

The NCT is determined based on simple open-loop experiment of the object. The construction of NCT will determine the h and m parameters in order to design the compensator. In order to construct the NCT, a simple open-loop experiment has to be conducted. In the experiment, an actuator of linear motion system is driven with a step input. Then, the displacement and velocity responses of the linear motion system are measured.

Figure 6.11 shows the stepwise input, velocity, and displacement responses of the linear motion system. The rated input to the actuator is 10 V which is driven by a stepwise input as shown in Fig. 6.11a. The velocity and the displacement responses due to the stepwise input are shown in Fig. 6.11b, c are measured, respectively. Then, from these responses the NCT response is constructed as shown in Fig. 6.12. The determined NCT has $-17.095\ \text{s}^{-1}$ inclinations near the origin of m and 143 mm/s maximum rate of the NCT h.

After NCT determination was done, the parameter of $\omega_n$and ξ was selected for PI compensator design. The influence of $\omega_n$and ξ was examined and the best of PI compensators were used for the experiment. Three PI compensator parameters as listed in Table 6.3 were tested by simulation and experiment. Different values of $\omega_n$and ξwere chosen as comparison results. Compensators PI-1 and PI-2 were

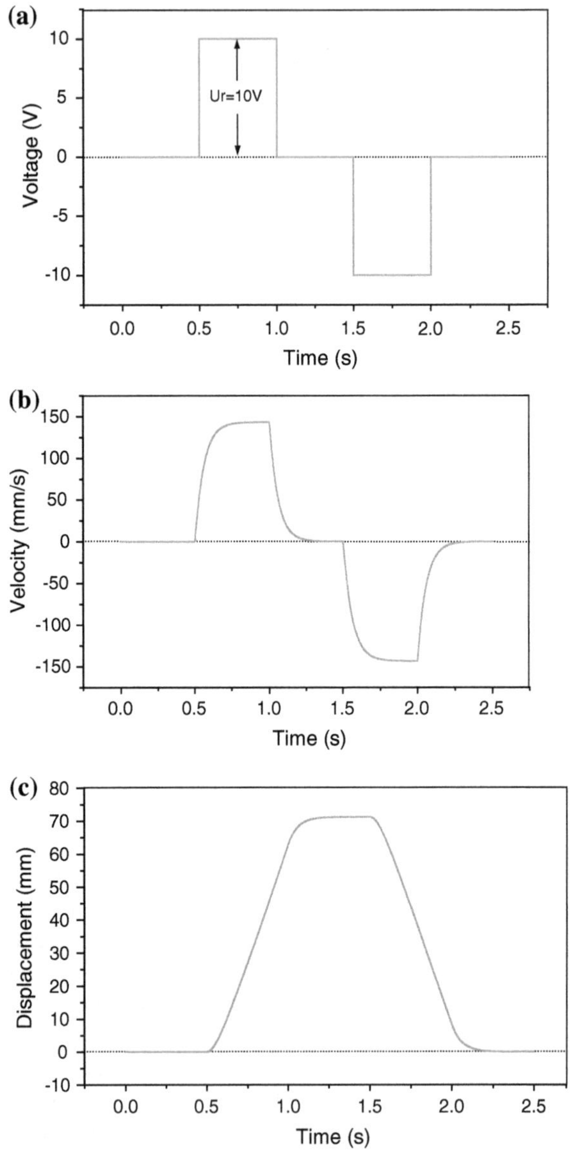

Fig. 6.11 Simple open-loop response of the linear motion system for NCT determination. **a** Step-wise input; **b** Velocity response; **c** Displacement response

designed using the value of ξ and ω_n was selected differently. Meanwhile, larger value ξ and lower value of ω_n were chosen for PI-3. Figure 6.13 shows the responses on the phase plane when the 50, 100, and 200 mm step input references were used, respectively.

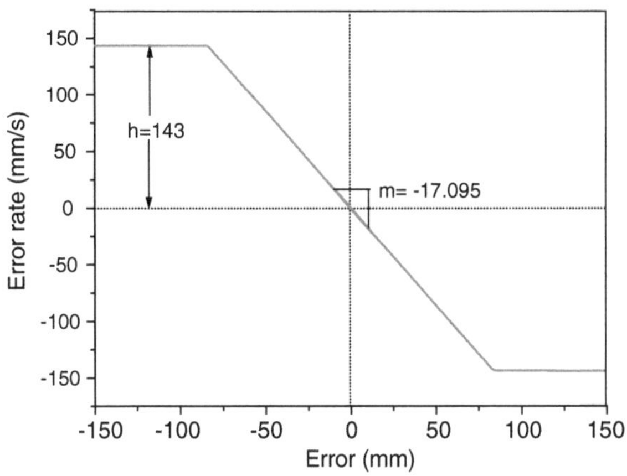

Fig. 6.12 Construction of NCT of linear motion system

On the other hand, Fig. 6.14 shows the position responses when 50, 100, and 200 mm step input references were used, respectively. The detailed position performances and performances based upon overshoot, settling time, rise time, and steady-state error are shown in Table 6.4. The results on the phase plane showed that the compensator of PI-3 made the object follow the NCT more quickly than PI-1 and PI-2 in the case of 50, 100, and 200 mm step input references. The results on the position responses showed that the compensator of PI-3 gave the smallest overshoot than PI-1 and PI-2 in the case of 50, 100, and 200 mm step input references. Although the compensator of PI-1 gave the quickest rise time, it gave the highest overshoot and slowest settling time. The compensator of PI-2 gave the slowest rise time and all compensators gave almost the similar zero steady-state error.

The results showed that the compensator PI-3 gave the best overall performances as compared with the compensators PI-1 and PI-2. Moreover, the results confirmed that the larger value ξ and lower value of ω_n were preferable in the PI compensator parameter design. The compensator PI-3 will be used for NCTF controller design for the linear motion system since it gave the best overall performances. Therefore, the design parameters $\xi = 5000$ and $\omega_n = 1$ are selected. The value of the compensator parameters are calculated as $K_p = 40.907$ and $K_i = 0.004$.

6.5.3 Some Performance Results

The performances of the NCTF controller were compared with the PV controller. Figure 6.15 shows the simulated and experimental responses of the NCTF and PV controller when the 50, 100, and 200 mm step references were used.

The results showed that the simulated and experimental responses of the NCTF controller are good and closed to each other compare than PV controller when a

Table 6.3 PI compensator parameters design

Compensator	ζ	ω_n	K_p	K_i
PI-1	25	100	20.453	40.907
PI-2	50	10	4.091	0.409
PI-3	5000	1	40.907	0.004

Table 6.4 Comparison of the performances of different compensators to the step input

	Step input (mm)	Overshoot (%)	Settling time (s)	Rise time (s)	Steady-state error
PI-1	50	3.9132	0.6430	0.1160	−0.0206
	100	7.0223	0.9390	0.1210	−0.0767
	200	12.9723	1.2810	0.1470	−0.3133
PI-2	50	1.2868	0.4070	0.2600	−0.4385
	100	1.2868	0.4070	0.2600	−0.8770
	200	1.4009	0.4180	0.2630	−1.9525
PI-3	50	0	0.2450	0.1310	0
	100	0	0.2710	0.1450	0
	200	0	0.3190	0.1740	−0.0020

50, 100, and 200 mm step input references were used. However, some discrepancies exist, chiefly due to modeling errors and the diverse nonlinear elements present in the actual system. Some of the possible reasons to explain them include modeling approximations and uncertainties in model parameter values; power amplifier (UPM) saturation (in voltage and/or current); DC motor dead-band effect; Coulomb (dry or static) friction effect on the cart; stiction effect; rack and pinion and gear backlash; noise and a cart system's track not properly leveled.

The detailed performances when step input references were used are shown in Table 6.5 for simulated results and Table 6.6 for experimental results. The simulation and experimental results showed that the NCTF controller gave almost zero overshoot than the PV controller. The simulation overshoot of PV controller gave more than 5 % but less than 8 %, while the experiment gave more than 10 % but less than 12 %. Although the PV controller gave slightly faster rise time than the NCTF controller, the settling time for both controllers is almost same. Moreover, both controllers gave zero steady-state error for simulation results, but by experiment, the NCTF controller gave better results than PV controller.

In order to show the continuous responses of both controllers, square input references were used. Figure 6.16 shows the response when −50 to 50 mm, −100 to 100 mm, and −200 to 200 mm square input references were used. The detailed results are shown in Table 6.7 for simulated results and Table 6.8 for experimental results.

The overall performance between NCTF and PV techniques is shown in Fig. 6.17. The simulated and experimental results showed that the NCTF controller gave consistent small overshoot than the PV controller. However, for the rise time performance, NCTF controller was slightly slower as compared to the PV controller.

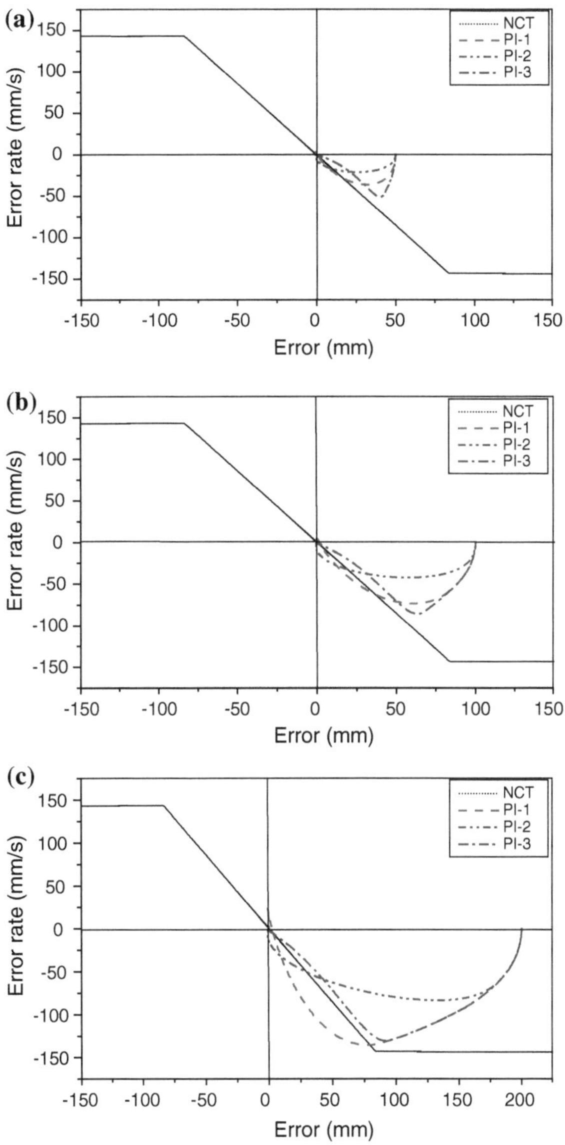

Fig. 6.13 Responses on phase plane. **a** The responses on the phase plane when 50 mm; **b** The responses on the phase plane when 100 mm; **c** The responses on the phase plane when 200 mm

The results confirmed that the NCTF controller successfully controlled the position better than the PV controller and performed more consistently at different desired positions. In the NCTF control system, the NCT governs the positioning response performance. The parameter of u_p represents the difference between the actual error

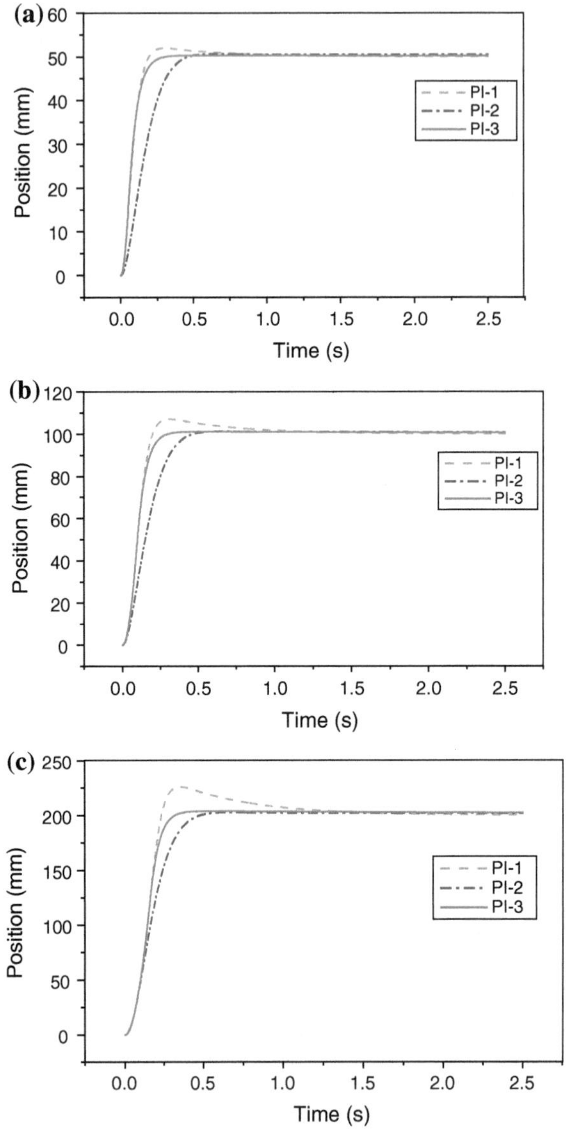

Fig. 6.14 Responses at different positions. **a** The position responses when 50 mm; **b** The position responses when 100 mm; **c** The position responses when 200 mm

rate and NCT. If the parameter of u_p is zero, the plant motion perfectly follows the NCT. Consequently, it causes a good performance of the NCTF controller. The NCTF controller is also effective to compensate for the effect of the friction which is the source of positioning inaccuracy [5, 17]. Non-model-based method of NCTF controller was proposed and designed. The effectiveness of the NCTF was examined

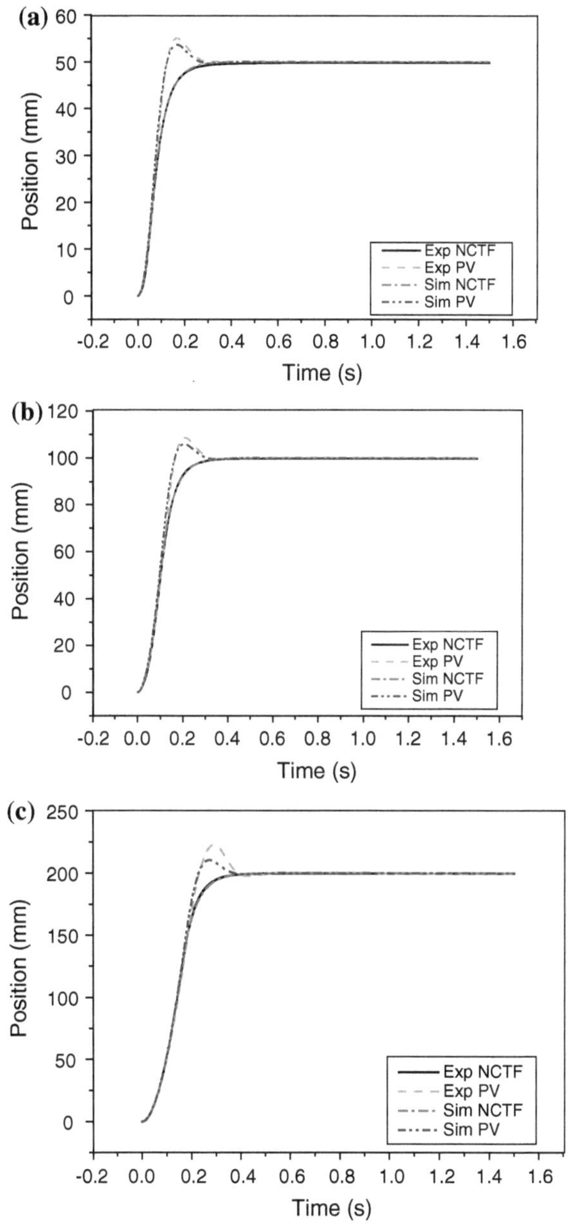

Fig. 6.15 The results for position responses to a step input. **a** Response to 50 mm step input; **b** Response to 100 mm step input; **c** Response to 200 mm step input

through the simulation and experiment of linear motion servo system. The results show that the NCTF controller gave better results for overshoot. The settling time and steady-state error are almost unchanged compared to the PV controller. However,

Table 6.5 Simulated positioning performances for the controllers when step input references were used

Controllers	Positions (mm)	Overshoot (%)	Performance		
			Settling time (s)	Rise time (s)	Steady-state error
NCTF	50	0	0.24	0.13	0
	100	0	0.27	0.14	0
	200	0	0.31	0.17	0
PV	50	7.32	0.24	0.08	0
	100	5.87	0.27	0.10	0
	200	5.27	0.33	0.14	0

the rise time for NCTF is a little bit higher than PV. Overall, the experimental results proved that the NCTF controller successfully controlled the position of the linear motion system than the PV controller.

6.6 Conclusions

The objective of this project was to design a practical positioning controller for linear motion servo system. The acquired controller for linear motion servo system must have high performance, simple structure, and be easy to design. The non-model-based method of NCTF controller was proposed. The NCTF consists of a nominal characteristic trajectory (NCT) and PI compensator. The NCTF control system was designed based on the simple open-loop experiment. The effectiveness of NCTF was evaluated and compared with existing PV controller through simulation and experiment. Both the simulation and experimental results showed that the NCTF controller is more effective for controlling position of linear motion servo system than the PV controller. The linear motion servo system consists of a cart driven by a DC motor, via a rack and pinion mechanism to ensure consistent and continuous traction. The cart is also equipped with a rotary joint to which a free turning rod (or pendulum) can be attached. The focus of this project is only to design a controller for the positioning of cart. The project reported here some limitations and can be further extended. Some of the recommendations for future development are as follows:

(i) Experimental robustness evaluation of the linear motion system.
(ii) Design and implement another controller for anti-swing of the pendulum.

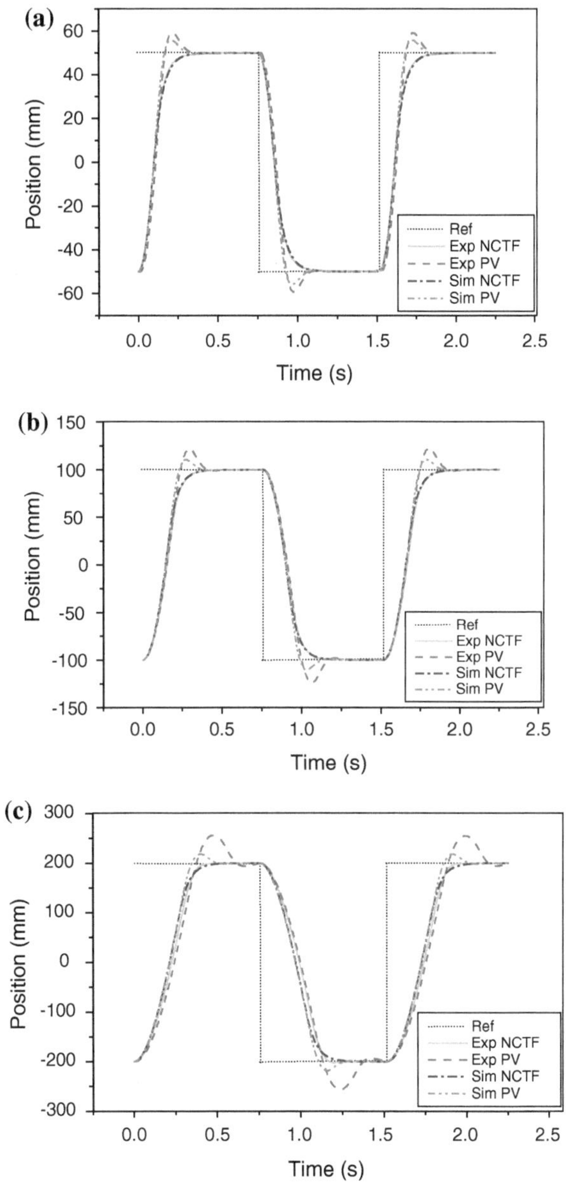

Fig. 6.16 The results for position responses to a square input. **a** Response to 50 mm square input; **b** Response to 100 mm square input; **c** Response to 200 mm square input

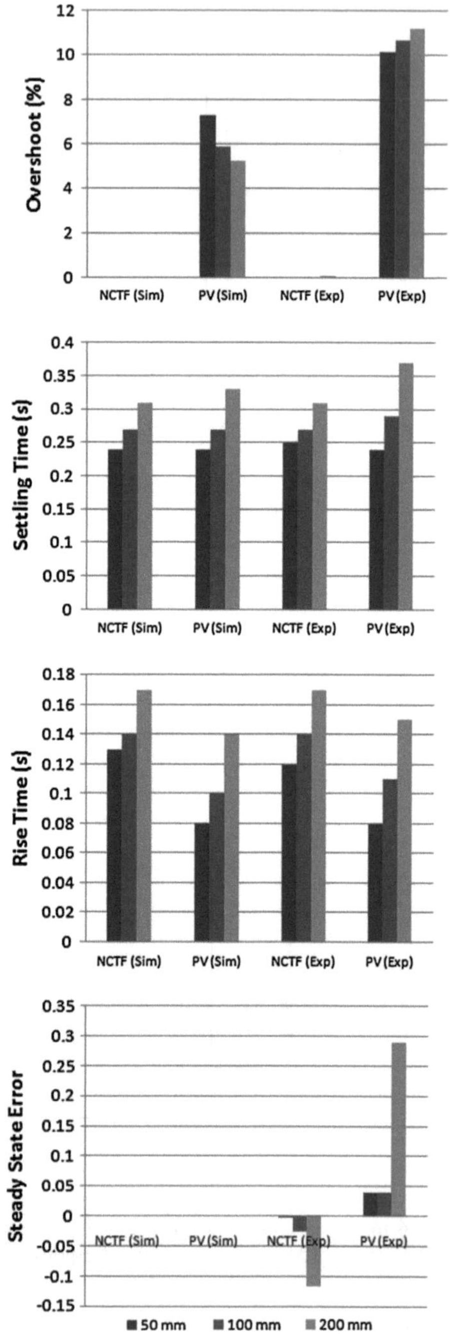

Fig. 6.17 Performances comparison between the controllers

Table 6.6 Experimental positioning performances for the controllers when step input references were used

Controllers	Positions (mm)	Overshoot (%)	Performance		
			Settling time (s)	Rise time (s)	Steady-state error
NCTF	50	0	0.25	0.12	−0.001
	100	0.07	0.27	0.14	−0.025
	200	0.09	0.31	0.17	−0.118
PV	50	10.19	0.24	0.08	0.04
	100	10.67	0.29	0.11	0.04
	200	11.24	0.37	0.15	0.29

Table 6.7 Simulated positioning performances for the controllers when square input references were used

Controllers		Performance	
	Positions (mm)	Overshoot (%)	Rise time (s)
NCTF	−50 to 50	0	0.12
	−100 to 100	0	0.12
	−200 to 200	0	0.14
PV	−50 to 50	11.74	0.05
	−100 to 100	10.53	0.07
	−200 to 200	9.13	0.11

Table 6.8 Experimental positioning performances for the controllers when square input references were used

Controllers		Performance	
	Positions (mm)	Overshoot (%)	Rise time (s)
NCTF	−50 to 50	0	0.12
	−100 to 100	0.07	0.12
	−200 to 200	0.09	0.14
PV	−50 to 50	14.61	0.05
	−100 to 100	15.33	0.08
	−200 to 200	15.10	0.12

References

1. Wahyudi (2002) New practical control of PTP positioning system. Ph.D Thesis, Tokyo Institute of Technology
2. Fitri MY, Wahyudi, Akmeliawati R, Wijaya AA (2009) NCTF control method of two mass system for PTP positioning. In: Proceedings of the international conference on man-machine system (ICoMMS), 2009
3. Al-Kasasbeh RT, Alshamasin MS, Ionescu F, Korenevskiy N (2012) Modelling and parameter estimation for operator intelligence in man-machine systems. Int J Model Ident Control, Indersciences 15(1):69–85

4. Omar HM (2003) Control of gantry and tower cranes. M.S. Thesis, Virginia Tech, Blacksburg
5. Wahyudi, Jalani J, Muhida R (2005) Design and implementation of classical PID controller for an automatic gantry crane system. In: Proceedings of the international conference on recent advances in mechanical and materials engineering, 2005
6. Alshamasin MS, Ionescu F, Al-Kasasbeh RT (2012) Modelling and simulation of a SCARA robot using solid dynamics and verification by MATLAB/Simulink. Int J Model Ident Control, Indersciences 15(1):28–38
7. Hua YJ, Shine YK (2006) Adaptive control for 3-D overhead crane systems. In: Proceedings of the 2006 American control conference, Minneapolis, 14–16 June 2006
8. Yu D, Guo Q, Hu Q (2006) Position control of linear servo system using intelligent feedback controller. In: Proceeding of the sixth international conference on intelligent systems design and application (ISDA'06): 0–7695-2528-8/06, 2006
9. Delibasi A, Turker T, Cansever G (2004) Real-time DC motor position control by fuzzy logic and PID controller using labview. Yildiz Technical University, Istanbul
10. Qu S, Tian Y, Chen C, Ai L (2012) A small intelligent car system based on fuzzy control and CCD camera. Int J Model Ident Control, Indersciences 15(1):48–54
11. Wahyudi, Jalani J (2005) Design and implementation of fuzzy logic controller for an intelligent gantry crane system. In: Proceedings of the 2nd international conference on mechatronics, pp 345–351
12. Wahyudi, Jalani J, Muhida R, Salami MJE (2007) Control strategy for sutomatic gantry crane systems: a practical and intelligent approach. Int J Adv Robot Syst, 4(4):447–456
13. Maeda GJ, Sato K (2008) Practical control method for ultra-precision positioning using a ballscrew mechanism. Precis Eng 32:309–318
14. Sato K, Maeda GJ (2009) A practical control method for precision motion: improvement of NCTF control method for continuous motion control. Precis Eng 33:175–186
15. Wahyudi, Sato K, Shimokohbe A (2003) Characteristic of practical position for point-to-point (PTP) positioning systems: effect of design parameters and actuator saturation on positioning performance. Precis Eng 27:157–169
16. Yakub MFM, Aminudin BA (2011) Improved NCTF control method for a two-mass rotary positioning systems. Intel Control Autom
17. Chong S-H, Sato K (2010) Practical controller design for precision positioning, independent of friction characteristic. Precis Eng 34:286–300

Chapter 7
A High Order PID-Sliding Mode Control: Simulation on a Torpedo

Ahmed Rhif, Zohra Kardous and Naceur Ben Hadj Braiek

Abstract Position and speed control of the torpedo present a real problem for the actuators because of the high level of system nonlinearity and because of the external disturbances. The nonlinear systems control is based on several different approaches, which include the sliding mode control. This chapter deals with the basic concepts, mathematics, and design aspects of a control for nonlinear systems that make the chattering effect lower. As solution to this problem we will adopt as a starting point the high order sliding mode approaches and then the PID-sliding surface. Simulation results show that this control strategy can attain excellent control performance with no chattering problem.

Keywords Sliding mode control · PID controller · Chattering phenomenon · Nonlinear system

7.1 Introduction

Modern torpedoes are the most effective marine weapons but they have a range much lower than antiship missiles. Torpedoes are propelled engine equipped with an explosive charge, and sometimes with an internal guidance system that controls the direction, speed, and depth. The typical shape of a torpedo is that of a cigar 6 m long with a diameter of 50 cm and weight of one ton. Torpedoes are the main weapons

A. Rhif (✉) · Z. Kardous · N. Ben Hadj Braiek
Laboratoire des Systèmes Avancés, Ecole Polytechnique de Tunisie, BP 743, 2078, La Marsa, Tunisia
e-mail: ahmed.rhif@issatso.rnu.tn

Z. Kardous
e-mail: zohra.kardous@enit.rnu.tn

N. Ben Hadj Braiek
e-mail: naceur.benhadj@ept.rnu.tn

L. Liu et al. (eds.), *Applied Methods and Techniques for Mechatronic Systems*, Lecture Notes in Control and Information Sciences 452,
DOI: 10.1007/978-3-642-36385-6_7,

of a submarine, but are also used by ships and by aircraft. They are increasingly wire-guided (cable connects the submarine for several thousand meters and makes it possible to reprogram or redirect the machine according to the evolution of the target). However, most modern torpedoes can be completely autonomous. They have active sonar, which make them able to direct themselves to the target they have been designated prior to launch. Other types of torpedoes, for example, are self-possessed, and especially during the second half of World War II, an acoustic sensor (passive sonar) allowed them to move to the noise emitted by the engines of the target. However, sometimes this kind of torpedo locks onto the engine noise of the submarine pitcher, so the standard procedure was to dive low speed after such a shot. Modern torpedoes are powered by steam or electricity. The former have speeds ranging from 25 to 45 knots, and their scope ranges from 4 to 27 km. They consist of four elements: the warhead, the air section, rear section, and tail section. The warhead is filled with explosives (181–363 kg). The steam-air section is about one-third of the torpedo and contains compressed air and fuel tanks and water for the propulsion system. The rear section contains the turbine propulsion systems with the guidance and control of depth. Finally, the tail section contains the rudders, exhaust valves, and propellers. Orders of a torpedo electric are similar to those of steam torpedoes, but the tank air is replaced by batteries and the turbines by an electric motor [1, 2].

The sliding mode control has proved its effectiveness through theoretical studies. Its principal scopes of application are robotics and electrical engines [3–7]. The advantage of such a control is its robustness and effectiveness versus the disturbances and the model uncertainties. Indeed, to make certain the convergence of the system to the desired state, a high level control is often requested. In addition, the discontinuous part of the control generates the chattering phenomenon which is harmful for the actuators. In fact, there are many solutions suggested for this problem. In the literature, sliding mode control with limiting band has been considered by replacing the discontinuous part of the control with a saturation function. Also, fuzzy control was proposed as a solution thanks to its robustness. On the other hand, the high order sliding mode consists in the sliding variable system derivation.

In the literature, different approaches have been proposed for the synthesis of nonlinear surfaces. In [8], the proposed area consists of two terms, a linear term that is defined by the Herwitz stability criteria and another nonlinear term used to improve transient performance. In [9], to measure the armature current of a DC motor, Zhang Li used the high order sliding mode as it is faster than the traditional methods such as vector control. To eliminate the static error that appears while parameter measurements one uses a PI controller [6]. Thus, the author has chosen to write the sliding surface in a transfer function of a proportional integral form while respecting the convergence properties of the system to this surface. The same problem of the static error was also treated by adding an integrator block just after the sliding mode control.

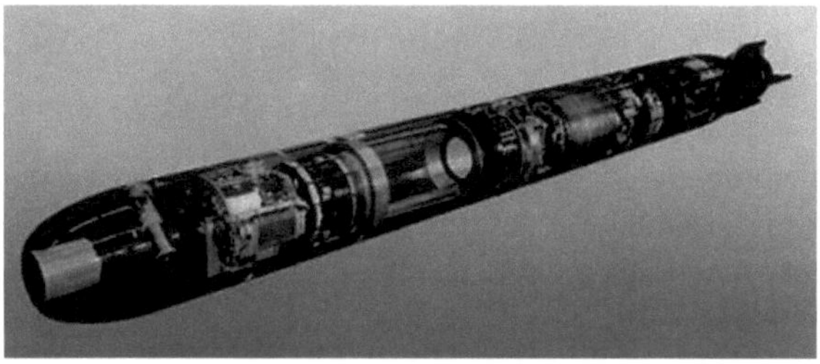

Fig. 7.1 Schematic of a torpedo

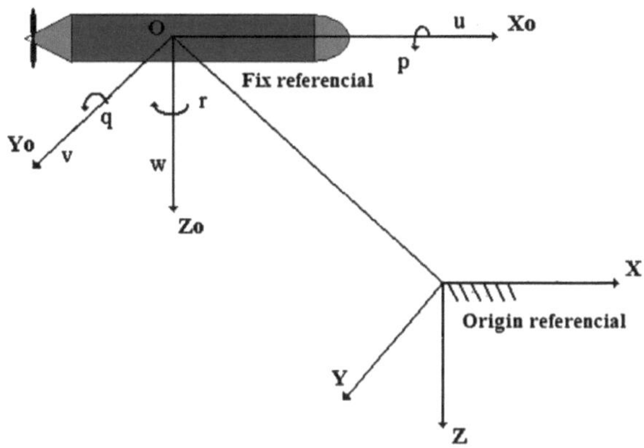

Fig. 7.2 Inertial frame and body-fixed frame

7.2 Process Modeling

Torpedoes (Fig. 7.1) are systems with strong nonlinearity and always subject to disturbances and model parameter uncertainties, which make their measurement and their control a hard task. Equation (7.1) represents the torpedo's motion's equation in six degrees of freedom. ***M*** is the matrix of inertia and added inertia, ***C*** is the matrix of Coriolis and centrifugal terms, ***D*** is the matrix of hydrodynamic damping terms, ***G*** is the vector of gravity and buoyant forces, and τ is the control input vector describing the efforts acting on the torpedo in the body-fixed frame. ***B*** is a nonlinear function depending on the actuators characteristics, and ***u*** is the control input vector.

$$\begin{aligned} &M\dot{v} + C(v)v + D(v)v + G(\eta) = \tau \\ &\tau = B(u). \end{aligned} \tag{7.1}$$

For the modeling of this system, two references are defined (Fig. 7.2): one fix reference related to the vehicle which is defined in an origin point: $R_0(X_0, Y_0, Z_0)$, the second reference is related to the Earth R(x, y, z). The torpedo presents a strong nonlinear aspect that appears when we describe the system in three dimensions (3D), so the state function will present a new term of disturbances as shown in (7.2).

$$\dot{X} = AX + Bu + \varphi(X, u) \tag{7.2}$$

with $\|\varphi(X, u)\| \leq MX$, M > 0.

As we consider only the linear movement in immersion phase, we need only four degrees of freedom for which we describe the system only in two dimensions (2D). All developments were done, and the resulting state space describing the system is given by (7.3) and (7.4).

$$\dot{X} = AX + Bu \tag{7.3}$$

$$\dot{\mathrm{X}} = \begin{bmatrix} \dot{\omega} \\ \dot{q} \\ \dot{\theta} \\ \dot{z} \end{bmatrix}, A = \begin{bmatrix} a_{11} & a_{12} & 0 & 0 \\ a_{21} & a_{22} & a_{23} & 0 \\ 0 & 1 & 0 & 0 \\ 1 & 0 & a_{43} & 0 \end{bmatrix} \text{ and } B = \begin{bmatrix} b_{11} \\ b_{21} \\ 0 \\ 0 \end{bmatrix}. \tag{7.4}$$

where,

$\boldsymbol{\omega}$ is linear velocity, **q** the angular velocity, $\boldsymbol{\theta}$ the angle of inclination, and **z** the depth. The system control is provided by ***u*** which presents the immersion deflection.

In this way, the system could be represented by two parts [2]: $H_1(p)$ the transfer function of inclination (7.5) and $H_2(p)$ the transfer function of immersion (7.6).

$$H_1(p) = \frac{7660}{p(p+40)} \tag{7.5}$$

$$H_2(p) = \frac{6514(p+6.85)}{p(p+1.91)(p+12.5)(p+40)}. \tag{7.6}$$

7.3 The Torpedo Controller Design: Sliding Mode Control

The appearance of the sliding mode approach occurred in the Soviet Union in the 1960s with the discovery of the discontinuous control and its effect on the system dynamics. This approach is classified in monitoring with Variable System Structure (VSS). The sliding mode is strongly recommended due to its facility of establishment, its robustness against disturbances, and model uncertainties. The principle of the sliding mode control is to force the system to converge toward a selected surface and then to remain there and to slide on in spite of uncertainties and disturbances. In general, the main purpose of the sliding mode control consists in bringing back the state trajectory toward the sliding surface and to make it move above this surface until reaching the equilibrium point. The sliding mode exists when commutations between two controls u_{max} and u_{min} remain until reaching the desired state. On the other hand, the sliding mode exists when $s\dot{s} < 0$. This condition is based on Lyapunov quadratic function. In fact, control algorithms based on Lyapunov method have proved effective for controlling linear and nonlinear systems subject to disturbances. In this way, the existence condition of sliding mode control can be satisfied by the candidate Lyapunov function: $V = \frac{1}{2}s^2$ [10–18].

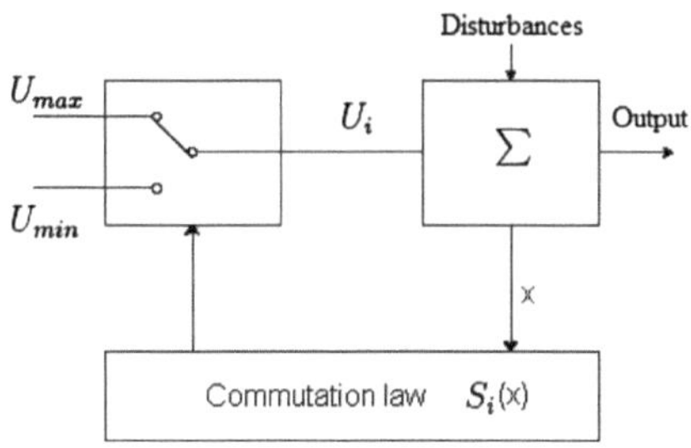

Fig. 7.3 Control unit commutation structure

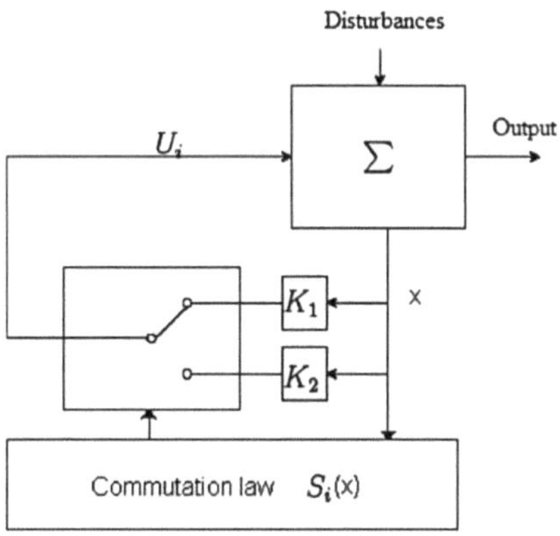

Fig. 7.4 Control unit with commutation on the state feedback

There are three different sliding mode structures: first, commutation takes place on the control unit (Fig. 7.3), the second structure uses commutation on the state feedback (Fig. 7.4), and finally, it is a structure by commutation on the control unit with addition of the equivalent control (Fig. 7.5).

In this study, we chose to use the first structure because it is the most solicited (Fig. 7.3).

To ensure the existence of the sliding mode, we must produce a high level of discontinuous control. For that we will use a relay which commutates between two extreme values of control [19]. Second, we have to define a first order sliding surface. In this study we will describe it as follows (7.7):

$$s = k_1 e + k_2 \dot{e}. \tag{7.7}$$

In the convergence phase to the sliding surface, we have to verify that:

$$\dot{V} = \frac{1}{2}\frac{\partial}{\partial t}\left(s^2\right) \leq -\eta\,|s| \tag{7.8}$$

with $\eta > 0$

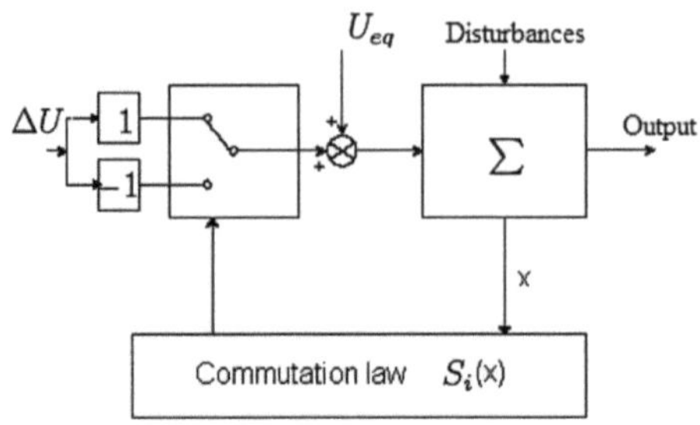

Fig. 7.5 Control unit with addition of the equivalent control

In this part of controlling, the control law of the sliding mode could be given by (7.9).

$$u = k\,sign\,(s) \tag{7.9}$$

with

$$sign\,(s) = \begin{cases} 1, & s > 0 \\ -1, & s < 0 \end{cases} \tag{7.10}$$

sign(.): is the sign function
k : a positive constant that represents the discontinuous control gain.

In this way, using the PID controller, the sliding surface will be represented as follows (7.11):

$$s = \alpha_1 e + \alpha_2 \dot{e} + \alpha_3 \int_0^t edt \tag{7.11}$$

with $\alpha_i > 0$
then,

$$\dot{s} = \alpha_1 \dot{e} + \alpha_2 \ddot{e} + \alpha_3 e \tag{7.12}$$

To reduce the chattering phenomenon, we use the saturation function which gives:

$$u = \lambda sat\left(\frac{s}{\phi}\right) \tag{7.13}$$

where,

$$u = \begin{cases} \lambda\,sign\,(s)\ if\ |s| \geq \phi \\ \lambda\left(\frac{s}{\phi}\right)\ if\ |s| \leq \phi \end{cases} \tag{7.14}$$

with λ and $\phi > 0$, ϕ defines the thickness of the boundary layer [16, 17].

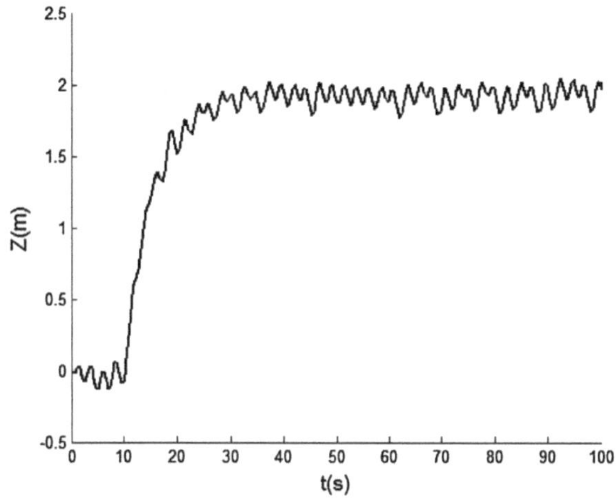

Fig. 7.6 System immersions by SMC1

7.4 Simulation Results

Simulations results are illustrated in Figs. 7.6, 7.7, 7.8, 7.9, 7.10, 7.11, 7.12, 7.13, 7.14, 7.15, 7.16 and 7.17. Figures 7.6 and 7.7 show that, with a first order sliding mode control (SMC1) using the sliding surface (7.7) with $k_1 = 1$ and $k_2 = 2.5$, we can reach the desired value in a short time but the control level ($u_= \pm 3$) and its switching frequency are high (Fig. 7.8). In addition, we notice that the reaching phase presents some commutations known as chattering effect. However, the second order sliding mode control (SMC2), with the sliding surface coefficients $\beta_1 = 2, \beta_2 = 5, \beta_3 = 2$, can reduce considerably the chattering phenomenon (Figs. 7.10 and 7.11) but the level of the control is always high ($u \approx \pm 1.8$) and its commutation frequency is even higher (Fig. 7.12).

As a solution to this problem, we apply the PID-SMC1 with sliding surface defined in (7.11) with $\alpha_1 = 1, \alpha_2 = 4$, and $\alpha_3 = 0.04$. To reach the sliding surface and to converge to zeros, we choose $\phi = 2$ and $\lambda = 1$. The simulation results of this approach are given in Figs. 7.14, 7.15, 7.16 and 7.17. We notice that the system error converges to zero (Figs. 7.14 and 7.15) and that we have reduced considerably the chattering effect relatively to the two last approaches simulated in this chapter. In other words, we notice that the control level (Fig. 7.16) has little commutation in the beginning of the system evolution, then it stabilizes in ($u = 0.4$) after a short period of time (~ 30 s). This excellent result is also validated in Figs. 7.9, 7.13, and 7.17 which present the pitching angle of the torpedo that seems very adequate with the PID-SMC1 approach (Fig. 7.17).

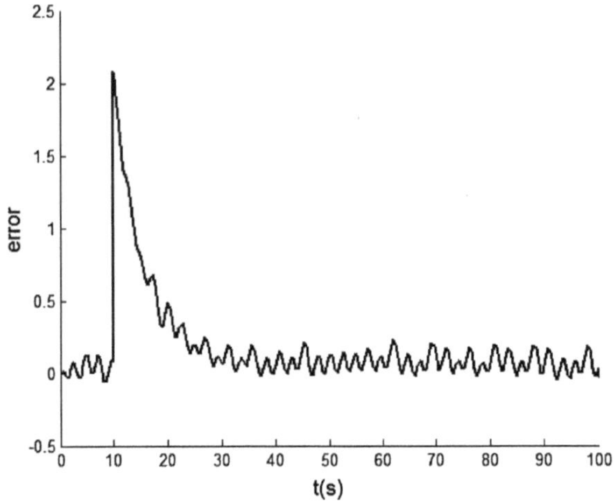

Fig. 7.7 System error by SMC1

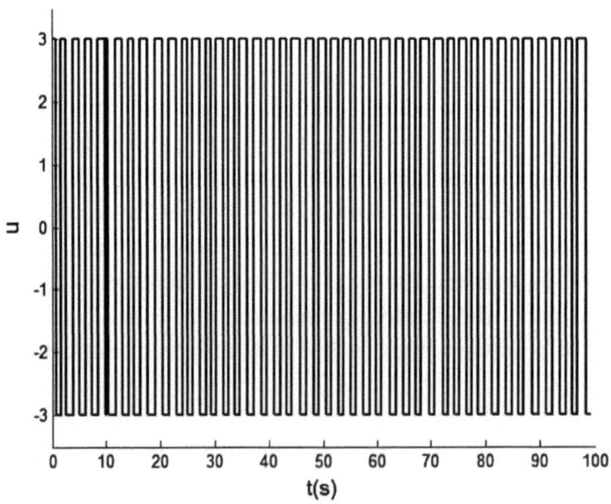

Fig. 7.8 Control evolutions by SMC1

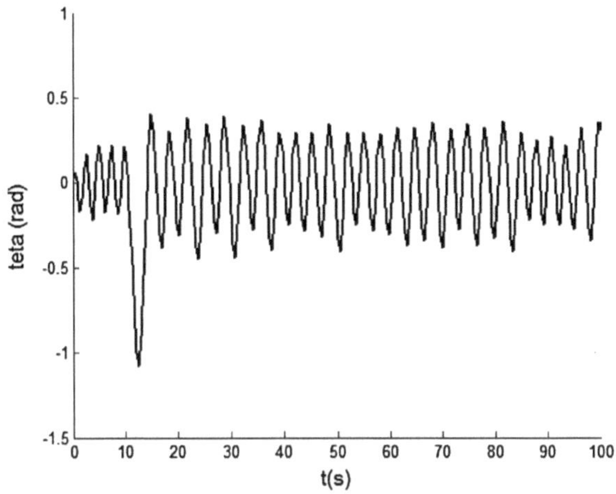

Fig. 7.9 Immersion angle by SMC1

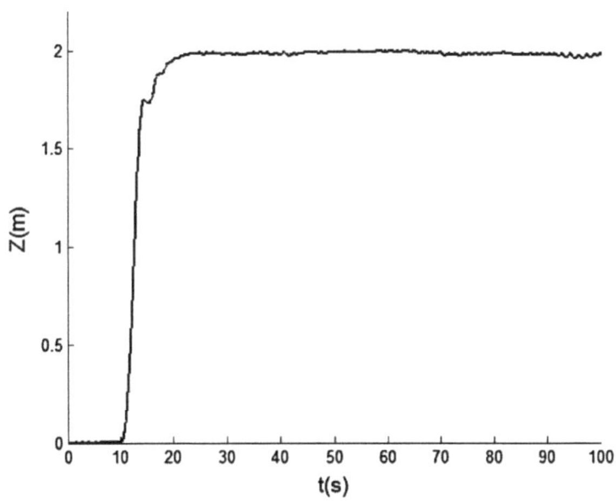

Fig. 7.10 System immersions by SMC2

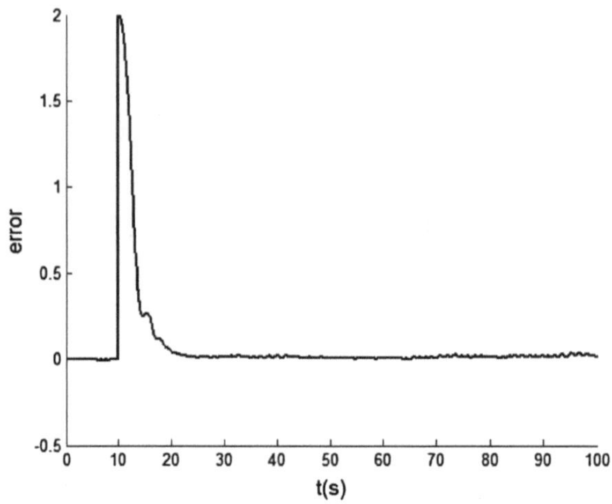

Fig. 7.11 System error by SMC2

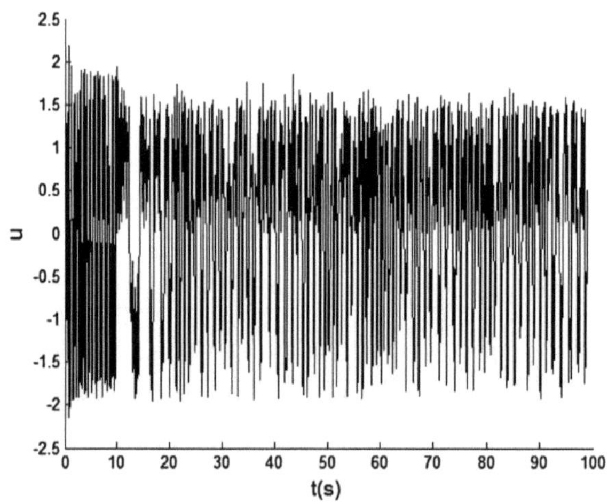

Fig. 7.12 Control evolutions by SMC2

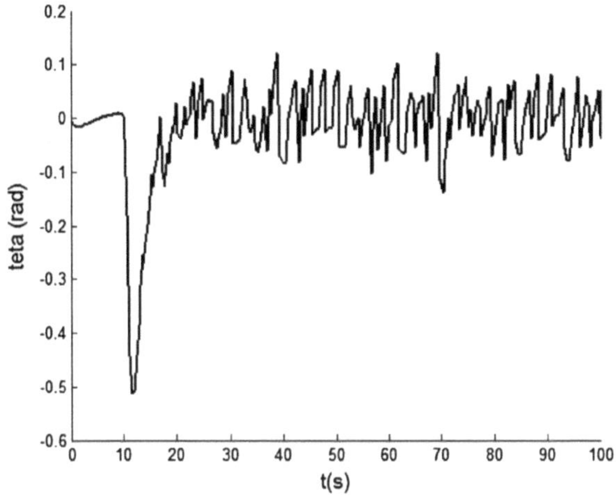

Fig. 7.13 Immersion angle by SMC2

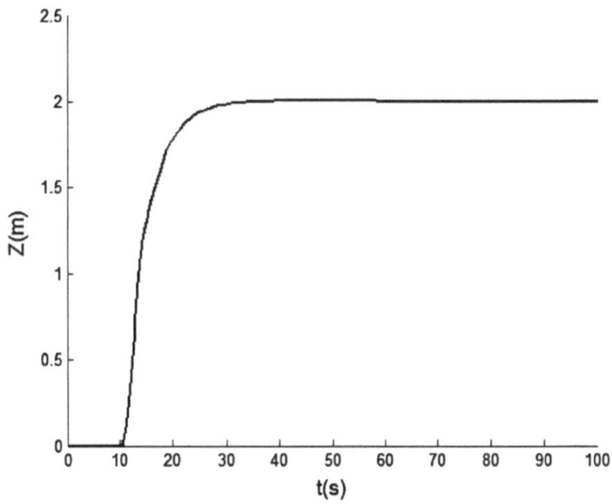

Fig. 7.14 System immersions by PID-SMC1

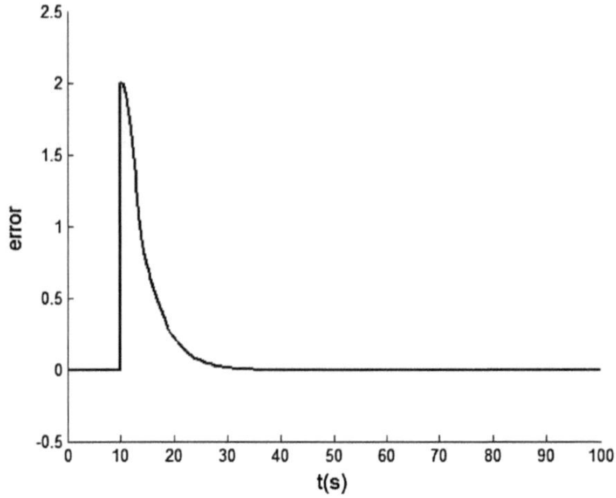

Fig. 7.15 System error by PID-SMC1

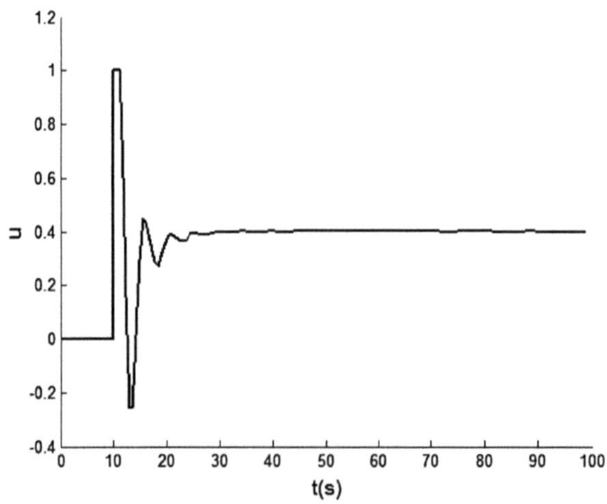

Fig. 7.16 Control evolutions by PID-SMC1

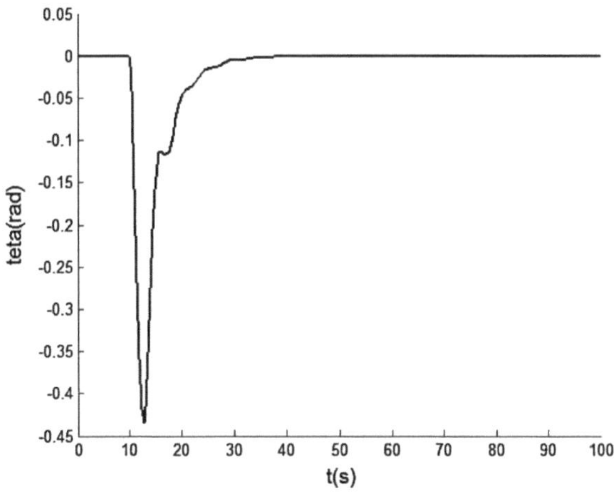

Fig. 7.17 Immersion angle by PID-SMC1

7.5 Conclusion

In this work, torpedoes controller has been presented, detailed, and justified by simulation results. This chapter could be a ready study for those who want to research on the sliding mode control. We approached the synthesis method of a control law by sliding mode using a nonlinear sliding surface. For the first time, we presented the class and the properties of this sliding surface adopted. Then, a sliding mode control using the sliding surface developed together with stability studies was elaborated. After that, second order sliding mode control was developed and tested by simulation on a torpedo. Finally, to reduce the static error and the number of derivatives that make the system inaccurate and that could represent some imperfections in real applications, a PID-sliding surface is developed and simulated. This last approach shows very effective qualities of control and robustness, especially in terms of the control level reduction and the sliding mode switching control minimization.

References

1. Miller J, Bevly DM (2013) A system for autonomous canine guidance. Int J Model Ident Control 20(1):33–46
2. Nam SQ, Kown SH, Yoo WS, Lee MH, Jeon WS (1993) Robust fuzzy control of o three fin torpedo. J Soc Nav Architects Jpn 173: 231–235
3. Rhif A, Kardous Z, Ben Hadj Braiek N (2011) A high order sliding mode-multimodel control of non linear system simulation on a submarine mobile. In: International multi-conference on systems, signals and devices, Tunisia, March 2011
4. Rhif A, Kardous Z, Hadj Braiek NB (2012) A sliding mode multimodel control for a sensorless photovoltaic system. J Sci Ind Res 71:418–424

5. Utkin VI (1977) Variable structure systems with sliding modes. IEEE Trans Autom Control 22(2):212–222
6. Eker İ (2005) Sliding mode control with PID sliding surface and experimental application to an electromechanical plant. ISA Trans 45(1):109–118
7. Rhif A (2011) Position control review for a photovoltaic system: dual axis sun tracker. IETE Tech Rev 28:479–485
8. Rhif A (2011) A review note for position control of an autonomous underwater vehicle. IETE Tech Rev 28:486–493
9. Li Z, Shui-sheng Q (2005) Analysis and experimental study of proportional-integral sliding mode control for DC/DC converter. J Electron Sci Technol Chin 3(2)
10. Lee DS, Youn MJ (1989) Controller design of variable structure systems with nonlinear sliding surface. Electron Lett 25(25):1715–1716
11. Emel'yanov SV (1963) On pecularities of variables structure control systems with discontinuous switching functions. Doklady ANSSR 153:776–778
12. Emel'yanov SV (1967) Variable structure control systems. Moscow, Nouka
13. Utkin VI, Young KD (1978) Methods for constructing discontinuity planes in multidimensional variable structure systems. Auto Remote control 166–170
14. Shi Y, Zhou C, Huang X, Yin Q (2012) Fault-tolerant sliding mode control for the interferometer system under the unanticipated faults. Int J Model Ident Control 16(4):353–362
15. Tzypkin YZ (1955) Theory of control relay systems. Gostekhizdat, Moscow
16. Anosov DV (1959) On stability of equilibrium points of relay systems. Autom Remote Control 2:135–149
17. Jie S, Yong Z, Chengliang Y (2012) Longitudinal brake control of hybrid electric bus using adaptive fuzzy sliding mode control. Int J Model Ident Control 15(3):147–155
18. Rhif A, Kardous Z, Ben Hadj Braiek N (2013) A sliding mode-multimodel control for torque evolution of a double feed asynchronous generator. In: International conference on electrical engineering and software applications, Hammamet, 2013
19. Vaidyanathan S (2013) Global chaos synchronization of Liu-Yang systems via sliding mode control. In: International conference on control, engineering and information technology (CEIT'14). Sousse, Tunisia

Chapter 8
Sliding Mode Control with Self-Turning Law for Uncertain Nonlinear Systems with Time-Delay and External Disturbances

Ran Zhen, Jinyong Chen, Xueli Wu, Quanmin Zhu, Hassan Nouri, Xiaojing Wu and Jianhua Zhang

Abstract This study proposes a novel sliding mode control with self-turning law for nonlinear systems with time-delay and external disturbances possessing uncertain parameters. The adjustable control gain and a bipolar sigmoid function are online tuned to force the tracking error to approach zero. The proposed control scheme provides good transient and steady-state performance. Moreover as the proposed controller, the chatting phenomenon can be avoided and the problem of time-delay and external disturbances are solved for a class of nonlinear systems. The closed-loop control system stability is proved to use the Lyapunov method. Steady-state system performance and chattering are considerably improved. Numerical simulation results are given to illustrate the effectiveness of the proposed procedure.

Keywords Sliding mode control · Adaptive control · Lyapunov method · Boundary layer

8.1 Introduction

A useful and effective control scheme to deal with uncertainties, time-varying properties, nonlinearities, and bounded externals disturbances is the sliding mode control (SMC) [1]. The SMC has been developed and applied widely in the industry

R. Zhen · J. Chen · X. Wu · X. Wu · J. Zhang
Department of Electrical Engineering, Hebei University of Science and Technology,
26 Yuxiang Street, Shijiazhuang, 050018, People's Republic of China

R. Zhen · Q. Zhu · H. Nouri
Department of Engineering Design and Mathematics, University of the West of England,
Frenchay Campus, Coldharbour Lane, Bristol, BS16 1QY, UK

X. Wu (✉)
Institute of Electrical Engineering, Yanshan University, 438 Hebei West Street,
Qinhuangdao, 06604, China
e-mail: xlwu0311@163.com

L. Liu et al. (eds.), *Applied Methods and Techniques for Mechatronic Systems*,
Lecture Notes in Control and Information Sciences 452,
DOI: 10.1007/978-3-642-36385-6_8,

[2–8]. For instance, the sliding mode control has been applied to DC–DC buck converters [9], the Ball-beam system [10], the control of a direct drive robot [11], 3 DOF planar robot manipulators [12], and so on. SMC is commonly favored as a powerful robust control method for its independence from parametric uncertainties and external disturbances under matching conditions. The first step in the SMC design is to choose a switching manifold so that the state variables restricted to the manifold have a desired dynamics. However, the conventional sliding mode control for uncertain nonlinear systems demands uncertainties bounded constraint and larger switch gain to ensure the robustness of uncertainties and external disturbances. But a large control gain is undesired for increased chances of input chattering. The boundary layer technique and switch gain adaptive mechanism are common techniques for alleviating input chattering [13–17]. In [13], the sliding mode controller is designed to prolong the stack life of the PEM fuel cell. In [14], the adaptive fuzzy SMC algorithms which were designed to eliminate the chattering action of the control signal were proposed. For a class of uncertain nonlinear MIMO system, the switch gain parameter adaptive methods, which can effectively eliminate chattering and does not know precisely the bounded uncertainties, were also proposed [15]. The [16] presents a sliding mode controller for a narrow tilting three-wheeled vehicle. In [17], adaptive fuzzy sliding mode control for a class of SISO continuous nonlinear systems with unknown dynamics and bounded disturbances is introduced, and is applied to the electro-hydraulic servo mechanism.

All the above mentioned, the sliding mode controllers still have a lot of problems. For instance, the time-delay and the external disturbances [18–20] are not discussed in the above chapters .The objective of this chapter is to propose a sliding mode controller for a class of nonlinear systems with time-delay and external disturbances. Based on this proposed method, a systematic sliding mode controller with self-tuning law (SSMC) is designed for the class of nonlinear system in this chapter. The proposed control scheme provides good transient and steady-state performance. Moreover as the proposed controller, the chatting phenomenon can be avoided and the problems of time-delay and external disturbances are solved for a class of nonlinear systems. Applying the Lyapunov stability method, an effective tuning law is developed to improve tracking precision. The stability and robustness of the proposed self-tuning SMC algorithm are proven. State trajectory is designed to reach the sliding surface. The self-tuning rate is defined as a function of the magnitude of the tracking error. Steady-state system performance and chattering are considerably improved. The novel design method that is proposed in this chapter is very good to solve the control problem of nonlinear systems with time-delay and external disturbances. Numerical simulation results are given to demonstrate the effectiveness and the feasibility of the proposed approach.

The rest of this chapter is organized as follows: Sect. 8.2 presents the system definitions and the classical SMC design method. Section 8.3 introduces our proposed SMC scheme. The simulation results are presented to show the effectiveness of the proposed control for a class of nonlinear systems with time-delay and external disturbances in Sect. 8.4. Section 8.5 provides the concluding remarks.

8.2 Sliding Mode Controller Design

Consider a class of nonlinear systems, which can be represented as

$$\begin{cases} \dot{x}_i = x_{i+1} \qquad (i = 1, 2, \ldots, n-1) \\ \dot{x}_n = f(x(t)) + b(x(t))\,u + g(x(t-d)) + w \\ y = x_1 \end{cases} \tag{8.1}$$

where $x = [x_1, x_2, \ldots, x_n]^T \in R^n$ is state vector and $u \in R^1$ is the control input, and $y \in R^1$ is the system output. w is the external disturbance, $f(x(t))$, $g(x(t-d))$, and $b(x(t))$ are time-varying nonlinear function and not precisely known. Let $f(x) = f_0(x) + \Delta f(x)$ and $g(x(t-d)) = g_0(x(t)) + \Delta g(x(t-d))$, where $f_0(x) = [x_2, \ldots, x_n, f_0]^T$ and $g_0(x(t)) = [0, \ldots, 0, g_0]^T$ are the nominal parts of $f(x)$ and $g(x)$, respectively. The parameter variations, $\Delta f(x) = [0, \ldots, 0, \Delta f_0]^T$ and $\Delta g(x(t-d)) = [0, \ldots, 0, \Delta g_0]^T$ are assumed to the bound and differentiable with respect to t. Assume the uncertainties in (8.1) satisfy the matching condition and all the states are quantifiable. In order to design controller must introduce the following hypothesis:

Assume 1 The parameter variations $\Delta f(x)$, $\Delta g(x(t-d))$ are assumed to the bound, $|\Delta f(x)| \le |F|$, $|\Delta g(x(t-d))| \le |G|$ $\quad F, G$ are constant.

Assume 2 Control gain b boundary is known, but the world itself is time-varying or dependent.

$$0 < b_{\min}(x(t)) \le b(x(t)) \le b_{\max}(x(t))$$

Assume 3 Interference w is bounded, $|w| \le W$ W is constant.

Defined the tracking error as $e = x_1 - x_{1d}$. The control objective is to drive to the output signal y to the desired y_d and the tracking error vector is

$$e = x - \left[x_{1d}, \dot{x}_{1d}, \ldots, x_{1d}^{(n-1)}\right]^T = \left[e, \dot{e}, \ldots, e^{(n-1)}\right]^T.$$

The first step in designing SSMC is defining the sliding function. Let the sliding function be

$$s = c_1 e + c_2 \dot{e} + \cdots + c_{n-1} e^{(n-2)} + e^{(n-1)}$$

The constants c_i are selected to be positive such that the polynomial $P(p) = c_1 + c_2 p + \cdots + c_{n-1} p^{(n-2)} + p^{(n-1)}$ is Hurwitz. p is Laplace operator. Notably, the choice of c_i determines the rate of decay of the tracking error. s can be rewritten into

$$s = x^{(n-1)} - x_d^{(n-1)} + \sum_{i=1}^{n-1} c_i e^{(i-1)}$$

An appropriate control input u is chosen to satisfy the condition

$$s\dot{s} \leq o \tag{8.2}$$

If (8.2) holds, eventually $s \to 0$ and $e \to 0$.
Let the control input u be formulated as

$$u = \hat{b}^{-1}\left[-f_0 - g_0 + x_d^{(n)} - \sum_{i=1}^{n-1} c_i e^{(i)} - k\text{sgn}(s) \right] = u_{\text{eq}} + u_{\text{vss}} \tag{8.3}$$

where

$$\begin{cases} k \geq \beta\left[(F + G + W + \eta) + (\beta - 1)\left|\hat{u}\right| \right] \\ u_{\text{eq}} = \hat{b}^{-1}\hat{u} \\ \hat{u} = -f_0 - g_0 + x_d^{(n)} - \sum\limits_{i=1}^{n-1} c_i e^{(i)} \\ u_{\text{vss}} = -\hat{b}^{-1}k\text{sgn}(s) \end{cases} \tag{8.4}$$

η is constant.

The validity of the preceding formula is given in Theorem 1.

Theorem 1 Consider the system (8.1). Given the sliding function and the control law (8.3) and (8.4), the condition (8.2) is guaranteed and the tracking error declines asymptotically.

Proof. Direct differentiation of s yields

$$\dot{s} = f + g + bu + w - x_d^{(n)} + \sum_{i=1}^{n-1} c_i e^{(i)} \tag{8.5}$$

Choose a Lyapunov function candidate, such as $V = \frac{1}{2}s^2$

$$\begin{aligned} \dot{V} &= s\dot{s} \\ &= s\left(f + g + bu + w - x_d^{(n)} + \sum_{i=1}^{n-1} c_i e^{(i)} \right) \\ &= s\left(f + g + b\hat{b}^{-1}\hat{u} - b\hat{b}^{-1}k\text{sgn}\,(\text{s}) + \text{w} - \text{y}_\text{d}^{(\text{n})} - \sum_{\text{i}=1}^{\text{n}-1} \text{c}_\text{i}\text{e}^{(\text{i})} \right) \\ &= s\,(f - f_0 + g - g_0 + w) + s\left(b\hat{b}^{-1} - 1 \right)\hat{u} - b\hat{b}^{-1}k\,|s| \end{aligned} \tag{8.6}$$

where $\hat{b} = [b_{\max}b_{\min}]^{1/2}$ $\beta^{-1} \leq b^{-1}\hat{b} \leq \beta$ $\beta = (b_{\max}b_{\min})^{1/2} + 2 \geq 3$. So

$$\begin{aligned} &\dot{V} \leq |s|\,(F{+}G{+}W){+}\,|s|\left|b\hat{b}^{-1} - 1\right|\left|\hat{u}\right| - \left[(F + G + W + \eta) + (\beta - 1)\left|\hat{u}\right| \right]|s| \\ &= |s|\left|b\hat{b}^{-1} - 1\right|\left|\hat{u}\right| - \left[\eta\,(\beta - 1)\left|\hat{u}\right| \right]|s| \end{aligned}$$

Since $\beta - 1 \geq 2$, $-2/3 \leq b\hat{b}^{-1} - 1 \leq \beta - 1$ and $\left| b\hat{b}^{-1} - 1 \right| \leq \beta - 1$, one obtains $\dot{V} \leq -\eta \left| s \right| \leq 0$

8.3 Self-Turning Law Design

The boundary layer technique is frequently utilized to eliminate the chattering typically found in conventional SMC. However, the control gain constant and the width of the boundary layer are often fixed numbers and there is no guarantee for the precise tracking. Here, a self-tuning method is proposed to solve the chattering problem, and guarantee precise tracking.

The self-tuning SMC law for system (8.1) is

$$u = \hat{b}^{-1} \left[-f_0 - g_0 + x_d^{(n)} - \sum_{i=1}^{n-1} c_i e^{(i)} - \hat{\beta} \phi \left(\hat{\lambda}, s \right) \right] \tag{8.7}$$

where $\varphi \left(\hat{\lambda}, s \right)$ is a bipolar sigmoid function,

$$\varphi \left(\hat{\lambda}, s \right) = \frac{\left(1 - \exp \left(-\hat{\lambda} s \right) \right)}{\left(1 + \exp \left(-\hat{\lambda} s \right) \right)} \tag{8.8}$$

The self-tuning law is designed as

$$\dot{\hat{\beta}} = \eta \hat{b} e \operatorname{sgn} \left(\frac{\partial x_1}{\partial u} \right) \left(\frac{1 - \exp \left(-\hat{\lambda} s \right)}{1 + \exp \left(-\hat{\lambda} s \right)} \right) \tag{8.9}$$

$$\dot{\hat{\lambda}} = \mu \hat{b} e \operatorname{sgn} \left(\frac{\partial x_1}{\partial u} \right) s \tag{8.10}$$

where η and μ are positive constants used to adjust the tuning rates. In discrete time implementation, $\frac{\partial x_1}{\partial u}$ is replaced with $\frac{\Delta x_1}{\Delta u}$, as the sampling time is small. The differences, Δx_1 and Δu, can be calculated using the current states and previous states.

Theorem 2 Consider the system (8.1). Given the sliding function (8.2), the control law (8.7) and (8.8), and the adaptation law (8.9) and (8.10), then (8.1) will be stabilized. The input chattering and the steady-state error are eventually eliminated.

Proof. Choose a Lyapunov function candidate, such as

$$V = \frac{1}{2} e^2 \tag{8.11}$$

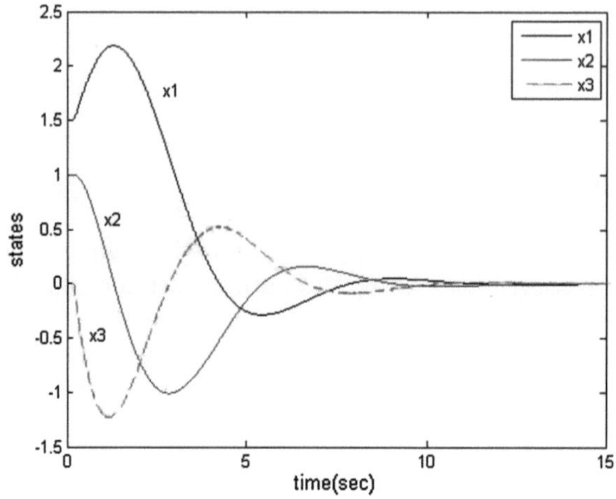

Fig. 8.1 State responses

By direct computation

$$\dot{V} = \frac{\partial V}{\partial e}\frac{\partial e}{\partial x_1}\frac{\partial x_1}{\partial u}\left(\frac{\partial u}{\partial \hat{\beta}}\frac{\partial \hat{\beta}}{\partial t} + \frac{\partial u}{\partial \hat{\lambda}}\frac{\partial \hat{\lambda}}{\partial t}\right) = \dot{V}_1 + \dot{V}_2. \tag{8.12}$$

Thus, the first term on the right-hand side of (8.12) becomes

$$\dot{V}_1 = \frac{\partial V}{\partial e}\frac{\partial e}{\partial x_1}\frac{\partial x_1}{\partial u}\frac{\partial u}{\partial \hat{\beta}}\frac{\partial \hat{\beta}}{\partial t} = -e\frac{\partial x_1}{\partial u}\hat{b}^{-1}\left(\frac{1-\exp\left(-\hat{\lambda}s\right)}{1+\exp\left(-\hat{\lambda}s\right)}\right)\dot{\hat{\beta}}. \tag{8.13}$$

Substituting (8.9) into (8.13) yields

$$\dot{V}_1 = -\eta e^2\left|\frac{\partial x_1}{\partial u}\right|\left(\frac{1-\exp\left(-\hat{\lambda}s\right)}{1+\exp\left(-\hat{\lambda}s\right)}\right)^2 \leq 0. \tag{8.14}$$

Similarly, the second term of the right-hand side of (12) becomes

$$\dot{V}_2 = \frac{\partial V}{\partial e}\frac{\partial e}{\partial x_1}\frac{\partial x_1}{\partial u}\frac{\partial u}{\partial \hat{\lambda}}\frac{\partial \hat{\lambda}}{\partial t} = -e\frac{\partial x_1}{\partial u}\hat{b}^{-1}\hat{\beta}\frac{2s\exp\left(-\hat{\lambda}s\right)}{1+\exp\left(-\hat{\lambda}s\right)^2}\dot{\hat{\lambda}}. \tag{8.15}$$

Substituting (8.10) into (8.15) yields

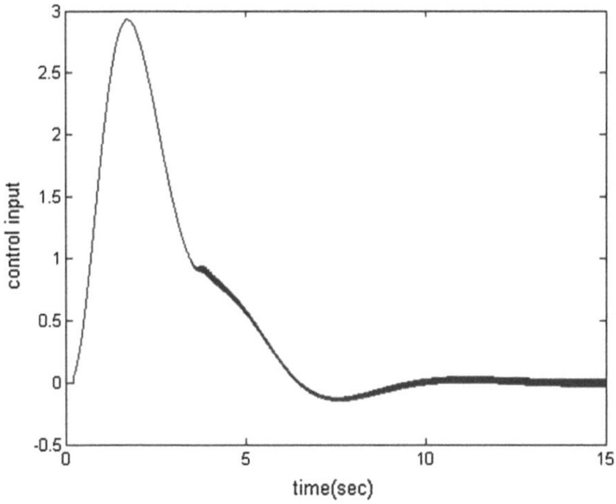

Fig. 8.2 Control input u

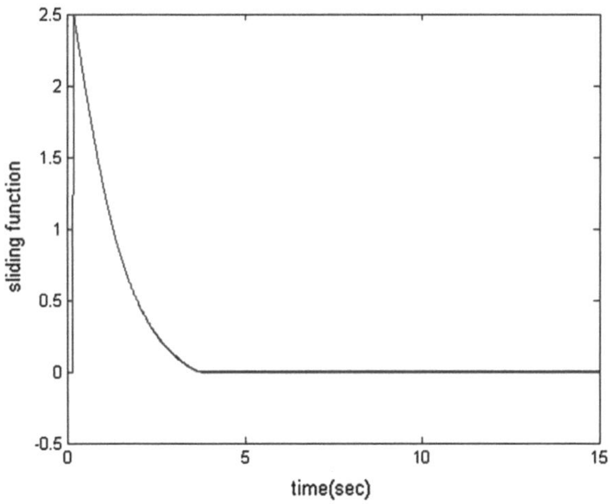

Fig. 8.3 The sliding function

$$\dot{V}_2 = -\mu e^2 \left| \frac{\partial x_1}{\partial u} \right| \frac{2s \exp\left(-\hat{\lambda} s\right)}{1 + \exp\left(-\hat{\lambda} s\right)^2} \hat{\beta} s^2 \leq 0. \tag{8.16}$$

The only possible condition for $\dot{V} = 0$ in (8.14) and (8.16) is $e = 0$ or $s = 0$. In either case, the tracking error is or approaches zero. Therefore, the elimination of tracking error can be guaranteed.

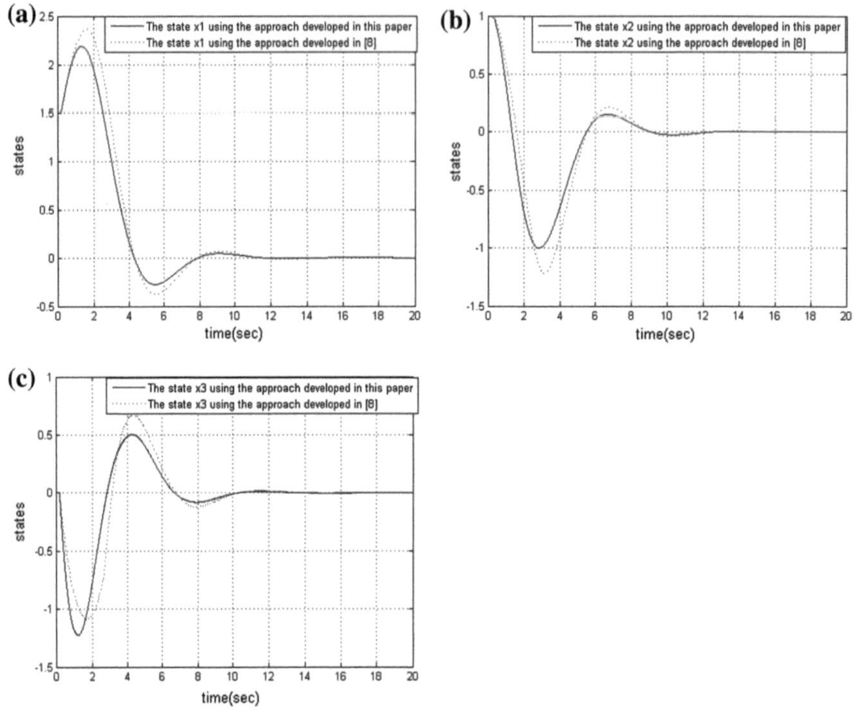

Fig. 8.4 The comparison of states responses

8.4 Illustrative Examples

$$\begin{cases} \dot{x}_1 = x_2 \\ \dot{x}_2 = x_3 \\ \dot{x}_3 = -(1 + 0.3\sin(t))x_1^2 - (1.5 + 0.2\cos(t))x_2 - (2 + 0.5\sin t)x_3 \\ \qquad +(1 + 2\cos(t-d))x_2(t) + u + 0.5\sin t \end{cases}$$

Let $f_o(x) = \left[x_2, x_3, -x_1^2 - 1.5x_2 - 2x_3\right]^T$, $\Delta f(x) = [0, 0, -0.3\sin(t)x_1^2 - 0, 2\cos(t)x_2 - 0.5\sin(t)x_3]^T$, $g_0(x(t)) = [0, 0, x_2(t)]^T$ and $\Delta g(x_2(t-d)) = [0, 0, 2\cos(t-d)x_2(t)]^T$. The initial state is $x = \left[1.5\ 1\ 0\right]^T$. The parameter c in the sliding function is chosen to be $c^T = \left[c_1\ c_2\ c_3\right]^T = \left[1\ 1\ 1\right]^T$. The initial parameter values are selected to be $\hat{\beta}(0) = 2$ and $\hat{\lambda}(0) = 2$. The desired state is $x_d = \left[0\ 0\ 0\right]^T$. The control strategy (8.7)–(8.10) is applied. The simulation results are shown in Fig. 8.1 through 3. Control input chattering does not occur, and the tracking error vanishes quickly. The proposed SSMC demonstrates effectiveness in tracking performance and obtaining system robustness.

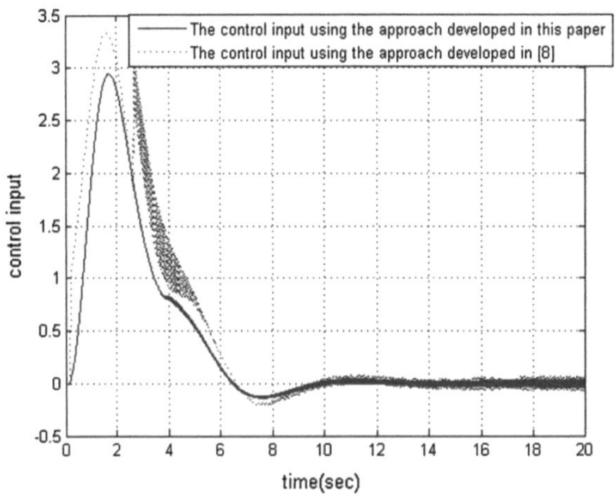

Fig. 8.5 The comparison of control input u

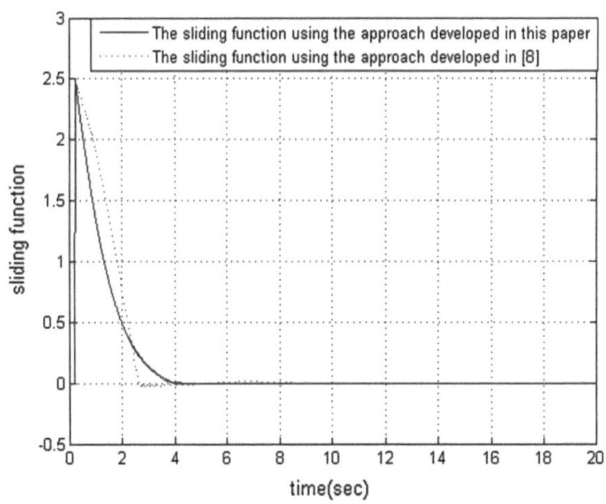

Fig. 8.6 The comparison of the sliding function

As can be seen from the above, the trajectories of the states' system $x_1(t)$, $x_2(t)$ and $x_3(t)$ are plotted in Fig. 8.1. As it can be seen, system (8.1) tracks the desired signals quickly. The sliding surface is shown in Fig. 8.3. In Fig. 8.2, it is demonstrated that there is no chattering in the control signal of the proposed method.

As it can be seen, the state of the system overshoot volume is less than [8] in Fig. 8.4. In Fig. 8.5, the convergence rate is increased and the magnitude of control input is less than in [8]. The sliding surface can be faster to balance position than [8] in Fig. 8.6.

8.5 Conclusion

This work proposes a novel control method for nonlinear systems with parametric uncertainties. Using the Lyapunov stability method, the proposed method guarantees robustness as well as tracking performance. A self-tuning law enables the automatic adaptation of the adjustable control gain parameter and the adjustable boundary layer width to the terminal values. The control input chattering is completely eliminated. Smooth control activity and excellent tracking performance are shown. Application of a planetary gear type inverted pendulum control system demonstrates that the proposed control method is effective.

Acknowledgments This chapter was supported by the Education Department fund of Hebei Province (Z2010136), and the Natural Science fund of Hebei Province (F2012208075).

References

1. Nekoukar V, Erfanian A (2011) Adaptive fuzzy terminal sliding mode control for a class of MIMO uncertain nonlinear systems. Fuzzy Sets Syst 179(1):34–49
2. Zhen R, Meng H, Meng FH (2010) Application of the fuzzy neural network with inverse identification structure on neutralization process. Int J Model Ident Control 9(1/2):53–58
3. Hu XX, Gang L, Gao HJ (2012) Adaptive sliding mode tracking control for a flexible air-breathing hypersonic vehicle. J Franklin Inst 349(2):559–577
4. Michael B, Pablo RR (2012) Sliding mode filter design for nonlinear polynomial systems with unmeasured states. Inform Sci 204(30):82–91
5. Dominguez JR (2013) Robust nested sliding mode integral control for anti-lock brake system. Int J Veh Des 62(2–4):188–205
6. Wu XL, Wu XJ, Du WX (2010) Adaptive fuzzy sliding mode control for uncertain nonlinear system with input nonlinearity. Int J Model Ident Control 11(3/4):186–192
7. Wu XJ, Wu XL, Zhen R, Luo XY, Zhu Q (2011) Adaptive control for time-delay non-linear systems with non-symmetric input non-linearity. Int J Model Ident Control 13(3):152–161
8. Kuo TC, Huang YJ, Chang SH (2008) Sliding mode control with self-tuning law for uncertain nonlinear systems. ISA Trans 47(2):171–178
9. Hasan K (2013) Non-singular: terminal sliding-mode control of DC-DC buck converters. Control Eng Pract 21(3):321–332
10. Qian DW, Liu XJ (2012) Adaptive hierarchical sliding mode control for ball-beam system. Int J Model Ident Control 4(5/6):205–211
11. Erbatur K, Calli B (2009) Fuzzy boundary layer tuning for sliding mode systems applied to the control of a direct drive robot. Soft Comput 13(11):1099–1111
12. Ahmed F (2011) Adaptive fuzzy sliding mode control using supervisory fuzzy control for 3 DOF planar robot manipulators. Appl Soft Comput 11(8):4943–4953
13. Seyed MRR (2012) Control of PEM fuel cell system via higher order sliding mode control. Int J Model Ident Control 6(3/4):310–329
14. Ho HF, Wong YK, Rad AB (2009) Adaptive fuzzy sliding mode control with chattering elimination for nonlinear SISO systems. Simul Model Pract Theory 17(7):1199–1210
15. Roopaei M, Balas VE (2009) Adaptive gain sliding mode control in uncertain MIMO systems. Soft Computing Applications, SOFA '09.3rd International Workshop on, pp. 77–82
16. Nestor R, Enric FC (2013) A sliding mode controlled three wheeled narrow vehicle. Int J Veh Des 62(2–4):123–146

17. Otto C (2012) Adaptive fuzzy sliding mode control for electro-hydraulic servo mechanism. Expert Syst Appl 39(11):10269–10277
18. Wu XL, Zhang, JH, Zhu QM (2010) A generalized procedure in designing recurrent neural network identification and control of time-varying-delayed nonlinear dynamic systems. Neurocomputing 73(7–9):1376–1383 (SCI)
19. Zhen R, Wu XL, Zhang JH (2010) Sliding model synchronization controller design for chaotic neural network with time-varying delay. Intelligent Control And Automation (WCICA), 2010 8th World Congress on, pp. 3914–3919
20. Zhang JH, Zhu QM, Wu XL, Li Y (2013) A generalized indirect adaptive neural networks backstepping control procedure for a class of non-affine nonlinear systems with pure-feedback prototype. Neurocomputing Available Online. Accessed 24 May 2013

Chapter 9
Applied Methods and Techniques for Modeling and Control on Micro-Blog Data Crawler

Kai Gao, Er-Liang Zhou and Steven Grover

Abstract Models can provide mechanisms to improve system performance. This chapter presents the applied methods and techniques for modeling and controlling on micro-blog crawler. With the rapid development of social studies and social network, millions of people present or comment or share their opinions on the platform everyday, and as a result, produce or spread their opinions and sentiments on different topics. The microblog has been an effective platform to know or mine social opinions. In order to do so, crawling the relevant microblog data is necessary. But it is hard for a traditional web crawler to crawl micro-blog data as usual, as by using Web 2.0 techniques such as AJAX, the micro-blog data is dynamically generated rapidly. As most microblogs' official platforms cannot offer some suitable tools or RPC interface to collect the big data effectively and efficiently, we present an algorithm on modeling and controlling on micro-blog data crawler based on simulating browsers' behaviors. This needs to analyze the simulated browsers' behaviors in order to obtain the requesting URLs to simulate and parse and analyze the sending URL requests according to the order of data sequence. The experimental results and the analysis show the feasibility of the approach. Further works are also presented at the end.

Keywords Models · Social networks · Micro-blog · Crawler

K. Gao (✉) · E.-L. Zhou
School of Information Science and Engineering, Hebei University of Science and Technology, No. 26 YuXiang Road, Shijiazhuang, 050000, Hebei, China
e-mail: gaokai@hebust.edu.cn

S. Grover
Comrise Company, Concord Center Building 2, Hazlet, NJ, 07730, USA
e-mail: steven@comrise.com

L. Liu et al. (eds.), *Applied Methods and Techniques for Mechatronic Systems*, Lecture Notes in Control and Information Sciences 452,
DOI: 10.1007/978-3-642-36385-6_9, © Springer-Verlag Berlin Heidelberg 2014

9.1 Background

Nowadays, most of the research in the fields of mechatronic systems or social studies have spent significant effort to find rules from various complicated phenomena by principles, observations, measured data, and logic derivations. The rules are normally summarized as concise and quantitative expressions or "models," and this can provide mechanisms to improve the system (represented by its model) performance. As for the social studies, the social network data (e.g., Twitter, Facebook, Sina_micro-blog, Tecent_micro-blog, etc.) have attracted millions of users and academic and industry researchers to research on modeling and mining the knowledge behind the magnanimity information, and as a result, there has been tremendous interest in social networks. Due to its fast development and wide usage, the microblog has attracted the attention of users, enterprises, governments, and researchers, and so applied methods and techniques for modeling and controlling in this field is very important.

As the foundation of the micro-blog data mining, data collection is the key phase, crawling or collecting the relevant micro-blog data effectively and efficiently is important. But the microblog has many differences compared with traditional web applications. For example, there are many online users, at the same time, its different interactive and displaying mode and login operation are needed, and AJAX technology is widely used, etc. Traditional web crawlers, for instance, can only get the corresponding web pages, but they cannot get the relevant structure and the corresponding social relationships as well as users' backgrounds and fans. That is to say, being different from traditional web application, there are some differences on micro-blog data's login operation, display way, privacy policy, data processing, etc. So the traditional web crawler is not suitable for micro-blog data crawling or collection.

This section presents some details on modeling of micro-blog data crawler based on simulating browsers' behaviors. On the basis of this method, we have collected several million blog data in a short time period.

9.2 Motivation

Although there has been some research on AJAX-based web pages, the technique is not suitable to the micro-blog application. Encouraging developers to develop applications on micro-blog services, some providers of micro-blog services usually offer some special APIs, which can provide developers with the probability of constructing uniform and universal architecture to utilize the APIs to automatically download and save these special data. But the mere APIs-based method has some limitations on rights, calling times, special policies, and so on, and some extra tasks cannot be done by only using these official APIs.

In this chapter, we present some strategies based on simulating browsers' behaviors to obtain the data from micro-blog platform. The main idea is to simulate browsers' behaviors by using the browser's (e.g., FireFox) core to get the corresponding data. This can solve the problem of parsing the JavaScript code, and can do special login operation, etc. In order to crawl the data effectively, we present the following strategies: (1) focused crawling on some special crowds; (2) meta-topic searching and crawling: that is to say, we crawl the special contents by using the microblog's searching function; (3) parallel crawling: based on big data processing by using the *Redia* and *MongoDB*, we use the multiprocessing technology to download and save the data simultaneously. The proposed crawler is composed of four modules, i.e., simulating module, data crawling module, data parsing module, and data persistence module. The experimental results and the analysis show the feasibility of the approach. Further works are also presented at the end.

9.3 Related Work

Online social networking technologies enable individuals to simultaneously share information with any number of peers. With the launch of Twitter in 2007, the microblog has become highly popular, and many researchers want to investigate the micro-blog information propagation patterns [1] or analyze structures of the micro-blog network to identify influential users [2]. Reference [3] discusses some of the ways in which earlier works used text content to analyze online networks, as well as background on language coordination and the exchange-theoretic notions of power from status and dependence. Reference [4] studies several long-standing questions in media communications research, in the context of the micro-blog service Twitter, regarding the production, flow, and consumption of information. A framework which enriches the semantics of Twitter Messages (i.e., tweets) and identifies topics and entities (e.g., persons, events, products) mentioned in tweets is present in reference [5]. Reference [6] conducts a study on recommending URLs posted in Twitter messages and compares strategies for selecting and ranking URLs by exploiting the social network of a user as well as the general popularity of the URLs in Twitter. Authors of reference [7] investigate the attributes and relative influence of 1.6M Twitter users by tracking 74 million diffusion events that took place on the Twitter follower graph over a 2-month interval, and they conclude that the word-of-mouth diffusion can only be harnessed reliably by targeting large numbers of potential influencers, thereby capturing average effects. Reference [8] examines the role of social networks in online information diffusion with a large-scale field experiment, and the authors further examine the relative role of strong and weak ties in information propagation. Although stronger ties are individually more influential, it is the more abundant weak ties that are responsible for the propagation of novel information, and the authors suggest that weak ties may play a more dominant role in the dissemination of information online than currently believed. In reference [9], authors address the problem of discovering topically meaningful communities from a social network, and authors propose a probabilistic scheme that incorporates topics, social relationships, and nature of posts for more effective community discovery, and then they demonstrate the effectiveness of the model and show that it performs better than existing community discovery models. Reference [10] examines the application of an event-driven sampling approach to the Live Journal social network, and the approach makes use of the "always on" atom feed provided by Live Journal that contains all public blog posts in near real-time to inform the sampling process of user friendship networks, and this has the effect of targeting sampling toward the public active users of the network. In addition to proposing models and algorithms for learning the model parameters and for testing the learned models to make predictions, reference [11] develops techniques for predicting the time by which a user may be expected to perform an action.

As for data crawling, in order to overcome the inherent bottlenecks with the traditional crawling, reference [12] proposes the design of a parallel migrating web crawler. Reference [13] proposes a dynamic data crawling methods, which include the sensitive checking of website changes and dynamic retrieving of pages from target websites, and the authors implement an application and compare the performance between the conventional static approaches and the proposed dynamic ones. In reference [14], authors present a novel URLs ordering system that relies on a cooperative approach between the crawlers and the web servers based on file system and web log information, and the proposed algorithm is based on file timestamps and web log internal and external counts. Reference [15] presents a micro-blog service crawler named as *MBCrawler*, which is designed on the APIs provided by micro-blog services, and the architecture is modular and scalable, so it can fit specific features of micro-blog services. Reference [16] presents a dynamic cooperation model for different crawlers' message exchanging, and both the experimental results and the application validate the feasibility of the algorithm.

As for the modeling methods, reference [17] presents the commonly used statistical modeling methods, such as stepwise regression, radial basis function partial least squares, partial robust M-regression, ridge regression, and principal component regression that can be applied in the proposed multicollinearity domain. The Viterbi algorithm, a widely used maximum likelihood estimating method, can be used in natural language processing, and reference [18] presents an effective search space reduction for human pose estimation with Viterbi algorithm.

Although the proposed algorithm has some relationship with the above related work, there are many differences. The proposed crawler works with simulating browser behavior to collect Sina_Micro-blog (http://weibo.com/) and Tecent_Micro-blog (http://t.qq.com/) data. The proposed algorithm is based on simulating browsers' behaviors. As for browser, reference [19] describes how to calculate various object-oriented metrics of three versions of Mozilla Firefox, and the neural network approach can predict high and medium severity errors more accurately than low severity errors.

The experimental results and the analysis show the feasibility of the proposed approach.

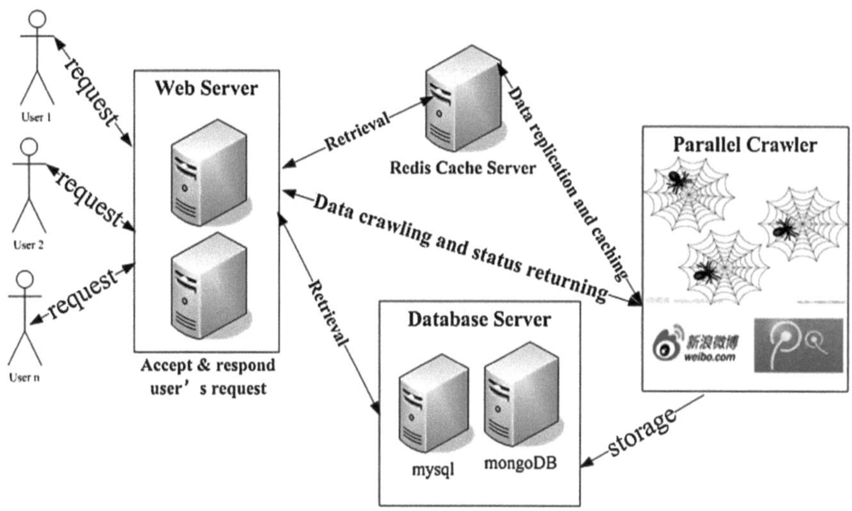

Fig. 9.1 System architecture

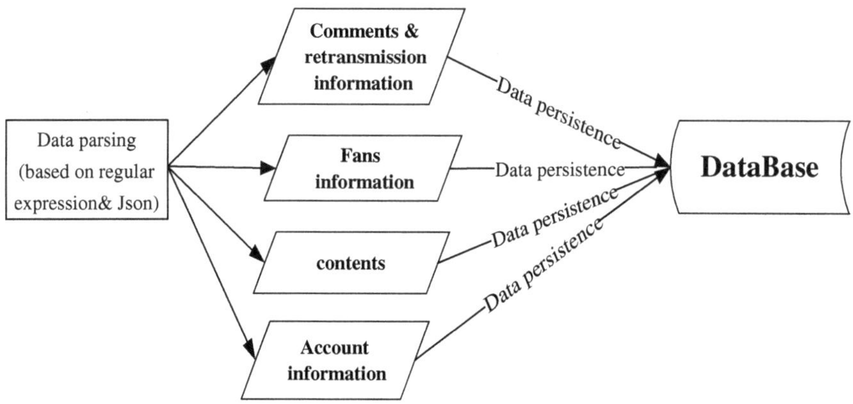

Fig. 9.2 Parsed data and its persistence

9.4 System Architecture

Social networks are often huge, and therefore crawling the micro-blog data could be both challenging and interesting. As microblog's big data properties, it is impossible to crawl all the micro-blog data. Instead, it is feasible to crawl some kinds of data (e.g., account information, contents or topics, attentions or fans, etc.). In this section, we propose the system architecture on parallel crawling. Figure 9.1 shows the architecture, and Fig. 9.2 shows the parsed data and its persistence.

In practice, we can use the *RDBMS* to store the parsed data, and the *Redis* is used as the cache server, so users' retrieval request can be done through the web server.

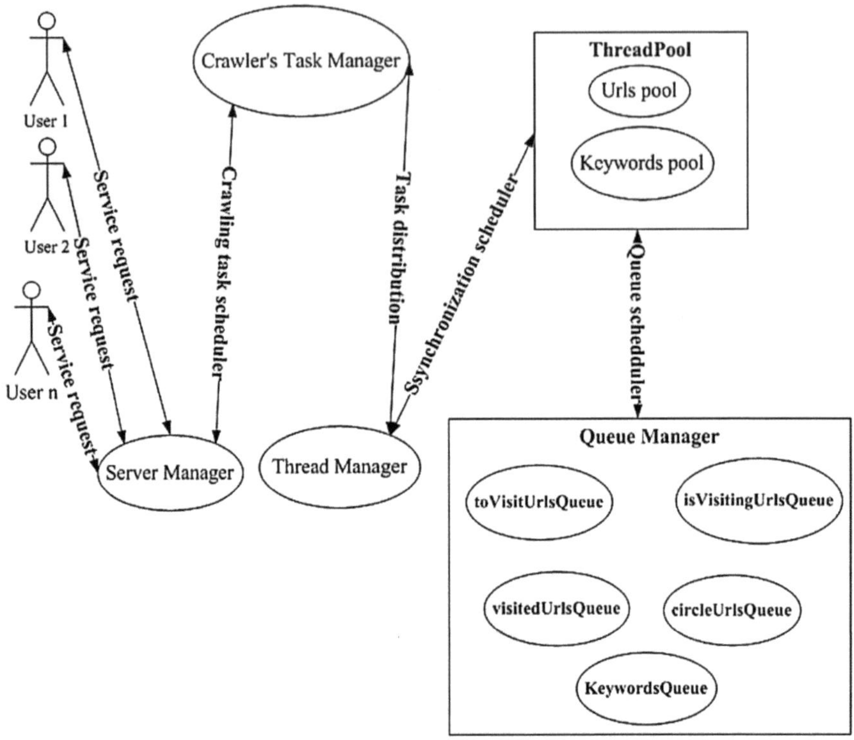

Fig. 9.3 Multi-thread-based parallel crawling

In detail, as for the multi-thread based parallel crawling, we use a thread pool and a queue manager to schedule the tasks. There are several different queues, including the *toVisitUrlsQueue*, *isVisitingUrlsQueue*, *visitedUrlsQueue*, *circleUrlsQueue*, *keywordsQueue*, etc., see Fig. 9.3.

9.5 Case Studies and Implementation of the Simulated Browser-Based Crawling

Instead of merely using the official APIs, we propose a simulated browser-based crawling, as the merely APIs-based method has some limitations on rights, calling times, and so on, and perhaps some extra tasks cannot be done by only using official APIs. We present some strategies based on simulating browsers' behaviors to obtain the micro-blog data, and the proposed crawler is composed of four modules, i.e., simulating module, data crawling module, data parsing module, and data persistence module.

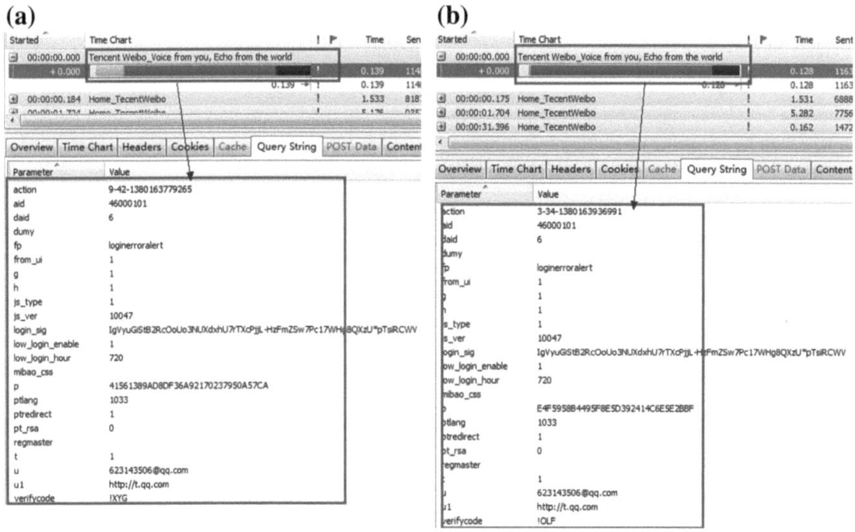

Fig. 9.4 Different situations with the same account

9.5.1 *Simulation of the Login Operation and Cookies Data Obtaining*

Commercial websites often use technologies (e.g., HTTP compression, SSL encryption and chunked encoding) to provide some reasonable levels of security and system performance. As for the micro-blog data effectively crawling, simulated login operation is necessary. Otherwise (for example, by only using official APIs-based crawling), only few data can be crawling. Here, the simulated login operation means this kind of operation allows the crawler to use some legal accounts and their corresponding passwords to login the corresponding micro-blog platforms, and the key phase is the encrypted data parsing. Here, we use the HttpWatch [20], which integrates with Internet Explorer or Mozilla Firefox to provide some unrivaled levels of HTTP monitoring, without the need for separately configured proxies or network sniffers. Simply interacting with a website, HttpWatch can display a log of requests and responses alongside the web page itself, and it can even show interactions between the browser and its cache. As a result, each HTTP transaction can be examined or parsed so as to see the values of headers, cookies, query strings, and other HTTP-related data. HttpWatch can work well with these technologies to provide a view of HTTP activity. By using HttpWatch, we can obtain 21 or more different parameters during the simulated login phase. But in practice, there usually exist some different situations during the login phase, and Fig. 9.4 shows the two situations when using the same account.

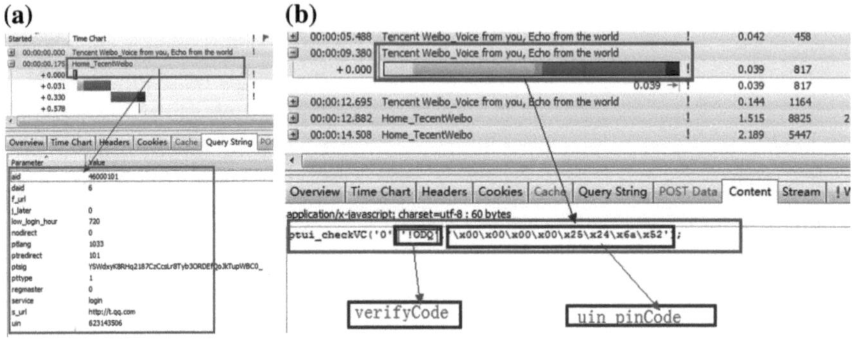

Fig. 9.5 Requested preliminary parameters (**a**) and the return values after the requested period (**b**)

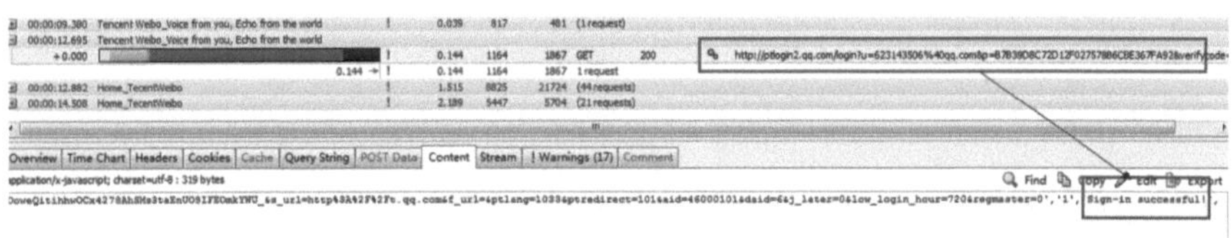

Fig. 9.6 Returned results

From the parsed results, we can conclude that the requested URLs usually contain static and dynamic parameters (e.g., the parameter *P* and *verifycode* in Fig. 9.4, and the *verifycode* parameter is usually used as the password encryption). As the microblogs' login passwords are usually multilevers and multilayer encrypted, it usually contains some other preliminary parameters, and the displayed parameter in Fig. 9.5b (i.e., "\x00\x00\x00\x00\x25\x24\x6a\x52") is the encrypted parameter in Fig. 9.4a. Now the encrypted resolving phase is finished, and the returned or parsed content is shown in Fig. 9.6.

As for judging whether the corresponding user is legal or illegal, it needs to analyze the cookies data. On the other hand, whether actually login or not, when requesting the server data, if the user can get the legal cookies, he or she can obtain the same data as if he or she really "login" the web server. In detail, in order to obtain the cookies data, it needs three steps. First, it needs to obtain the *verifycode* and *uin* parameters, see algorithm 1 below. Second, by using the JavaScript analysis engine and invoking the encrypted function, we can obtain the parsed parameters, see Fig. 9.7. Last, it needs to merge the relevant parameters to obtain the corresponding cookie data, which is the result of the simulating login phase, see algorithm 2 below, and the parsed cookies data result is shown in Fig. 9.8.

```
//parsing the verifyCode and uin parameters
Algorithm 1 (i.e., void getCheckVC())
Input:
(1) preLoginUrl//requested URLs before the login operation;
(2) username
(3) host//parameter on requesting the preLoginUrl
Output:
(1) retJson: returned parameters on preLoginUrl request
(2) verifyCode
(3) uin
Step1. retJson=getPreLoginJson(preLoginUrl,username,host);
Step2. verifyCode=parseVerifyCode(retJson);
Step3. uin=parseUin(retJson);
```

```
//obtaining the cookie data
Algorithm 2 (i.e., void getCookies())
Input:
(1) loginUrl // Url to login the corresponding micro-blog page
(2) username
(3) password
(4) verifyCode //verification code obtained from the former step
(5) host
(6) referrer// means the source urls to do the login operation
Output: cookie data
Step1. cookie=getCookies(loginUrl,username,password,verifyCode,
host,referer);
Step2.cookie=checkValidate(cookie) //verify the cookie
Step3.transmitCookie(cookie)//transfer the obtained cookies to
the corresponding threads
```

9.5.2 Data Parsing and Persistence

After collecting the corresponding data, it needs to be parsed and stored. As for the content, there are some differences between data within the blogger's main page and other common pages. As usual, the main page returns data in a traditional way, while other common pages usually use AJAX [21] and JSON [22] technology to return data to client in order to enhance the performance or optimize the user experience. By using JavaScript, these returned data can be parsed and filled into the corresponding sites. Figure 9.9 shows the private crawled data, and algorithm 3 shows the main steps of this processing step.

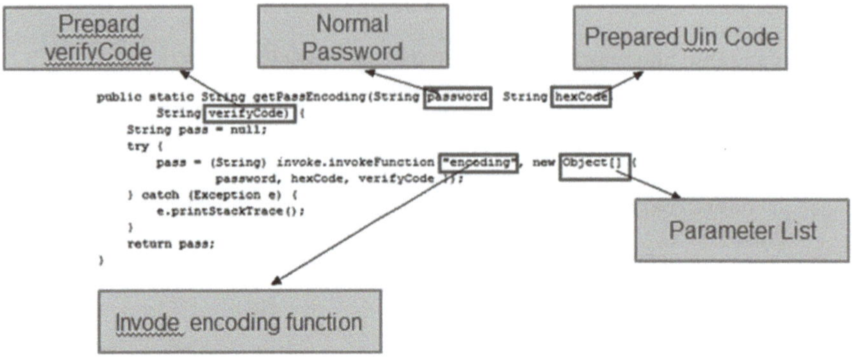

Fig. 9.7 The parsed parameters

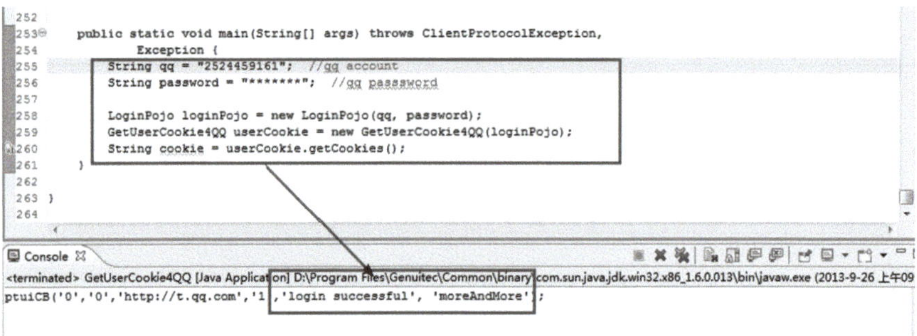

Fig. 9.8 The cookies data of the simulating login phase

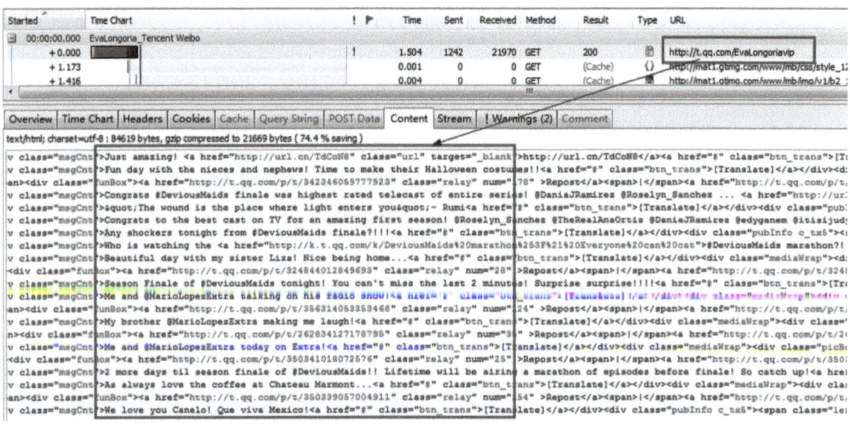

Fig. 9.9 The crawled content

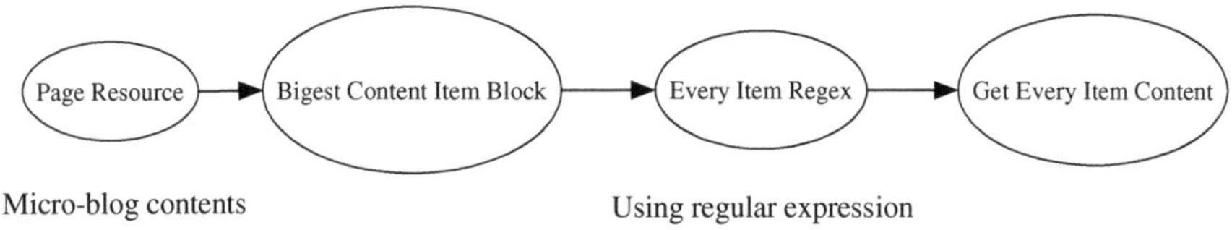

Fig. 9.10 The processing flow

```
// Get the content of corresponding Urls
Algorithm 3 (i.e., grabOnePageArticle(String url, boolean
isFirst))
Input:
(1) Url //pages to be crawled;
(2) Url_Refer //one of the header parameter within the HTTP
request;
(3) Url_Host //one of the header parameter such the port
number, etc.
Output:
(1) content.
Step1. Url=getCompleteUrl(time,currentPage,aid,uid) // Fill in
the Url parameters dynamically, including the current time,
currentPage, aid parameter (i.e., id parameters of the micro-
blog contents, see Fig.\,\ref{fig:9}), uid parameter
(i.e., user name).
Step2.  content=grabPageSource4QQ.grabPageSourceOfQQ(Url,
Url_Refer, Url_Host);
```

The data parsing module needs to process the crawled data and parse the micro-blog content. The crawled data can be classified into two corpus, i.e., plaintext and cipher text. As the plaintext is regular and uniform, we use the regular expression to extract the real contents (e.g., the micro-blog contents, Url, id, published time, IP address, reviewed or commented number, forwarded number, etc.). The processing flow is shown in Fig. 9.10.

9.6 Experimental Results and Analysis

9.6.1 About the Testing Data Set and the Experimental Environment

As for these micro-blog big data, its persistence is an important issue. We use the *MongoDB, Redis* to store and cache them in our real application, and *Mysql* is

(a)

url	name	sex
http://t.qq.com/EvaLongoriavip	EvaLongoria	femail
http://t.qq.com/BeverleyM	Beverley	femail
http://t.qq.com/PaulScheer	PaulScheer	mail
http://t.qq.com/neilgaimanvip	NeilGaiman	mail
http://t.qq.com/JohnStamos	JohnStamos	mail
http://e.t.qq.com/KaylaCollins	KaylaCollins	femail
http://t.qq.com/SnoopDogg	SnoopDogg	mail
http://t.qq.com/Nacho_Polo	Nacho_Polo	mail
http://t.qq.com/JeffKGilburg	JeffKGilburg	mail
http://t.qq.com/CarolynHennesy	CarolynHennesy	femail
http://t.qq.com/NikkiReed	NikkiReed	femail
http://t.qq.com/AbigailSpencer	AbigailSpencer	femail
http://t.qq.com/BonnieJillLaflin	BonnieJillLaflin	femail
http://t.qq.com/LukeBilyk	LukeBilyk	mail
http://t.qq.com/JamesPhelan	JamesPhelan	femail
http://t.qq.com/winds_official	winds_official	mail
http://t.qq.com/PeytonList	PeytonList	femail
http://t.qq.com/jamiefoxxvip	jamiefoxxvip	mail
http://t.qq.com/CarrieAnnInaba	CarrieAnnInaba	femail

(b)

sendUrl	docUrl	article	person_id
http://t.qq.com/EvaLongoriavip	http://t.qq.com/p/t/163872010546035	Just amazing! http://url.cn/TdCoN8[Translate]	EvaLongoriavip
http://t.qq.com/EvaLongoriavip	http://t.qq.com/p/t/342346059777923	Fun day with the nieces and nephews! Time to make their Hallo	EvaLongoriavip
http://t.qq.com/EvaLongoriavip	http://t.qq.com/p/t/339346042208598	Congrats #DeviousMaids finale was highest rated telecast of er	EvaLongoriavip
http://t.qq.com/PaulScheer	http://t.qq.com/p/t/348849124784899	.@AnthonySHead is a very bad man who hates San Diego on th	PaulScheer
http://t.qq.com/PaulScheer	http://t.qq.com/p/t/337348059180032	My wife with @MsJuneDiane (New Girl, Flight of the Conchords,	PaulScheer
http://t.qq.com/PaulScheer	http://t.qq.com/p/t/333848030426112	Meeting an important Hollywood friend to discuss a project.[Tr	PaulScheer
http://t.qq.com/PaulScheer	http://t.qq.com/p/t/310387092696122	NTSF:SD:SUV:: is going to London (for Real!) for our TV Movie.	PaulScheer
http://t.qq.com/OrlandoJones	http://t.qq.com/p/t/34997129217626	Huh... well... The more you know. :/ http://url.cn/K53Cjt[Translat	OrlandoJones
http://t.qq.com/OrlandoJones	http://t.qq.com/p/t/258643107676980	Seriously Congress... can't you get your freaking priorities strai	OrlandoJones
http://t.qq.com/AliciaWitt	http://t.qq.com/p/t/227077002802688	10 minutes to showtime! http://url.cn/P9Txu3[Translate]	AliciaWitt
http://t.qq.com/DarrenRovell	http://t.qq.com/p/t/323345120290114	Nov. 13, 1994: Last time the Bengals, Browns and Lions won o	DarrenRovell
http://t.qq.com/DarrenRovell	http://t.qq.com/p/t/321342042720847	Today's payouts: Northwestern paying Maine $450K, Nebraska	DarrenRovell
http://t.qq.com/cocorochaonline	http://t.qq.com/p/t/329349120476115	New cover for all my Chinese friends!!! Harpers Bazaar China -	cocorochaonline
http://t.qq.com/cocorochaonline	http://t.qq.com/p/t/339348036456824	Wilhelmina Models takes you behind-the-scenes on Coco Rocl	cocorochaonline
http://t.qq.com/cocorochaonline	http://t.qq.com/p/t/276017069296076	Watch with Allure Magazine as I chop off my long hair into a pi	cocorochaonline
http://t.qq.com/cocorochaonline	http://t.qq.com/p/t/243927112474143	So excited to be at the #MetropolitanOpera for opening night!	cocorochaonline

Fig. 9.11 The parsed results, **a** account information **b** other parsed content

only used as the experimental platform. Figure 9.11 shows the *MySql*-based parsed Tecent_Micro-blog data.

In order to evaluate the algorithm's performance, we classify the following tested data set according to their authorities into three classes. The reason we use the following three classes is that the micro-blog official platform usually presents the following three different classes, and the three classes have different data sizes. In detail, the first class has less data while the two others have more. In detail, the first class is the ordinary (i.e., minor authorities) microbloggers, whose propagative scope is only limited within old friends or classmates; the second class is those medium authorities' microbloggers (e.g., network magazines microblog), whose contents can be followed or spread into all kinds of users; the last class is famous persons' microblogs. We will test the performance on the above different data environment.

9.6.2 Ordinary Microblogger's Performance Evaluation

In this section, we use someone's microblog as an example. Figure 9.12a shows the original microblog, and b shows the crawled and parsed data, respectively. It is clear that there is no difference between them, and all contents have been obtained correctly.

9.6.3 Medium Authorities' Microblogger Performance Evaluation

Here, we use some medium authorities' microbloggers as the experimental platform. Figure 9.13a shows the micro-blog interface, while b shows the crawled and parsed data.

Fig. 9.12 Ordinary and the parsed results of the ordinary microblogger's, **a** ordinary data, **b** parsed results

9.6.4 *Famous Persons' Microblog*

As for these famous persons' microblogs, we use the famous actor Tom Cruise's microblog as an example. Figure 9.14a shows his micro-blog interface, while b shows the crawled and parsed data.

9.6.5 *Performance Evaluation*

Sometimes micro-blog services provide some APIs. Through these services, the well-structured data can be easily obtained, so it can provide us the probability of constructing uniform and universal software architecture to utilize the provided APIs to automatically download data. However, there are usually some limits and

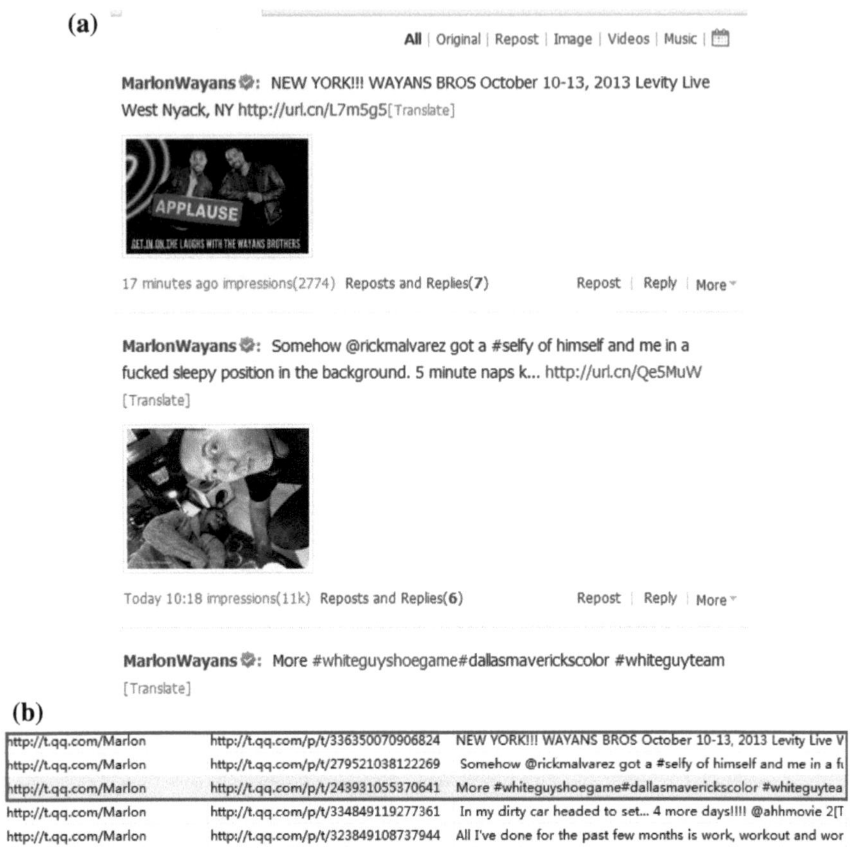

Fig. 9.13 Medium authorities' microblogger and the corresponding parsed result, **a** ordinary data, **b** parsed results

obstacles. In order to evaluate the performance, we present the comparison between APIs-based crawling and the simulating browser behavior approach, see Table 9.1.

From the above comparison, as for the proposed method, it is clear that the parsed data's degree of integrity and the accuracy or scope is higher. But, as shown before, if the template or the main framework of the microblog has been changed, the accuracy of the parsed data is lower than usual. Fortunately, these changes occur rarely. If we can track or analyze the parsed data periodically, it is easy to find the changes and then revise some special rules to parse the corresponding data.

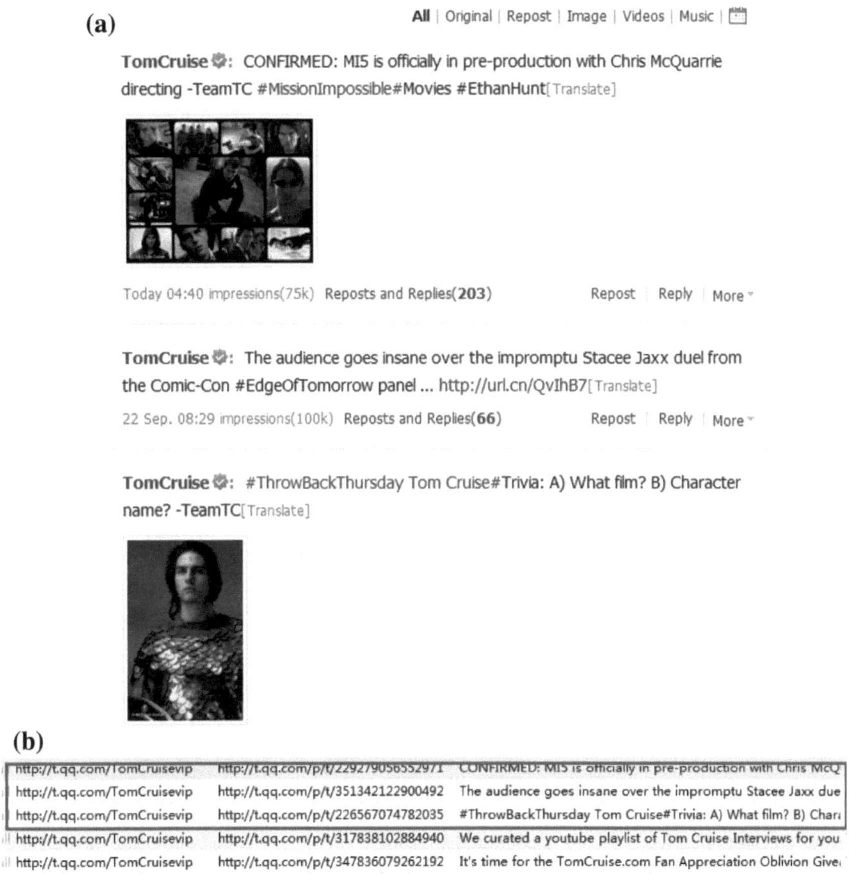

Fig. 9.14 Famous persons' micro-blog and the parsed results, **a** ordinary data, **b** parsed results

9.7 Conclusion

It is hard for a traditional web page crawler to crawl micro-blog data as usual, and most microblogs' official platforms cannot offer some suitable tools or RPC interfaces to collect the data effectively and efficiently. This chapter presents some algorithms and strategies on crawling and parsing micro-blog data effectively based on simulating browsers' behaviors. This needs to analyze the simulated browsing behavior in order to obtain the requesting URLs, to simulate and analyze the sending URLs requests according to the order of data sequence. It needs to focus on crawling on some special crowds and crawl some special contents by using the microblog's searching function. Parallel crawling and the multiprocessing technology are also used to download the data simultaneously. The experimental results and the analysis show the feasibility of the approach. Existing works are also presented at the end.

Table 9.1 The performance comparison of the two approaches

Performance	Algorithms	
	The proposed simulating browser behavior-based crawling	Official APIs-based crawling
About the micro-blog data entrance	Micro-blog platform account	(1) Micro-blog platform account (2) AppKey (3) AccessToken
About the data usability	Flexible and easy to use	(1) Existing more limitations such as fixed or lower crawling frequency, it is not flexible (2) The parsed contents are usually fixed and researchers cannot choose some special categories from them
About the multi-threads performance	Better	Poor
About the crawling frequency	Flexible, and the only limitation is the bandwidth	Not flexible, as there are more limitations such as API invoking rights or authorization
About the stability	Better. The only limitation is the variation of the Official's platform, but the variation is rare	Better
About the data integrity	Better, and almost all needed data can be parsed	Poor, and some data cannot be parsed such as the *docUrl*, etc. *DocUrl* parameter can be seen from the Fig. 9.12b
About visiting or data requested frequency	Larger than 60 times (per minutes)	Less than 1 times (per minutes)
About the degree of data integrity	100 %	95 %, lacking some materials such as *docUrl*, etc
About the data accuracy	99 %	100 %
About the data scope	Widely, and almost all the data appearing on the micro-blog platform can be crawled and parsed	Narrowly, and the only small new parts, such as microbar, microgroup, etc., can be obtained

Acknowledgments Some earlier works were done in Beijing Institute of Technology with the help of Dr. Hua-ping Zhang and Prof. Yin-ping Zhao. This work is sponsored by the National Science Foundation of Hebei Province (No. F2013208105) and the National Science Foundation of China (No. 61272362). It is also sponsored by Hebei Province Scientific and Technical Key Task (No. 12213516D).

References

1. Kwak H, Lee C, Park H, Moon S (2010) What is Twitter, a social network or a news media? In: 19th international conference on world wide web. ACM Press, USA, pp 591–600
2. Weng J, Lim EP, Jiang J, He Q (2010) TwitterRank: finding topic-sensitive influential twitterers. In: 3rd international conference on web search and web data mining. ACM Press, USA, pp 261–270
3. Cristian DNM, Lee L, Bo P, Kleinberg J (2012) Echoes of power: language effects and power differences in social interaction. In: 21th international conference on world wide web. ACM Press, France, pp 699–708
4. Wu S, Hofman JM, Mason WA, Watts DJ (2011) Who says what to whom on Twitter. In: 20th international conference on the world wide web. ACM Press, India, pp 705–714
5. Abel F, Gao Q, Houben GJ, Tao K (2011) Analyzing user modeling on Twitter for personalized news recommendations. In: International conference on user modeling, adaptation and personalization. LNCS, vol 6787. Springer, Spain, pp 1–12
6. Chen J, Nairn R, Nelson L, Bernstein M, Chi E (2010) Short and tweet: experiments on recommending content from information streams. In: 28th international conference on human factors in computing systems. ACM Press, USA, pp 1185–1194
7. Bakshy E, Hofman JM, Mason WA, Watts DJ (2011) Everyone's an influencer: quantifying influence on Twitter. In: 3rd international conference on web search and data mining. ACM Press, Hong Kong, pp 65–74
8. Bakshy E, Rosenn I, Marlow C, Marlow C (2012) The role of social networks in information diffusion. In: International conference on world wide web. ACM Press, France, pp 519–528
9. Sachan M, Contractor D, Tanveer AF, Subramaniam LV (2012) Using content and interactions for discovering communities in social networks. In: International conference on world wide web. ACM Press, France, pp 331–340
10. Dan C, Shipman FM (2009) Capturing on-line social network link dynamics using event-driven sampling. In: International conference on computational science and engineering, vol 4. Vancouver, Canada, pp 284–291
11. Goyal A, Bonchi F, Lakshmanan LV (2010) Learning influence probabilities in social networks. In: 3th international conference on web search and data mining. ACM Press, USA, pp 241–250
12. Agarwal A, Durgesh S, Pandey AKA, Goel V (2012) Design of a parallel migrating web crawler. J Adv Res Comput Sci Softw Eng 2(4):147–153
13. Kim KS, Kim KY, Lee KH, Kim TK, Cho WS (2012) Design and implementation of web crawler based on dynamic web collection cycle. In: International conference on information networking (ICOIN). Bali, Indonesia, pp 562–566
14. Chandramouli A, Gauch S, Eno J (2012) A cooperative approach to web crawler URL ordering, human–computer systems interaction: backgrounds and applications. J Adv Intell Soft Comput 98:343–357
15. Lu G, Liu S, Lü K (2013) MBCrawler: a software architecture for micro-blog crawler. In: International conference on information technology and software engineering. Lecture Notes in Electrical Engineering, vol 212. Springer, Berlin, Heidelberg, pp 119–127
16. Gao K, Li SW (2010) The cooperation model for multi agents and the identification on replicated collections for web crawler. Int J Model Identif Control 11(3–4):224–231

17. Garg A, Tai K (2013) Comparison of statistical and machine learning methods in modelling of data with multicollinearity. Int J Model Identif Control 18(4):295–312
18. Han G, Zhu H, Ge J (2013) Effective search space reduction for human pose estimation with Viterbi recurrence algorithm. Int J Model Identif Control 18(4):341–348
19. Singh S, Mittal P, Kahlon KS (2013) Empirical model for predicting high, medium and low severity faults using object oriented metrics in Mozilla Firefox. Int J Comput Appl Technol 47(2/3):110–124
20. HttpWatch: Introduction to HttpWatch 8.x (2013). http://help.httpwatch.com/#introduction.html
21. Ajax: Introduction to Ajax (2013). http://api.jquery.com/category/ajax/
22. Json: Introduction to Json (2013). http://www.json.org/index.html

Chapter 10
Development of an Improved Genetic Algorithm for Resolving Inverse Kinematics of Virtual Human's Upper Limb Kinematics Chain

Gangfeng Deng, Xianxiang Huang, Qinhe Gao, Ying Zhan and Quanmin Zhu

Abstract Inverse kinematics is the key technique in virtual human motion control and it is difficult to obtain the solutions by using geometric, algebraic, or iterative algorithms. In this chapter, an Improved Genetic Algorithm (IGA) is proposed to resolve the inverse kinematics problem in upper limb kinematics chain (ULKC). First, the joint-units of ULKC and its mathematical models are constructed by using D–H method; then population diversity and population initialization are accomplished by simulating human population, and the adaptive operators for mutation are designed. The simulation results show that compared with the Standard Genetic Algorithm (SGA), the IGA can provide higher precise solutions in searching process and avoid "premature" stop or inefficient searching in later stage with high probability.

Keywords Upper limb kinematics chain · Inverse kinematics · D–H method · Genetic algorithm · Population initialization

10.1 Introduction

Control of joint angles is the key technique in virtual human motion control. The popular approaches include geometric control, behavior control, physical control, motion capture control, and synthesis control [1], which have different characteristics

G. Deng (✉) · X. Huang · Q. Gao · Y. Zhan
Xi'an Research Institute of High Technology, Xi'an, 710025, People's Republic of China
e-mail: 20378481@qq.com

Q. Zhu
College of Chemical Engineering, China University of Petroleum, Qingdao, 266580, People's Republic of China

Q. Zhu
Department of Engineering Design and Mathematics, University of the West of England, Coldharbour Lane, Bristol, BS16 1QY, UK

L. Liu et al. (eds.), *Applied Methods and Techniques for Mechatronic Systems*,
Lecture Notes in Control and Information Sciences 452,
DOI: 10.1007/978-3-642-36385-6_10,

[2, 3]. The kinematics control, one of the geometric control methods, has been the most widely used in human motion control while the end-effector's position and orientation are determined.

Kinematics control algorithm contains forward kinematics (FK) algorithm and inverse kinematics (IK) algorithm. Calculating the position and orientation of the end-effector from the joint Cartesian space is known as FK problem and calculating the joint angles from position and orientation of the end-effector is called IK problem. Many traditional methods have been used to resolve the IK problem such as geometric [4], algebraic [5], and iterative methods [6]. However, these methods have their own demerits in solving the IK problem of the mechanical structure or the body structure.

In recent years, Fuzzy [7], Artificial Neural Networks (ANN) [8, 9] and Genetic Algorithm (GA)-based evolutionary approaches have been applied to solve the IK problem for many kinds of manipulators, and great progress has been achieved. Satish [10] uses a 3-layer perceptron neural network to resolve the IK problem in a two degree-of-freedom (DOF) serial chain manipulator, as the training data subdivide degree is only 1.125 degrees, the precision of solution is not good enough. Bassam [11] uses a 4-layer perceptron neural network to resolve the IK problem in 2 DOF manipulator, which can avoid the emergence of redundant solutions and improve the precision by dividing the solution space into many sub-spaces, but the complexity of the algorithm is increased on one hand. Reference [12] researches the multi-layer perceptron and functional link artificial neural network, respectively, to compute 2 and 3 DOF serial manipulators' inverse solution. When the robotic arm is 3 DOF, the solution's precision is less than satisfactory. As the training sample's subdivision degree determines the precision of the solution in neural network, when the body has more than 3 DOF, due to the limitation in the number of training samples, it is difficult to obtain a high precise solution. The GA is widely applied to solve the IK problem due to its excellent characteristics of evolutionary optimization; Saleh [13] obtains the IK solutions of the three joints which decide the end-effector position of PUMA560 robot by using the genetic algorithm based on the niche and clustering technique. The method can search four optimal solutions at a time but it is time-consuming—every generation computation spent 4 s; Banga [14] combines GA and Analytical Hierarchy Process (AHP) to resolve three nodes manipulator's IK problem, AHP is used to select a different fitness function to prefer a one robotic arm indicator (such as attitude, friction, adjust time), and GA optimizes the inverse solution according to the fitness function selected by AHP. Due to the fitness function selection mechanism, the method can meet multiple-objectives optimization, but it also increases the probability of obtaining suboptimal solutions. Due to the random selection of initial population and the large searching domain, the computational time of GA optimization is usually long and the algorithm is prone to "premature" and later search slowly. So, scholars study another approach, using GA to optimize the parameters of neural network, and then use this neural network to compute the IK solutions. Liu [15] uses GA to optimize weights and thresholds of BP neural network by applying partition coding method; then the BP neural network is used to compute the 2 DOF robot's IK solution. Carlos [16] combines Fuzzy neural network (FNN) algorithm and GA for solving the inverse kinematics problem of a 2 DOF manipulator. This method uses GA to optimize the network structure, the set of rules and membership functions of the FNN, and then this FNN algorithm is used to compute the two joint angle's values of manipulator. These methods will accelerate the optimization process of the inverse solution when the algorithm completes the parameter optimization. The disadvantage is that the precision of the solution depends on the neural network, and the accuracy of the neural network's output depends on its structure, parameters, and training samples. When the dimension of the input and output increases, the number of parameters also will be exponentially increased. And it requires a lot of training samples, which would make the training inefficient or even unattainable. As a simple neural network algorithm, the method mentioned above is only suitable for solving the IK problem in less than 3 DOF; when DOF is equal to or larger than 3, the precision of solutions will be far from satisfactory.

The objective of this study is to resolve the IK problem of virtual human's ULKC which is the most complex, flexible, and commonly used agent of the human body which has 6 DOF. This study proposes an improved GA method to resolve the IK problem of ULKC, which can avoid GA falling into "premature" stop or inefficient searching in later stage with high probability.

10.2 Mathematical Model of ULKC

In this section, we introduce the basic principle of D–H method, construct the kinematics chain of ULKC, and analyze the difficulty in obtaining solutions from the kinematics equation.

10.2.1 Introduction of D–H Method

At present, the D–H method is usually used to construct the coordinate of kinematics chain whose mathematical model is used to represent by matrices. The D–H method is based on the transformation of the homogeneous coordinates in three-dimension space, which can link the movement, transformation, and mapping with matrix calculation.

(1) The posture description of kinematics chain

a. Position description

In the three-dimension Cartesian coordinate system, the position of point is usually described by a 3×1 vector. Figure 10.1 shows the point P in the coordinate system {A}. Its homogeneous coordinates are $P = (P_x, P_y, P_z, 1)^T$.

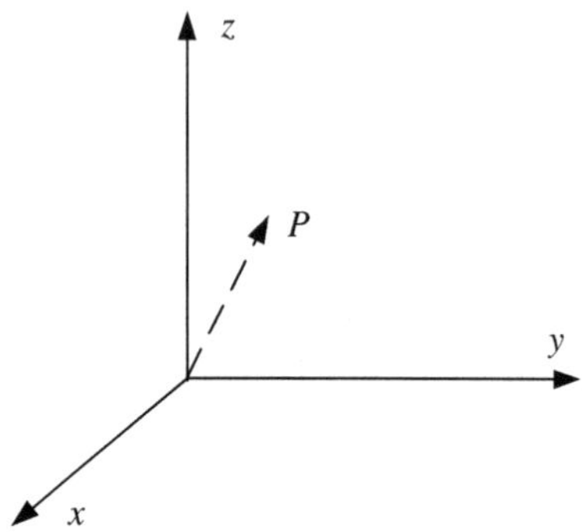

Fig. 10.1 Point P in the coordinate system

b. Orientation description

In order to study the kinematics chain's movement and operation, both the object's position and orientation need to be described. 4×4 homogeneous matrix ${}_A^B\boldsymbol{R}$ represents the orientation of the coordinate system {B} relative to the coordinate system {A}. $n_x,\ n_y,\ n_z,\ o_x,\ o_y,\ o_z,\ a_x,\ a_y,\ a_z$ are the direction cosine values that three unit vectors $\vec{x},\ \vec{y},\ \vec{z}$ of the coordinate system {B} to the coordinate system {A}

$$ {}_A^B\boldsymbol{R} = \begin{bmatrix} n_x & o_x & a_x & 0 \\ n_y & o_y & a_y & 0 \\ n_z & o_z & a_z & 0 \\ 0 & 0 & 0 & 1 \end{bmatrix}. \tag{10.1} $$

${}_A^B\boldsymbol{R}$ represents the orientation of the coordinate system {B} to the coordinate system {A}.

Rotate θ around the axis x, y, and z, respectively, and their corresponding rotation matrices can be expressed as follows:

$$ \boldsymbol{Rot}(x,\theta) = \begin{bmatrix} 1 & 0 & 0 & 0 \\ 0 & \cos\theta & -\sin\theta & 0 \\ 0 & \sin\theta & \cos\theta & 0 \\ 0 & 0 & 0 & 1 \end{bmatrix}, \tag{10.2} $$

$$ \boldsymbol{Rot}(y,\theta) = \begin{bmatrix} \cos\theta & 0 & \sin\theta & 0 \\ 0 & 1 & 0 & 0 \\ -\sin\theta & 0 & \cos\theta & 0 \\ 0 & 0 & 0 & 1 \end{bmatrix}, \tag{10.3} $$

$$ \boldsymbol{Rot}(z,\theta) = \begin{bmatrix} \cos\theta & -\sin\theta & 0 & 0 \\ \sin\theta & \cos\theta & 0 & 0 \\ 0 & 0 & 1 & 0 \\ 0 & 0 & 0 & 1 \end{bmatrix}. \tag{10.4} $$

Spatially, when the coordinate systems {A} and {B} have the same orientation and different origins, the translation matrix is as follows:

$$ {}_{A}^{B}\boldsymbol{trans}(B_x, B_y, B_z) = \begin{bmatrix} 1\,0\,0\,B_x \\ 0\,1\,0\,B_y \\ 0\,0\,1\,B_z \\ 0\,0\,0\,1 \end{bmatrix}. \tag{10.5} $$

It represents the origin position of the coordinate system {B} in the coordinate system {A}.

c. The expression of position and orientation of objects in space

When coordinates system {A} and {B} have different origins and different orientations, the matrix ${}_{A}^{B}\boldsymbol{T}$ represents the position and orientation of B in the coordinate system {A}

$$ {}_{A}^{B}\boldsymbol{T} = {}_{A}^{B}\boldsymbol{trans}(B_x, B_y, B_z){}_{A}^{B}\boldsymbol{R} = \begin{bmatrix} n_x & o_x & a_x & B_x \\ n_y & o_y & a_y & B_y \\ n_z & o_z & a_z & B_z \\ 0 & 0 & 0 & 1 \end{bmatrix}. \tag{10.6} $$

It is proved that every posture matrix can be decomposed to many rotation matrices and translation matrices.

d. The spatial transformation between different coordinate systems by the posture matrix

If ${}^{A}P = ({}^{A}Px, {}^{A}Py, {}^{A}Pz, 1)^T$ and ${}^{B}P = ({}^{B}Px, {}^{B}Py, {}^{B}Pz, 1)^T$ are homogeneous coordinates of the point P in the coordinates {A} and {B}, respectively, then, ${}^{A}P = {}_{A}^{B}\boldsymbol{T}{}^{B}P$.

The transformation between different coordinate systems can be conducted on the basis of the above formula.

(2) The nature of matrices in D–H method

a. The nature of translation matrix

Proposition 1: if $\{\boldsymbol{trans}(a, b, c)|\{a, b, c \in \boldsymbol{R}\}$ represents the exchange group on matrix multiplication, then,

$$ \boldsymbol{trans}(0, 0, 0) = \boldsymbol{I} \tag{10.7} $$

$$ \boldsymbol{trans}(a, b, c) \cdot \boldsymbol{trans}(a, b, c)^{-1} = \boldsymbol{trans}(-a, -b, -c). \tag{10.8} $$

Property 1:

$$ \boldsymbol{trans}(a_1, b_1, c_1) \cdot \boldsymbol{trans}(a_2, b_2, c_2) = \boldsymbol{trans}(a_1 + a_2, b_1 + b_2, c_1 + c_2). \tag{10.9} $$

Corollary 1:

$$\boldsymbol{trans}(a, b, c)^{-1} = \boldsymbol{trans}(-a, -b, -c). \tag{10.10}$$

b. The nature of rotation matrix

Proposition 2: if $\{\boldsymbol{Rot}(x, \theta)|\theta \in \boldsymbol{R}\}$ represents the exchange group on matrix multiplication, so does to $\{\boldsymbol{Rot}(y, \theta)|\theta \in \boldsymbol{R}\}$ and $\{\boldsymbol{Rot}(z, \theta)|\theta \in \boldsymbol{R}\}$.

Property 2:

$$\begin{aligned} \boldsymbol{Rot}(x, \theta_1) \cdot \boldsymbol{Rot}(x, \theta_2) &= \boldsymbol{Rot}(x, \theta_1 + \theta_2) \\ \boldsymbol{Rot}(y, \theta_1) \cdot \boldsymbol{Rot}(y, \theta_2) &= \boldsymbol{Rot}(y, \theta_1 + \theta_2) \\ \boldsymbol{Rot}(z, \theta_1) \cdot \boldsymbol{Rot}(z, \theta_2) &= \boldsymbol{Rot}(z, \theta_1 + \theta_2). \end{aligned} \tag{10.11}$$

Corollary 2:

$$\begin{aligned} \boldsymbol{Rot}(x, \theta)^{-1} &= \boldsymbol{Rot}(x, \theta)^T = \boldsymbol{Rot}(x, -\theta) \\ \boldsymbol{Rot}(y, \theta)^{-1} &= \boldsymbol{Rot}(y, \theta)^T = \boldsymbol{Rot}(y, -\theta) \\ \boldsymbol{Rot}(z, \theta)^{-1} &= \boldsymbol{Rot}(z, \theta)^T = \boldsymbol{Rot}(z, -\theta). \end{aligned} \tag{10.12}$$

c. The nature of the posture matrix

If the posture matrix ${}^m_i\boldsymbol{T}$ represents the transformation from coordinate system $\{m\}$ to coordinate system $\{i\}$, and the posture matrix ${}^k_m\boldsymbol{T}$ represents the conversion from coordinate system $\{k\}$ to coordinate system $\{m\}$; then, the posture matrix, from coordinate system $\{k\}$ to coordinate system $\{i\}$, can be computed by the following formula:

$${}^k_i\boldsymbol{T} = {}^m_i\boldsymbol{T}{}^k_m\boldsymbol{T}. \tag{10.13}$$

Then, the transformation from ith joint-unit to jth joint-unit can be described as:

$${}^j_i\boldsymbol{T} = {}^{i+1}_i\boldsymbol{T}{}^{i+2}_{i+1}\boldsymbol{T} \cdots {}^j_{j-1}\boldsymbol{T}. \tag{10.14}$$

10.2.2 Structure of ULKC

The human ULKC contains root joint, chest joint, shoulder joint, elbow joint, wrist joint, and palm joints. Usually the palm is regarded as an independent research object as it has a lot of joints. Therefore, the wrist is the end-effector of the ULKC.

Human joints can be approximately treated as the rotary joint which is classified into P-type (shaft parallel to the link) and V-type (shaft perpendicular to the link). This study considers each joint-unit a combination of P-type and V-type. While the joint-unit only consists of P-type or V-type, this study adds another type which is

Fig. 10.2 The joint-units of the ULKC

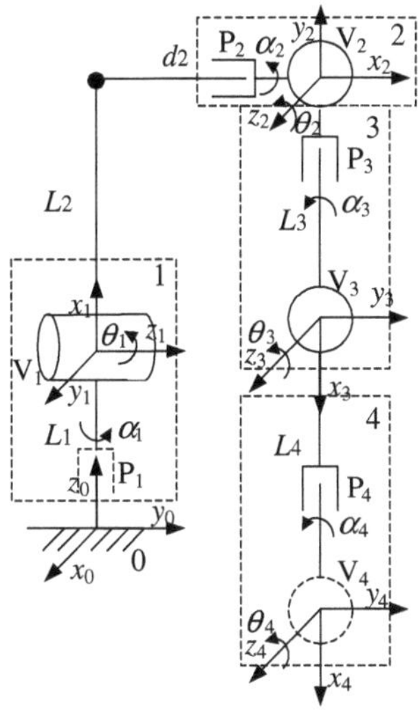

within the ZERO rotation range. The rotation shaft of the P-type joint is the x-axis of the local coordinate system of the joint-unit, and the rotation shaft of the V-type joint is the z-axis of the local coordinate system of the joint-unit, then the y-axis can be determined by the right-handed screw rule. According to this specification, the local coordinate systems of human ULKC's joint-units are established, as shown in Fig. 10.2.

The 0th joint-unit is the human root joint-unit whose coordinate is relatively fixed, and it is also the basic coordinate of the entire kinematics chain. The 1st joint-unit is the chest joint-unit which contains a real V-type joint and a complementary P-type joint; thus, it can only rotate around z_1. The 2nd joint-unit is the shoulder joint-unit, which consists of a V-type joint and a P-type joint. The 3rd joint-unit is the elbow joint-unit which contains a V-type joint and a P-type joint. The 4th joint-unit is the wrist joint-unit which consists of a complementary V-type joint and a real P-type joint. It should be noted that the joints P_2, V_2, P_3 are derived from the shoulder joint which is a spherical joint. Placing P_3 under the elbow unit is conducive to joint grouping and mathematical description of the kinematics chain. α_i represents the rotation angle of P_i, and θ_i the rotation angle of V_i. α_1 and θ_4 are always equal to 0.

According to the parameter values of adult body in GB10000-88 (Chinese), set values for L_1, L_2, L_3, L_4, d_2 and the ranges of $\theta_1, \theta_2, \theta_3, \alpha_2, \alpha_3, \alpha_4$. They are listed in Table 10.1.

10.2.3 *Posture Transformation of ULKC*

With reference to Sects. 10.2.1 and 10.2.2, the transformation from first joint-unit to zeroth joint-unit of ULKC can be described as follows:

Table 10.1 The parameters of the ULKC

Joint-unit	L_i/mm	$\theta_i/°$	$\alpha_i/°$	d_i/mm
1	19.5	$0 \sim 60$	0	0
2	31.5	$-30 \sim 180$	$-180 \sim 45$	19.85
3	33.3	$-140 \sim 0$	$-45 \sim 90$	0
4	25.3	0	$-90 \sim 90$	0

(1) Translate L_1 along the axis z_0, then the origin of the two coordinate systems are coincided;
(2) Rotate $-90°$ around the axis x_0, then the shaft z_0 and z_1 are collinear;
(3) Rotate $-90°$ around the axis z_0, then the shaft x_0 and x_1 are collinear.

Therefore, the transformation matrix is:

$$ {}_0^1\boldsymbol{T} = \boldsymbol{trans}(0, 0, L_1)\boldsymbol{Rot}(x, -90°)\boldsymbol{Rot}(z, -90°). \tag{10.15}$$

According to Eqs. (10.2–10.5),

$$\boldsymbol{trans}(0, 0, L_1) = \begin{bmatrix} 1 & 0 & 0 & 0 \\ 0 & 1 & 0 & 0 \\ 0 & 0 & 1 & L_1 \\ 0 & 0 & 0 & 1 \end{bmatrix}, \tag{10.16}$$

$$\boldsymbol{Rot}(x, -90°) = \begin{bmatrix} 1 & 0 & 0 & 0 \\ 0 & 0 & 1 & 0 \\ 0 & -1 & 0 & 0 \\ 0 & 0 & 0 & 1 \end{bmatrix}, \tag{10.17}$$

$$\boldsymbol{Rot}(z, -90°) = \begin{bmatrix} 0 & 1 & 0 & 0 \\ -1 & 0 & 0 & 0 \\ 0 & 0 & 1 & 0 \\ 0 & 0 & 0 & 1 \end{bmatrix}. \tag{10.18}$$

Substitute Eqs. (10.16–10.18) into Eq. (10.15):

$${}_0^1\boldsymbol{T} = \begin{bmatrix} 0 & 1 & 0 & 0 \\ 0 & 0 & 1 & 0 \\ 1 & 0 & 0 & L_1 \\ 0 & 0 & 0 & 1 \end{bmatrix}. \tag{10.19}$$

Similarly, we can obtain the transformation matrices of other adjacent joints' local coordinate systems:

$$ {}_1^2\boldsymbol{T} = \begin{bmatrix} 0 & C\alpha_2 C\theta_1 - S\alpha_2 S\theta_1 & -C\alpha_2 S\theta_1 - S\alpha_2 C\theta_1 & L_2 C\theta_1 \\ 0 & C\alpha_2 S\theta_1 + S\alpha_2 C\theta_1 & C\alpha_2 C\theta_1 - S\alpha_2 S\theta_1 & L_2 S\theta_1 \\ 1 & 0 & 0 & d_2 \\ 0 & 0 & 0 & 1 \end{bmatrix}, \tag{10.20} $$

$$ {}_2^3\boldsymbol{T} = \begin{bmatrix} C(\theta_2 - \pi/2) & -S(\theta_2 - \pi/2)C\alpha_3 & S(\theta_2 - \pi/2)S\alpha_3 & L_3 C(\theta_2 - \pi/2) \\ S(\theta_2 - \pi/2) & C(\theta_2 - \pi/2)C\alpha_3 & -C(\theta_2 - \pi/2)S\alpha_3 & L_3 S(\theta_2 - \pi/2) \\ 0 & S\alpha_3 & C\alpha_3 & 0 \\ 0 & 0 & 0 & 1 \end{bmatrix}, \tag{10.21} $$

$$ {}_3^4\boldsymbol{T} = \begin{bmatrix} C\theta_3 & -C\alpha_4 S\theta_3 & S\alpha_4 S\theta_3 & L_4 C\theta_3 \\ S\theta_3 & C\alpha_4 C\theta_3 & -S\alpha_4 C\theta_3 & L_4 S\theta_3 \\ 0 & S\alpha_4 & C\alpha_4 & 0 \\ 0 & 0 & 0 & 1 \end{bmatrix}. \tag{10.22} $$

In the equations above, S represents sine operator; C represents cosine operator.

According to Eq. (10.14), the posture transformation matrix of wrist joint (end-effector) relative to the root joint can be computed as follows:

$$ \begin{aligned} {}_0^4\boldsymbol{T} &= {}_0^1\boldsymbol{T}\,{}_1^2\boldsymbol{T}(\theta_1, \alpha_2)\,{}_2^3\boldsymbol{T}(\theta_2, \alpha_3)\,{}_3^4\boldsymbol{T}(\theta_3, \alpha_4) \\ &= \begin{bmatrix} {}_0^4n_x & {}_0^4o_x & {}_0^4a_x & {}_0^4B_x \\ {}_0^4n_y & {}_0^4o_y & {}_0^4a_y & {}_0^4B_y \\ {}_0^4n_z & {}_0^4o_z & {}_0^4a_z & {}_0^4B_z \\ 0 & 0 & 0 & 1 \end{bmatrix} \end{aligned} \tag{10.23} $$

where,

$$ \begin{aligned} {}_0^4n_x &= S(\theta_2 - \pi/2)C(\theta_3)(C\alpha_2 C\theta_1 + S\alpha_2 C\theta_1) - S\theta_3(S\alpha_3(S\alpha_2 S\theta_1 - C\alpha_2 C\theta_1) \\ &\quad - C(\theta_2 - \pi/2)C(\alpha_3)(C(\alpha_2)S(\theta_1) + S(\alpha_2)C\theta_1)) \end{aligned} $$

$$ {}_0^4n_y = C(\theta_2 - \pi/2)C\theta_3 - S(\theta_2 - \pi/2)C\alpha_3 S\theta_3 $$

$$ \begin{aligned} {}_0^4n_z &= -S\theta_3(S\alpha_3(C\alpha_2 S\theta_1 + S\alpha_2 C\theta_1) + C(\theta_2 - \pi/2)C\alpha_3(S\alpha_2 S\theta_1 - C\alpha_2 C\theta_1)) \\ &\quad - S(\theta_2 - \pi/2)C\theta_3(S\alpha_2 S\theta_1 - C\alpha_2 C\theta_1) \end{aligned} $$

$$ \begin{aligned} {}_0^4o_x &= -S\alpha_4(C\alpha_3(S\alpha_2 S\theta_1 - C\alpha_2 C\theta_1) + C(\theta_2 - \pi/2)S\alpha_3(C\alpha_2 S\theta_1 + S\alpha_2 C\theta_1)) \\ &\quad - C\alpha_4 C\alpha_3(S\alpha_3(C\alpha_2 S\theta_1 - C\alpha_2 C\theta_1) - C(\theta_2 - \pi/2)C\alpha_3(C\alpha_2 S\theta_1 + S\alpha_2 C\theta_1)) \\ &\quad - S(\theta_2 - \pi/2)C\alpha_4 S\theta_3(C\alpha_2 S\theta_1 + S\alpha_2 C\theta_1) \end{aligned} $$

$$ {}_0^4o_y = S(\theta_2 - \pi/2)S\alpha_3 S\alpha_4 - C(\theta_2 - \pi/2)C\alpha_4 S\theta_3 - S(\theta_2 - \pi/2)C\alpha_3 C\alpha_4 C\theta_3 $$

$$ {}_0^4o_z = S(\theta_2 - \pi/2)C\alpha_4 S\theta_3(S\alpha_2 S\theta_1 - C\alpha_2 C\theta_1) - C\alpha_4 C\theta_3(S\alpha_3(C\alpha_2 S\theta_1 + S\alpha_2 C\theta_1) $$

$$
\begin{aligned}
&+C(\theta_2-\pi/2)C\alpha_3(S\alpha_2S\theta_1-C\alpha_2C\theta_1))-S\alpha_4(C\alpha_3(C\alpha_2S\theta_1+S\alpha_2C\theta_1)\\
&-C(\theta_2-\pi/2)S\alpha_3(S\alpha_2S\theta_1-C\alpha_2C\theta_1))\\
{}_0^4a_x &= S\alpha_4C\theta_3(S\alpha_3(S\alpha_2S\theta_1-C\alpha_2C\theta_1)-C(\theta_2-\pi/2)C\alpha_3(C\alpha_2S\theta_1+S\alpha_2C\theta_1))\\
&-C\alpha_4(C\alpha_3(S\alpha_2S\theta_1-C\alpha_2C\theta_1)+C(\theta_2-\pi/2)S\alpha_3(C\alpha_2S\theta_1+S\alpha_2C\theta_1))\\
&+S(\theta_2-\pi/2)S\alpha_4S\theta_3(C\alpha_2S\theta_1+S\alpha_2C\theta_1)\\
{}_0^4a_y &= S(\theta_2-\pi/2)C\alpha_4S\alpha_3+C(\theta_2-\pi/2)S\alpha_4S\theta_3+S(\theta_2-\pi/2)C\alpha_3S\alpha_4C\theta_3\\
{}_0^4a_z &= S\alpha_4C\theta_3(S\alpha_3(C\alpha_2S\theta_1+S\alpha_2C\theta_1)+C(\theta_2-\pi/2)C\alpha_3(S\alpha_2S\theta_1-C\alpha_2C\theta_1))\\
&-C\alpha_4(C\alpha_3(C\alpha_2S\theta_1+S\alpha_2C\theta_1)-C(\theta_2-\pi/2)S\alpha_3(S\alpha_2S\theta_1-C\alpha_2C\theta_1))\\
&-S(\theta_2-\pi/2)S\alpha_4S\theta_3(S\alpha_2S\theta_1-C\alpha_2C\theta_1)\\
{}_0^4B_x &= L_2S\theta_1+L_3S(\alpha_2-\pi/2)(C\alpha_2S\theta_1+S\alpha_2C\theta_1)-L_4S\theta_3(S\alpha_3(S\alpha_2S\theta_1-C\alpha_2C\theta_1)\\
&-C(\theta_2-\pi/2)C\alpha_3(C\alpha_2S\theta_1+S\alpha_2C\theta_1))+L_4S(\theta_2-\pi/2)C\theta_3(C\alpha_2S\theta_1+S\alpha_2C\theta_1)\\
{}_0^4B_y &= d_2+L_3C(\theta_2-\pi/2)+L4C(\theta_2-\pi/2)C\theta_3-L4S(\theta_2-\pi/2)C\alpha_3S\theta_3\\
{}_0^4B_z &= L_1+L_2C\theta_1-L_3S(\theta_2-\pi/2)(S\alpha_2S\theta_1-C\alpha_2C\theta_1)-L_4S\theta_3(S\alpha_3(C\alpha_2S\theta_1+S\alpha_2C\theta_1)\\
&+C(\theta_2-\pi/2)C\alpha_3(S\alpha_2S\theta_1-C\alpha_2C\theta_1))-L_4S(\theta_2-\pi/2)C\theta_3(S\alpha_2S\theta_1-C\alpha_2C\theta_1).
\end{aligned}
\tag{10.24}
$$

It is difficult to compute the inverse solutions from such a complex posture transformation matrix like Eq. (10.24) by using the analytical method or geometric method. Also, as there are six variables θ_1, θ_2, θ_3, α_2, α_3, and α_4 in the matrix and the end-effector of the ULKC is 6 DOF, it is also difficult to compute the precise solutions quickly by simply using the neural network algorithm or the standard genetic algorithm.

10.2.4 Summary

In this section, D–H method is used to describe the kinematics equations of ULKC. The basic principle of D–H method is that transformation of points in different coordinate systems can be conducted according to transitivity of transformation matrices. ULKC is a kinematics chain. Its IK problem is to resolve the joint angles when the posture of end-effector is known. We derive the transformation matrices of adjacent local coordinates systems and the kinematics Eq. (10.24) of ULKC. It is difficult to resolve Eq. (10.24) by basic geometric method or analytic method due to its complex and nonlinear features. From the next section, GA will be improved to resolve Eq. (10.24).

10.3 Improvement of Standard Generation Algorithm

In this section, an improved GA is put forward with the assistance of an adaptive genetic operator.

10.3.1 Introduction of SGA

Genetic algorithm is an efficient and global searching algorithm based on natural selection and genetic theory. The possible solution in the problem domain is seen as an individual or chrome in the population. The SGA is a group operation algorithm which only uses the standard genetic operators, such as selection operator, crossover operator, and mutation operator. It is the foundation of other improved GA.

(1) The mathematical model of SGA

The SGA can be expressed as:
SGA=$(C, E, P_0, M, \phi, \Gamma, \Psi, T)$

where,

C coding method;
E the fitness function of individual;
P_0 the initialization population;
M the amount of population;
O selection operator;
Γ crossover operator;
Ψ mutation operator;
T the termination condition;

(2) The process of the SGA

a. Chromosome coding and decoding (Fig. 10.3)

In the SGA, the individuals in the population are represented by fixed-length binary strings, whose allele consists of the binary digit $\{0, 1\}$. The genes of each individual in the initial population can be generated via the random numbers distributed uniformly. For example, X = 100010111011 can be used to depict an individual, whose chromosome length is 16.

Coding: a parameter, with the range of [U1, U2], can be expressed by the binary code symbol with the length of K, which generate 2^k kinds of code. Thus the corresponding relationship between the parameters and coding is as follows:

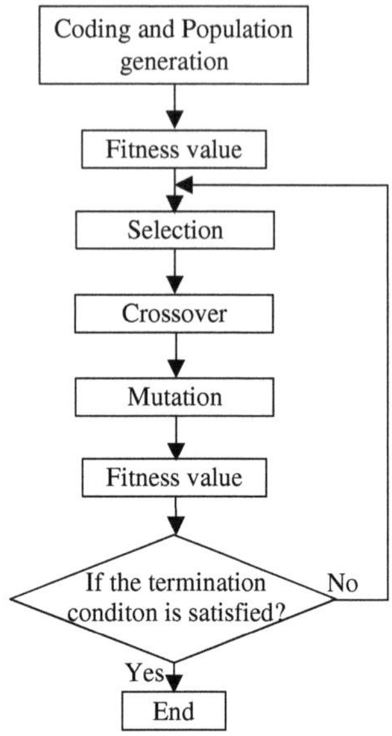

Fig. 10.3 The flowchart of SGA

$$\begin{aligned} &000000\cdots0000 = 0 \to U_1 \\ &000000\cdots0001 = 1 \to U_1 + \delta \\ &000000\cdots0002 = 2 \to U_1 + 2\delta \\ &\quad\vdots \qquad\qquad\quad \vdots \qquad \vdots \\ &111111\cdots1111 = 2^k - 1 \to U_2 \end{aligned} \tag{10.25}$$

where, $\delta = \frac{U_2 - U_1}{2^k - 1}$.

Decoding: if the code of an individual is $b_k b_{k-1} b_{k-2} \ldots b_2 b_1$, the corresponding decoding formula is as follows:

$$X = U_1 + \left(\sum_{i=1}^{k} b_i \cdot 2^{i-1} \right) \cdot \frac{U_2 - U_1}{2^k - 1}. \tag{10.26}$$

b. The evaluation of individuals' fitness

In the SGA, the probability of individual's genetic opportunity is decided by its fitness value. All the individual fitness values must be nonnegative for evaluating the genetic probability correctly. So, we need to determine the transformation principle from target function to fitness function, especially when the target value is negative.

c. Genetic operators

Three kinds of genetic operators used by the SGA are as follows:

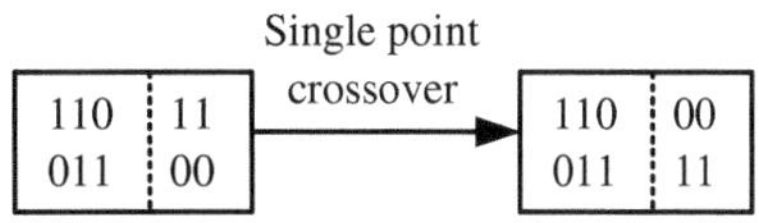

Fig. 10.4 The process of crossover

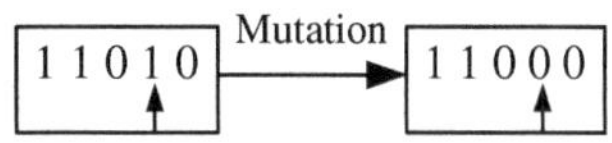

Fig. 10.5 The process of mutation

1. Use the proportional selection operator to select the father individuals. The proportional selection factor refers to the genetic probability. If the number of individuals is M, the fitness of ith individual is f_i, then, the probability of ith individual is:

$$P_i = f_i \Big/ \sum_{k=1}^{M} f_k. \tag{10.27}$$

If the probability of individual selection is given, uniform random numbers between [0, 1] are generated to determine which individual is to mate. The individuals, who have a large selection probability, can be repeatedly selected and their genes are expanded in the population, while the small ones would eliminate gradually.

2. The single-point crossover operator is applied to crossover operation in SGA. Only one cross point exists in the single-point crossover operator, the two individuals are chosen as cross objects, then, a crossover point is generated randomly. The chosen individuals exchange their genes at the cross point and generate two children individuals (Fig. 10.4).

3. The mutation operation uses the basic bit mutation operator and the uniform mutation operator. In order to avoid the "premature" convergence problem, there is a small probability for the individuals composed of binary gene code to shift the genetic code from 0 to 1 or 1 to 0 (Fig. 10.5).

4. The parameters of the SGA

There are four parameters of the SGA need to be set; they are M, T, P_c, and P_m.

M represents the size of the population, which refers to the individual number of the population, usually ranging from 20 to 100;

T represents the terminal generation times, which usually ranges from 100 to 500;

P_c stands for the crossover probability, usually 0.4–0.99;

P_m stands for the mutation probability, normally 0.0001–0.1.

10.3.2 Principles of Improvement

Since the intact structure and theory of GA was systemically proposed by Holland in 1975, numerous scholars have devoted to developing GA. The coding method, the control parameters, and crossover mechanism are deeply studied. Therefore, various improved GA are proposed, such as Hierarchic Genetic Algorithm [18, 19], CHC Algorithm [20, 21], Messy Genetic Algorithm [22, 23], Niched Genetic Algorithm [24], Adaptive Genetic Algorithm [25, 26], and Parallel Genetic Algorithm [27–29], whose characteristics and performances are compared in article [30].

The present research focuses on the improvement of GA from the perspectives of population initialization and its diversity. An improved GA is put forward in this chapter with the assistance of adaptive genetic operator.

We operate on the individuals from the perspectives of selection, crossover, and mutation. The SGA contains some demerits such as "premature" and inefficient searching in later stage, which result from the following aspects, such as single population, random selection of initial population, and no-adaptive crossover and mutation probabilities. This research will improve the SGA on the aspects of population diversity, the selection strategy for initial population, and the adaptive genetic operators to overcome the "premature" and inefficient searching in later stage.

a. Population diversity. In the SGA, all the N individuals in a population, ***Chrom***, the differences among individuals will be reduced with the increase of generation times, which leads the searching process to the "premature" convergence. We introduce the population migration: the population ***Chrom*** is divided into p independent subpopulations, $\boldsymbol{SubPop}_i (i = 1, 2, \ldots, p)$; then each sub population has $m = N/p$ individuals. The genetic operations are conducted in the isolated subpopulation in the normal condition; we let the individuals migrate to other subpopulations with the probability P_{mig}, while it meets the needs of certain conditions (such as get the certain generation times). It can ensure the diversity of individuals in population, and avoid the "premature" convergence.
b. Population initialization. Randomly selecting the individuals will reduce the differences among individuals significantly after several genetic operations, which results in inefficient searching in later stage. To solve this problem, Rasit [17] proposes a method in which the initial population is generated by three trained Elman neural networks. Three outputs from three networks can be obtained at a time. After putting these outputs into the initial population, the GA is used to optimize the solution. The disadvantages of this method are: complex structure, which consists of 155 neurons in the hidden layer of the three Elman neural networks; the small number of sample (no more than 3); other individuals are randomly generated. Based on the research of the phenomenon of human populations, we found there is at least a significant difference in the human populations—color. The populations with different colors certainly have different chromosomes, and the colors of individuals within the same population are similar but different; on the other hand, there are no exactly same individuals in the same human population. These ensure the diversity of human genes, thus we propose three principles for initial population:

   1. There is at least one gene different between subpopulations;
   2. There is at least one gene similar (relative) between individuals within the same subpopulation;
   3. There are no two same individuals within the same subpopulation.

These three principles can be summarized as "the individuals are similar but different within the same sub population; the individuals are different but similar between sub populations."

c. The adaptive genetic operators. In order to improve the capability of GA to avoid the "premature" convergence and inefficient searching in later stage, the genetic operators should be adaptive. That means the crossover probability P_c and mutation probability P_m would be increased with the similarity of individuals increased. We design the adaptive operators as follows:

$$P_c = P_{c0} + \mu\beta_i \tag{10.28}$$

$$P_m = P_{m0} + \nu\beta_i \tag{10.29}$$

where, P_{c0} is the initial crossover probability and P_{m0} is the initial mutation probability; $\beta_i$is the proportion of the maximum fitness individuals account to the total number of individuals in ith generation; μ and ν are the two constant factors which are used to make P_c and P_m not more than 1.

The improved GA has the features of diversity population, the similar but different initial population, the adaptive genetic operators, which will be helpful to search the high-accuracy solutions quickly.

10.3.3 Process Flow of IGA

The process flow of the GA contains coding, population initialization, fitness value evaluation, selection, crossover, mutation, and judge the terminal condition. According to the above principles, we design the IGA as the following steps:

a. Coding. The fixed-length binary string is used to represent the individual in the SGA, where allele is {0, 1}. Binary coding will greatly increase gene length, and must shift between the coding and decoding repeatedly. It is not suitable for the high dimension. We use the real coding to represent each individual whose chromosome concludes six genes $\{\theta_1, \theta_2, \theta_3, \alpha_2, \alpha_3, \alpha_4\}$. The variable domains are shown in Table 10.1.
b. Population initialization. The population initialization contains population diversity and individuals' initialization. The population diversity is realized by setting subpopulations. We divide the population ***Chrom*** into p independent subpopulations $\boldsymbol{SubPop}_i (i = 1, 2, \ldots, p)$, each subpopulation has m individuals, then the entire population has $N = pm$ individuals. The genetic operations are independently conducted within the subpopulations. We let the individuals of every subpopulation migrate to other subpopulations with probability P_{mig} every g generations, which will ensure the individual diversity within populations.

According to the principle "the individuals are similar but different within the same sub population, the individuals are different but similar between sub populations," we initiate the population: using θ_1 as feature gene to distinguish the different subpopulations; we divide the interval $(0, \pi/3)$ of variable domain of θ_1 into p equal length intervals, and the length of each interval is:

$$\sigma = \frac{\pi/3}{p}. \tag{10.30}$$

Then, the ith interval is $((i-1)\sigma, i\sigma)$ from which the first gene θ_1 of the ith subpopulation are generated, and the other five genes are generated from their own variable domain randomly. This method can ensure: a. the first genes of individuals from the same subpopulation are similar but different; b. the other five genes are all the same, which is a small probability event; c. the individuals from different subpopulations are not possibly the same.

c. Fitness value evaluation. The only requirement for the fitness function is that, for each chromosome, we can calculate a corresponding comparable nonnegative result. According to the posture matrix ${}_0^4\boldsymbol{T}'$ computed by individual and the target posture matrix ${}_0^4\boldsymbol{T}$ except the last line (the last line of the all posture matrix is the same), we define the error function:

$$F(\theta_1, \theta_2, \theta_3, \alpha_1, \alpha_2, \alpha_3) = \frac{1}{3 \times 4}\sqrt{\sum_{i=1}^{3}\sum_{j=1}^{4}({}_0^4\boldsymbol{T}'_{i,j} - {}_0^4\boldsymbol{T}_{i,j})^2}. \tag{10.31}$$

The fitness function is designed as follows:

$$\mathrm{Fit}(F) = \frac{1}{1 + F(\theta_1, \theta_2, \theta_3, \alpha_1, \alpha_2, \alpha_3)}. \tag{10.32}$$

The individual's fitness value of $\mathrm{Fit}(F)$ will be larger when the error $F(\theta_1, \theta_2, \theta_3, \alpha_1, \alpha_2, \alpha_3)$ is smaller.

d. Selecting the father individuals. The randomized competitive selection method is used to select the father individuals. Each time, we select a pair of individuals by using roulette selection mechanism, and then make the two individuals compete. The one who has the higher fitness will be selected. Repeat this process until the population is full.
e. Crossover. We use the two-point crossover method which needs to randomly select two integers k_1 and k_2, and exchange the (k_1+1)th to (k_2)th genes of two father individuals with probability P_c; the value of P_c is derived by Eq. (10.28).
f. Mutation. Mutation means that one or several genes of each individual are changed with mutation probability P_m; the P_m is derived by Eq. (10.29).
g. Termination conditions. When the error computed by Eq. (10.31) is equal to or smaller than the given value ε, or the GA operation goes to the maximum generation times, the computational process comes to the end.

10.3.4 Summary

In this section, the principle, "the individuals are similar but different within the same sub population; the individuals are different but similar between sub populations," is proposed to direct the initialization and diversity of the population. The population is divided into some independent subpopulations which are marked by the first gene. Thus, the chromosomes of individuals from different subpopulations are absolutely different, and the chromosomes of individuals from same subpopulations are similar. And the migration strategy is also used to keep diversity. Formulas (10.28) and (10.29) are designed to make the crossover operator and mutation operator adaptive. These methods can keep the IGA diversity and avoid "premature" convergence and inefficient searching in later stage. Besides, error function (10.31) and fitness function (10.32) are designed.

10.4 Simulation Studies

In this section, SGA and IGA are compared in terms of optimizing capability in solving the IK problem of ULKC. Then, the IGA is applied in the control of ULKC of a virtual human.

10.4.1 Simulation of SGA

We set 120 individuals in the population; crossover probability $P_c = 0.7$, mutation probability $P_m = 0.131$; the termination condition was the error $F \leq 0.01$ or the number of generation time was up to 200.

For example, a target posture matrix was generated randomly. See the following matrix:

$$ {}_0^4\boldsymbol{T} = \begin{bmatrix} 0.87814 & -0.02665 & 0.47767 & 46.3554 \\ 0.07128 & 0.99459 & -0.07554 & 54.6941 \\ -0.47307 & 0.10039 & 0.87529 & 33.5743 \\ 0 & 0 & 0 & 1 \end{bmatrix}. \tag{10.33} $$

The result of the SGA operation is shown in Fig. 10.6.

The solid line represents the error of the optimal individual in the population, and the dotted line represents the average error of all the individuals. The computation fell into "premature" convergence when the 31st generations, and the error was 0.0329 when generation time was 200, which did not reach the target value 0.01.

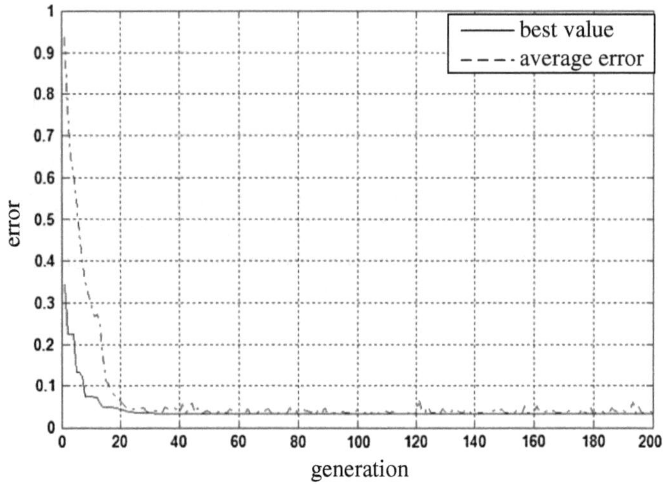

Fig. 10.6 Computational curves of SGA

10.4.2 Simulation of IGA

We divided the population into three subpopulations, each with 40 individuals;

We let the individuals migrate to other subpopulations with the probability of 0.2 every 20 generation times;

Crossover probability and mutation probability are:

$$P_{ci} = 0.6 + 5\beta_i \tag{10.34}$$

$$P_m = 0.125 + 2\beta_i \tag{10.35}$$

The termination condition was the error$F \leq 0.01$;

The target posture matrix was expressed as matrix (10.33).

The result of the IGA operation was shown in Fig. 10.7.

The error was 0.0094 when the generation time was 30, which reached the target value 0.01, and then the computation came to an end. The best individual was:

Chr_best = {34.380 79.038 – 76.716 – 124.272 29.286 – 0.738}

According to Chr_best, we obtained the actual posture matrix:

$$ {}_0^4\boldsymbol{T}' = \begin{bmatrix} 0.87611 & -0.01764 & 0.48180 & 46.2860 \\ 0.06416 & 0.99471 & -0.08025 & 54.1649 \\ -0.47783 & 0.10122 & 0.87260 & 33.3975 \\ 0 & 0 & 0 & 1 \end{bmatrix}. \tag{10.36}$$

Comparing the ${}_0^4\boldsymbol{T}'$and ${}_0^4\boldsymbol{T}$, we saw that the solution was highly accurate.

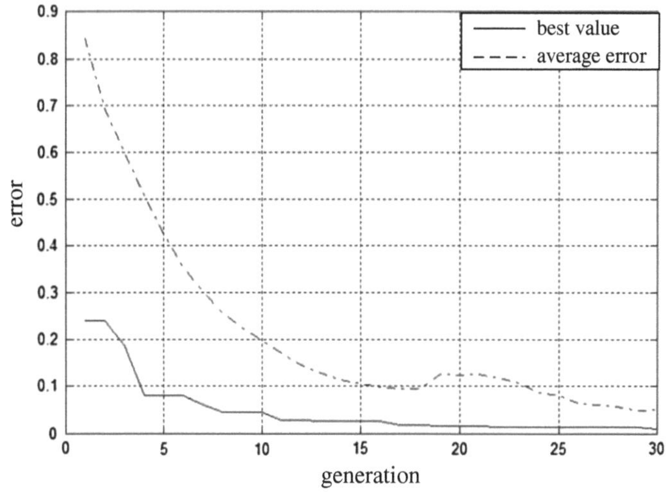

Fig. 10.7 Computational curves of IGA

Table 10.2 Results of simulation

No.	The generation times when error is up to 0.01	
	IGA	SGA
1.	38	41
2.	28	–
3.	43	–
4.	21	25
5.	25	40
6.	30	55
7.	35	–
8.	32	32
9.	59	–
10.	31	86

Note The line means the error has not reached 0.01 after 200 generation times

10.4.3 Comparison of SGA and IGA

We compared the performance differences of SGA and IGA by 10 times simulation to resolve the ULKC IK problem. The results are shown in Table 10.2.

By inspection of Table 10.2, it can be seen that all 10 computations based on IGA meet the target error 0.01 within 200 generations. And SGA cannot meet the target error after 200 generations for four times. The results showed that compared with SGA, IGA falls into the "premature" convergence with low probability.

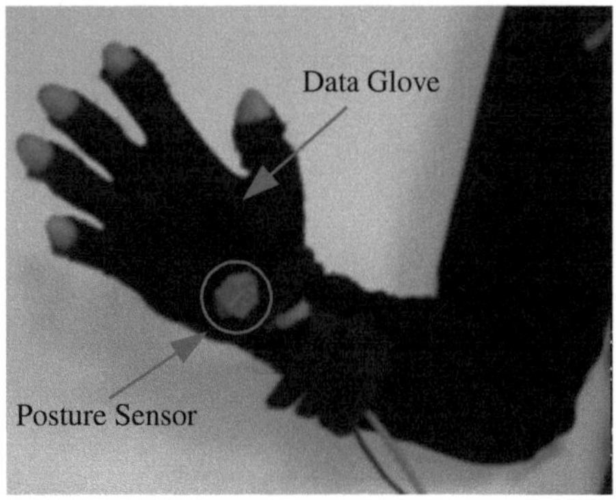

Fig. 10.8 Hardware of the system

10.4.4 Application

We applied the IGA in a virtual maintenance system. Posture sensor was used to measure the wrist joint's posture of operator relative to its root joint, and the position of operator's root was fixed. The postures of root and sensor of operator are mapped into the postures of root joint and end-effector of virtual human in the virtual circumstance. The data glove was used to control the virtual human's behaviors like grasp and release. The hardware is shown in Fig. 10.8.

The termination condition is $F < 0.05$ or the generation is 100 times. We set the scan period 100 ms.

The operator stretched or shrank his hand in the real world. The posture sensor measured the posture of wrist joint relative to the root; thus the posture of end-effector relative to the root joint of virtual human was obtained. Compute the inverse solution with IGA and get the six joint angles. Then the form of virtual human's ULKC was displayed on the screen. When the hand of virtual human touched the part, the data glove was used to control the hand and catch it. The operator and virtual maintenance scene are shown in Fig. 10.9.

The computer measurements indicated that it cost 0.068 s to compute one genetic operation with Intel Core i3 (clocked at 2.93 GHz) processor. And in most cases, the IGA approach can obtain the required solution within 20 generations. The experimental results showed that the IGA approach was reliable in application.

10.4.5 Summary

In this section, SGA and IGA have been tested to resolve the IK problem of ULKC. The same target posture matrix has been resolved by both SGA and IGA which consists of the same number of individuals. The simulation curves are shown in Figs. 10.6 and 10.7, respectively. Obviously, the simulation result of IGA is better

Fig. 10.9 Operator and virtual human

than that of SGA. A number of simulations have been conducted for the comparison, which is shown in Table 10.2. It can be concluded that the optimizing capability of IGA is much better than that of SGA. The IGA has been tested to control the ULKC of virtual human and to provide demonstrations for the other potential applications.

10.5 Conclusions

In this research, we have proposed a new IGA approach to resolve IK problem and obtained the precise inverse solution of ULKC. Compared with SGA, the IGA can avoid "premature" convergence and inefficient searching occurred with much high probability in the later stage of the computation. We have utilized D–H method to structure the mathematic model of ULKC, and improved the GA's population diversity, population initialization, and adaptive genetic operators. Therefore the proposed procedure provides high precise solutions in a quick search route. The simulation bench tests have demonstrated the concepts and algorithms. In addition, the IGA can be applied to resolve many other chain's inverse kinematic problems.

Acknowledgments This work is partially supported by the National Nature Science Foundation of China under Grant 61102170 and 61273188, and the National Advanced Research Foundation of China under Grant 9140A27040112JB47081.

References

1. Wang X, Sun S-Q, Chai C-L (2009) An overview of 3D human motion editing and synthesis. J Image Graph 14(12):233–242
2. Liu G-D, Pan Z-G, Cheng X, Li L, Zhang M-M (2010) A survey on machine learning in the synthesis of human motions. J Comput Aided Des Comput Graph 22(9):1619–1627

3. Xia S-H, Wei Y, Wang Z-Q (2010) A survey of physics-based human motion simulation. J Comput Res Devel 47(8):1354–1361
4. Chen P, Liu L, Yu F (2012) A geometrical method for inverse kinematics of a kind of humanoid manipulator. Robot 34(2):211–216
5. Wang H, Cai Y-F, Zhang W-G (2011) Analytical algorithm of inverse kinematics model for 7DOF manipulator. J JingSu Univ (Nat Sci Ed) 32(3):254–259
6. Zhang X, Wang Z-Y, Wang Z-S (2009) A real-time inverse kinematics algorithm for human motion modeling. J Comput Aided Des Comput Graph 21(6):853–860
7. Qu S-C, Liu X-B, Wang Q-Q (2013) A brief fuzzy controller for an intelligent tracking system. Int J Model Ident Control 19(2):171–178
8. Zhang B-G, Zhang R-Z, Wang G (2012) Breakout prediction for continuous casting using genetic algorithm-based back propagation neural network model. Int J Model Ident Control 16(3): 199–205
9. Golea N, Debbache G, Golea A (2012) Neural network-based adaptive sliding mode control for uncertain non-linear MIMO systems. Int J Model Ident Control 16(4):334–344
10. Satish K, Kashif I (2012) Implementation of artificial neural network applied for the solution of inverse kinematics of 2-link serial chain manipulator. Int J Eng Sci Technol 4(9):4012–4024
11. Bassam D, Shadi K, Mohamed A (2010) Applying neural network architecture for inverse kinematics problem in robotics. Softw Eng Appl 3:230–239
12. Santosh KN, Swetalina P, Subudhi PRS (2012) A novel application of artificial neural network for the solution of inverse kinematics controls of robotic manipulators. Intell Syst Appl 9:81–91
13. Saleh T, Christopher C, William M (2006) A genetic algorithm approach to solve for multiple solutions of inverse kinematics using adaptive niching and clustering. IEEE Proc Evol Comput 1815–1822
14. Banga VK, Singh Y, Kumar R (2007) Simulation of robotic arm using genetic algorithm and AHP. World Acad Sci Eng Technol 5:95–101
15. Liu D-H, Yuan S-C, Wang J-Y (2008) Neural networks based on the genetic algorithm and its application in mechanical engineering. J XiDian Univ 35(1):152–156
16. Carlos K, Maria LC (2012) Robot arm fuzzy control by a neuro-genetic algorithm. http://citeseerx.ist.psu.edu/viewdoc/summary?doi=10.1.1.135.4167
17. Rasit K (2011) A neuro-genetic approach to the inverse kinematics solution of robotic manipulators. Sci Res Essays 6(13):2784–2794
18. Wang J, Li J-H, Ni N (2011) Restricted searching area hierarchical genetic algorithm for UAV path planning. J Detect Control 33(4):39–43
19. Cheng W, Jin Q-R (2013) Optimum base station frequency allocation based on hierarchical genetic algorithms. Comput Digital Eng 41(2):168–170
20. Qi Y, Qin H-L, Shen S-T, Li Y-H (2004) A study of the optimization for fuzzy diagnostic rules based on the reformative CHC algorithm. Acta Aeronautica Et Astronautica Sinica 25(04):362–367
21. Huang W-P, Wang X-Y (2007) Structural damage diagnose is based on improved CHC algorithm. J Vib Measur Diagn 27(3):232–235
22. Goldberg DE, Korb B, Deb K (1989) Messy genetic algorithms: motivation, analysis and first results. Complex Syst 3:493–530
23. Zhang Z-Y, Xie G, Xie K-M (2005) Definition of node numbers of hidden layer of feed-forward neural network by messy genetic algorithm. J TaiYuan Univ Technol 36(4):392–394
24. Saleh T, Christopher C, William M (2006) A genetic algorithm approach to solve for multiple solutions of inverse kinematics using adaptive niching and clustering. IEEE Congr Evol Comput Sheraton Vancouver Wall Centre Hotel 1815–1822
25. Lv X-Q, Chen S-G, Lin J (2013) Adaptive genetic annealing algorithm of solving 0/1 knapsack. J Chongqing Univ Posts Telecommun (Natl Sci Ed) 25(1):138–142
26. Chen K, Hu X-G (2013) Method of relay routing based on genetic adaptive ant colony system algorithm. J Cent South Univ (Sci Technol) 44(2):572–579

27. Topping B-H, Sziveri J, Bahreinejad A (1998) Parallel processing, neural networks and genetic algorithms. Adv Eng Softw 29(10):763–786
28. Matsumura T, Nakamura M, Okech J (1998) A parallel and distributed genetic algorithm on loosely-coupled multiprocessor system. IEICE Trans Fundam Electron Commun Comput Sci 81(4): 540–546
29. Mayer MK (1999) A network parallel genetic algorithm for the one machine sequencing problem. Comput Math Appl 37(3):71–78
30. Liu G, Cao Y (2007) Performance comparison of several improved genetic algorithm. Microcomput Inf 30:190–192

Chapter 11
Sliding Mode Control for Nonlinear Discrete Time Systems with Matching Perturbations

Yang Li, Quanmin Zhu, Xueli Wu and Jianhua Zhang

Abstract This chapter considers sliding mode control of nonlinear discrete time systems with matching perturbations. The nonlinear sliding mode controller, whose parameters assure the closed-loop system stable, is designed in order to drive the state trajectories toward to a small bounded region. The controller is approximated by a polynomial equation in current control term $u(k)$ according to Taylor series expansion. The algebraic solutions can be obtained by resolving a polynomial equation in the latest control term $u(k)$. The integrated procedure provides a straightforward methodology to apply sliding mode control design technique for nonlinear systems. The simulation results are provided to illustrate the effectiveness of the proposed scheme.

Keywords Nonlinear discrete time system · Nonlinear controller · Sliding mode control · Matching perturbations

11.1 Introduction

The discrete time sliding mode control algorithm is very important when it is implemented by digital controller [1]. However, the discrete time sliding mode controller is not easily designed from the continuous counterpart through simple equivalence. So it is necessary to study the discrete time sliding mode control algorithm.

X. Wu · J. Zhang
Hebei University of Science and Technology, Shijiazhuang, 050054, China

Y. Li (✉) · X. Wu
Yanshan University, Qinhuangdao, 066004, China
e-mail: zhuchanzhuyi@126.com

Q. Zhu
University of the West of England, Coldharbour Lane, Bristol, BS16 1QY, UK

L. Liu et al. (eds.), *Applied Methods and Techniques for Mechatronic Systems*,
Lecture Notes in Control and Information Sciences 452,
DOI: 10.1007/978-3-642-36385-6_11, © Springer-Verlag Berlin Heidelberg 2014

Reference [2] proposed the necessary reaching condition of the discrete time sliding mode control system. Reference [3] proposed the sufficient and necessary reaching condition of the discrete time sliding mode control system. Reference [4] proposed the discrete reaching condition based on Lyapunov function by using equivalent form in Ref. [3]. Reference [1] proposed the "reaching law approach" which is the equivalence of the reaching condition in inequality form and defined notions of reaching condition. References [4] and [5] developed the idea of an equivalent control and the sector of sliding mode. Reference [6] proposed discrete-time equivalent controller in the prescribed boundary layer for Sampled-Data Systems. Reference [7] proposed a simple methodology for designing sliding mode control that can eliminate chattering for discrete time systems with matching perturbations. However, these methods cannot assign the desired closed-loop eigenvalues directly. Reference [8] proposed the output feedback sliding mode controller for a sampled data linear systems with matching disturbances which can assign the desired closed-loop eigenvalues directly. The applications of the sliding mode control method have been extensively studied, Refs. [9, 10] considered the application of the sliding mode control for discrete time systems.

On the other hand, a number of reports for discrete time nonlinear controller have been presented from application demand. However, it is not easy for the description of the controller nonlinearities because the lack of general modeling framework for a wide range of nonlinearities. Some researches into the nonlinear controllers have been presented, such as Ref. [11] proposed a control-oriented model to represent a wide range of nonlinear discrete-time dynamic plants, a pole placement controller is designed for providing a straightforward methodology when designing systems with nonlinear controller. Reference [12] proposed state-dependent parameter (SDP) models to deal with the nonlinearities of system states and system controllers. In all of the above methods with sliding mode control, there is no direct method to handle the nonlinearities of the discrete time system. This is the motivation to propose the new study in which a direct nonlinear sliding mode controller is proposed to handle the nonlinearities in such control system design.

With reference to some previous results, a simple design technique of nonlinear sliding mode controller for discrete time nonlinear system with matching perturbations is discussed in this chapter. The controller is designed by using the sliding mode control concept, but the controller proposed in this chapter is more general than Ref. [8]. The nonlinear controller is approximated by a polynomial equation in current control term $u(k)$ according to Taylor series expansion [13]. The algebraic solutions can be obtained by resolving a polynomial equation in the latest control term $u(k)$. It provides a straightforward method to deal with the sliding mode nonlinear controller for the discrete time nonlinear system.

The organization of this chapter is: in Sect. 11.2, the problem formulation is proposed. In Sect. 11.3, the structure of the general controller is designed to drive the state trajectories into a small region with respect to the bound of perturbations. In Sect. 11.4, the parameters of the general controller are designed according to a set of preassigned eigenvalues. In Sect. 11.5, the direct sliding mode controller is obtained by using Newton–Raphson algorithm to resolve the general controller polynomial equation. In Sect. 11.6, two examples are given to illustrate the effectiveness of the proposed procedure.

11.2 Preliminaries

The nonlinear discrete time system can be described as follows:

$$x(k+1) = Ax(k) + B(\varphi(u(k)) + v(k,x,u)), \tag{11.1}$$

where $x \in R^n$ is the state vector, $\varphi(u(k)) \in R^m$ is the nonlinear controller, $u(k)$ is the control input, where $m \leq n$, $v(k,x,u)$ is bounded matching perturbations, A and B are constant matrixes with appropriate dimensions.

Lemma 1 [8]. The sufficient discrete time sliding condition $|\overline{\sigma}_i(k+1)| < |\overline{\sigma}_i(k)|$ is held if the following inequality satisfies:

$$\sigma(k)\,\Delta\sigma(k+1) < -\frac{1}{2}(\sigma(k+1))^2, \quad \sigma(k) \neq 0, \tag{11.2}$$

where $\Delta\sigma(k+1) = \sigma(k+1) - \sigma(k)$.

Assumption 1. The norm of the matching perturbation is bounded as follows:

$$\|v(k,x,u)\|_p \leq \delta, \tag{11.3}$$

where $p = 1, 2, \infty$.

The sliding coefficient matrix $S \in R^{m\times n}$ of the sliding function is chosen such that SB is nonsingular matrix. The sliding function is designed as

$$\sigma(k) = Sx(k). \tag{11.4}$$

11.3 Design of Controller Structure

The controller structure is designed in this section. The sliding mode control idea is inspired by Ref. [8], but the nonlinear sliding mode controller is introduced in this section.

Theorem 1. Given the system equation described in (11.1), and the sliding function described in (11.4). Let $\overline{\sigma} = (SB)^{-1}\sigma = \left[\overline{\sigma}_1\,\overline{\sigma}_2\cdots\overline{\sigma}_m\right]^T$. If the sliding mode controller is designed as

$$\varphi\left(u\left(k\right)\right) = \varphi\left(u_e\left(k\right)\right) + \varphi\left(u_\sigma\left(k\right)\right), \tag{11.5}$$

where

$$\varphi\left(u_e\left(k\right)\right) = \left(SB\right)^{-1}\left(\beta S - SA\right)x\left(k\right). \tag{11.6}$$

$$\varphi\left(u_\sigma\left(k\right)\right) = \left[\,\varphi\left(u_{\sigma 1}\right)\,\varphi\left(u_{\sigma 2}\right)\,\cdots\,\varphi\left(u_{\sigma m}\right)\right]^T = -K_0\overline{\sigma}\left(k\right), \tag{11.7}$$

where β is constant, and $K_0 = \mathrm{diag}\left[\,k_{01}\;k_{02}\;\cdots\;k_{0m}\,\right]$.

Then the following cases can be obtained:
(A) The state trajectories of the controlled system whose controller is described by (11.5) is driven into the following region:

$$\Re_A = \left\{\begin{array}{l} z\left(k\right) \in R^n : \left\|z\left(k\right)\right\|_p \\ \le \dfrac{\left\|P^{-1}B\right\|_p \delta}{1-\left\|P^{-1}\left(A+B(SB)^{-1}(\beta S-SA)-BK_0(SB)^{-1}S\right)P\right\|_p} \end{array}\right\}, \tag{11.8}$$

where $z\left(k\right) = P^{-1}x\left(k\right)$, P is diagonal transformation matrix.
(B) The discrete sliding condition $\left|\overline{\sigma}_i\left(k+1\right)\right| < \left|\overline{\sigma}_i\left(k\right)\right|$ will be satisfied outside the following region:

$$\Re_B = \left\{z\left(k\right) \in R^n : \left|\sum_{j=1}^{n} l_{ij} z_j\left(k\right)\right| \le \max\left[\frac{\rho\left(k\right)}{k_{0i}}, \frac{\rho\left(k\right)}{2-k_{0i}}\right]\right\}, \tag{11.9}$$

where $0 < k_{0i} < 2$, $L = \left[l_{ij}\right] = \left(SB\right)^{-1}SP$, and $\left\|v\left(k\right) + \left(\beta - 1\right)Lz\left(k\right)\right\|_p \le \rho\left(k\right)$.

Proof. (A) The system equation with nonlinear controller described by (11.10) can be obtained by substituting (11.5) into (11.1).

$$\begin{aligned} x\left(k+1\right) = &\left(A + B\left(SB\right)^{-1}\left(\beta S - SA\right) - BK_0\left(SB\right)^{-1}S\right)x\left(k\right) \\ &+ Bv\left(k\right). \end{aligned} \tag{11.10}$$

Equation (11.10) can be transformed into the following equation by using the transformation $z\left(k\right) = P^{-1}x\left(k\right)$:

$$\begin{aligned} z\left(k+1\right) = &P^{-1}\left(A + B\left(SB\right)^{-1}\left(\beta S - SA\right) - BK_0\left(SB\right)^{-1}S\right)Pz\left(k\right) \\ &+ P^{-1}Bv\left(k\right), \end{aligned} \tag{11.11}$$

where P is the transformation matrix.

If the state trajectories satisfy:

$$\|z\|_p > \frac{\left\|P^{-1}B\right\|_p \delta}{1 - \left\|P^{-1}\left(A + B\,(SB)^{-1}\,(\beta S - SA) - BK_0\,(SB)^{-1}\,S\right)P\right\|_p}. \tag{11.12}$$

Then

$$\begin{aligned}
&\|z\,(k+1)\|_p \\
&= \left\|P^{-1}\left(A + B\,(SB)^{-1}\,(\beta S - SA) - BK_0\,(SB)^{-1}\,S\right)Pz\,(k) + P^{-1}Bv\,(k)\right\|_p \\
&\le \left(\left\|P^{-1}\left(A + B\,(SB)^{-1}\,(\beta S - SA) - BK_0\,(SB)^{-1}\,S\right)P\right\|_p \right.\\
&\quad \left. + \frac{\left\|P^{-1}B\right\|_p \delta}{\|z\|_p}\right)\|z\|_p \\
&< (\left\|P^{-1}\left(A + B\,(SB)^{-1}\,(\beta S - SA) - BK_0\,(SB)^{-1}\,S\right)P\right\|_p \\
&\quad + 1 - \left\|P^{-1}\left(A + B\,(SB)^{-1}\,(\beta S - SA) - BK_0\,(SB)^{-1}\,S\right)P\right\|_p)\,\|z\|_p \\
&\le \|z\,(k)\|_p\,.
\end{aligned} \tag{11.13}$$

Therefore, it can indicate that the system state trajectories are driven into the region as follows:

$$\Re_A = \left\{\begin{array}{l} z\,(k) \in R^n : \|z\,(k)\|_p \le \\ \dfrac{\left\|P^{-1}B\right\|_p \delta}{1 - \left\|P^{-1}\left(A + B(SB)^{-1}(\beta S - SA) - BK_0(SB)^{-1}S\right)P\right\|_p} \end{array}\right\}. \tag{11.14}$$

Which means the system state trajectories can be driven into a small closed and bounded region by the controller. The bound of the region is determined by the system parameters and the magnitude of perturbation.
(B) The following equation can be obtained by substituting (11.3) into (11.10):

$$\begin{aligned}
\Delta\sigma\,(k+1) &= \sigma\,(k+1) - \sigma\,(k) \\
&= SB\varphi\,(u_\sigma\,(k)) + SBv\,(k) + (\beta - 1)\,Sx\,(k)\,.
\end{aligned} \tag{11.15}$$

Equation (11.15) indicates that

$$\Delta\overline{\sigma} = (SB)^{-1}\Delta\sigma = \varphi(u_\sigma) + v + (\beta - 1)Lz, \tag{11.16}$$

where $\overline{\sigma} = (SB)^{-1}\sigma = \left[\overline{\sigma}_1\,\overline{\sigma}_2 \cdots \overline{\sigma}_m\right]^T$ and $L = [l_{ij}] = (SB)^{-1}\,SP$.

If the system (11.11) is stabilized, then $z\,(k)$ is bounded. Because $v(k)$ is bounded, so $\bar{v}\,(k) = v\,(k) + (\beta - 1)\,Lz\,(k)$ is bounded, too. Therefore, it gives $\|\bar{v}\,(k)\|_p \le$

$\rho(k)$. Equation (11.16) implies that

$$\Delta\overline{\sigma}_i = \varphi(u_{\sigma i}) + \bar{v}_i = -k_{oi}\overline{\sigma}_i + \bar{v}_i, \quad i = 1, 2, \ldots, m. \tag{11.17}$$

Now, suppose that the following equation is satisfied with

$$|\overline{\sigma}_i(k)| > \max\left\{\frac{\rho(k)}{k_{0i}}, \frac{\rho(k)}{2 - k_{0i}}\right\} = \frac{\rho(k)}{2 - k_{0i}}, \tag{11.18}$$

where $0 < k_{0i} < 2$.
Equation (11.18) can lead to the following two cases:

(a) $\overline{\sigma}_i > 0$. Equation (11.18) implies that $2\overline{\sigma}_i(k) - \overline{\sigma}_i(k)k_{0i} > \rho(k)$, which means that $1 > \frac{\rho(k)+\overline{\sigma}_i(k)k_{0i}}{2\overline{\sigma}_i(k)} > \frac{\bar{v}_i+\overline{\sigma}_i(k)k_{0i}}{2\overline{\sigma}_i(k)}$, that means $-1 < \frac{-k_{oi}\overline{\sigma}_i+\bar{v}_i}{2\overline{\sigma}_i}$.

(b) $\overline{\sigma}_i < 0$. Equation (11.18) implies that $-2\overline{\sigma}_i(k)+\overline{\sigma}_i(k)k_{0i} > \rho(k)$, which means that $1 > \frac{\rho(k)-\overline{\sigma}_i(k)k_{0i}}{-2\overline{\sigma}_i(k)} > \frac{\bar{v}_i-\overline{\sigma}_i(k)k_{0i}}{-2\overline{\sigma}_i(k)}$, that means $-1 < \frac{-k_{oi}\overline{\sigma}_i+\bar{v}_i}{2\overline{\sigma}_i}$.

On the other hand, Eq. (11.18) also implies $|\overline{\sigma}_i| > \frac{\rho}{k_{0i}}$, which indicates that

$$-k_{oi} + \frac{\bar{v}_i}{\overline{\sigma}_i} \leq -k_{oi} + \frac{|\bar{v}_i|}{|\overline{\sigma}_i|} \leq -k_{oi} + \frac{\rho}{|\overline{\sigma}_i|} < 0. \tag{11.19}$$

According to case (a), case (b), and (11.19), it gives

$$-1 < (-k_{oi}\overline{\sigma}_i + \bar{v}_i)/2\overline{\sigma}_i < 0. \tag{11.20}$$

According to (11.17) it gives

$$-\infty < \frac{2\overline{\sigma}_i}{\Delta\overline{\sigma}_i} < -1, \tag{11.21}$$

which also implies $\Delta\overline{\sigma}_i\overline{\sigma}_i < -\frac{1}{2}(\Delta\overline{\sigma}_i)^2$.
where $\overline{\sigma} = (SB)^{-1}Sx = Lz$. Therefore, from Lemma 1, it can be obtained that $|\overline{\sigma}_i(k+1)| < |\overline{\sigma}_i(k)|$ which also means that the sliding function $\overline{\sigma}_i(k)$ is decreasing outside $\Re_B$ when the nonlinear sliding mode controller is used.

Remark 1. It should be noted that the obtained controller in Theorem 1 is not strict sliding mode controller due to the nonlinearities. We can use the U model approach to solve it in Sect. 11.5.

Remark 2. There are two advantages of the controller. First, the upper bound of the perturbations need not be known before the controller implement. Second, the chattering phenomenon will never happen because there is no switching action in the controller. A multiple robotic manipulators system (MRMS) is composed of n robotic manipulators. An MRMS containing four robotic manipulators is shown in Fig. 11.1.

11.4 Design of Controller Parameters

The controller parameters are determined in this section. The design method has been proved by Ref. [14] for single-input systems, and has been proved by Ref. [8] for multi-input systems.

(A) Determination of the controller parameter β

If the sliding coefficient matrix S is full rank matrix and $K_0 = I$, then $n - m$ eigenvalues of the system described in Eq. (11.10) is determined by the following reduced order system:

$$x(k+1) = \left(A - B(SB)^{-1}SA\right)x(k) = \bar{A}x(k) \tag{11.22}$$

$$\sigma(k) = Sx(k) = 0. \tag{11.23}$$

And the rest m eigenvalues are $\beta - 1$.

(B) Determination of the controller parameter S

First, the eigenvector matrix W is determined. This method has been proved by Ref. [14]. Consider the following system:

$$\begin{aligned} \dot{x} &= Ax + Bu \\ \sigma &= Sx, \end{aligned} \tag{11.24}$$

where $A \in R, u \in R, y \in R$, and SB is nonsingular, the feedback system eigenvalue assignment question is that

$$(A + BK)W = WJ, \tag{11.25}$$

where K is an $m \times n$ feedback matrix and it is chosen to yield the desired closed-loop poles specified by the eigenvalues of J.

The problem of arbitrary eigenvector assignment has been tackled by Ref. [15], where it has been shown that, in general, it is possible to specify all components of any one eigenvector arbitrarily using state feedback method. In matrix form, (11.25) is equivalent to

$$AW - WJ = BL, \tag{11.26}$$

where L is an arbitrary $m \times (n - m)$ matrix and it is chosen to provide linear combinations of the columns of matrix B.

Second, the eigenvector matrix S is determined.

Let the matrix S satisfy

$$SB = X, \tag{11.27}$$

where X is an arbitrary $m \times m$ nonsingular matrix and

$$SW = 0. \tag{11.28}$$

A solution to (11.28) always exists, since B is full rank, giving the particular solution

$$S = XB^{-1}. \tag{11.29}$$

This solution also satisfies (11.28), since it is required from $B^{-1}W = 0$. A systematic method of finding B^{-1} which will always satisfy $B^{-1}W = 0$ is by constructing $\begin{bmatrix} B & W \end{bmatrix}^{-1}$. The first m rows of this inverse matrix is B^{-1} that satisfies $B^{-1}W = 0$.

11.5 Solution of the Nonlinear Controller

The nonlinear controller is solved in this section. The method has been proved by Ref. [11]. According to Taylor series (1721), analytic functions of the Taylor series at a given point are finite order functions of its Taylor's series, which completely determines the function in some neighborhood of the point. So the following polynomial equation in the current control term $u(k)$, was proposed to approximate the nonlinear controller $\varphi\,(u\,(k))$ that is described by (11.5):

$$\varphi\,(u\,(k)) = \sum_{j=0}^{M} \alpha_j\,(k)\,u^j\,(k-1) \tag{11.30}$$

The control input $u\,(k)$ can be obtained by Newton–Raphson algorithm.

Remark 4. As far as the authors know, there is almost no straightforward approach for nonlinear system control [16–22], the method of Ref. [11] provides the straightforward approach.

11.6 Algorithm for Implement

In this section, a step-by-step procedure is listed to implement the control scheme.

Step 1. Calculate the controller parameter β and sliding matrix S.

Step 2. Assign initial values of the state $x\,(1)$, the ideal state of the system and assign $i = 1$.

Step 3. Calculate the controller $\varphi\,(u\,(i))$ based on Eqs. (11.5) to (11.7) and initial values of the state $x\,(i)$.

Step 4. Calculate the controller $u\,(i)$ of (11.30) based on the Newton–Raphson algorithm.

Step 5. $i = i + 1$ go to Step 3.

This is an online algorithm for sliding model control of discrete time nonlinear dynamic system.

11.7 Simulation

Example 1. The selected discrete time nonlinear system is described as follows:

$$x(k+1) = \begin{bmatrix} 1 & 0.01 & 0 \\ 0 & 1 & 0.01 \\ -0.01 & 0.02 & 0.99 \end{bmatrix} x(k) + \begin{bmatrix} 0.01 & -0.01 \\ 0 & 0.01 \\ 0.01 & 0 \end{bmatrix} \left(v(k) + \begin{bmatrix} \varphi(u_1(k)) \\ \varphi(u_2(k)) \end{bmatrix} \right). \tag{11.31}$$

The nonlinear controller outputs are expressed in terms of (11.30):

$$\begin{bmatrix} \varphi(u_1(k)) \\ \varphi(u_2(k)) \end{bmatrix} = \begin{bmatrix} u_1^3(k) + u_1^2(k) + u_1(k) + 1 \\ u_2^3(k) + u_2^2(k) + u_2(k) + 1 \end{bmatrix}. \tag{11.32}$$

The matching perturbations of the system can be described as follows:

$$v(k) = \begin{bmatrix} x_1 x_2 - 0.2x_3 + 0.1x_1 u_1 - x_2 \cos(k) + 0.2 \sin(k) \\ x_2^2 - x_3 - 2x_3^2 + 0.1x_1 + 0.01x_2 u_2 + 0.5 \cos(k) \end{bmatrix}. \tag{11.33}$$

The eigenvalues of the discrete time nonlinear system (11.31) are assigned to {0.6,0.6,0.8}, and $\beta = 1.6$, and x_d is the ideal value of the state, and y_d are the desired output value of the system.

The sliding coefficient matrix S is obtained according to Ref. [13].

$$S = \begin{bmatrix} 2111.8 & 2111.8 & -2011.8 \\ 2117.6 & 2217.6 & -2117.6 \end{bmatrix}. \tag{11.34}$$

The equivalent control law is designed as

$$\varphi(u_e) = \begin{bmatrix} 1247 & 1286.2 & -1248.3 \\ 1249.4 & 1331.7 & -1313.9 \end{bmatrix} x. \tag{11.35}$$

And,

$$\begin{aligned} \varphi(u_\sigma) &= -\overline{\sigma}(k) = (SB)^{-1} \sigma \\ &= -\begin{bmatrix} 2111.8 & 2111.8 & -2011.8 \\ 2117.6 & 2.2176 & -2117.6 \end{bmatrix} x, \end{aligned} \tag{11.36}$$

where $x_1(1) = 1, x_2(1) = -0.5, x_3(1) = 0.5$.

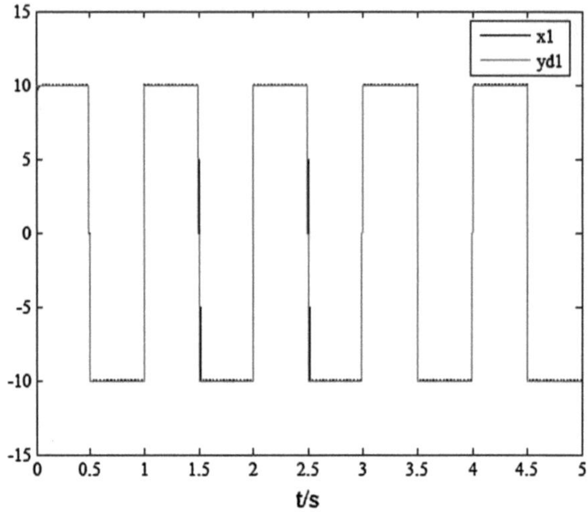

Fig. 11.1 State variable x_1 and desired output y_{d1}

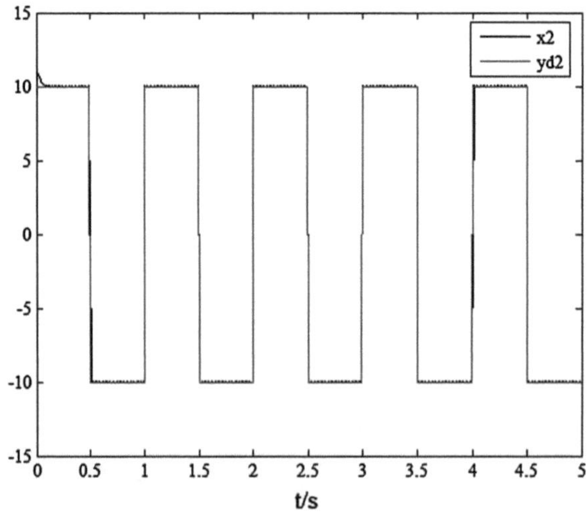

Fig. 11.2 State variable x_2 and the desired output y_{d2}

Figures 11.1, 11.2, and 11.3 show that the state trajectories are all tracked to the desired output y_d rapidly. Figures 11.4 and 11.5 show the sliding function σ also into small bounded region rapidly. Figures 11.6 and 11.7 show that the controller output trajectories obtained by Newton–Raphson algorithm.

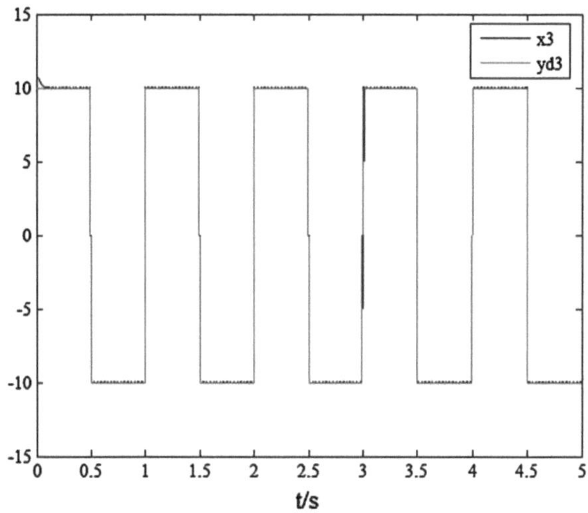

Fig. 11.3 State variable x_3 and desired output y_{d3}

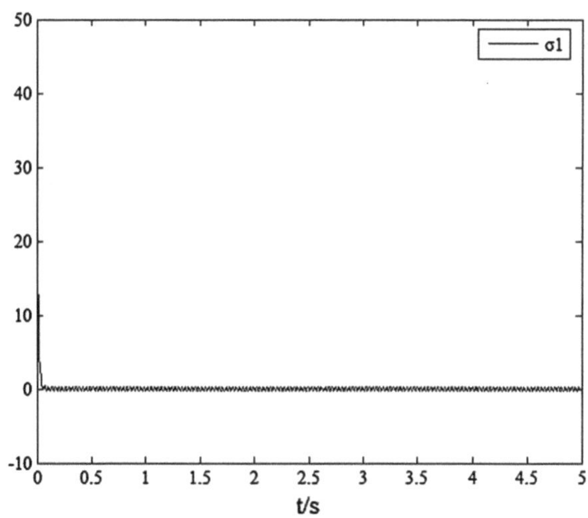

Fig. 11.4 Sliding surface σ_1

Note that the controllers proposed in Ref. [7] can be used for discrete time linear system with matching perturbations, but it does not contain controller nonlinearities. Therefore, the controller proposed in Ref. [8] cannot be used directly for those system with controller nonlinearities.

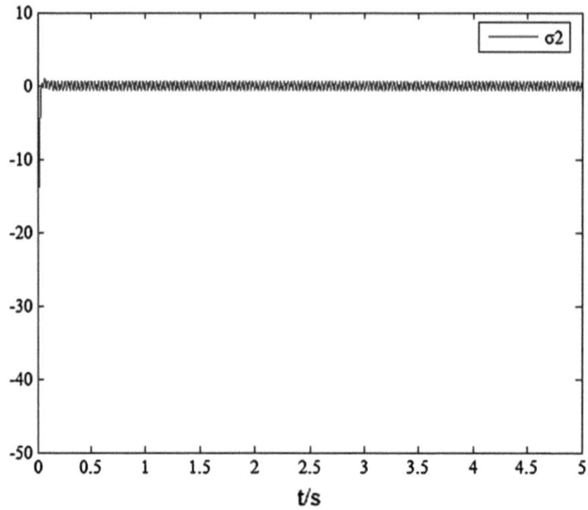

Fig. 11.5 Sliding surface σ_2

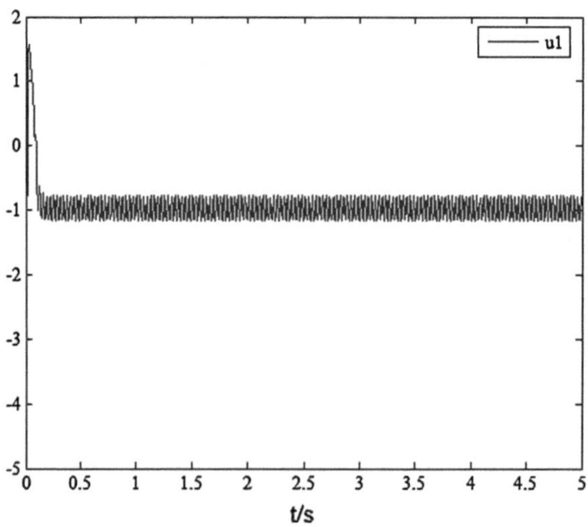

Fig. 11.6 controller output u_1

Example 2. Comparison with the PID method.

Figures 11.8, 11.9, and 11.10 show that the state trajectories are all tracked to the desired output y_d by using the PID method. The plots show very clearly that the peak overshoot and the settling time can be minimized and the system performance improves significantly by using the sliding mode controller proposed in this chapter.

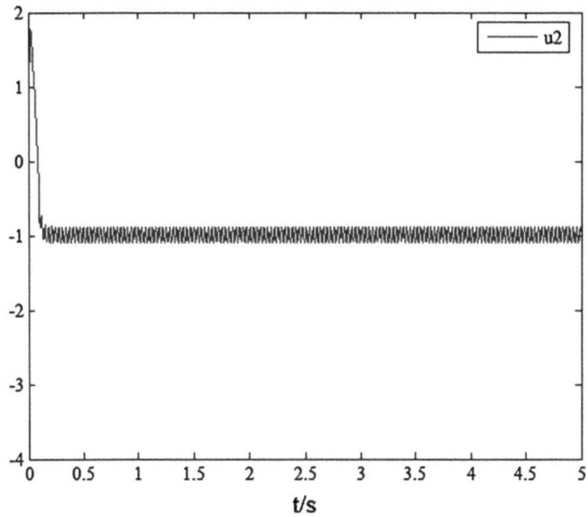

Fig. 11.7 controller output u_2

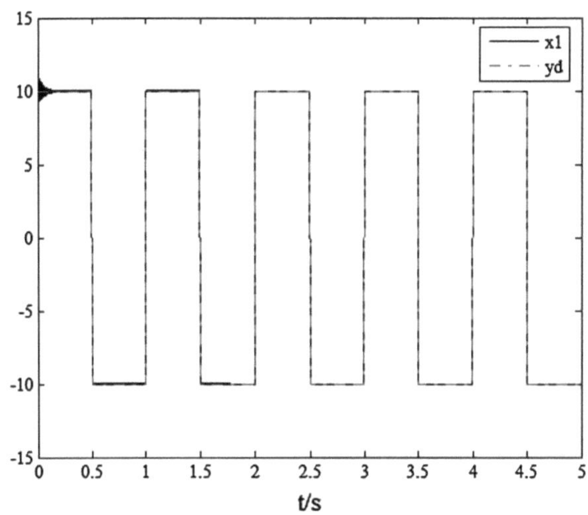

Fig. 11.8 State variable x_1 and desired output y_{d1}

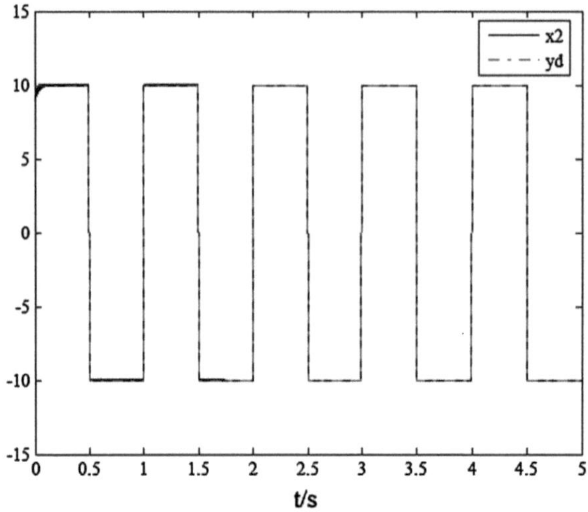

Fig. 11.9 State variable x_2 and the desired output y_{d2}

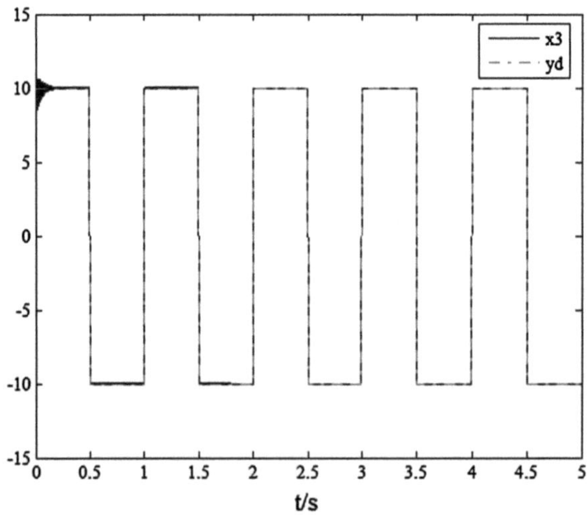

Fig. 11.10 State variable x_3 and desired output y_{d3}

11.8 Conclusions

The nonlinear sliding mode controller is designed for the discrete time nonlinear systems. The controlled system state trajectories are driven into small bound region, and the system is stable by the determination of the nonlinear controller parameters. The nonlinear controllers are represented by a polynomial equation, and the algebraic solutions can be obtained by Newton–Raphson algorithm. The method proposed in this chapter provides a straightward methodology to use sliding mode control design techniques when nonlinearities embedded in the controller. Here, only the sliding mode control of nonlinear systems is investigated, but it is strongly believed that the idea of this chapter is effective for most other classes of discrete time control systems. Further studies on the developed methodology, such as discrete time switched systems, discrete time neutral systems, time delay systems, will be conducted to provide a comprehensive framework in designing discrete time nonlinear control systems.

Acknowledgments The authors thank the support from the National Natural Science Foundation of China (Grant No. 61273188) and National Natural Science Foundation of Hebei Province (F2012208075) and Taishan Scholar Construction Engineering Special Funding. The authors, hereby, gratefully acknowledge this support.

References

1. Gao W, Wang Y, Homaifa A (1995) Discrete-time variable structure control systems. IEEE Trans Ind Electron 42(2):117–122
2. Milosavljevic D (1985) General conditions for the existence of a quasi-sliding mode on the switching hyperplane in discrete variable structure systems. Autom Remote Control 46(30):307–314
3. Sarpturk SZ, Istefanopulos Y, Kaynak O (1987) On the stability of discrete variable structure systems. IEEE Trans Autom Control 32(10):930–932
4. Furuta K (1990) Sliding mode control of a discrete system. Syst Control Lett 14:145–152
5. Wang WJ, Wu GH, Yang DC (1994) Variable structure control design for uncertain discrete time systems. IEEE Trans Autom Control 39(1):99–102
6. Utkin V, Drakunov S (1989) On discrete-time sliding mode control. In: Proceedings of IFAC symposium on nonlinear control system, Capri, Italy, pp 484–489
7. Su WC, Drakunov SV, Üzgüner Ü (2000) An $O\left(T^2\right)$ boundary layer in sliding mode for sampled-data systems. IEEE Trans Autom Control 45(3):482–485
8. Cheng C-C, Lin M-H, Hsiao J-M (2000) Sliding mode controllers design for linear discrete-time systems with matching perturbations. Automatica 36(8):1205–1211
9. Corradini ML, Orlando G (1998) Variable structure control of discretized continuous-time systems. IEEE Trans Autom Control 43:1329–1334
10. Utkin V, Gulder J, Shi J (2009) Sliding mode control in electro-mechanical systems, 2nd edn. Taylor and Francis, London
11. Zhu QM, Guo LZ (2010) A pole placement controller for non-linear dynamic plants. Proc Instn Mech Eng Part I: J Syst Control Eng 216:467–476
12. Taylora CJ, Chotaib A, Young PC (2009) Non-linear control by input-output state variable feedback pole assignment. Int J Control 82(6):1029–1044
13. Apostol TM (1963) Mathematical analysis. Addison-Wesley, Reading

14. Pan Y, Furuta K (1994) Vss control design for discrete-time system. Control-Theory Adv Technol, Part 1 10(4):669–687
15. El-Ghezawi ME, Zinober ASI, Billings SA (1983) Analysis and design of variable structure systems using a geometric approach. Int J Control 38(3):657–671
16. Zhang J, Li H, Guan X, Wu X (2010) Stability of cellular neural networks with time varying delay. Int J Model Identif Control 9(1/2):59–64
17. Wu X, Zhang J, Zhu Q (2010) A generalized procedure in designing recurrent neural network identification and control of time-varying-delayed nonlinear dynamic systems. Neurocomputing 73:1376–13833
18. Wu XL, Zhang JH, Guan XP (2010) Delay-dependent asymptotic stability of BAM neural networks with time delay. Kybernetes 39(8):1313–1321
19. Zhang J, Zhu Q, Wu X, Li Y (2013) A generalized indirect adaptive neural networks backstepping control procedure for a class of non-affine nonlinear systems with pure-feedback prototype. Neurocomputing (in press)
20. Wu X, Yang L, Zhang J (2012) Synchronisation of unified chaotic systems with uncertain parameters in finite time. Int J Model Identif Control 17(4):295–301
21. Furze JM, Zhu QM, Qiao F, Hill J (2013) Linking and implementation of fuzzy logic control to ordinate plant strategies. Int J Model Identif Control (in press)
22. Yoshimura T (2012) Adaptive sliding mode control for uncertain discrete-time systems using an improved reaching law. Int J Model Identif Control 16(4):380–391

Chapter 12
Type-2 Fuzzy Wavelet Neural Network Controller Design Based on an Adaptive Gradient Descent Method for Nonlinear Dynamic Systems

Hamidreza Abbasi, Ali Akbar Safavi and Maryam Salimifard

Abstract The integration of fuzzy systems, Wavelet theory, and neural networks has recently become a popular approach in the engineering fields for control of nonlinear systems. Therefore, the application of Fuzzy Wavelet Neural Network controllers is clearly obvious to investigators. A lot of research has been done in the control of nonlinear systems by using the models based on type-1 Fuzzy Logic Systems (FLS). However, they are regularly unable to handle uncertainties in the rules. This chapter develops a novel structure of Type-2 Fuzzy Wavelet Neural Networks (T2FWNN) to control a nonlinear system. This has been performed by invoking some of the specific advantages of wavelets, such as dynamic compatibility, compression, and step parameter adaptation along with a combination of type-2 fuzzy concepts regarding the neural networks abilities. The proposed network is constructed based on a set of TSK fuzzy rules that includes a wavelet function in the consequent part of each rule. This can provide appropriate tools on adaptation of plant output signal to follow a desired one. In this regard, the merits of utilizing wavelets and type-2 FLS simultaneously have been discussed and explored to efficiently handle the uncertainties. It is worth mentioning that the stability of the system is effectively dependent on the learning procedure and the initial values of the network parameters. Here, an adaptive gradient descent strategy is used to adjust the unknown parameters. Furthermore, the performance of the proposed T2FWNN is compared with the type-1 FLS networks. As investigated, this method has gained considerably high levels of accuracy with the reasonable number of parameters. Finally, the efficiency of the proposed approach is demonstrated via the simulation results of two nonlinear case studies.

Keywords Type-2 fuzzy neural networks · Nonlinear systems · Control · Wavelets

H. Abbasi (✉)
School of Electrical learning, Shiraz University, Shiraz, Iran
e-mail: hamid_abs62@yahoo.com

A. A. Safavi · M. Salimifard
Department of Power and Control, School of Electrical and Computer Engineering, Shiraz University, Shiraz, Iran

L. Liu et al. (eds.), *Applied Methods and Techniques for Mechatronic Systems*, Lecture Notes in Control and Information Sciences 452,
DOI: 10.1007/978-3-642-36385-6_12,

12.1 Introduction

Control is an important research and application area. It can be applied to multiple different fields, such as industrial processes, business, engineering, weather, safety issues, signal processing, and many other fields. In this regard, it is known that uncertainty can rule as an unwanted item which affects final desired signals and emerges in multiple forms. The concept of information is inherently associated with the concept of uncertainty [1, 2]. The most fundamental aspect of this relation is that the uncertainty unwantedly involved in any problem-solving situation is a result of some information deficiency, which may be imprecise, incomplete, fragmentary, vague, contradictory, or deficient in other aspects; these will lead to facing a not fully reliable system. Uncertainty is an adhesive part of information [3]. The template of fuzzy conceptual reasoning allows handling a number of these uncertainties spontaneously and therefore fuzzy systems employ type-1 fuzzy sets, which represent uncertainty by numbers in the range [0, 1] when an entity is uncertain. In this regard, consider a specific measurement procedure, it is difficult to determine the measured value exactly, and of course type-1 fuzzy sets make more sense than traditional sets subsequently [3]. On the other hand, neural networks (NNs) are considered for modeling the desired control structures. In this chapter, the integration of NN and wavelet function is proposed. Wavelet function is a waveform that has limited duration and an average value of zero. The integration of the localization properties of wavelets and the learning abilities of NN shows advantages of wavelet neural networks (WNNs) over NN in complex nonlinear system modeling. A WNN that uses wavelet functions has been proposed by researchers for solving approximation and classification problems [4]. The wavelet analysis can approximate different types of functions at different levels of resolutions. The synthesis of fuzzy wavelet neural inference system includes the finding of the optimal definitions of the premise and consequent part of fuzzy IF–THEN rules through the training capability of wavelet neural networks, evaluating the error response of the system. A combination of fuzzy technology and WNN has been considered for solving signal-processing and control problems [5]. Fuzzy systems with linear combination of the basis function are discussed in [5].

Nowadays, Fuzzy Wavelet Neural Networks (FWNN) are proposed to combine wavelet theory, fuzzy logic, and neural networks at the same time for a particular purpose. Hence, fuzzy wavelet neural systems are strikingly applied to solve control, identification, and prediction problems and also the FWNN-based controller is developed for the control of dynamic plants [6–9]. The FWNN structure that is constructed on the base of a set of fuzzy rules is proposed and used for approximating nonlinear functions [10]. The combination of wavelet network and fuzzy logic allows us to develop a system that has fast training speed and to describe nonlinear objects that are characterized with uncertainty. Wavelet transform has the ability to analyze nonstationary signals to discover their local details. Moreover, Fuzzy logic allows us to reduce the complexity of the data and to deal with uncertainties. Besides, neural networks have some self-learning characteristics that increase the accuracy of the control methods. Most of the mentioned structures are based on type-1 fuzzy systems, that is the antecedent and/or the consequent parts of the fuzzy rules which are commonly selected of type-1 fuzzy membership functions. In such cases, the type-2 fuzzy sets are employed to reduce the effects of the uncertainties in the rule base. The uncertainties in type-2 fuzzy set are deriving from different sources [11]. All in all, the amount of uncertainty in a system can be decreased by means of type-2 fuzzy logic because this method provides better capabilities to handle linguistic uncertainties by modeling vagueness and data unreliability [12].

First, the type-2 fuzzy sets introduced by Zadeh as an extension of the type-1 fuzzy sets [3]. The theoretical methods of type-2 fuzzy system and relative design rules are interpreted in [11, 13, 14]. In addition, type-2 fuzzy logic systems have spread in multiple applications such as for forecasting of time-series [14], pattern recognition [15], and robot control issues [16]. Interval type-2 fuzzy logic is vastly applied to minimize the effects of uncertainty which is produced by the instrumentation active parts, environmental noise, and other external disturbances as described in [17]. The stability of a type-2 fuzzy controller is described in [18]. In addition, there are records of some experiments showing some enhancements in analyzing the accuracy of fuzzy models of type-2 over their type-1 counterparts [19, 20]. In general, the methods for designing a type-2 fuzzy model based on experimental data can be classified into two main categories. The first class of methods assumes that an optimal type-1 fuzzy model has already been designed and afterward a type-2 fuzzy model is constructed through some augmentation of the existing model. The second category of design methods is concerned with the construction of the type-2 fuzzy model directly from experimental data. In both cases, an optimization method could help to obtain the optimal type-2 fuzzy model for the particular application.

Mamdani and TSK-type IF–THEN rules are the general methods for designing Fuzzy systems. In the former type, antecedent and either the consequent parts assigned with fuzzy values. On the other hand, in TSK-type fuzzy rules, the antecedent part is formed by fuzzy values, and consequent parts are built on crisp values or often linear functions. Many investigation works have been gaining better performance by utilizing TSK-type fuzzy neural systems rather than the Mamdani-type fuzzy neural systems in learning accuracy [21]. A recent work [22] discusses the design of a type-2 fuzzy neural network (T2FNN). The derived training algorithm in this chapter is given in [23]. Designing a recent structure of type-2 fuzzy neural network for modeling nonlinear systems is dedicated in [24]. However, this has subsequently enhanced in [25] for nonlinear system identification and nonlinear time-varying channel equalization based on clustering and gradient techniques. In other approaches, other kinds of membership functions such as type-2 Gaussians containing fixed standard deviations and hesitant means have been employed in the antecedent part, and the weighting interval sets with unified membership rating, i.e., an interval type-1 fuzzy set is used in the consequent parts of these systems [22]. As mentioned above, uncertainties can be interpreted as the dynamic nature of the plants as of its relevant coefficients and can affect them in some unwanted cases. Hence, to reduce the uncertainties effects, here we proposed to design a type-2 TSK Fuzzy system based on wavelet neural network system structure (T2FWNN) which is applied for control application of nonlinear dynamic plants

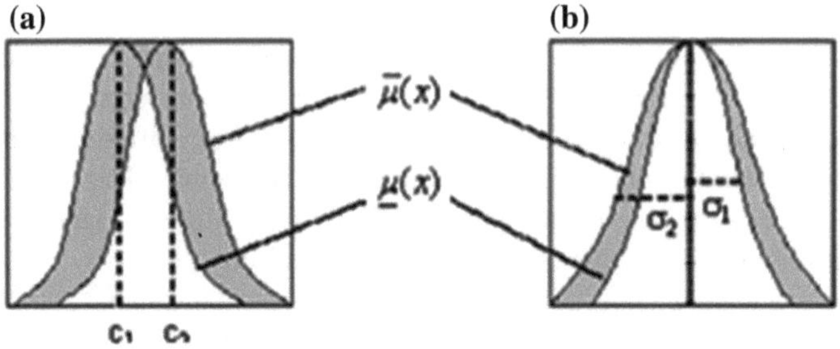

Fig. 12.1 Gaussian type-2 fuzzy set with (**a**) uncertain mean and (**b**) uncertain STD [26]

The chapter is organized as follows: In the following section, type-2 fuzzy sets and the inference mechanism of type-2 fuzzy TSK systems are briefly introduced. In Sect. 12.3, the structure of the proposed system is presented. In Sect. 12.4, the parameter update rule of type-2 FWNN based on gradient learning algorithm is derived. Simulation studies for the control approach of sample dynamic plant are carried out in Sect. 12.5. Finally, the conclusions are drawn in the last section.

12.2 Type-2 Fuzzy Systems

Type-2 fuzzy systems are intrinsically determined by fuzzy IF–THEN rules. Consequently, defined parameters in the antecedent and the consequent parts of the fuzzy rules consist of type-2 fuzzy membership functions.

Uncertainties could be classified into the STD and the mean in the Gaussian type-2 fuzzy sets. Gaussian type-2 fuzzy sets with uncertain mean and uncertain STD are shown in Fig. 12.1a and b, respectively. Relative mathematics equations for the membership function are expressed as follows:

$$\mu(x) = \exp\left(-\frac{1}{2}\frac{(x-c)^2}{\sigma^2}\right), \tag{12.1}$$

where c and σ are the center and the widths of the membership function, x is the input vector. Generally, both c and σ may be uncertain in the intervals $\sigma \in [\sigma 1, \sigma 2]$ and $c \in [c1, c2]$. In this chapter, the uncertainty in mean values is just considered in the membership function and no uncertainty is involved by STD, i.e., $c \in [c1, c2]$. The kernel of a type-2 fuzzy inference system is the fuzzy knowledge base. The relations between the input and output data points in a fuzzy knowledge base are expressed by fuzzy rules which have the IF–THEN form. A symbolic fuzzy rule is illustrated in the following format; here m and n are the number of inputs and outputs, respectively:

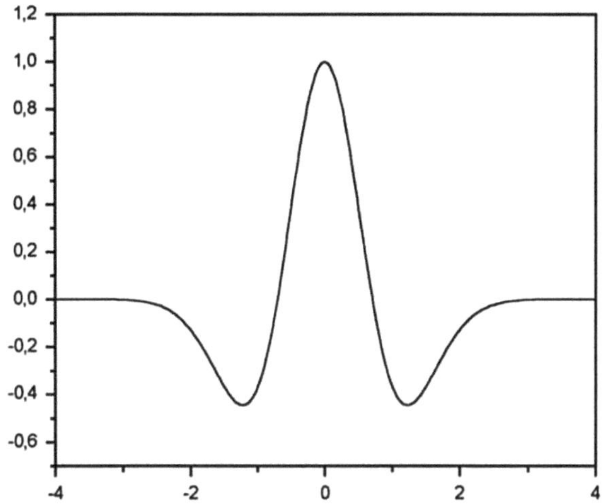

Fig. 12.2 The Mexican hat wavelet function

$$if\, x_1\, is\, A_1^l\, and\, X_2\, is\, A_2^l\, and \ldots and\, x_m\, is\, A_m^l$$
$$Then\, y_1\, is\, B_1^l\, and\, y_2\, is\, B_2^l\, and \ldots\, and\, y_n\, is\, B_n^l, \quad (12.2)$$

where x_i and $y_j(i = 1, \ldots, mj = 1, \ldots, n)$ are the input and output signals. A_i^l and B_j^l are the membership functions of the antecedent and consequent parts which are related to the *i*th input and *j*th output. It is worth mentioning that the antecedent and the consequent parts of a type-2 fuzzy system can be type-2 fuzzy sets or only one side can involve type-2 sets. In many chapters, the membership functions in the consequent part are taken as type 1 fuzzy sets. The design of fuzzy systems based on the fuzzy rules of (12.2) is given in [16].

The rules used in this chapter are such that the consequent parts are of TSK type, as indicated in the following:

$$If\, x_1\, is\, \tilde{A}_{j1}\, and\, x_2\, is\, \tilde{A}_{j2}\, and \ldots and\, x_m\, is\, \tilde{A}_{jm}$$
$$Then\, y_j = \sum_{j=1}^{n} w_{ji}(1 - z_{ji}^2)e^{-\frac{z_{ji}^2}{2}}, \quad (12.3)$$

where $x_1, x_2, \ldots, x_m$ are the input variables, $y_j(j = 1, \ldots, n)$ is the output variable, $\tilde{A}_{ji}$ is type-2 membership function for the *j*th rule of the *i*th input assumed as a Gaussian membership function. The concluding parts of the rules contain Mexican Hat WFs (Fig. 12.2). The use of wavelets with different dilation and translation values allows us to capture different behaviors and the essential features of the nonlinear model under these fuzzy rules. The proper fuzzy model that is described by the set of IF–THEN rules can be obtained by learning the dilation and translation parameters of the conclusion parts and the parameters of the membership function of the premise parts. Here, because of the use of wavelets, the computational strength and the generalization ability of T2FWNN are improved, and T2FWNN can describe the nonlinear process with an optimized accuracy.

It is to be noted that the use of interval type-2 fuzzy sets in the consequent parts of (12.3) increases the number of parameters to be updated with a possible increase in the approximation capability of the system. In the simulation section, a comparison of the performance of the type-2 FWNN used in this chapter and the other structures seen in the literature is given.

Actually, the antecedent parts of the fuzzy rules divide the input space into a set of fuzzy regions, and the consequent parts of the rules describe the system behavior in those regions, it could be assumed as the problem domain. Here, the major problem is to determining a certain category of data from a large dataset in the antecedent part such that a concise representation of the systems behavior is produced in the consequent part.

Assuming the antecedent uncertainties, the output of the type-2 fuzzy rules will have uncertainties consequently. Here, we assume the interval type-2 sets are used in the antecedents. The membership functions of the antecedent part are included from an upper and lower membership functions. They are denoted as $\bar{\mu}(x)$ and $\underline{\mu}(x)$, or $\bar{A}(x)$ and $\underline{A}(x)$, i.e.,

$$\mu_{\tilde{A}_k^i}(x_k) = [\underline{\mu}_{\tilde{A}_k^i}(x_k), \bar{\mu}_{\tilde{A}_k^i}(x_k)] = [\underline{\mu}^i, \bar{\mu}^i]. \tag{12.4}$$

$$y_j = w_j \psi_{j_m}(z). \tag{12.5}$$

$$\psi j(z) = \sum_{i=1}^{m} |a_{ij}|^{-\frac{1}{2}} (1 - z_{ij}^2) e^{-\frac{z_{ij}^2}{2}}. \tag{12.6}$$

Here, $z_{ij} = \frac{(x_i - b_{ij})}{a_{ij}}$ and a_{ij} and b_{ij} are the parameters of the WF between the$i$th input (i = 1,…, m) and the jth output (j = 1,…, n) of the wavelet.

The best choices for the implication operator of inference engine are "min" or "prod" t-norms. In this chapter, the second one is proposed to calculate the firing strengths as shown below:

$$\bar{f} = \bar{\mu}_{\tilde{A}_1}(x_1) * \ldots . * \bar{\mu}_{\tilde{A}_n}(x_n)$$

$$\underline{f} = \underline{\mu}_{\tilde{A}_1}(x_1) * \ldots . * \underline{\mu}_{\tilde{A}_n}(x_n). \tag{12.7}$$

Then, it is time to reduce the type and defuzzify the results. By employing equations in (12.5), the firing strengths of rules are determined. Afterward, the defuzzified output of the type-2 TSK fuzzy system is calculated. The inference engine of type-2 TSK FNS is assumed in [27]. The proposed inference engine given in [28] is used to determine the output of type-2 TSK FWNN, i.e.,

$$u = \frac{q\sum_{j=i}^{N} \underline{f_i} y_j}{\sum_{j=1}^{N} \underline{f_j}} + \frac{(1-q)\sum_{j=1}^{N} \bar{f_j} y_j}{\sum_{j=1}^{N} \bar{f_j}}, \quad (12.8)$$

where N is number of active rules, y_j is determined using (12.5), and the parameter q is a design factor that enables to adjust the lower or upper portions in the final output, depending on the level of certainty of the system. For a more detailed description of type-2 fuzzy systems, the reader is referred to [27]. In Sect. 12.3, the structure of the type-2 FWNN proposed in this chapter is described.

12.3 Structure of Type-2 FWNN

Wavelets are defined in the following form:

$$\Psi_j(x) = |a_j|^{-\frac{1}{2}} \psi\left(\frac{x - b_j}{a_j}\right) \qquad a_j \neq 0, \quad (12.9)$$

where $\Psi_j(x)$ represents the family of wavelets obtained from the single $\psi(x)$ function by dilations and translations, where $a_j = \{a_{1j}, a_{2j}, ..., a_{mj}\}$ and $b_j = \{b_{1j}, b_{2j}, ..., b_{mj}\}$ are the dilation and translation parameters, respectively. $x = \{x_1, x_2, \ldots, x_m\}$ are the input signals, $j = 1, \ldots, n$. $\psi(x)$ is localized in both time space and frequency space and is called a mother wavelet. Wavelet networks include WFs in the neurons of the hidden layer of the network. The output of WNN is calculated as

$$y = \sum_{j=1}^{k} w_j \Psi_j(x) = \sum_{j=1}^{k} w_j |a_j|^{-\frac{1}{2}} \psi\left(\frac{x - b_j}{a_j}\right), \quad (12.10)$$

where $\Psi_j(x)$ is the WF of the j th unit of the hidden layer, w_j is weight coefficients between the input and the hidden layers, a_j and b_j are the parameters of WF as described above. As it is obvious to researchers, WNN has good generalization abilities, can be easily trained than other networks, and therefore can approximate complex functions to some precision very compactly, such as multilayer perceptrons and radial-based networks [4, 10]. One of the important issues in the FWNNs and WNN is the initialization of the parameters of the networks which can lead to obtain fast output convergence. A number of methods are discussed in the literature for the initialization of wavelets, such as the orthogonal least square procedure [4]. An optimal initial choice of the dilation and translation parameters of the wavelet increase the training speed and result in fast convergence. The approximation and convergence properties of WNN are presented in [29].

In this chapter, the use of wavelet (rather than linear) functions are proposed to improve the computational power of the Type-2 fuzzy-neuro networks.

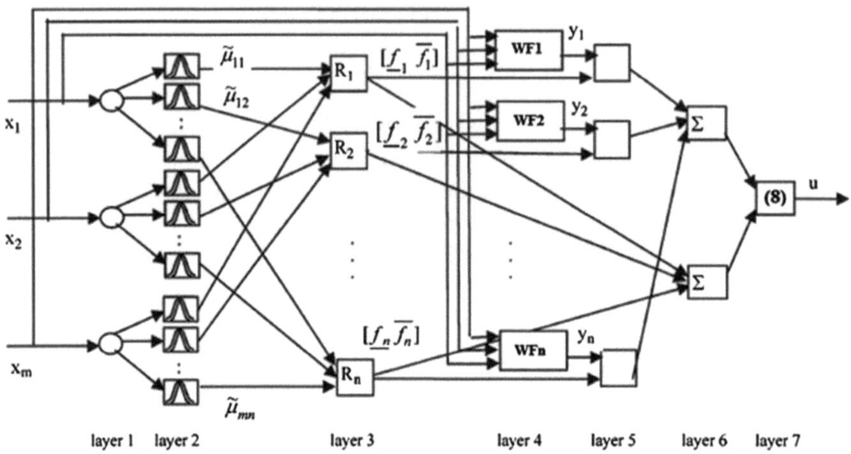

Fig. 12.3 Structure of type-2 TSK fuzzy wavelet neural network system with m inputs, n rules, and one final output u

The structure of the proposed multiple-input–single-output type-2 FWNN which is used in this chapter is shown in Fig. 12.3. In the current structure, network input signals are $X = \{x_1, x_2, \ldots, x_m\}$. As mentioned before, type-2 TSK fuzzy rules shown in (12.3) are used here. The development of the type-2 FWNN includes the determination of the structure and the proper values of the unknown parameters of the antecedent and the consequent parts of each rule. In this chapter, we assume the design of a type-2 FWNN when the membership functions of the antecedent part are Gaussian as in (12.1) and the membership functions of the consequent are WF. Under the uncertain considerations for both (c and σ) parameters of the Gaussian function (within certain intervals), the parameter space of the system would be expanded to a very large scale. In this chapter, both these parameters are considered to be uncertain and the learning rules are derived for both items. It means that the STD and mean parameters are both uncertain.

Assume that the parameters of membership functions are represented by uncertain mean and STD. In the first layer of Fig. 12.3, the input signals are inserted. In the second layer, each node corresponds to one linguistic term. Here, for input signals, the membership degrees $\underline{\mu}$ and $\bar{\mu}$ in which the input value belongs to a fuzzy set are determined using Eq. (12.1). The third layer performs the product operation according to (12.7). The fourth layer determines the outputs of the wavelet functions $y_j (j = 1, \ldots, n)$ (shown as WF in the blocks) using (12.10) in the consequent parts. In the fifth, sixth, and seventh layers, the type reduction and the defuzzification operations are made to calculate the output of the whole network. In this layer, the contribution of each wavelet to the output of the T2FWNN is determined. The product of membership degrees $\underline{f}$ and $\bar{f}$ and wavelet functions, y_j calculated in layer 5.

Two summation parts are considered in Layer 6. The calculation process of the sum of the output signals of layer 5 executed in one of these blocks [the numerator of

(12.8)], and the sum of the output signals of layer 4 calculated in the other block [the denominator of (12.8)]. By applying the Eq. (12.8), the whole output of the network will be calculated in Layer 7.

The design of the type-2 FWNN includes the determination of the structure and the proper values of the unknown coefficients in the antecedent and consequent parts of the fuzzy IF–THEN rules. Structure selection includes the determination of the number of fuzzy rules (layer 3) from the training data and the number of fuzzy sets (layer 2) for each of the input variables. These two parameters are closely dependent. That is, the selection of the value of the first parameter depends on the selection the value of the latter one. There are a number of approaches for structure selection. One of these approaches is the Interval Fuzzy type 2.

The number of parameters N to be updated in the proposed T2FWNN structure is given by

N = Number of parameters of the Type2 Gaussians
+ Number of parameters of the wavelets
+ Number of weights (w)

The constructed T2FWNN structure has the following advantages rather than other works: The T2FWNN is constructed on the basis of fuzzy rules that include WFs in the consequent parts of the rules. A wavelet is a nonlinear function of input linguistic variables that have local property. The use of such a structure allows the T2FWNN to approximate complex functions more effectively. In addition, the T2FWNN model has an acceptable network size and faster learning speed than other modeling structures.

12.4 Parameter Update Rules for Learning

After the design of the antecedent parts by using interval fuzzy type 2, an adaptive gradient descent strategy is used to adjust the unknown parameters. This algorithm implicitly implies that an adaptive learning rate is applied to design the consequent parts of the fuzzy rules. The latter guarantees the convergence and speeds up the learning of the network. In what follows, the parameter update rules are derived for a type-2 TSK FWNN with uncertain mean and STD and both will be updated. In the first stage, the output error (cost function) is assessed as follows:

$$E = \frac{1}{2}\sum_{i=1}^{0}\left(u_i^d - u_i\right)^2. \tag{12.11}$$

Here, u_i^d, and u_i are the desired and the current output values of the network, respectively and O would be considered as the number of output signals of the network (in our case, $O = 1$). Besides, the parameters $w_i, b_{ij}, a_{ij}, c1_{ij}, c2_{ij}$ and $\sigma_{ij}(i = 1, \ldots, mj = 1, \ldots, n)$ should be updated and must to be adjusted by means of the following formulas:

$$w_j(t+1) = w_j(t) - \gamma \frac{\partial E}{\partial w_j} + \lambda\left(w_j(t) - w_j(t-1)\right)$$
$$a_{ij}(t+1) = a_{ij}(t) - \gamma \frac{\partial E}{\partial a_{ij}} + \lambda\left(a_{ij}(t) - a_{ij}(t-1)\right)$$
$$b_{ij}(t+1) = b_{ij}(t) - \gamma \frac{\partial E}{\partial b_{ij}} + \lambda\left(b_{ij}(t) - b_{ij}(t-1)\right). \quad (12.12)$$

$$c1_{ij}(t+1) = c1_{ij}(t) - \gamma \frac{\partial E}{\partial \underline{c}_{ij}}$$

$$c2_{ij}(t+1) = c2_{ij}(t) - \gamma \frac{\partial E}{\partial \bar{c}_{ij}}$$
$$\sigma_{ij}(t+1) = \sigma_{ij}(t) - \gamma \frac{\partial E}{\partial \sigma_{ij}}. \quad (12.13)$$

where γ is the learning rate, λ is the momentum, which is used to speedup the learning process. m is the number of input signals of the network (input neurons), and n is the number of rules (the hidden neurons). The initial values are generated randomly.

The values of derivatives in (12.12) and (12.13) can be calculated by the following formulas:

$$\frac{\partial E}{\partial w_j} = \frac{\partial E}{\partial u}\frac{\partial u}{\partial y_j}\frac{\partial y_j}{\partial w_j}$$
$$= \left(u(t) - u^d(t)\right) \times \left(\frac{q \times \underline{f_i}}{\sum_{j=n}^{m} \underline{f_i}} + \frac{(1-q) \times \bar{f_i}}{\sum_{j=n}^{m} \bar{f_i}}\right) \times \sum_{j=1}^{n} \Psi_j(x). \quad (12.14)$$

$$\delta_j = \frac{\partial E}{\partial u}\frac{\partial u}{\partial y_j}\frac{\partial y_j}{\partial \Psi_j} = \left(u(t) - u^d(t)\right) \cdot \left(\frac{q \times \underline{f_i}}{\sum_{j=n}^{m} \underline{f_i}} + \frac{(1-q) \times \bar{f_i}}{\sum_{j=n}^{m} \bar{f_i}}\right) .w_i. \quad (12.15)$$

$$\frac{\partial E}{\partial a_{ij}} = \frac{\partial E}{\partial u}\frac{\partial u}{\partial y_i}\frac{\partial y_i}{\partial \psi_i}\frac{\partial \psi_i}{\partial z_{ij}}\frac{\partial z_{ij}}{\partial a_{ij}} = \delta_j(3z_{ij}^1 - z_{ij}^4)e^{-\frac{z_{ij}^2}{z}} / \left(\sqrt{a_{ij}^3}\right)$$
$$\frac{\partial E}{\partial b_{ij}} = \frac{\partial E}{\partial u}\frac{\partial u}{\partial y_i}\frac{\partial y_i}{\partial \psi_i}\frac{\partial \psi_i}{\partial z_{ij}}\frac{\partial z_{ij}}{\partial b_{ij}} = \delta_j(3z_{ij} - z_{ij}^2)e^{-\frac{z_{ij}^2}{z}} / \left(\sqrt{a_{ij}^2}\right). \quad (12.16)$$
$$\frac{\partial E}{\partial a_{ij}} = \frac{\partial E}{\partial u}\left(\frac{\partial u}{\partial \underline{f_i}}\frac{\partial \underline{f_i}}{\partial \underline{\mu}_{ij}}\frac{\partial \underline{\mu}_{ij}}{\partial \sigma_{ij}} + \frac{\partial u}{\partial \bar{f_i}}\frac{\partial \bar{f_i}}{\partial \bar{\mu}_{ij}}\frac{\partial \bar{\mu}_{ij}}{\partial \sigma_{ij}}\right). \quad (12.17)$$

$$\frac{\partial E}{\partial c1_{ij}} = \frac{\partial E}{\partial u}\left(\frac{\partial u}{\partial \underline{f_i}}\frac{\partial \underline{f_i}}{\partial \underline{\mu}_{ij}}\frac{\partial \underline{\mu}_{ij}}{\partial c1_{ij}} + \frac{\partial u}{\partial \bar{f_i}}\frac{\partial \bar{f_i}}{\partial \bar{\mu}_{ij}}\frac{\partial \bar{\mu}_{ij}}{\partial c1_{ij}}\right). \tag{12.18}$$

$$\frac{\partial E}{\partial c2_{ij}} = \frac{\partial E}{\partial u}\left(\frac{\partial u}{\partial \underline{f_j}}\frac{\partial \underline{f_j}}{\partial \underline{\mu}_{ij}}\frac{\partial \underline{\mu}_{ij}}{\partial c2_{ij}} + \frac{\partial u}{\partial \bar{f_i}}\frac{\partial \bar{f_j}}{\partial \bar{\mu}_{ij}}\frac{\partial \bar{\mu}_{ij}}{\partial c2_{ij}}\right). \tag{12.19}$$

Here,

$$\frac{\partial E}{\partial u} = (u(t) - u^d(t)), \quad \frac{\partial u}{\partial \underline{f_j}} = \frac{q \times (y_j - \underline{u})}{\sum_{j=1}^{n} \underline{f_j}}, \qquad \frac{\partial u}{\partial \bar{f_j}} = \frac{(1-q) \times (y_j - \bar{u})}{\sum_{j=1}^{n} \bar{f_j}}$$

$$\underline{u} = \frac{\sum_{j=1}^{n} \underline{f_j} y_j}{\sum_{j=1}^{n} \underline{f_j}}, \qquad \bar{u} = \frac{\sum_{j=1}^{n} \bar{f_j} y_j}{\sum_{j=1}^{n} \bar{f_j}}. \tag{12.20}$$

Then, by employing t-norm "prod" operator:

$$\frac{\partial \underline{f_j}}{\partial \underline{\mu}_{ij}} = \prod_{\substack{k=1\\k=i}}^{N1} \underline{\mu}_{kj}, \qquad \frac{\partial \bar{f_j}}{\partial \bar{\mu}_{ij}} = \prod_{\substack{k=1\\k=i}}^{N1} \bar{\mu}_{kj}, \tag{12.21}$$

where $i = 1, \ldots, \mathrm{N1}$, $\mathrm{k} = 1, \ldots, \mathrm{N1}$, and $j = 1, \ldots, \mathrm{N2}$.

Upper and lower membership functions between the ith input and the jth hidden neurons of layer 3 can be written as follows (Fig. 12.1a).

First, $G(c_{ij}, \sigma_{ij}, x_i)$ define as:

$$G(c_{ij}, \sigma_{ij}, x_i) = \exp\left(-\frac{1}{2}\frac{(x_i - c_{ij})^2}{\sigma_{ij}^2}\right). \tag{12.22}$$

$$\underline{\mu}_{ij}(x) = \begin{cases} G(c2_{ij}, \sigma_{ij}, x_i), & x_i \le \dfrac{c1_{ij} + c2_{ij}}{2} \\ G(c1_{ij}, \sigma_{ij}, x_i), & x_i > \dfrac{c1_{ij} + c2_{ij}}{2} \end{cases}. \tag{12.23}$$

$$\bar{\mu}_{ij}(x) = \begin{cases} G(c1_{ij}, \sigma_{ij}, x_i), & x_i < c1_{ij} \\ 1 & c1_{ij} < x_i < c2_{ij} \\ G(c1_{ij}, \sigma_{ij}, x_i), & x_i < c2_{ij} \end{cases}. \tag{12.24}$$

Then,

$$\frac{\partial \bar{\mu}_j(x_i)}{\partial c1_{ij}} = \begin{cases} G(c1_{ij}, \sigma_{ij}, x_i)\dfrac{(x_i - c1_{ij})}{\sigma_{ij}^2}, & x_i < c1_{ij} \\ 0, & c1_{ij} < x_i < c2_{ij} \\ 0, & x_i > c2_{ij} \end{cases}. \tag{12.25}$$

$$\frac{\partial \underline{\mu}_j(x_i)}{\partial c1_{ij}} = \begin{cases} 0, & x_i \geq \dfrac{c1_{ij} + c2_{ij}}{2} \\ G(c1_{ij}, \sigma_{ij}, x_i)\dfrac{(x_i - c1_{ij})}{\sigma_{ij}^2}, & x_i > \dfrac{c1_{ij} + c2_{ij}}{2} \end{cases}. \tag{12.26}$$

$$\frac{\partial \bar{\mu}_j(x_i)}{\partial c2_{ij}} = \begin{cases} 0, & x_i < c1_{ij} \\ 0, & c1_{ij} < x_i < c2_{ij} \\ G(c2_{ij}, \sigma_{ij}, x_i)\dfrac{(x_i - c2_{ij})}{\sigma_{ij}^2} & x_i > c2_{ij} \end{cases}. \tag{12.27}$$

$$\frac{\partial \underline{\mu}_j(x_i)}{\partial c2_{ij}} = \begin{cases} G(c2_{ij}, \sigma_{ij}, x_i)\frac{(x_i - c2_{ij})}{\sigma_{ij}^2}, & x_i \leq \frac{c1_{ij}+c2_{ij}}{2}, \\ 0 & x_i \geq \frac{c1_{ij}+c2_{ij}}{2} \end{cases} \tag{12.28}$$

$$\frac{\partial \underline{\mu}_j(x_i)}{\partial c2_{ij}} = \begin{cases} G(c1_{ij}, \sigma_{ij}, x_i)\frac{(x_i - c1_{ij}^2)}{\sigma_{ij}^3}, & x_i < c1_{ij} \\ 0, & c1_{ij} < x_i < c2_{ij} \\ G(c2_{ij}, \sigma_{ij}, x_i)\frac{(x_i - c2_{ij}^2)}{\sigma_{ij}^3}, & x_i < c2_{ij} \end{cases}. \tag{12.29}$$

$$\frac{\partial \underline{\mu}_j(x_i)}{\partial \sigma_{ij}} = \begin{cases} G(c2_{ij}, \sigma_{ij}, x_i)\frac{(x_i - c2_{ij})^2}{\sigma_{ij}^3}, & x_i \leq \frac{c1_{ij}+c2_{ij}}{2} \\ G(c1_{ij}, \sigma_{ij}, x_i)\frac{(x_i - c1_{ij})^2}{\sigma_{ij}^3}, & x_i > \frac{c1_{ij}+c2_{ij}}{2} \end{cases}. \tag{12.30}$$

The parameters of the type-2 FWNN can thus be updated using (12.12) and (12.13) considering Eqs. (12.14–12.30). Also, the parameter q in (12.8) is to adjust the lower or upper portions in the final network output. In addition, the value of q should be optimized using an adequate learning rule from an initial value (e.g., 0.5);

$$q(t+1) = q(t) - \gamma \frac{\partial E}{\partial q}. \tag{12.31}$$

$$\frac{\partial E}{\partial q} = (u(t) - u^d(t)).\left(\frac{\underline{f}_j}{\sum_{j=1}^m \underline{f}_j} + \frac{\bar{f}_j}{\sum_{ij}^m \bar{f}_j}\right). \tag{12.32}$$

Updating of the parameters of the T2FWNN is carried out using Eqs. (12.11–12.32).

In all learning algorithms a major problem is convergence. While we employed the gradient descent method, the convergence depends on the suitable initial values of the learning rate and the momentum term. In most neural networks problems, the best selection is in the interval [0,1]. An unstable learning could be achieved by choosing a large value of the learning rate. On the other hand, a slow learning speed undoubtedly would be expected while using small values for the learning rate. In this chapter, the parameter updating procedure is performed by means of an intelligent adaptive approach. At the beginning, a small value of the learning rate γ is considered to start the parameter learning of the type-2 FWNN system. Then, based on the change of error $\Delta E = E(t) - E(t + 1)$, γ will be increased or decreased if the ΔE is positive or negative respectively, while learning. A small value of γ may slow down the learning process and a large value of γ may cause the learning process to a fluctuating condition. Moreover, it could lead to instability. So the γ changes regarding the value of ΔE. This strategy leads to stable learning of the T2FWNN, and the convergence of the system would be guaranteed and speeds up by means of the presented learning procedure. In addition, a momentum term (λ) is used to speedup the learning processes. A type-2 fuzzy system is more accurate in the modeling of systems rather than a type 1 fuzzy system. An increase in the computational load (as compared to type 1) may not be the case as a type-2 FWNN may be able to describe the process using less fuzzy rules than a type 1 fuzzy system. This is demonstrated in Sect. 12.5.

12.5 Simulation Studies

In this section, the performance of the described T2FWNN structure is evaluated by two simulation studies for control applications; the dynamic plant models are directly selected from the literature in order to enable us to make a direct performance comparison.

The plant model implicit uncertainties cannot properly be described by deterministic models, and therefore, regular control approaches based on such models are unable to result in the desired performance. Thus, soft computing methodologies can be helpful as an alternative choice. Considering the inherent characteristics of the dynamic industrial plants and the effect of disturbances on such plants, the uncertainties of the plant coefficients could be considered as the dynamic character of the plants. It is known that, these types of uncertainties could be handled by employing type-2 fuzzy sets. Furthermore, the overall type-2 FWNN structure described in Sect. 12.3 is used as the adaptive controller for the control of plants. In the following, the simulation results are presented and contrasted with those procured using other methods proposed in the literature.

In the following, we will discuss the performance of the proposed type-2 FWNN for control approaches and therefore investigate its benefits. The scheme of the Type-2 FWNN-based control system is shown in Fig. 12.4. Here, g(k) is the input set-point

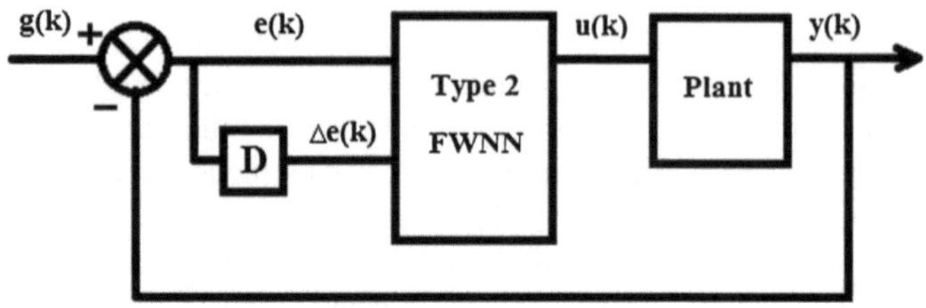

Fig. 12.4 Structure of Type-2 FWNN-based control system

signal, y(k) is the output signal of the plant, e(k), Δe(k) are the Type-2 FWNN inputs which are called error and the change in error, respectively. Operator D is used to demonstrate the differentiation operation. In the current investigation, the error and change in error are the inputs of type-2 FWNN controller. Parameter learning procedure of the proposed control system is carried out by using gradient descent algorithm in a closed-loop manner. Consequently, the IF–THEN rules of the controller are generated. Finally, the controller output signal is to be applied to the plant.

In the first example, the discussed T2FWNN scheme is employed to control a proposed dynamic plant which is described by the following difference equation:

$$y(k) = \frac{y(k-1)y(k-2)(y(k-1)+2.5)}{1+y(k-1)^2+y(k-2)^2} + u(k), \tag{12.33}$$

where y(k), y(k-1), and y(k-2) are the current, one-step, and two-step delayed outputs of the plant, respectively, and u(k) is the output type-2 FWNN controller. The following reference signal is used as set point for the control system:

$$g(k) = \begin{cases} 10, & 0 < k \le 50 \\ 15, & 50 < k \le 100 \\ 10, & 100 < k \le 150 \\ 15, & 150 < k \le 200 \end{cases}. \tag{12.34}$$

The initial values of the parameters of the proposed type-2 FWNN, i.e., the parameters of the antecedent membership functions in the second layer (c1, c2, and σ) and the parameters of the consequent part (a, b, and w) are generated randomly in the interval [−10, 10].

An adaptive approach is used for updating these parameters according to the parameter updating rules derived before in (12.11–12.32). Hence, learning of the parameter values of type-2 FWNN is carried out for the given input signals. In addition, as a suitable performance criterion the following RMSE is used:

$$RMSE = \sqrt{\frac{1}{k}\sum_{i=1}^{k}(y(k) - y_N(k))^2}. \tag{12.35}$$

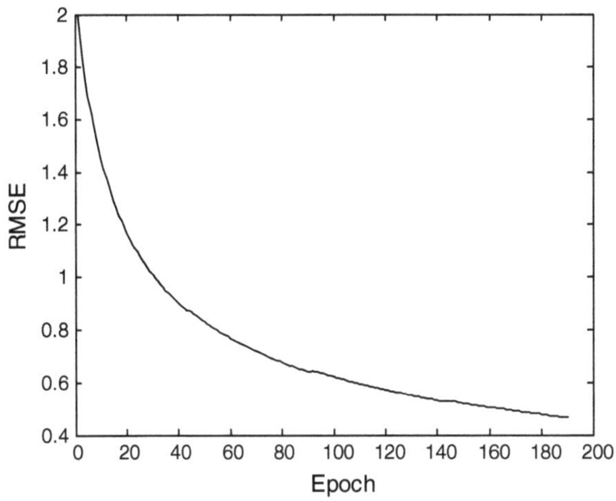

Fig. 12.5 RMSE values while learning

Here, $y_N(k)$ is the output of the type-2 FWNN and k is equal to 200, it is equal to say that the number of data points used for training is 200. The training of the type-2 FWNN system with three rules (33 parameters) is performed for 200 data points. Figure 12.5 depicts the relative RMSE curve. In Fig. 12.6, the time response characteristic of the control system is shown. For more detail, it is worth mentioning that the current structure has 33 parameters to be updated. The best and the mean RMSE values over the time steps were 0.0772 and 0.1521, respectively. By increasing the number of rules, the number of parameters will grow considerably. In this case, the best and mean RMSE values of the type-2 TSK fuzzy system decrease slightly more. It is completely evident by the simulation results obtained for different numbers of rules that the performance of type-2 FWNN with some higher number of the rules is better than that of the type 1 FWNN. On the other hand, type-2 FWNN has a better capability to describe the process using less fuzzy rules; however, it should not be forgotten that in this case, the proposed method has less parameters and, therefore, is computationally less expensive. The above is compared in Table 12.1 with the results of TRFN-G [28], type-2 FNS [26], and type 1 fuzzy wavelet neural network (FWNN). It is noteworthy to say that the indicated results in Table 12.1 are for the test data. As is obvious, the RMSE value for the type-2 FWNN model is less than that of the other approaches.

As the last example, the T2FWNN control system is considered to control a nonlinear plant. The dynamic plant is described as:

$$y(k) = 0.72y(k-1) + 0.025y(k-2)u(k-1) + 0.01u^2(k-2) + 0.2u(k-3). \tag{12.36}$$

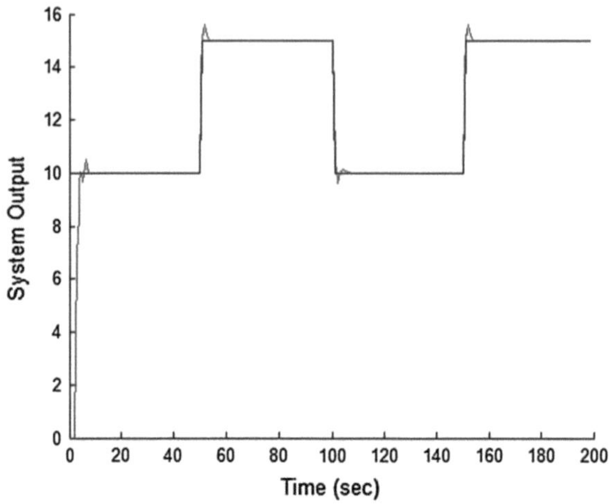

Fig. 12.6 Time response of control systems for two different values of set points

Table 12.1 Comparison table for different models of dynamic plant control

Model type	Number of rules	Number of parameters	RMSE	
			Mean	Best
TRFN-G [28]	3	33	1.086	0.887
Type-1 FWNN	3	27	0.35159	0.31812
Type-2 TSK FNS [26]	3	39	0.1802	0.0967
	9	117	0.1469	0.0725
Type-2 FWNN	3	33	0.1521	0.0772
	5	53	0.1115	0.0658

Similar to the previous example, the mathematical model of dynamic plant is extracted from the literature and is the same as that used in prior works. As is obvious, the current output of plant in each step depends on the previous input and output signals directly. However, only the current state of system and the control signal is fed into the T2FWNN controller as its inputs.

Also, for the control of the plant, we used the previous excitation signal in (12.34). The parameters of the T2FWNN are updated using different set-point signals for 200 data points. The performance of the T2FWNN for this example is shown in Figs. 12.7 and 12.8, which illustrate the relative RMSE curve and the time response characteristics. Table 12.2 compares the best and mean RMSE values with the other approaches. As can be seen, the performance of the T2FWNN is dramatically better than those that can be obtained with FWNN and TRFN-G [28]. Eliminating the fluctuations of the system response to the set point can also be considered as a significant benefit of the proposed structure.

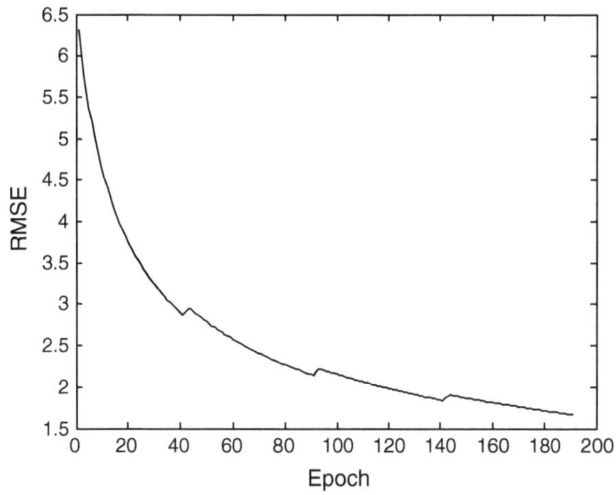

Fig. 12.7 RMSE values while learning

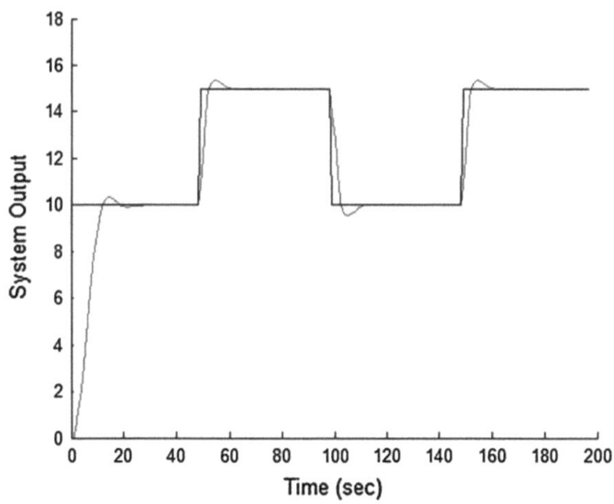

Fig. 12.8 Time response of control systems for two different values of set points

Table 12.2 Comparison table for different models of dynamic plant control

Model type	Number of parameters	RMSE	
		Mean	Best
TRFN-G [28]	33	1.2353	1.4394
Type-1 FWNN	27	1.2043	1.13402
Type-2 FWNN	33	0.9592	0.8834

12.6 Conclusion

In this chapter, novel adaptive configuration is designed for control approaches of nonlinear dynamic plants that combines the advantages of wavelet functions, neural networks, and type-2 fuzzy logic as well. The proposed type-2 FWNN structure based on TSK fuzzy rules has been used as a controller. Here, employing of type-2 fuzzy sets enhance the capability of the system to face with the uncertainties and to handle uncertain information properly. The parameters update rules of the structure are derived based on the interval type-2 fuzzy systems and gradient descent algorithm. Also, two simulation studies are performed to compare control approaches results of the proposed structure with the literature. To facilitate the decision-making on different model performance comparison, the plant models are taken from the literature. It is seen that T2FWNN can converge faster and is more adaptive to be set by new set points. It yields better performance in control cases, resulting in smaller RMSE values with regard to having the justifiable number of parameters.

Future investigations on the topic could be included for either employing other learning algorithm or utilizing fuzzy type 2 on the wavelet functions of the consequent part. On the other hand, optimization of the structure of the T2FWNN and the extending of research for the control of other plants could be considered as another viewpoint in future works.

References

1. Melin P, Castillo O (2001) A new method for adaptive model-based control of nonlinear dynamic plants using a neuro-fuzzy-fractal approach. J Soft Comput 5:171–177
2. Mendel JM (2000) Uncertainty, fuzzy logic, and signal processing. Sig Proc J 80:913–933
3. Zadeh LA (1975) The concept of a linguistic variable and its application to approximate reasoning. Inf Sci 8:43–80
4. Zhang Q, Benviste A (1992) Wavelet networks. IEEE Trans Neural Networts 3(6):889–898
5. Lin Y, Wang FY (2005) Predicting chaotic time series using adaptive wavelet-fuzzy inference system. In: Proceeding of IEEE intelligent vehicles symposium, Las Vegas, USA, pp 888–893
6. Abiyev RH, Kaynak O (2008) Fuzzy wavelet neural networks for identification and control of dynamic plants-a novel structure and a comparative study. IEEE Trans Industr Electron 55(8):3133–31408
7. Mousavi SH, Noroozi N, Safavi AA, Ebadat A (2010) Modeling and control of nonlinear systems using novel fuzzy wavelet networks: the output adaptive control approach. In: IEEE CDC, 2010, USA
8. Ebadat A, Noroozi N, Safavi AA, Mousavi SH (2011) New fuzzy wavelet network for modeling and control: the modeling approach. Commun Nonlinear Sci Numer Simul 16:3385–3396
9. Safavi AA, Romagnoli JA (1997) Application of wavelet based neural networks to modeling and optimization of an experimental distillation column. IFAC J Eng Appl Artif Intell 10(3): 301–313
10. Zekri M, Sadri S, Sheikholeslam F (2008) Adaptive fuzzy wavelet network control design for nonlinear systems. Fuzzy Sets Syst 159:2668–2695
11. Daniel WCH, Ping-An Z, Jinhua X (2001) Fuzzy wavelet networks for function learning. IEEE Trans Fuzzy Syst 9(1):200–211
12. Mendel JM, Bob RI, (2002) Type-2 fuzzy sets made simple. IEEE Trans Fuzzy Syst 10:117–127

13. Wagenknecht M, Hartmann K (1988) Application of fuzzy sets of type 2 to the solution of fuzzy equations systems. Fuzzy Sets Syst 25:183–190
14. Mendel JM, John RI, Liu F (2006) Interval type-2 fuzzy logic systems made simple. IEEE Trans Fuzzy Syst 14(6):808–821
15. Karnik NN, Mendel JM (1999) Application of type-2 fuzzy logic systems to forecasting of time-series. Inf Sci 120:89–111
16. Hwang C, Rhee FC-H (2007) Uncertain fuzzy clustering: interval type-2 fuzzy approach to C-means. IEEE Trans Fuzzy Syst 15(1):107–120
17. Hagras HA (2004) A hierarchical type-2 fuzzy logic control architecture for autonomous mobile robots. IEEE Trans Fuzzy Syst 12(4):524–539
18. Castillo O, Melin P (2008) Intelligent systems with interval type-2 fuzzy logic. Int J Innovative Comput Inf Control 4(2):771–783
19. Castillo O, Aguilar L, Cázarez N, Cárdenas S (2008) Systematic design of a stable type-2 fuzzy logic controller. Appl Soft Comput 8(3):1274–1279
20. Castro JR, Castillo O, Melin P, Rodriguez-Diaz A (2009) A hybrid learning algorithm for a class of interval type-2 fuzzy neural networks. Inf Sci 179:2175–2193
21. Dereli T, Baykasoglu A, Altun K, Durmusoglu A, Turksen IB (2011) Industrial appli-cations of type-2 fuzzy sets and systems: a concise review. Comput Ind 62:125–137
22. Jang J-SR, Sun Ch-T, Muzutani E (1997) Neuro-fuzzy and soft computing: a computational approach to learning and machine intelligence. Prentice Hall, Upper Saddle River
23. Wang CH, Cheng CS, Lee TT (2004) Dynamical optimal training for interval type-2 fuzzy neural network. IEEE Trans Syst Man Cybern 34(3):1462–1477
24. Hagras HA (2006) Comments on "dynamical optimal training for interval type-2 fuzzy neural network (T2FNN)". IEEE Trans Syst Man Cybern 36(5):1206–1209
25. Juang C-F (2002) A TSK-type recurrent fuzzy network for dynamic systems processing by neural network and genetic algorithms. IEEE Trans Fuzzy Syst 10(2):155–170
26. Zhang J, Walter GG, Miao Y, Lee WNW (1995) Wavelet neural networks for function learning. IEEE Trans Signal Process 43(6):1485–1497
27. Abiyev RH, Kaynak O (2011) A type-2 neuro-fuzzy system based on clustering and gradient techniques applied to system identification and channel equalization. Appl Soft Comput 11:1396–1406
28. Abiyev RH, Kaynak O (2010) Type 2 fuzzy neural structure for identification and control of time-varying plants. IEEE Trans Industr Electron 57(12):4147–4159
29. Begian MB, Melek WW, Mendel JM (2008) Parametric design of stable type-2 TSK fuzzy systems. In: Proceedings of North American fuzzy information processing systems, pp 1–6

Chapter 13
Multivariable Closed-Loop Identification and Its Application to Boiler-Turbine System of Power Unit

Shi-he Chen, Ya-gang Wang, Xi Zhang and Xiao-feng Li

Abstract In this chapter, a new technique for multivariable closed-loop identification is presented. On the basis of process input and output data in the control loops, the process frequency-response matrix is estimated with signal decomposition and frequency spectrum analysis, and then a transform-function matrix is identified by least square method. The required input and output data are obtained while the processes are still in normal closed-loop operation. The closed-loop identification is applied to the boiler-turbine coordinated control system of power unit. Simulation example is given to show both effectiveness and accuracy of the identification method for boiler-turbine unit.

Keywords Closed-loop identification · Frequency response · Multivariable system · Boiler-turbine unit

13.1 Introduction

Most modern multivariable controls, such as inverse Nyquist array and characteristic locus method, require a full model of the process in the form of a transfer-function matrix or a frequency-response matrix over the entire working frequency range [1]. In many cases, such a model is not available and physical modeling may require a prohibitive engineering effort. Thus, practical and effective estimation of the full model of the processes becomes appealing.

There are two ways to identify a multivariable process for control application, i.e., open-loop and closed-loop ones. In any case, an excitation of the process is

S. Chen · X. Zhang · X. Li
Guangdong Electric Power Research Institute, Guangzhou, 510600, China

Y. Wang(✉)
University of Shanghai for Science and Technology, Shanghai, 200093, China
e-mail: ygwang@usst.edu.cn

L. Liu et al. (eds.), *Applied Methods and Techniques for Mechatronic Systems*,
Lecture Notes in Control and Information Sciences 452,
DOI: 10.1007/978-3-642-36385-6_13,

needed to extract useful information about process dynamics. For open-loop transient response experiments, step or pulse excitation signals are commonly injected at the process input, and the response is measured. To perform such experiment, the process must be stable and the process is in equilibrium when the experiments begin. Some time-domain methods based on the step test can be adopted for the multivariable process identification [2, 5, 18]. The main advantage of the step test is that it requires little prior knowledge. However, it is an open-loop method and quite sensitive to nonlinearity in the system [11, 14].

Closed-loop testing is preferred to open-loop [3, 17], since the former causes less perturbation to the process. A majority of existing techniques for closed-loop process identification are in the frequency domain, while the frequency range of interest for such applications is usually from zero up to the process critical frequency [8]. In process control practice, the closed-loop relay feedback test provides several interesting advantages [2]. This identification technique has been extended to determine the dynamics of unknown multivariable plants. A straightforward extension of relay feedback test to multivariable processes is sequential loop test [9, 15]. In addition, there are several attempts for simultaneous loop test [10, 13, 20]. Unlike the single loop counterpart, however, the method has certain drawbacks when applied to multivariable system: (1) the sequential loop test ignores the combined dynamics; (2) for simultaneous loop test, it is hard to find the critical point for the multivariable process; (3) they partially identify the frequency-response matrix at one or two points, which are insufficient to model the process accurately for designing better controllers.

Boiler-turbine units in coal-fired power plants are typical multivariable systems, which have large time-delay, time-variant, nonlinear, strong coupling, and different dynamic characteristics. Coordinated control system for boiler-turbine unit is the most complex automation system in the power plant, which makes unit load follow the load demand and keeps steam pressure stable as well. The coordinated control strategy is generally adopted to control boiler, turbine, and generator as a whole [6, 12]. Currently, most of the control methods are based on the traditional PID strategy and synthesize characteristics of boiler-turbine, such as feedforward control, decoupling control, inner-model control, predictive control, etc. [7, 16, 19]. Because boiler-turbine unit is a nonlinear system, the dynamic characteristics of system are variable in different load ranges. Although controller parameters are well obtained at normal load-point, system performance becomes poor when load-point changes. Thus, the parameter tuning and load adaptation of coordinated control system become appealing.

This chapter proposes a closed-loop identification method based on response data of process input and output during normal operation. The process frequency-response matrix is first estimated by data decomposition and frequency spectrum analysis, and finally a transform-function matrix is identified. The closed-loop multivariable identification can be applied to the boiler-turbine unit, which improves the adaptation of the coordinated control system.

13.2 Multivariable Closed-Loop Identification

13.2.1 Signal Decomposition

The signals that vary from one steady state to another one are commonly encountered in industrial processes, such as process output due to set-point change. The variation between two steady states includes the dynamics character of our interest in the frequency domain. However, the signal usually cannot be dealt directly with Fourier analysis since they are not absolutely integrable if the two steady states are at different values.

To overcome this difficulty, we consider Laplace transform defined as

$$F(s) = \int_0^{\infty} f(t)e^{-st}\mathrm{d}t. \tag{13.1}$$

Decompose $f(t)$ into transient part $\Delta f(t)$

$$\Delta f(t) = f(t) - f(\infty), \tag{13.2}$$

and stationary part $f_s(t)$

$$f_s(t) = f(\infty) \cdot 1(t), \tag{13.3}$$

where $1(t)$ is unit step function, and function $f(t)$ can be expressed as

$$f(t) = \Delta f(t) + f_s(t). \tag{13.4}$$

The stationary part $f_s(t)$ is a step function and its Laplace-transform $F_s(s)$ is given by

$$F_s(s) = \int_0^{\infty} f_s(t)e^{-st}\mathrm{d}t = \frac{f(\infty)}{s}. \tag{13.5}$$

As transient part $\Delta f(t)$ will decay to zero after certain time, the Laplace-transform $F\ (s)$ of $f(t)$ can then be expressed as

$$F(s) = \Delta F(s) + F_s(s) = \int_0^{\infty} \Delta f(t)e^{-st}\mathrm{d}t + \frac{f(\infty)}{s}, \tag{13.6}$$

where $\Delta F(s)$ is the Laplace transforms of the transient part $\Delta f(t)$.

Suppose that at $t = T_f$, $f(t)$ have reached the steady-state value and $\Delta f(t)$ is approximately zero after $t = T_f$, then (13.6) becomes

$$F(s) \approx \int_0^{T_f} \Delta f(t)e^{-st}\mathrm{d}t + \frac{f(\infty)}{s}. \tag{13.7}$$

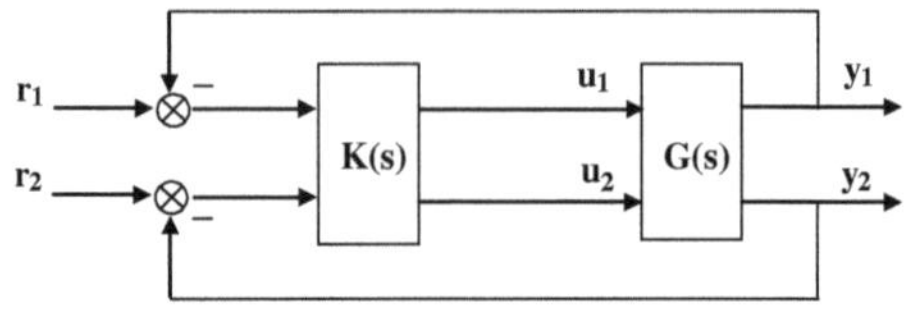

Fig. 13.1 Block diagram of multivariable closed-loop control system

The treatment of signal decomposition although straightforward, provides a useful tool for the following system identification in the frequency domain.

13.2.2 Process Frequency Response

Considering 2×2 multivariable closed-loop control system, shown in Fig. 13.1, where $K(s)$ is controller, and $G(s)$ is process.

If any set-point changes, the process outputs and control variables will change from the previous steady state to the new one, and these signals can be decomposed into

$$y_i(t) = \Delta y_i(t) + y_i(\infty) \tag{13.8}$$

$$u_i(t) = \Delta u_i(t) + u_i(\infty), \tag{13.9}$$

where $\Delta y_i(t)$, $y_i(\infty)$, $\Delta u_i(t)$, $u_i(\infty)$, $i = 1, 2$ are transient parts and steady-state parts of y_i(t) and u_i(t), respectively.

To identify the dual-input dual-output process, we first change the set-point r_1 and keep the set-point r_2 unchanged, and record the process inputs $u_i(t)$ and outputs $y_i(t)$ until they reach new steady states at t_{ui} and t_{yi}, respectively. Then $Y^1(s)$ and $U^1(s)$ can be expressed as

$$Y^1(s) = \begin{bmatrix} \int_0^{t_{y1}{}^1} \Delta y_1{}^1(t)e^{-st}\mathrm{d}t + \dfrac{y_1{}^1(\infty)}{s} \\ \\ \int_0^{t_{y2}{}^1} \Delta y_2{}^1(t)e^{-st}\mathrm{d}t + \dfrac{y_2{}^1(\infty)}{s} \end{bmatrix} \tag{13.10}$$

$$U^1(s) = \begin{bmatrix} \int_0^{t_{u1}{}^1} \Delta u_1{}^1(t)e^{-st}\mathrm{d}t + \dfrac{u_1{}^1(\infty)}{s} \\ \\ \int_0^{t_{u2}{}^1} \Delta u_2{}^1(t)e^{-st}\mathrm{d}t + \dfrac{u_2{}^1(\infty)}{s} \end{bmatrix}, \tag{13.11}$$

which follow the relation

$$Y^1(s) = G(s)U^1(s). \tag{13.12}$$

Since Eq. (13.12) is in vector form, it is not sufficient to determine $G(s)$ from $Y^1(s)$ and $U^1(s)$ alone. Next, we change set-point r_2 and keep set-point r_1 unchanged, and then we can obtain the following equation:

$$\left[Y^1(s)\ Y^2(s)\right] = G(s)\left[U^1(s)\ U^2(s)\right]. \tag{13.13}$$

As there is always a set-point change for each test, the $U^k(s), k = 1, 2$ are linearly independent. The matrix $G(s)$ are determined from (13.13) as

$$G(s) = \left[Y^1(s)\ Y^2(s)\right] \times \left[U^1(s)\ U^2(s)\right]^{-1}. \tag{13.14}$$

Substituting $U^k(s)$ and $Y^k(s)$ in (13.14), we have

$$G(s) = \begin{bmatrix} \int_0^{t_{y1}{}^1} \Delta y_1{}^1(t)e^{-st}\mathrm{d}t + \frac{y_1{}^1(\infty)}{s} & \int_0^{t_{y1}{}^2} \Delta y_1{}^2(t)e^{-st}\mathrm{d}t + \frac{y_1{}^2(\infty)}{s} \\ \int_0^{t_{y2}{}^1} \Delta y_2{}^1(t)e^{-st}\mathrm{d}t + \frac{y_2{}^1(\infty)}{s} & \int_0^{t_{y2}{}^2} \Delta y_2{}^2(t)e^{-st}\mathrm{d}t + \frac{y_2{}^2(\infty)}{s} \end{bmatrix} \times \begin{bmatrix} \int_0^{t_{u1}{}^1} \Delta u_1{}^1(t)e^{-st}\mathrm{d}t + \frac{u_1{}^1(\infty)}{s} & \int_0^{t_{u1}{}^2} \Delta u_1{}^2(t)e^{-st}\mathrm{d}t + \frac{u_1{}^2(\infty)}{s} \\ \int_0^{t_{u2}{}^1} \Delta u_2{}^1(t)e^{-st}\mathrm{d}t + \frac{u_2{}^1(\infty)}{s} & \int_0^{t_{u2}{}^2} \Delta u_2{}^2(t)e^{-st}\mathrm{d}t + \frac{u_2{}^2(\infty)}{s} \end{bmatrix}^{-1}, \tag{13.15}$$

then

$$G(s) = \begin{bmatrix} s\int_0^{t_{y1}{}^1} \Delta y_1{}^1(t)e^{-st}\mathrm{d}t + y_1{}^1(\infty) & s\int_0^{t_{y1}{}^2} \Delta y_1{}^2(t)e^{-st}\mathrm{d}t + y_1{}^2(\infty) \\ s\int_0^{t_{y2}{}^1} \Delta y_2{}^1(t)e^{-st}\mathrm{d}t + y_2{}^1(\infty) & s\int_0^{t_{y2}{}^2} \Delta y_2{}^2(t)e^{-st}\mathrm{d}t + y_2{}^2(\infty) \end{bmatrix} \times \begin{bmatrix} s\int_0^{t_{u1}{}^1} \Delta u_1{}^1(t)e^{-st}\mathrm{d}t + u_1{}^1(\infty) & s\int_0^{t_{u1}{}^2} \Delta u_1{}^2(t)e^{-st}\mathrm{d}t + u_1{}^2(\infty) \\ s\int_0^{t_{u2}{}^1} \Delta u_2{}^1(t)e^{-st}\mathrm{d}t + u_2{}^1(\infty) & s\int_0^{t_{u2}{}^2} \Delta u_2{}^2(t)e^{-st}\mathrm{d}t + u_2{}^2(\infty) \end{bmatrix}^{-1}. \tag{13.16}$$

Replace s $= j\omega$ in (13.16), we have

$$G(j\omega)=\begin{bmatrix} j\omega\int_0^{t_{y1}{}^1}\Delta y_1{}^1(t)e^{-j\omega t}dt+y_1{}^1(\infty) & j\omega\int_0^{t_{y1}{}^2}\Delta y_1{}^2(t)e^{-j\omega t}dt+y_1{}^2(\infty) \\ j\omega\int_0^{t_{y2}{}^1}\Delta y_2{}^1(t)e^{-j\omega t}dt+y_2{}^1(\infty) & j\omega\int_0^{t_{y2}{}^2}\Delta y_2{}^2(t)e^{-j\omega t}dt+y_2{}^2(\infty) \end{bmatrix} \times \begin{bmatrix} j\omega\int_0^{t_{u1}{}^1}\Delta u_1{}^1(t)e^{-j\omega t}dt+u_1{}^1(\infty) & j\omega\int_0^{t_{u1}{}^2}\Delta u_1{}^2(t)e^{-j\omega t}dt+u_1{}^2(\infty) \\ j\omega\int_0^{t_{u2}{}^1}\Delta u_2{}^1(t)e^{-j\omega t}dt+u_2{}^1(\infty) & j\omega\int_0^{t_{u2}{}^2}\Delta u_2{}^2(t)e^{-j\omega t}dt+u_2{}^2(\infty) \end{bmatrix}^{-1}. \tag{13.17}$$

Suppose that the number of the frequency response to be identified in the frequency range (0, $\omega_{\max}$) is M, we can estimate the process frequency response $G(j\omega_l)$ at the discrete frequencies 0, $\Delta\omega$, $2\Delta\omega$,..., $(M-1)\Delta\omega$, where $\Delta\omega = \omega_{\max}/M \cdot \omega_{\max}$ is defined as the maximum of frequencies corresponding to π phase lag in dual loops, which can be determined by the following iterative formula:

$$\omega_{n+1} = \omega_n - (\pi + \phi_n)\frac{\omega_n - \omega_{n-1}}{\phi_n - \phi_{n-1}} \tag{13.18}$$

$$\phi_n = \min\{\arg[\mathrm{G}(j\omega_n)]\}, \tag{13.19}$$

where the frequency responses $\mathrm{G}(j\omega_n)$, can be calculated by (13.17). The initial values for ω_{n-1} and ϕ_{n-1} are set to zero, and ω_n is recommended to be as small as possible, e.g., 10^{-3}. The iterative formula has a quadratic convergence rate near the solution. Simulation results show that 99 % accuracy for $\omega_{\max}$ can be quickly achieved within several iterations.

For $\omega = 0$, the matrix (13.17) will be

$$G(0)=\begin{bmatrix} y_1{}^1(\infty) & y_1{}^2(\infty) \\ y_2{}^1(\infty) & y_2{}^2(\infty) \end{bmatrix} \times \begin{bmatrix} u_1{}^1(\infty) & u_1{}^2(\infty) \\ u_2{}^1(\infty) & u_2{}^2(\infty) \end{bmatrix}^{-1}. \tag{13.20}$$

13.2.3 Transfer-Function Matrix

From the process frequency response $G(j\omega_l), l = 1, 2, \ldots M$, we can obtain the transfer-function matrix. The second-order plus dead-time model is adopted for each element of transfer-function matrix

$$g'(s) = \frac{b_1 s + b_0}{a_2 s^2 + a_1 s + 1} e^{-Ls}, \tag{13.21}$$

which can represent both monotonic, oscillatory, and non-minimum phase processes. Its parameters may be determined by matching $g'(j\omega)$ to $g(j\omega)$ at multiple points $\omega_l, l = 1, 2, \ldots, M$, which means

$$g(j\omega_l) = \frac{j\omega_l b_1 + b_0}{(j\omega_l)^2 a_2 + j\omega_l a_1 + 1} e^{-j\omega_l L}, l = 1, 2, \ldots, M. \tag{13.22}$$

The parameters a_1, a_2, b_1, b_0, and L in (13.21) are determined by both magnitude and phase conditions of (13.22). For the frequency ω_l, the magnitude and phase values of (13.22) are given as

$$-\omega_l{}^4 \left|g(j\omega_l)\right|^2 a_2{}^2 - \omega_l{}^2 \left|g(j\omega_l)\right|^2 (a_1{}^2 - 2a_2) + \omega_l{}^2 b_1{}^2 + b_0{}^2 = \left|g(j\omega_l)\right|^2 \tag{13.23}$$

$$-\arg\left[g(j\omega_l)\right] + \tan^{-1}\left(\frac{b_1}{b_0}\right) - \tan^{-1}\left(\frac{a_1\omega_l}{1 - a_2\omega_l^2}\right) = \omega_l L, \tag{13.24}$$

for $l = 1, 2, \ldots, M$, the magnitude equation of (13.23) can be arranged in a matrix form

$$\Phi\theta = \Gamma, \tag{13.25}$$

where

$$\Phi = \begin{bmatrix} -\omega_1{}^4 \left|g(j\omega_1)\right|^2 & -\omega_1{}^2 \left|g(j\omega_1)\right|^2 & \omega_1{}^2 & 1 \\ -\omega_2{}^4 \left|g(j\omega_2)\right|^2 & -\omega_2{}^2 \left|g(j\omega_2)\right|^2 & \omega_2{}^2 & 1 \\ \vdots & \vdots & \vdots & \vdots \\ -\omega_M{}^4 \left|g(j\omega_M)\right|^2 & -\omega_M{}^2 \left|g(j\omega_M)\right|^2 & \omega_M{}^2 & 1 \end{bmatrix}$$

$$\Gamma = \begin{bmatrix} \left|g(j\omega_1)\right|^2 \\ \left|g(j\omega_2)\right|^2 \\ \vdots \\ \left|g(j\omega_M)\right|^2 \end{bmatrix} \theta = \begin{bmatrix} \theta_1 \\ \theta_2 \\ \theta_3 \\ \theta_4 \end{bmatrix} = \begin{bmatrix} a_2{}^2 \\ a_1{}^2 - 2a_2 \\ b_1{}^2 \\ b_0{}^2 \end{bmatrix}.$$

θ in (13.25) can be solved using the linear least-squares method

$$\theta = (\Phi^T \Phi)^{-1} \Phi^T \Gamma, \tag{13.26}$$

and the original model parameters can then be recovered from θ

$$\begin{bmatrix} a_2 \\ a_1 \\ b_1 \\ b_0 \end{bmatrix} = \begin{bmatrix} \sqrt{\theta_1} \\ \sqrt{\theta_2 + 2\sqrt{\theta_1}} \\ \pm\sqrt{\theta_3} \\ \pm\sqrt{\theta_4} \end{bmatrix}. \tag{13.27}$$

If $G(0) > 0$, then $b_0 = \sqrt{\theta_4} > 0$. A negative b_1 corresponds to an inverse response, i.e., the output step response will first move in an opposite direction to its final value, and a positive b_1 corresponds to a minimum phase process. Thus, by observing the output time response, b_1 is able to be determined by

$$b_1 = \begin{cases} -\sqrt{\theta_3}, & \text{if a inverse response is detected;} \\ \sqrt{\theta_3}, & \text{otherwise} \end{cases}$$

If $G(0) < 0$, reverse the signs of b_0 and b_1 accordingly.

After model parameters a_1, a_2, b_1, b_0 are determined, the delay L can be estimated using the phase relation in (13.24)

$$\begin{bmatrix} \omega_1 \\ \omega_2 \\ \vdots \\ \omega_M \end{bmatrix} L = \begin{bmatrix} -\arg\left[g(j\omega_1)\right] + \tan^{-1}\left(\dfrac{b_1}{b_0}\right) - \tan^{-1}\left(\frac{a_1\omega_1}{1-a_2\omega_1^2}\right) \\ -\arg\left[g(j\omega_2)\right] + \tan^{-1}\left(\dfrac{b_1}{b_0}\right) - \tan^{-1}\left(\frac{a_1\omega_2}{1-a_2\omega_2^2}\right) \\ \vdots \\ -\arg\left[g(j\omega_M)\right] + \tan^{-1}\left(\dfrac{b_1}{b_0}\right) - \tan^{-1}\left(\frac{a_1\omega_M}{1-a_2\omega_M^2}\right) \end{bmatrix}, \quad (13.28)$$

L can thus be obtained, again using the least-squares method.

13.3 Application to Power Unit

13.3.1 Dynamic Model of Power Unit

The power unit is composed of boiler, turbine, and generator. The schematic diagram of the coal-fired boiler-turbine unit is shown in Fig. 13.2, and the primary variables are marked in the figure. There are strong couplings between the throttle pressure P_T and electricity generated N.

The energy conversion and heat transfer of power unit can be classified into three processes, such as furnace combustion and heat transfer, pipeline transfer, turbine working. The mass, energy, and momentum balance equations can be listed based on the three processes, and then a block diagram of structure can be obtained to describe the dynamic characteristic of boiler-turbine unit, where there exist crossed relationship and coupling between inputs and outputs, shown in Fig. 13.3. This is a 2×2 dynamic nonlinear model, which reflects the energy balance and the inner nonlinearity of system as well. The B is fuel quantity, the N is electricity generated, the P_T is throttle pressure, and the μ is governor valve position.

Considering the minor disturbance on the rated operation condition, the model of power unit can be simplified into dual-input dual-output linear dynamic model around the operation point, shown in Fig. 13.4.

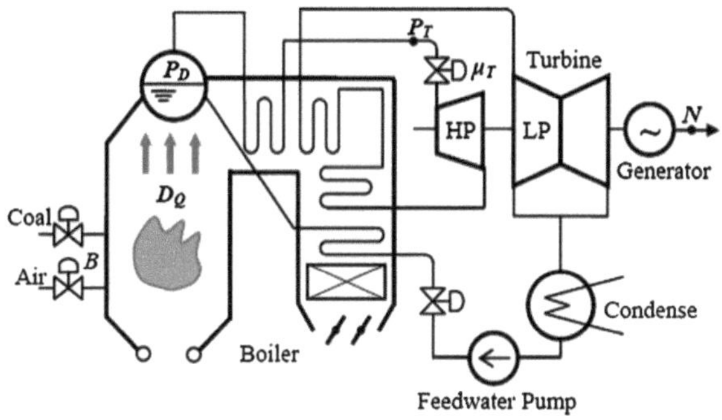

Fig. 13.2 Schematic diagram of the coal-fired boiler-turbine unit

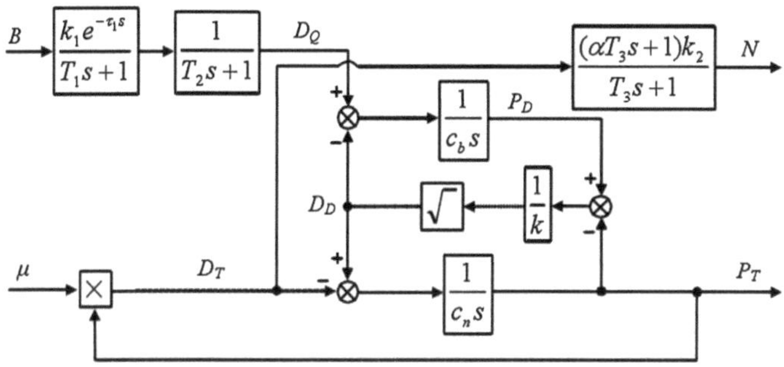

Fig. 13.3 Nonlinear model of boiler-turbine unit

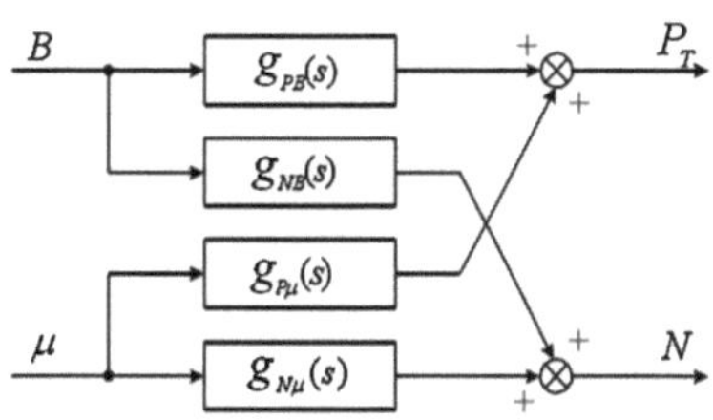

Fig. 13.4 Process model of boiler-turbine unit

13.3.2 Simulation Example

In order to show the accuracy of the proposed method in a noisy environment, the noise-to-signal ratio (NSR), defined as

$$\text{NSR} = \frac{\text{mean (abs(noise))}}{\text{mean (abs(signal))}}$$

is introduced in simulation.

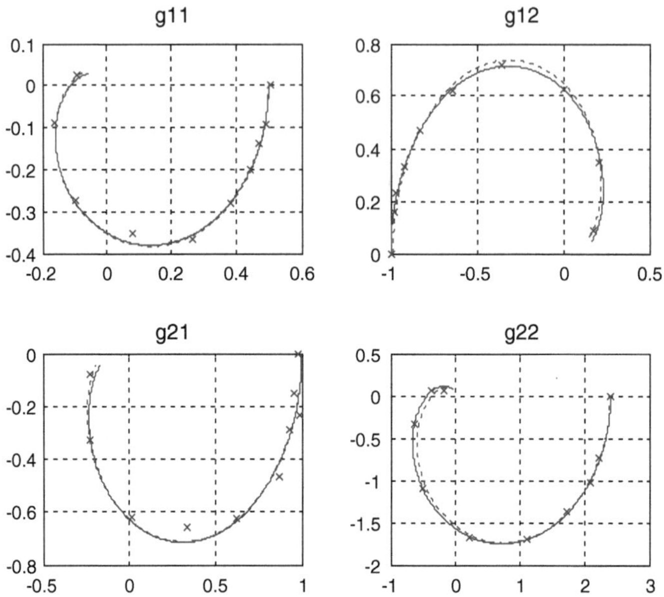

Fig. 13.5 Nyquist plots of estimated frequency responses

Consider a dynamic model of boiler-turbine unit:

$$G(s) = \begin{bmatrix} \dfrac{0.5}{(0.1s+1)^2(0.2s+1)^2} & \dfrac{-1}{(0.1s+1)(0.2s+1)^2} \\ \dfrac{1}{(0.1s+1)(0.2s+1)^2} & \dfrac{2.4}{(0.1s+1)(0.2s+1)^2(0.5s+1)} \end{bmatrix}.$$

On the basis of these two set-point responses, the frequency-response matrix is computed, and the second-order model is then calculated as

$$G'(s) = \begin{bmatrix} \dfrac{0.5006}{0.0694s^2+0.4898s+1}e^{-0.1326s} & \dfrac{-0.9875}{0.0691s^2+0.4409s+1}e^{-0.0376s} \\ \dfrac{0.0362s+0.9987}{0.0536s^2+0.4592s+1}e^{-0.1071s} & \dfrac{0.0743s+2.3899}{0.1937s^2+0.8343s+1}e^{-0.1668s} \end{bmatrix}.$$

Nyquist curve of the estimated results are shown in Fig. 13.5 (Solid line: Actual process, $\times$: Estimated frequency responses at 20 % NSR, Dashed line: Estimated model).

Simulation results show that the proposed identification method is practical and accurate though in the noisy environment. The obtained frequency response and second-order transfer-function matrix is adequate and ready for parameter tuning and retuning of coordinated control system of power units.

13.4 Conclusions

In this chapter, a multivariable closed-loop identification method has been presented. The proposed method requires no additional test signals and prior knowledge of the processes, and frequency-response at multiple points over the important frequency range can be identified, finally the second-order transfer-function matrix is obtained. The identification method is applied to the boiler-turbine control system of power unit and simulation example demonstrates the effectiveness and practicability of the method for boiler-turbine units.

Acknowledgments We thank the financing support from the National Natural Science Foundation of China under Grant 61074087 and the Innovation Program of Shanghai Municipal Education Commission of China under Grant 12ZZ144.

References

1. Amairi M, Aoun M, Najar S, Abdelkrim MN (2012) Guaranteed frequency-domain identification of fractional order systems: application to a real system. Int J Model Ident Control 17(1):32–42
2. Astrom KJ, Hagglund T (1984) Automatic tuning of simple regulators with specifications on phase and amplitude margins. Automatica 20:645–651
3. Astrom KJ, Hagglund T (1988) Automatic tuning of PID controllers. Instrument Society of America, Research Triangle Park, NC
4. Astrom KJ, Hagglund T (1995) PID controllers: theory, design, and tuning. Instrument Society of America, Research Triangle Park, NC
5. Bi Q, Cai WJ et al (1999) Robust identification of first-order plus dead-time model from step response. Control Eng Pract 7:71–77
6. Fang F, Liu JZ, Tan W (2004) Multivariable IMC-PID design for the coordinated control system of coal-fired power units. Dyn Eng 24(3):360–365
7. Fang F, Tan W, Liu JZ (2005) Tuning of coordinated controllers for boiler-turbine units. Acta Automatica Sinica 31(2):291–296
8. Hagglund T (1991) Process Control in Practice. Studentlitteratur and Chartwell-Bratt, Sweden
9. Loh AP, Hang CC, Quek CK, Vasnani VU (1993) Autotuning of multiloop proportional-integral controllers using relay feedback. Ind Eng Chem Res 32:1102–1107
10. Loh AP, Tan WW, Vasnani VU (1994) Relay feedback of multivariable systems and its use for auto-tuning of multi-loop PI controllers. In: Control'94, pp 21–24
11. Luyben WL (1990) Process modeling, simulation and control for chemical engineers. McGraw-Hill, New York
12. Mahmoud MS, Qureshi A (2012) Model identification and analysis of small-power wind turbines. Int J Model Ident Control 17(1):19–31
13. Palmor ZJ, Halevi Y, Krasney N (1995) Automatic tuning of decentralized PID controllers for TITO processes. Automatica 31(7):1001–1010
14. Privara S, Cigler J, Vana Z, Ferkl L (2012) Incorporation of system steady state properties into subspace identification algorithm. Int J Model Ident Control 16(2):159–167
15. Shih HS, Cheng CY (1994) Use of relay feedback test for automatic tuning of multivariable systems. AIChE J 40:627–646
16. Sun L, Lv JH, Wei J (2008) Research on AGC optimization control strategy of large thermal power unit. Jiangsu Electr Eng 27(1):5–8

17. Tarhouni M, Zidi S, Laabidi K, Ksouri-Lahmari M (2012) Least squares support kernel machines (LS-SKM) for identification. Int J Model Ident Control 17(1):68–77
18. Wang QG, Zhang Y (2001) Robust identification of continuous systems with dead-time from step responses. Automatica 37:377–390
19. Zhao H, Li W, Taft C, Benstman J (1999) Robust controller design for simultaneous control of throttle pressure and magawatt output in a power plant unit. Int J Robust Nonlinear Control 9(7):425–446
20. Zhuang M, Atherton DP (1994) PID controller design for a TITO system. Proc IEE Part D 141:111–120

Chapter 14
Tracking and Statistics Method Based on LBTM for Traffic Car Flow

Jian Liu, Zhiheng Gong, Xin Wang, Enyang Gao and Zexian Xu

Abstract Since limitations have been encountered in the achievement of the traditional targets tracking algorithms, a novel tracking method based on local block-graphs targets matching is proposed. Car flow tracking is applied to prove the effectiveness of the method. First, the images containing the targets are captured by the video frames to get the targets block-graphs. Second, the block-graphs can be used to achieve targets matching, and the detection process can be achieved by targets matching to get the optimal targets set. Finally, the targets set can be used to achieve the minimum deviation forecasting of all frames, and the targets tracking can be achieved. A junction video is selected as the experimental data, and a large number of experiments have been done. The results show that the proposed method not only has a great effect to track cars, but also has a better detection rate and tracking accuracy. The car flow statistics can be completed effectively.

J. Liu · Z. Gong (✉) · X. Wang · E. Gao
College of Information and Control Engineering, Shenyang Jianzhu University, Shenyang, 110168, China
e-mail: gongzhiheng@126.com

J. Liu
e-mail: jeanliu10@163.com

X. Wang
e-mail: wangxin7988@163.com

E. Gao
e-mail: gaoenyang@sia.cn

Z. Xu
School of Computing, National University of Singapore, Singapore, 117417, Singapore
e-mail: zexianxu@hotmail.com

L. Liu et al. (eds.), *Applied Methods and Techniques for Mechatronic Systems*,
Lecture Notes in Control and Information Sciences 452,
DOI: 10.1007/978-3-642-36385-6_14,

14.1 Introduction

In recent years, targets tracking has become one of the most topical issues in the field of pattern recognition. The process can be divided into targets detection and targets tracking. We work within the paradigm of detect-then-track, where an object detector is run on each frame to hypothesize the object of interest, followed by a local block-graphs targets matching stage to link detection into multi-frame trajectories. The targets detection is the basis for the targets tracking, which is usually combined with the Support Vector Machine (SVM) [1]. Some detection algorithms can be established to solve the problem of targets detection, such as Scale Invariant Feature Transform (SIFT) [2] and Histograms of Oriented Gradient (HOG) [3]. The essence of these algorithms is that the features of targets data are extracted, and then the targets data can be classified. Random Forest algorithm [4] is proposed by Breiman, and is effectively applied to the field of the targets detection.

The multi-targets tracking relates to the data association and the image analysis. More recent methods have attempted to solve the multi-targets tracking. Blei proposed the CTM algorithm by the scientific topic models [5]. Zhao proposed the TUMP algorithm to achieve the multi-targets tracking and statistics [6]. Kratz proposed the LSMP algorithm according to the modified TUMP [7]. Butt proposed the Lagrange relaxation cost network to solve the problem of multi-targets tracking [8]. The above algorithms obtain better tracking effects according to the analysis of all frames. However, the tremendous amount of calculation and the real-time tracking impact of the frames' pixels are the disadvantage factors.

We believe that the targets tracking problem should convert local analysis into global analysis. A novel method (Local Block-graphs Targets Matching, LBTM) is proposed by the local image processing and the data association analysis [9–13]. Our method optimizes the image from the frames to the local block-graphs, and we use the local block-graphs to achieve the targets matching. By making better use of global image information over each frame, our method is potentially capable of finding better solutions. The disadvantages of the traditional method can be overcome, and in order to prove the feasibility of method, we applied it to the car flow tracking and statistics [14–16].

The contents of the paper consist of five sections. In Sect. 14.2, the prerequisite knowledge is introduced, and the minimum loss of data association is achieved based on the maximum probability. In Sect. 14.3, our method is introduced. It is considered to solve the targets tracking and statistics. In Sect. 14.4, a number of experiments have been performed to prove the method feasibility. In Sect. 14.5, the research conclusions are presented.

14.2 Prerequisite Knowledge

Our method is based on the data association [17–19], and the LBTM is one of the data association. Data association is the problem of finding the detections corresponding to the targets in different frames of a video. The input to the LBTM problem in a sequence can ideally be represented by the block-graphs in which all the detections in each frame are connected to all the other detections in the other frames. Similarly, the results can be ideally represented by several subgraphs of the input.

14.2.1 Multi-targets Data Association

The multi-targets data association can be considered as the data mapping into a Cost-Flow Network (CFN), then we can use the CFN to solve the data association of minimum loss. The mapping principle is based on the movement trajectory and trend of the nonoverlapping targets.

We assume $X = \{\mathbf{x_i}\}$ is a generalized set of targets data, $\mathbf{x_i}$ consists of x_i, s_i, and a_i. x_i is the target coordinate, s_i is the target size, and a_i is the edge coordinate of target. It can be written as

$$\mathbf{x_i} = (x_i, s_i, a_i). \tag{14.1}$$

According to Eq. (14.1), the multi-targets set can be defined by

$$T_k = \left(\mathbf{x_{k_1}}, \mathbf{x_{k_2}}, \ldots, \mathbf{x_{k_m}}\right), \mathbf{x_{k_i}} \in X \tag{14.2}$$

We assume $T = \left\{T_k^i\right\}$ is a generalized set of multi-targets. According to Eqs. (14.1) and (14.2), we can convert the multi-targets data association into the issue of maximize probability, and the expressions are given by

$$\hat{T} = \arg\max_T P\left(T_k | X\right) \tag{14.3}$$

$$\hat{T} = \arg\max_T \prod_i P\left(\mathbf{x_i} | T\right) P\left(T\right). \tag{14.4}$$

Note that in order to get optimum result, we optimize Eq. (14.4). The T_k^i can be used to optimize Eq. (14.4), which is a multi-targets set. The limited condition can be given by

$$T_k^i \cap T_k^j = \emptyset, i \neq j \tag{14.5}$$

Equation (14.4) can be optimized by Eq. (14.5), and the results can be given by

$$\hat{T} = \arg\max_{T} \prod_{i} P(\mathbf{x_i}|\, T) \prod_{T_k^i \in T} P\left(T_k^i\right) \tag{14.6}$$

$$P(\mathbf{x_i}|\, T) = \begin{cases} 1-\beta_i & T_k^i \in T,\ \mathbf{x_i} \in T_k \\ \beta_i & \text{otherwise} \end{cases} \tag{14.7}$$

$$P\left(T_k^i\right) = P\left(\left\{\mathbf{x_{k_0}}, \mathbf{x_{k_1}}, \ldots, \mathbf{x_{k_n}}\right\}\right). \tag{14.8}$$

If we do not consider the mapping loss, the optimum result T can be obtained according to Eqs. (14.6), (14.7), and (14.8).

14.2.2 Minimum Loss of Data Association

In order to solve the mapping loss, we can convert the target function into the binarization according to the definition of CFN. It is given by

$$f_{en,i} = \begin{cases} 1 & T_k^i \in T, \mathbf{x_i} \in T_k \\ 0 & \text{otherwise} \end{cases} \tag{14.9}$$

$$f_{en,j} = \begin{cases} 1 & T_k^j \in T, \mathbf{x_j} \in T_k \\ 0 & \text{otherwise} \end{cases} \tag{14.10}$$

$$f_{i,j} = \begin{cases} 1 & T_k^i, T_k^j \in T, \mathbf{x_i},\ \mathbf{x_j} \in T_k \\ 0 & \text{otherwise} \end{cases}. \tag{14.11}$$

The restrictive condition for minimum loss of data association can be obtained by Eqs. (14.9), (14.10), and (14.11).

$$f_{en,i} + \sum_j f_{j,i} = f_{en,j} + \sum_j f_{i,j}. \tag{14.12}$$

Equation (14.16) can be written again by Eq. (14.12), and it is given by

$$\hat{T} = \arg\min_{T} \sum_{T_k^i \in T} -\log P\left(T_k^i\right) + \sum_i -\log P(\mathbf{x_i}|\, T) \tag{14.13}$$

$$\hat{T} = \arg\min_{T} \sum_i C_{en,i}, i f_{en,i} + \sum_{i,j} C_{i,j} f_{i,j} + \sum_j C_{en,j}, j f_{en,j}, \tag{14.14}$$

where $C_{en,i} = -\log P(\mathbf{x_i})$, $C_{en,j} = -\log P\left(\mathbf{x_j}\right)$, $C_{i,j} = -\log \frac{\beta_i}{1-\beta_i}$.

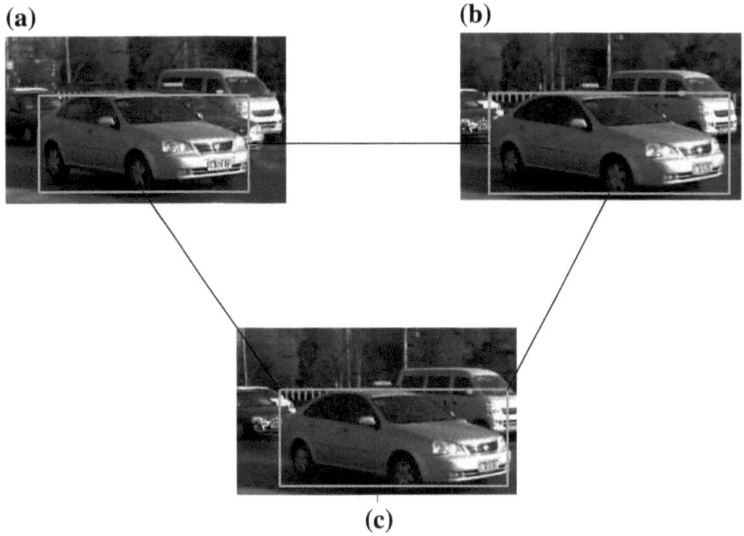

Fig. 14.1 Single target matching results

The generalized multi-targets set for minimum mapping loss can be obtained by Eq. (14.14), and we can achieve the multi-targets data association of minimum error.

14.3 Our Method

We propose an algorithm that can use LBTM to complete the detection [20] and tracking. Unlike [5–8], which analyze all the frames to carry out the tracking, our process is local. We apply the maximum probability to forecast targets trajectory. The mathematical expressions of target detection and tracking are presented in the this section, followed by rigorous problem formulation in subsequent sections.

14.3.1 Local Block-Graphs Targets Matching

It is assumed that a, b, and c are the three local block-graphs, and the cars are contained by it. The matching results of single target is shown in Fig. 14.1.

Figure 14.1 shows that we use the local block-graphs matching to achieve the detection of single target, which is the essence of our method.

The multi-targets detection can be achieved by the superimposition of single target detection. Note that we use Fig. 14.1 as an example to achieve multi-targets matching. The ideal result is shown in Fig. 14.2.

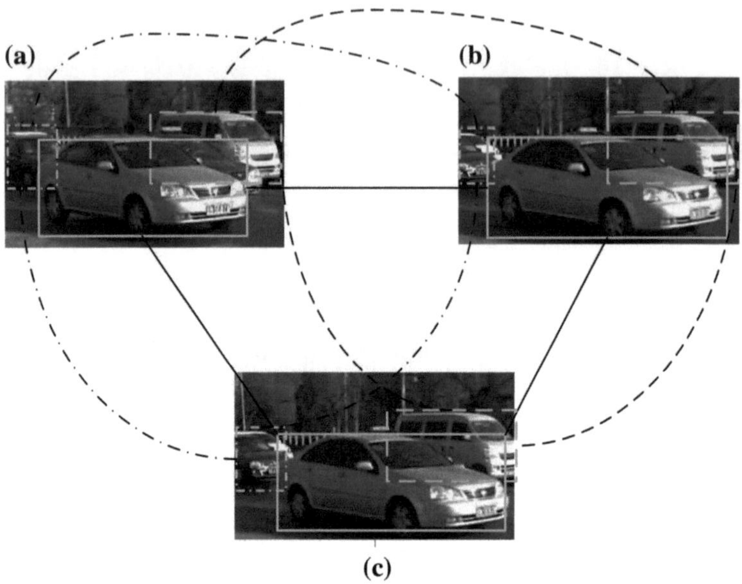

Fig. 14.2 Multi-target matching results

Note that the ideal targets matching is independent, and each process of targets matching is based on the block-graphs according to the Fig. 14.2. We consider adjacent frames as the matching frames, and the block-graphs are extracted from the frames. We strictly consider the overlapping influence when doing the targets matching, therefore the targets matching can be completed. The targets detection also can be achieved.

14.3.2 Targets Detection

The targets detection can be considered as the process of data association according to Sect. 14.2.1, and the targets are contained in some frames. It is assumed that the targets set V is contained in a generalized image G, and G consists of V, E, and ω. E is the edge set of G, ω is the weight of the edge set. G can be given by

$$G = (V, E, \omega). \tag{14.15}$$

The targets set V can be divided into f block-graphs, and we define a generalized set C_i, $1 \leq i \leq f$. It is assumed that the υ_m^i is the mth target of the ith block-graphs, and the C_i can be given by

$$C_i = \left\{\upsilon_1^i, \upsilon_2^i, \upsilon_3^i, \ldots\right\}. \tag{14.16}$$

Fig. 14.3 Divided results of car body region

According to Eq. (14.16), the E can be given by

$$E = \left\{ \left(\upsilon_m^i, \upsilon_n^j \right) \middle| i \neq j \right\}. \tag{14.17}$$

The car body can be divided into nine regions according to the different characteristics location, which consists of the car windows (1, 2), the back wheels (3, 4), the front wheels (5, 6), the headlights (7, 8), and windshield (9). It is shown in Fig. 14.3.

The characteristics regions can be converted into color vectors according to the images. The Φ_{ml}^i is the the lth color vectors of the mth target in the ith block-graph, and the weight of the edge set ω can be given by

$$\omega \left(\upsilon_m^i, \upsilon_n^j \right) = \sum_{l=1}^{9} k \left(\Phi_{ml}^i, \Phi_{nl}^j \right), \tag{14.18}$$

where K is the RBF kernel function of color vectors.

We set up a subimage G_s of G, given by

$$G_s = (V_s, E_s, \omega_s). \tag{14.19}$$

Here, V_s can be defined by the generalized set C_i, and it can be given by

$$V_s = \left\{ \upsilon_a^1, \upsilon_b^2, \upsilon_c^3, \ldots \right\}, \tag{14.20}$$

where υ_a^1 and υ_b^2 represent the ath target of the first block-graph and bth target of the second block-graph, respectively.

According to Eqs. (14.17), (14.18), and (14.20), E_s, and ω_s can be given by

$$E_s = \{ E(p, q) \,|\, p \in V_s, q \in V_s \} \tag{14.21}$$

$$\omega_s = \{ \omega(p, q) \,|\, p \in V_s, q \in V_s \}. \tag{14.22}$$

We consider the detection of targets set V_s as an event γ according to Sect. 14.2.2, and when the event is completed, the loss can be defined as $\gamma_a(V_s)$. We use the minimum loss of data association to get $\gamma_a(V_s)$, given by

$$\gamma_a(V_s) = \frac{1}{2}\left(\sum_{i=1}^{f}\sum_{j=1,j\neq i}^{f}\omega\left(V_s(i), V_s(j)\right)\right). \tag{14.23}$$

We expect that the targets set V_s can be detected perfectly, and $\gamma_a(V_s)$ is the minimum value. According to Eq. (14.23), the target function can be given by

$$\widehat{V_s} = \arg\min_{V_s}\left(\gamma_a(V_s)\right). \tag{14.24}$$

$\widehat{V_s}$ is the result of targets detection according to the Eq. (14.24).

14.3.3 Targets Tracking

It is assumed that the s block-graphs are contained in all frames, and the targets tracking is another event γ'. The loss can be defined as $\gamma_s(V_s)$, and it is given by

$$\gamma_s(V_s) = \gamma_a(V_s) + \alpha\gamma_m(V_s), \tag{14.25}$$

where α is weight coefficient. According to the way of space vectors expression, V_s can be defined by

$$\dot{X}_s = X_s(i+1) - X_s(i), 1 \le i \le (s-1), \tag{14.26}$$

where $\dot{X}_s$ is the targets deviation between the $(i+1)$th block-graph and the ith block-graph. According to Eq. (14.26), $\gamma_m(V_s)$ can be given by

$$\gamma_m(V_s) = \sum_{i=1}^{s}\sum_{j=1}^{s-1}\overbrace{\left|X_s(i) - \underbrace{\left(X_s(j) + \dot{X}_s(j)(i-j)\right)}_{\text{prediction}}\right|}^{\text{deviation}}. \tag{14.27}$$

Note that $X_s(j) + \dot{X}_s(j)(i-j)$ is the prediction of the targets tracking, and $\left|X_s(i) - \left(X_s(j) + \dot{X}_s(j)(i-j)\right)\right|$ is the deviation of the target tracking. According to Eqs. (14.25) and (14.26), the optimum result of targets tracking $\widehat{V_s}'$ can be given by

$$\widehat{V_s}' = \arg\min_{V_s}\left(\gamma_a(V_s) + \alpha\gamma_m(V_s)\right). \tag{14.28}$$

We achieve the target tracking by the above proposed algorithm in which the processes of tracking have three steps.

Step 1: we capture some frames, which contain the targets block-graphs. We extract the targets block-graphs from those frames, and use it to achieve the targets matching.

Step 2: according to Sect. 14.3.2, we establish the target function. The targets set V_s can be obtained. The targets detection can be achieved among the captured frames.

Step 3: according to Sect. 14.3.3, we use the targets set V_s to calculate V_s'. Note that we use V_s' to forecast targets trajectory and trend among all frames, and the tracking can be achieve by above steps.

14.4 Experimental Evaluations

We use the simulation experiments to reveal the feasibility and superiority of the method, and the CPU of computer is Inter(R) Core(TM) Duo-E7500, the size of internal storage is 6 G. Our code is implemented in MATLAB. We divide our experimental results into three sections. Section 14.4.1 shows that our method is applied as a instrument for cars detection. Section 14.4.2 shows that we use our method to solve the problem of the cars tracking, and compare our method with other algorithms. Section 14.4.3 shows that car flow statistics can be achieved by our method.

14.4.1 Cars Detection Experiment

In order to prove the accuracy of targets detection, we choose two images as the experimental images, which contain many cars. We use the process of Sect. 14.3.2 to carry out the cars detection, the results are shown in Fig. 14.4.

Figure 14.4 shows that the cars can be detected accurately, and Fig. 14.4a and b is the demo graphs and the demo results, respectively. Note that Fig. 14.4b shows the results of detection according to Fig. 14.4a. So that for bigger cars in Fig. 14.4a, the detection results of our method are more accurate. Note that Fig. 14.4a contains some small cars. When the cars are too small, our method merely find out the "cars group".

14.4.2 Cars Tracking Experiment

In order to prove the superiority of targets tracking, we choose a video as the experimental data. We choose five cars as the tracking targets, which are marked from 1 to 5. These cars' statuses are contained from the 87th frame to the 264th frame, and the results are shown in Fig. 14.5.

Fig. 14.4 Detection results of the cars

Figure 14.5 shows that the five cars are tracked effectively from the 87th frame. Note that car-1 can be tracked effectively from the 87th frame to the 211th frame,

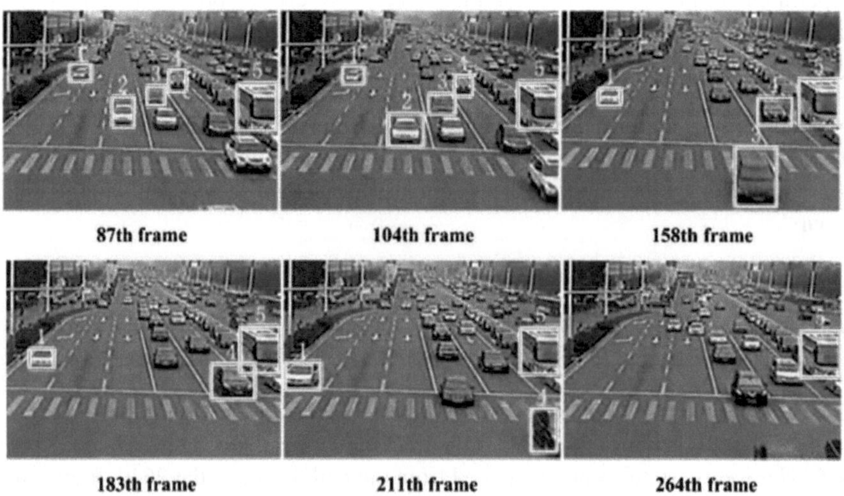

Fig. 14.5 Results of tracking experiments

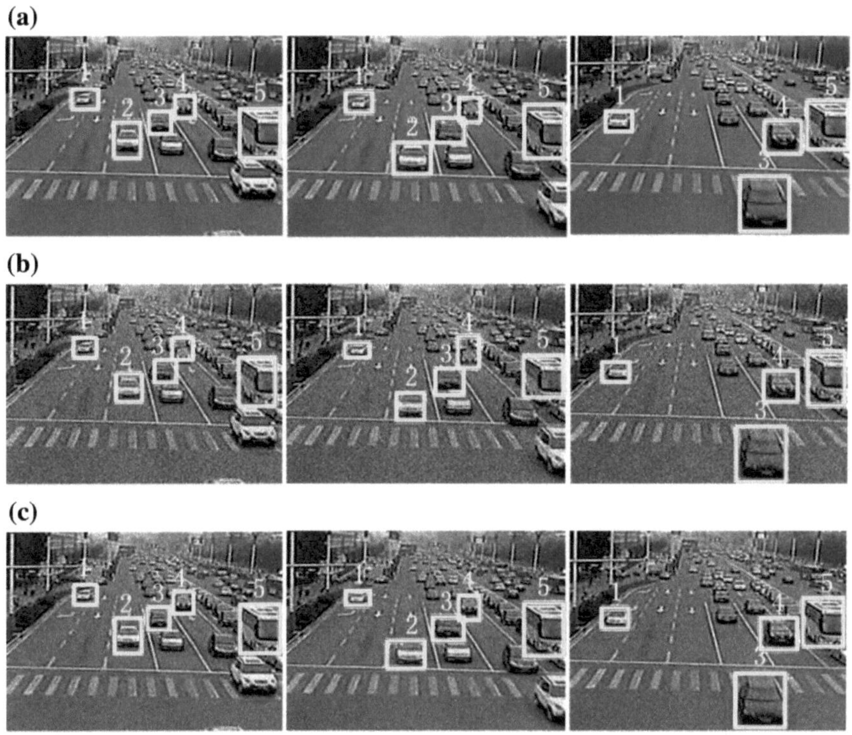

Fig. 14.6 This figure shows the tracking results in different noise intensity. Note that **a** is the tracking result in the Intensity-1, **b** is the tracking result in the Intensity-2, **c** is the tracking result in the Intensity-3

which disappears from the 211th frame. We consider car-1 passes through the crossing, and the process of car-1 tracking has been completed by our method. Note that car-2 disappears from the 158th frames. Note that car-3 and car-4 disappear from the 183th frame and the 264th frame, respectively, and our algorithm has the perfect tracking accuracy. Note that car-5 is motionless from the 87th frame to the 264th frame. The cars overlapping is a problem in the tracking process, and our method effectively overcomes the problem because our method depends on the maximum probability to carry out the tracking prediction, rather than merely analyzing the images.

To evaluate the tracking performance of our proposed method in different noise tasks, and to compare with different noise intensity specifically devised for the tracking accuracy, we add the different White Gaussian Noise (WGN) to the video, and the noise intensity can be defined by the distribution of WGN (0–10 is the noise intensity-1, 0–20 is the noise intensity-2, 0–30 is the noise intensity-3). The tracking results are shown in Fig. 14.6.

Table 14.1 Schedule of car flow statistics

Time	Numbers of the car flow	Error (%)
9:00	87	2.32
11:00	59	1.74
13:00	34	1.66
15:00	47	3.17
17:00	104	1.91

Figure 14.6 shows that the influence of WGN can be overcome by our algorithm. When the noise of Intensity-1 and Intensity-2 is added to the video, our method basically maintains the accuracy. Note that when the Intensity-3 is added to the video, the tracking results have the deviation because the pixel values change tremendously, and the images are too blurry.

14.4.3 Car Flow Statistics Experiment

We extract the two video (A and B) from the experimental video of Sect. 14.4.2, and the time of extracted video (A and B) is 120 s. We carry out the statistics once every 20 s, and we compare the experimental results with the results of CTM, TUMP, and LSMP. The results are shown in Fig. 14.7.

Figure 14.7 shows that our method has well-statistical results, which has the lower undetection rate than the CTM, TUMP, and LSMP. So that for a longer time, the number of car flow is much more. Figure 14.7a shows the correct value of car flow is 17 in the 20 s, and the results of our method, LSMP, TUMP, and CTM are 16, 15, 13, and 12, respectively. When the correct value of car flow is 92 in the 120 s, the results of our method, LSMP, TUMP, and CTM are 89, 84, 76, and 67, respectively. Figure 14.7b shows when the correct value of car flow is 97 in the 120 s, the results of our method, LSMP, TUMP, and CTM are 95, 86, 78, and 71, respectively. Note that our method has the better superiority in the statistics of cars flow.

In order to reveal the peak and low ebb of cars flow, we carry out the statistics once every 2 h from 9 o' clock to 18 o'clock, and the statistics time is 120 s. It is assumed that the number of cars for experimental result is N_t, and the standard value is N_b. The error σ can be defined by

$$\sigma = \frac{|N_t - N_b|}{N_b} \times 100\,\%, \tag{14.29}$$

σ can be calculated by Eq. (14.29), and the results are shown in Table 14.1.

Table 14.1 shows that 9 o'clock and 17 o'clock are the peak of cars flow, and 13 o'clock is the low ebb of cars flow. The number of cars flow has decreased from 11 o'clock to 13 o'clock, and has increased from 13 o'clock to 15 o'clock. The average

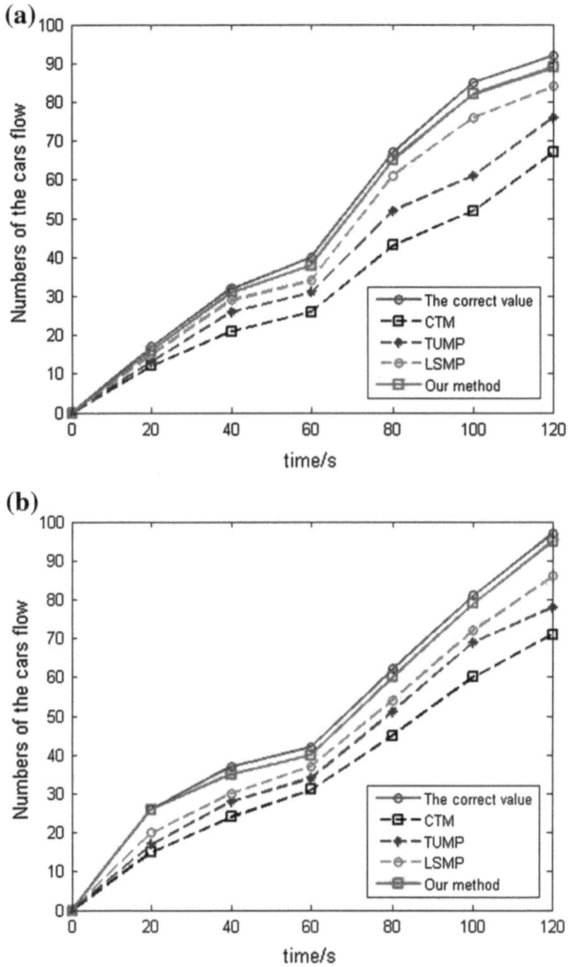

Fig. 14.7 This figure shows the statistics comparison results of car flow. Note that **a** and **b** is the result of videos *A* and *B*, respectively

error of our method maintains at about 2 %, which has a preferable statistical capacity in the different cars flow.

14.5 Conclusion

In this chapter, we make an in-depth study and analysis, and a novel algorithm is proposed based on the block-graph matching. We apply the algorithm to car flow tracking and statistics. The algorithm captures video frame images which contains targets, and then intercepts the images into blocks. Next, the block-graphs are processed using targets matching to achieve the targets detection. We use the targets set of detection to carry out the targets forecast of minimum deviation in all frames to complete the tracking.

The simulation experiments show that our method has the strong application capacity. Comparing the method with CTM, TUMP, and LSMP, it has the better accuracy and capacity of antijamming. The future work is to study the interference factors. Improving the accuracy of method will be a challenging topic for future study.

Acknowledgments The authors would like to thank for the supports from the National Natural Science Foundation of China (61272253) and the Ministry of Housing and Urban-Rural Development, China (2010-K9-22).

References

1. Mittag F, Saad M, Jahn A (2012) Use of support vector machines for disease risk prediction in genome-wide association studies: concerns and opportunities. Hum Mutat 33(12):1708–1718
2. Cruz-Mota J, Bogdanova I, Paquier B (2012) Scale invariant feature transform on the sphere: theory and applications. Int J Comput Vision 98(2):1–25
3. Dalai N, Triggs B (2005) Histograms of oriented gradients for human detection. Proceedings of IEEE conference on computer vision and, pattern recognition. Washington, USA, pp 886–893
4. Breiman L (2001) Random forests. Mach Learn 45(1):5–32
5. Blei DM, Lafferty JD (2007) A correlated topic model of science. Ann Appl Stat 1(1):17–35
6. Zhao XM, Gong D, Medioni G (2012) Tracking using motion patterns for very crowded scenes. In: Proceedings of European conference on computer vision, Florence, pp 315–328.
7. Kratz L, Nishino K (2012) Tracking pedestrians using local spatio-temporal motion patterns in extremely crowded scenes. IEEE Trans Pattern Anal Mach Intell 34(5):987–1002
8. Butt AA, Collins RT (2013) Multi-target tracking by Lagrangian relaxation to min-cost network flow. Proceedings of IEEE conference on computer vision and, pattern recognition. Portland, USA, pp 1–8
9. Andriyenko A, Schindler K (2011) Multi-target tracking by continuous energy minimization. Proceedings of IEEE conference on computer vision and, pattern recognition. CO, USA, pp 1265–1272
10. Benfold B, Reid I (2011) Stable multi-target tracking in realtime surveillance video. Proceedings of IEEE conference on computer vision and, pattern recognition. CO, USA, pp 3457–3464
11. Breitenstein MD, Reichlin F, Leibe B (2009) Robust tracking-by-detection using a detector confidence particle filter. In: Proceedings of IEEE international conference on computer vision, pp 1515–1522.
12. Henriques JF, Caseiro R, Batista J (2011) Globally optimal solution to multi-object tracking with merged measurements. Proceedings of IEEE international conference on computer vision. Coimbra, USA, pp 2470–2477
13. Shu G, Dehghan A, Oreifej O (2012) Part-based multiple-person tracking with partial occlusion handling. Proceedings of IEEE conference on computer vision and, pattern recognition. CO, USA, pp 1815–1821
14. Ren W, Gu Q, He DX, Zhao J (2013) Modelling and implementation of a car-like mobile robot for trajectory-tracking. Int J Model Iden Control 19(2):150–160
15. Qu SC, Liu XB, Wang QQ, Lei ZQ, Che YW (2013) A brief fuzzy controller for an intelligent tracking system. Int J Model Ident Control 19(2):171–178
16. Pang B, Shao C (2013) Modified newton-based ILC solving specific target tracking problems for non-linear batch reactor. Int J Model Ident Control 18(2):158–165

17. Zhang L, Li Y, Nevatia R (2008) Global data association for multi-object tracking using network flows. Proceedings of IEEE conference on computer vision and, pattern recognition. Anchorage, USA, pp 1–8
18. Yu Q, Medioni G, Cohen I (2007) Multiple target tracking using spatio-temporal markov chain monte carlo data association. Proceedings of IEEE conference on computer vision and, pattern recognition. Minneapolis, USA, pp 1–8
19. Goldberg AV (1997) An efficient implementation of a scaling minimum-cost flow algorithm. J Algorithms 22(1):1–29
20. Liu J, Gong ZH, Gao EY, Liu YN (2013) Research on electrical symbols recognition of HOG. J Shenyang Jianzhu Univ (Nat Sci) 29(3):571–576

Chapter 15
Object Manipulation Strategy Analysis and Realization for a Humanoid Robot

Qinjun Du, Hongzhe Sha, Fei Jia, Lina Liu and Xinghua Wu

Abstract It is hoped that the humanoid robot will be able to move flexibly and have excellent object operation capability like human beings. Based on people operation object strategy, this chapter proposes a humanoid robot object operation strategy that integrates the visual feedforward control strategy and the visual feedback control strategy. Using the visual feedforward control strategy, the humanoid robot can walk approaching the target object and make its arm close to the target object; using the visual feedback control strategy, the humanoid robot achieves the robot hand fine alignment with the target object. Based on the visual feedforward control strategy the time of the humanoid robot hand approaching the target object has been reduced; based on the visual feedback control strategy the accuracy of the operation has been improved. Based on the control strategy of this chapter proposed, the humanoid robot will be able to walk and operate objects independently.

Keywords Humanoid robot · Object manipulation · Control strategy

Q. Du (✉) · H. Sha · F. Jia · L. Liu · X. Wu
School of Electrical and Electronic Engineering, Shandong University of Technology,
12 Zhangzhou road, Shandong Zibo, 255091, China
e-mail: qinjundu@sohu.com

H. Sha
e-mail: 303398867@qq.com

F. Jia
e-mail: 2443226188@qq.com

L. Liu
e-mail: linaliu-126@163.com

X. Wu
e-mail: zblzwxh@126.com

L. Liu et al. (eds.), *Applied Methods and Techniques for Mechatronic Systems*,
Lecture Notes in Control and Information Sciences 452,
DOI: 10.1007/978-3-642-36385-6_15, © Springer-Verlag Berlin Heidelberg 2014

15.1 Introduction

The development purpose of the humanoid robot is to make the humanoid robot work in complex and dangerous environments in place of people, and serve people in the home, hospital, etc. [1, 2]. Humanoid robot is an intelligent machine, which can get to the region to operate objects through body movement, and can complete operation tasks using arm and hand with the help of various sensor information [3–5] such as grasp object operation. The humanoid robot is different from the industrial robot that needs to find the target object initiatively, reaches to the object by walk, and then completes the operation for the object. In this work process, humanoid robot not only needs to control the end effector, but also needs to control the movement of the head and body of the humanoid robot [6–8].

People hope that the humanoid robot will have the walk and operation capability of people [9–11]. The grasp operation of human beings can be divided into two steps: reach and grasp. The operation of reach is where the people get to the region in which the people can operate the target object. The operation of grasp is the people taking the object by hand. Human beings accomplish the operation object through the coordination of these two actions. The literature [12, 13] has described the human beings Reach-to-Grasp experiments in detail.

Human beings accomplish movement, at first, by a rough location of the target object through visual information; based on the visual information people complete movement with visual feedforward; second, based on the real-time visual feedback motion control, the hand reaches the vicinity of the target object and makes fine alignment with the target object.

Human beings realize the reach to grasp movement process using a combination control strategy of visual feedforward and visual feedback. The visual feedforward control strategy guides the hand close to the target object; the visual feedback strategy increases the accuracy of operation. The human beings' hand-eye system is based on the location information of visual feedback servo system. Although the binocular vision system is not necessary, it increases the efficiency of movement.

Visual servo control strategy is widely used in the robot hand-eye system [14–16]. In the field of industrial robots, based on the visual servo the robot grasp operation has achieved mature application. The end effector of most industrial robots is fixed on the base; the base of the robot is stationary. The target objects operated are generally located within the grasp region of the manipulator; industrial robots usually complete the target object operation without movement of the body of the robot.

In the process of humanoid robot walk, the joints of the head and hand deviate from the standard zero-position; therefore, the humanoid robot hand-eye system strict significance calibration is very difficult, or even impossible, greatly increasing the difficulty of the humanoid robot for object manipulation. Especially the standard zero-position offset of the head and hand joints have great impact on the visual servo system of the humanoid robot. Therefore, strict hand-eye system calibration is very difficult for the humanoid robot.

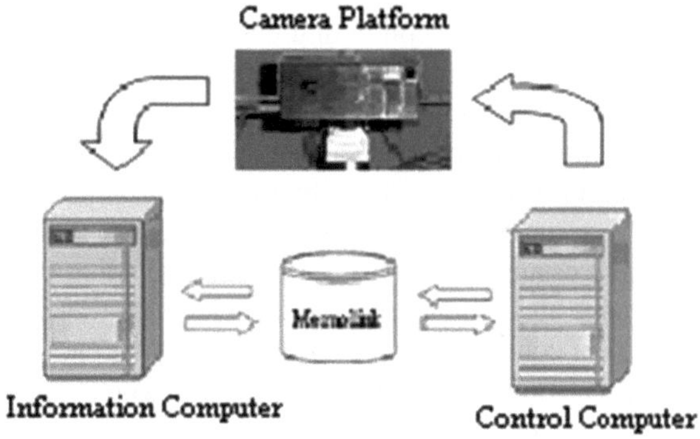

Fig. 15.1 Visual servo system architecture of the humanoid robot

In this chapter, based on the visual feedforward and visual feedback control strategy, the humanoid robot completes object operation. Using visual feedforward information, the hand of the humanoid robot approaches the target object; using the visual feedback control strategy, the hand of the humanoid robot achieves fine alignment with the target object. A non-strict hand-eye calibration system based on visual feedback is a reasonable way to solve humanoid robot object manipulation.

15.2 Visual Servo System Architecture

Visual servo control is widely used in the robot hand-eye system. In the field of industrial robots, the object operation based on the visual servo has achieved mature application [17].

To complete the operation of the target objects, the appropriate camera configuration is essential for the humanoid robot. In real life, human beings find and track objects through the rotation of the head and eyes. Humanoid robot must be able to turn its head to find, locate, and track the target objects, including moving objects and stationary objects. The head of the humanoid robot is designed the same as the human head, as shown in Fig. 15.1. The visual platform of the head has two degrees of freedom, turning left and right and rotates up and down, with two degrees of freedom to ensure the humanoid robot head points to anywhere in space.

We have built humanoid robot hand-eye system model in the literature [18]. The humanoid robot's arm and hand mimic the shape, structure, and operation functions of human beings, making it have the same operating flexibility as that of human beings. Therefore, the humanoid robot arm has six degrees of freedom and the hand has three degrees of freedom; in the planning and servo control of the control system, the humanoid robot imitates the human to move and grasp objects within the effective range which the robot arm can reach.

The whole humanoid robot perception and control system can be divided into visual information processing system and motion control system. The motion control system is responsible for the movement of each joint of the body; besides control of the walk, it needs to control the body balance and also the coordination of the hand with foot. Therefore, motion control part is a multitask control system and the control task is very heavy. The visual information processing part is mainly responsible for the acquisition and processing of visual information, including image acquisition and preprocessing, image analysis and search, target identification and tracking, and more. Visual information contains a huge amount of information.

In order to achieve real-time visual positioning and tracking, the chapter proposes perception and control system of a dual computer, in which one computer is responsible for the processing of visual information, and another computer is responsible for robot motion control; two computers communicate through Memolink which make the two computers share memory information as shown in Fig. 15.1.

In the real-time motion control system, the motion control computer communicates with DSP controllers via CAN bus. Each joint of the robot is driven by a Maxon DC motor with harmonic reducer. The DC motors are driven by PWM servo drivers. The main characteristics of this servo driver are small and light, high power, single power supply operation, unipolar control voltage input, and quick position control response. It has three control loops: current control loop, speed control loop, and position control loop.

15.3 Target Object Tracking

In the humanoid robot active vision system, the object image segmentation and tracking have no strict demarcation, but usually parallel processing.

15.3.1 The Target Object Tracking in the Video Image

The target object tracking process is divided into four cycle steps: search—match—refresh—forecasting. There are different algorithms of target object tracking, which according to the image feature information of the tracking process is how to achieve the matching steps and to use the model in the motion estimation and prediction. After the visual system has been initialized, the humanoid robot will find the target object based on the image information given. Subsequently, the target object tracking begins. After each frame image acquisition, a large number of image information are analyzed in order to locate the matching part of the image with the desired characteristics. This characteristics information is selected according to the tracking target object, so that when the matching part of the image acquisition with the desired characteristics is found, the visual system locates the position of the target object.

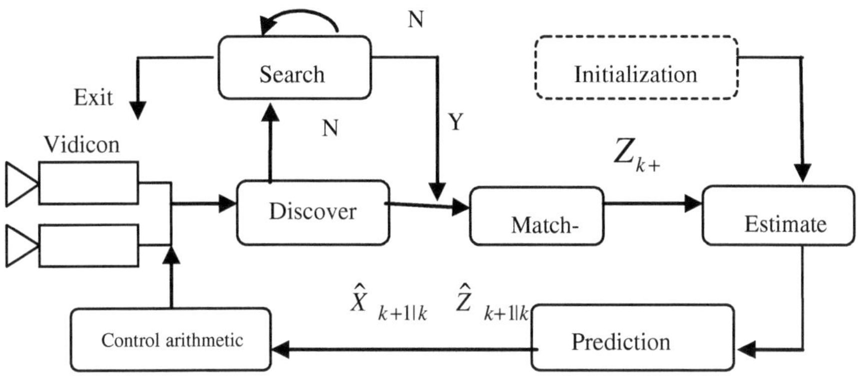

Fig. 15.2 Schematic of the target object tracking system

Figure 15.2 is a schematic of the target object tracking system, where X_k represents the target object state of the k frame image which includes the position and velocity information; and $X_{k+1|k}$ represents the target state of the $k + 1$ frame image. The X_k is the actual state of the target object, the $X_{k+1|k}$ is predicted through the motion model. Therefore, in the next frame, the visual system can quickly find and track the target object. Meanwhile, the predicted position information will be used to control the movement of the robot's head. If the target object is partially hidden from view or temporarily leaves the robot's field of view, the target object will be lost from the visual tracking system, the visual tracking system will continue to find the target object until the target object is rediscovered; in other words, in this case, the time of the visual system looking for the target object will be more than the predetermined time.

In real-time tracking system, motion estimation and prediction play very important roles. First, prediction of the movement condition, which includes position and velocity, can reduce the search area in the next frame, therefore reducing the amount of computation; second, the motion prediction reduces the time delay between the two systems of image processing and motion control loop, so that the tracking system responds more quickly.

The classic motion estimation is completed by the Kalman filter. The state of the Kalman filter is described by a four-dimensional space vector, which is expressed as

$$X(k) = (p(k), p^t(k))^T \tag{15.1}$$

$$p(k) = (x, y)^T \tag{15.2}$$

$$p'(k) = (x', y')^T, \tag{15.3}$$

where, $p(k)$ and $p'(k)$ are the position and velocity of the target object in the k frame image. Here, the movement of the target object is divided into a number of small time segments. Each segment can be approximated as a constant speed linear motion if the acceleration is zero. The expression formula uses Kalman filter for motion estimation and forecasting as the formula:

$$X_{k+1} = FX_k + Gv_k \tag{15.4}$$

$$Z_k = HX_k + w_k, \tag{15.5}$$

where $F = \begin{bmatrix} 1 & 0 & \Delta T & 0 \\ 0 & 1 & 0 & \Delta T \\ 0 & 0 & 1 & 0 \\ 0 & 0 & 0 & 1 \end{bmatrix}$, $G = \begin{bmatrix} \frac{\Delta T^2}{2} & 0 \\ 0 & \frac{\Delta T^2}{2} \\ \Delta T & 0 \\ 0 & -\Delta T \end{bmatrix}$, $H^T = \begin{bmatrix} 1 & 0 \\ 0 & 1 \\ 0 & 0 \\ 0 & 0 \end{bmatrix}$

X_{k+1}, X_k represent state vector position and velocity of the time k, respectively; Z_k represents output vector of the time k; v_k represents the acceleration of the time k; w_k represents the white noise information of the time k; ΔT represents the time periods between time k and $k+1$.

In the actual motion estimation and prediction process, we used the $\alpha - \beta$ filter, such as the formula (15.6)

$$\begin{cases} \hat{X}_{k+1|k} = F\hat{X}_{k|k} \\ \hat{Z}_{k+1|k} = H\hat{X}_{k+1|k} \\ \hat{X}_{k+1|k+1} = \hat{X}_{k+1|k} + W[Z_{k+1} - \hat{Z}_{k+1|k}] \end{cases} \tag{15.6}$$

where $W = \begin{bmatrix} \alpha & 0 & \frac{\beta}{\Delta T} & 0 \\ 0 & \alpha & 0 & \frac{\beta}{\Delta T} \end{bmatrix}^T$,

the $\hat{X}_{k|k}$ represents the estimate of the X_k when $Z_1, Z_2, \ldots, Z_k$ are known; the $\hat{X}_{k+1|k}$ represents the estimate of the X_{k+1} when $Z_1, Z_2, \ldots, Z_k$ are known; the $\hat{X}_{k+1|k+1}$ represents the estimate of the X_{k+1} when $Z_1, Z_2, \ldots, Z_{k+1}$ are known; the $\hat{Z}_{k+1|k}$ represents the estimate of the Z_{k+1} when $Z_1, Z_2, \ldots, Z_k$ are known.

15.3.2 The Realization of Vision Tracking

Visual servo control process is a cycle of real-time data acquisition, real-time decision-making, and real-time control. In the humanoid robot visual servo control system, the basis of the chosen device, it is assumed that a complete control cycle takes about m milliseconds. In visual information processing system, the processing of an image requires an average of about n milliseconds. Due to the different task characteristics of visual information processing and motion control, $n \gg m$, the visual information processing cycle is far greater than the cycle of motion control. In a visual information processing cycle, the humanoid robot system can complete a multicycle of motion control. Therefore, after a visual information processing cycle,

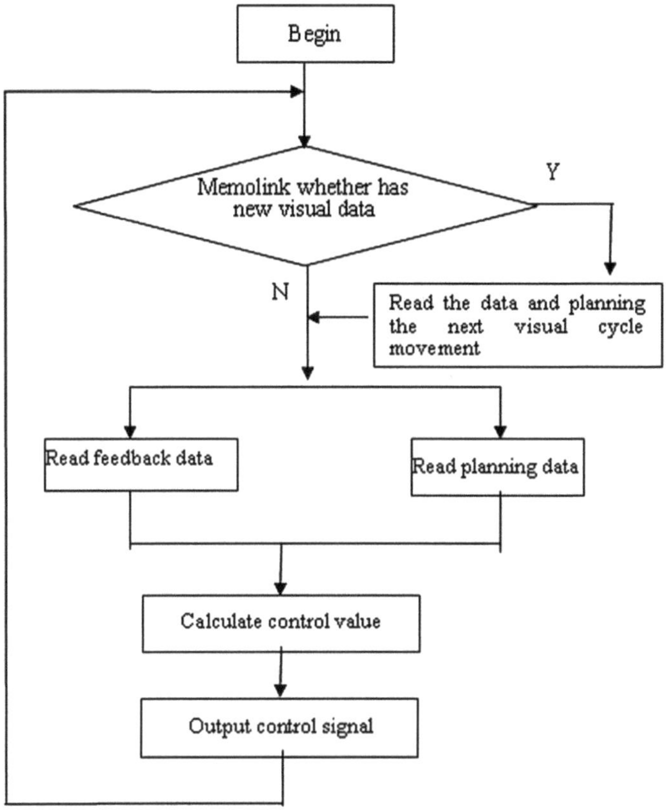

Fig. 15.3 Robot motion control software processes

the humanoid robot system should plan the motion within the next visual information processing cycle, that is, the multiple cycles motion planning should be done well, so as to ensure the robot's head is at a uniform, smooth, and accurate speed to track the target object.

Visual servo control process is shown in Fig. 15.3. In each motion control period, the motion control system needs to view Memolink at first, whether there is a new result of the visual information processing system by Memolink transmitted to the motion control system. If no new result, the program is according to the planning motion to control the robot movement; if there is a new result, the program must update the motion planning according to the result of the visual information system processing. In order to make the robot's head move smoothly, the time that we set in every motion planning cycle is slightly larger than the average of visual information processing cycle. We can ensure the original motion planning has not executed before the arrival of each new visual information processing result, which can make the robot head track the target object smoothly and stably.

Subsequently, according to the difference between the motion planning and feedback, the humanoid robot system calculates the control amount and then sends a control signal to control the rotation of the robot head.

Using traditional PID algorithm to calculate the amount of control, set $t(k)$ is the time of the k motion control cycle. At the time of $t(k)$, the system output is y_k, the value of motion planning is x_k, according to the PID algorithm the system output is y_{k+1} at the time of $t(k+1)$.

$$y_{k+1} = K_p(x_k - y_k) + K_i \sum (x_k - y_k) + K_d(x_k - y_k - x_{k-1} + y_{k-1}). \quad (15.7)$$

The above formula K_p, K_i, K_d are proportional coefficient, integral coefficient, and differential coefficient, respectively. In a control system, a certain integral coefficient can make the system with no accumulation error, but which will reduce the speed of response; a certain proportional coefficient can speed up the response speed of the system, and can respond in advance according to the change in input, but may cause system instability. Therefore, in the case the results are acceptable, we should only use the proportion coefficient, if the results fail to meet requirements, then use the integral coefficient and the differential coefficient.

Visual information system processing result is actually the result of the visual cycle processing, so that the control system input will always lag behind the actual target object position of a visual cycle. Therefore, if the motion planning for the robot head is only based on this input, the robot head will always lag the target object of a visual cycle. In order to reduce the impact on the results and the system lag, and also in order to speed up the response speed of the system, we use proportional derivative control in the system, here $K_i = 0$.

$$y_{k+1} = K_p(x_k - y_k) + K_d(x_k - y_k - x_{k-1} + y_{k-1}) \quad (15.8)$$

Using the formula, the control value y_{k+1} can be calculated. According to the control numerical number, the driver controls the motor running, then the humanoid robot head can be controlled to track the target object.

In practical applications, as the motion control differential coefficient is not zero, the robot head will have a certain overshoot and cause a swing of the head when tracking the target object to get to the equilibrium position. Meanwhile, since the target object image interferes with the environment, the robot head is also liable to cause a swing in the position of equilibrium. Therefore, the equilibrium position is not defined as the angle toward the target object, but is defined as a small angle range with the center of angle, as long as the target object is within the region corresponding in this small range, the robot head can be considered toward the target and need not adjust the direction of the robot head.

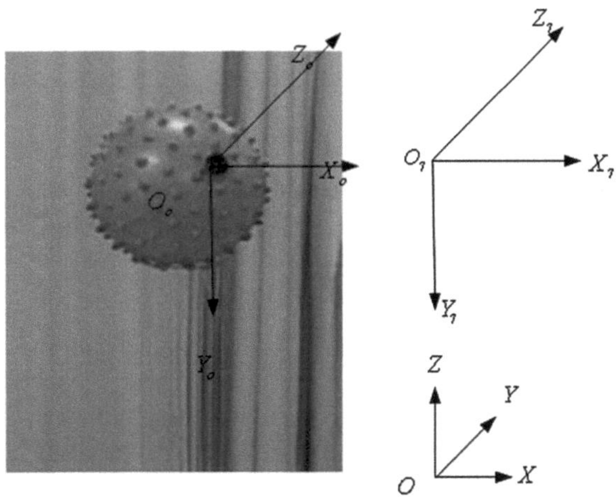

Fig. 15.4 The posture of target ball and hand

15.4 The Control Strategy of Approaching the Object

The literature [18] proposed a segmentation method based on multi-image information about the target object by using the depth, color, and shape information of the object to segment the target object image. It achieved the positioning of the target object and the robot hand, and the tracking of the target object.

The hand and arm of the humanoid robot imitates human arm movement to grasp the object according to the planning and servo control of the control system within the effective range which the robot arm can touch the object.

When the target object is not within the region that the robot arm can touch, the humanoid robot walks to approach the target object, using the movement of the shoulder and elbow. The robot hand approaches close to the target object, and then by the wrist and finger movements accomplishes grasp action. Through the two actions of walking and grasping, the human robot can complete the action of approach and grasp the target object.

This chapter selects a red ball as the target object. From grasp analysis, the target ball has the same posture whatever to place the target ball. As shown in Fig. 15.4, $O - XYZ$ is the world coordinates system, $O_o - X_oY_oZ_o$ is the coordinates system for the target ball. In order to grasp the target ball accurately, the end-effector coordinates system $O_7 - X_7Y_7Z_7$ of the center of the hand is established before the implementation of the grasp operation.

$[O_x, O_y, O_z]^T$ is the three-dimensional spatial position of the target ball, and $[P_x, P_y, P_z]^T$ is the desired position of the humanoid robot hand. The desired posture equation of the humanoid robot hand is as shown in Eq. (15.9), the translation amounts between the target balls coordinates system $O_o - X_oY_oZ_o$ and the end-effector coordinates system $O_7 - X_7Y_7Z_7$ are $[\Delta x, \Delta y, \Delta z]^T$.

$$T = \begin{bmatrix} 1 & 0 & 0 & P_x \\ 0 & 0 & 1 & P_y \\ 0 & -1 & 0 & P_z \\ 0 & 0 & 0 & 1 \end{bmatrix} = \begin{bmatrix} 1 & 0 & 0 & O_x + \Delta x \\ 0 & 0 & 1 & O_y + \Delta y \\ 0 & -1 & 0 & O_z + \Delta z \\ 0 & 0 & 0 & 1 \end{bmatrix} \tag{15.9}$$

The grasping object process of humanoid robot can be divided into five steps: find and coarse positioning target object, walking close to the target object, based on the visual feedforward coarse alignment, fine alignment based on visual feedback, grasp. Each step requires a different control strategy.

(1) The humanoid robot turns the head to look for the object ball, and roughly locates the position of the target ball.
(2) According to the three-dimensional position information of the target ball, the humanoid robot selects a suitable gait to walk close to the target ball, and ensures the target ball is located within the range of the humanoid robot arm operation region. Then standing before the target ball, the humanoid robot controls head alignment with the target ball and accurately measures three-dimensional information of the target ball.
(3) According to the target ball three-dimensional information, the humanoid robot controls the hand to reach the region which is near the target ball based on the visual feedforward control strategy, which realizes the end-effector coarse alignment with the target ball. At this time, the target ball and the hand of the robot are within the field of robot view.
(4) The three-dimensional information of the robot hand and the object ball are measured at the same time, the use of visual feedback achieves the hand fine alignment with the target ball.
(5) When the distance between the robot hand and the object ball is within grasp region, the control system of humanoid robot controls the hand to perform the grasp action.

In the coarse alignment process of the hand with the target ball, the humanoid robot uses the visual feedforward control strategy, which is commonly referred to the control of Look-then-Move. Figure 15.5 is the control process schematic diagram of Look-then-Move. It can be seen from the diagram that the entire control process is a serial process. Vision has not been introduced to the robot feedback, therefore, the control accuracy will be subject to the precision of the robot model, the camera model accuracy, precision of image feature extraction, accuracy of the posture estimation, where any error may affect the final positioning error. In this chapter, the coarse alignment of the robot hand with the object ball has been realized based on visual feedforward, guiding humanoid robot hand approaching the target ball, the hand, and the object ball within the robot vision field. At the same time, the model error of the system will not affect the final result, but can save time in approaching the target ball.

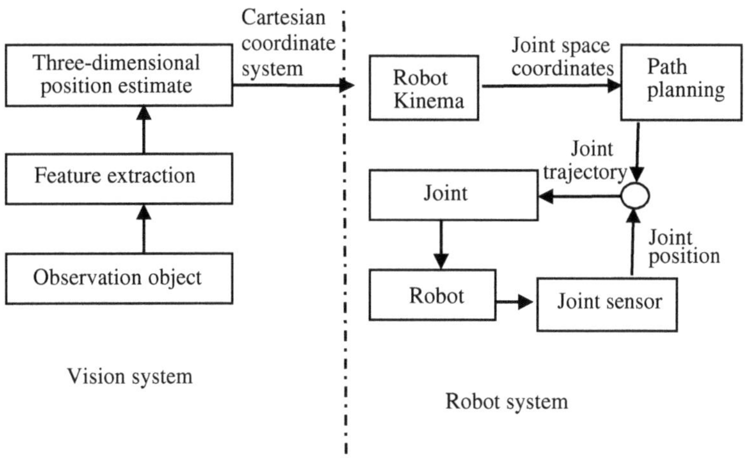

Fig. 15.5 Visual servo process based on the visual forward

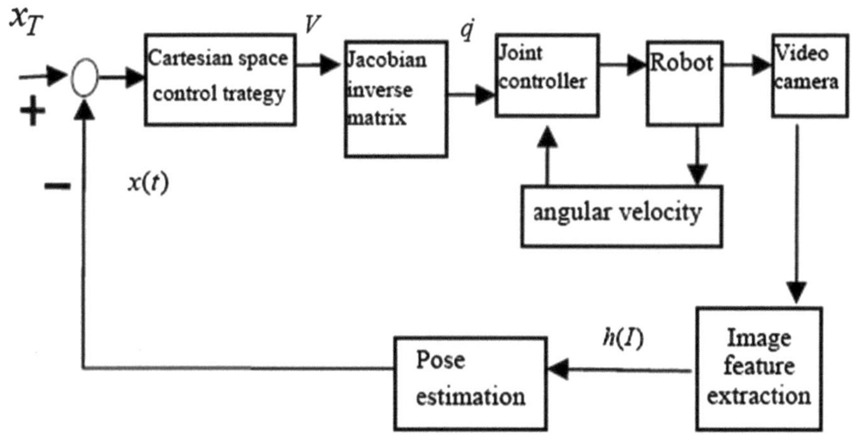

Fig. 15.6 The control process based on the location

15.5 The Control Strategy of Grasp Object Accurately

Based on the visual feedforward motion control strategy, the robot hand reaches the vicinity of the target object, the target object, and robot hand within the region of humanoid robot vision. Then, using the visual servo control based on location, the humanoid robot makes its hand fine alignment with the target object and implements grasp.

The control process based on the location as shown in Fig. 15.6. Visual servo control based on location uses visual feedback to compensate for hand-eye model error. Visual servo task is described by kinematic error function $E(x)$ of the end effect and the target object. The purpose of visual servo control is to make the error function $E(x = x^T) = 0$, where $x(t)$ represents three-dimensional position information of the hand center, and x^T is the current location information of the target ball.

$$E(x) = x_T - x(t). \tag{15.10}$$

The proportional controller is to ensure the error index decrease, the design of the proportional controller as shown in Eq. (15.11)

$$\dot{x}(t) = k \times E(x(t)) = k(x_T - x(t)). \tag{15.11}$$

The desired speed of the hand is proportional to the distance between the hand and the target object.

The whole process can be divided into the following steps:

(1) Each cycle, the robot vision system obtains the target ball and hand three-dimensional visual information, $x(t)$ is the three-dimensional position of the hand, and x_T is the three-dimensional position of the object ball.
(2) The humanoid robot control system calculates the kinematics error function $E(x) = x_T - x(t)$, and calculates the hand velocity according to the visual controller. This chapter uses a proportional controller to calculate velocity of the hand in the three-dimensional, calculated method as shown in Eq. (15.11).
(3) According to $[D] = [J][D_\theta]$, the three-dimensional differential movement of the hand can be translated into the differential movement of the joint space, and the differential velocity of every joint can be calculated. Control system makes suitable trajectory according to the differential motion of the joint space and executes the movement, and each joint movement in accordance with the joint control strategy.
(4) When the robot hand movements are close to the target object, the kinematic error function $E(x) = x_T - x(t)$ tends to zero; the whole servo process is over. If the error function $E(x) = x_T - x(t)$ is not equal to zero, the cycle continues.

15.6 Experiments

Based on the chapter that proposes control strategies, the humanoid robot grasps a red target ball of diameter 80 mm. Figure 15.7 is the video sequence of the humanoid robot grasping the target ball, the image segmentation of the target ball and hand in the whole process is stable. Figure (a) is the exact location of the target ball; figure (b) and (c) are the coarse alignment process based on visual feedforward; humanoid robot walks close to the target ball which is within the region of the humanoid robot grasp; figure (d), (e), and (f) are the fine alignment process based on the visual feedback; the humanoid uses the visual feedback to achieve precise alignment of the target object with the hand, and then grasps the red ball.

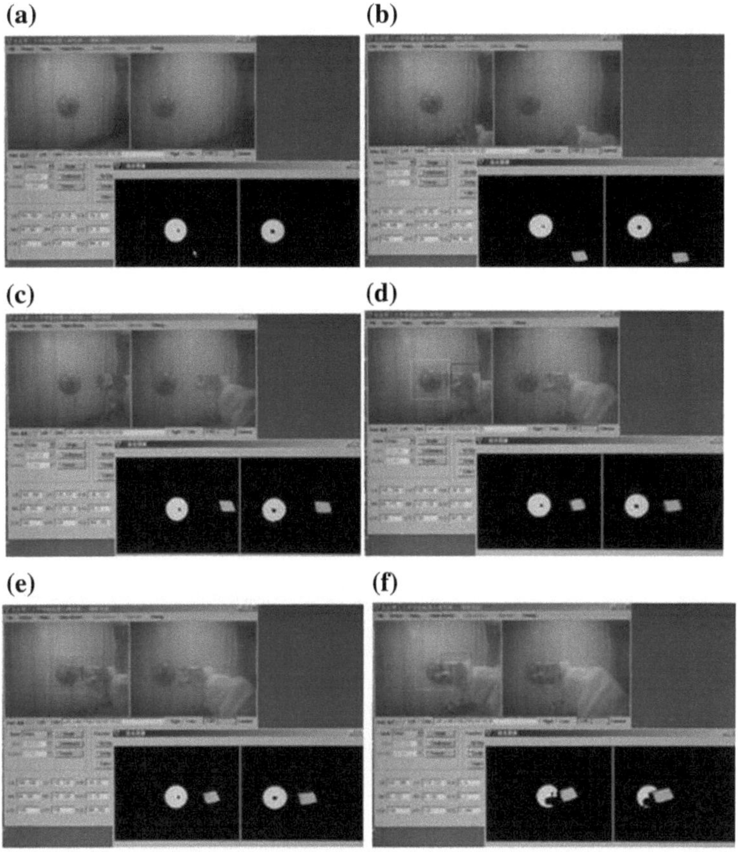

Fig. 15.7 The video sequence of the humanoid robot grasps the target red ball

In the experiments, the humanoid robot walks close to the target object, which is the target ball within the grasp region, and then turns its head aiming at the ball, at this time, the ball is located in the center of the left camera. The three-dimensional position of the object ball is measured by the humanoid robot vision system is $(-15.4, 384.7, 72.5)$. The humanoid robot controls the right arm to move close to the target ball according to the three-dimensional location information of the target ball, which achieves the coarse alignment of the hand with the target ball. The three-dimensional position of the hand is $(68.2, 335.9, 95.8)$ after completing the coarse alignment. Then, the humanoid robot adjusts the position of the hand to achieve the precise alignment with the target ball. The three-dimensional position of the hand is $(10.5, 329.5, 87.5)$ after completing fine alignment. The humanoid robot controls its finger to grasp the object ball after precise alignment.

Figure 15.8 is the second joint trajectory of the humanoid robot right arm grasp process. From the figure it can be seen that the trajectory is divided into two stages

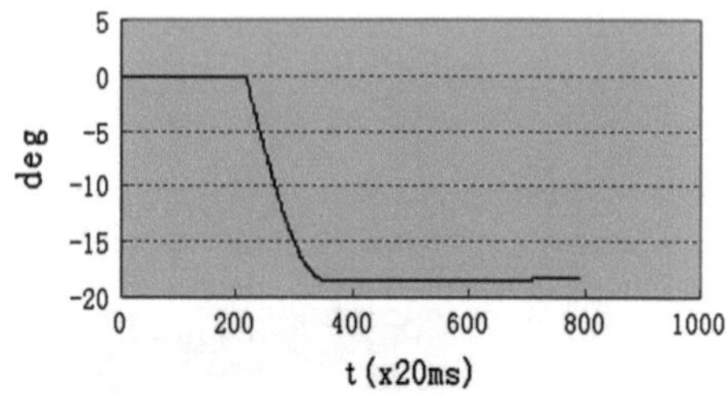

Fig. 15.8 The grasp process of the humanoid robot right arm second joint

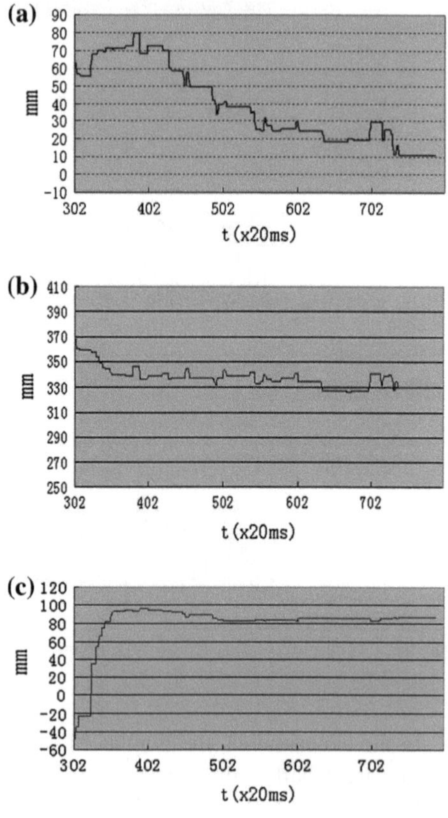

Fig. 15.9 The right hand trajectory of grasp process

of rough alignment and precise alignment . Approximately 200 to 400 cycles are the coarse alignment stage in the abscissa; the arm moves fast in this stage. From 400 to 800 cycles, the robot completes the fine alignment of the hand with the target ball; each joint of the robot adjusts in a small range in this stage.

Figure 15.9 is the robot hand trajectory measurement by the visual system in the world coordinates system. In order to ensure the robot safety, the waiting time is set in the control program to monitor the operating state of the robot. Figure 15.9 abscissa represents the 302–400 cycles for the coarse alignment stage; from 400 to 800 cycles for the precise alignment stage. In the coarse alignment stage, the direction of the hand movement is lifted up; in the robot fine alignment stage, the hand moves left and right or moves forward and backward. Owing to visual platform systematic errors, there is a big error between the hand and the target ball after the coarse alignment; based on visual feedback, the humanoid robot realizes the precise alignment of the robot hand with the target ball. Visual system measurement period is greater than the robot servo cycle, and therefore the curves are ladder shaped sometimes as shown in Figs. 15.8 and 15.9.

15.7 Conclusions

This chapter proposes visual servo control strategy integrating visual feedforward and visual feedback for non-strict calibration humanoid robot hand-eye system. Based on feedforward visual information the robot arm is guided to approach the target object, and then based on visual feedback control strategy the precise alignment of the humanoid robot hand with the target object is achieved. The visual feedforward reduces the time of the hand close to the target object, the visual feedback improves the accuracy of alignment. With this visual servo control strategy, the humanoid robot can realize the target object location, walking close to the target object; it grasps the target object in the laboratory environment.

Acknowledgments This research was supported by a grant from the National Natural Science Foundation of China, No: 61175090, the Natural Science Foundation of Shandong Province, No: ZR2010FL001.

References

1. Tao G, Mark L, Martin H (2011) Transferring human grasping synergies to robot. Mechatronics 21(1):272–284
2. Young SC, Kang GK, Ji YL et al (2011) MAHRU-M: a mobile humanoid robot platform based on a dual-network control system and coordinated task execution. Robot Auton Syst 59(6):354–366
3. Bedi P, Qayumi K, Kaur T (2012) Home security surveillance system using multi-robot system. Int J Comput Appl Technol 45(4):272–279
4. Henmi T, Deng MC, Lnoue A (2009) Swing-up controller for the acrobot using skill of human movements. Int J Model Ident Control 6(3):222–229
5. Dahiya RS, Metta G, Valle M, Sandini G (2010) Actilesensing: from humans to humanoids. IEEE Trans Robot 26(1):1–20
6. Du QJ, Shi X, Lu EQ et al (2012) Motion control system analysis and design for a humanoid robot. Int J Model Ident Control 17(2):158–165
7. Jia DY, Huang Q, Tian Y et al (2009) Object manipulation of a humanoid robot based on visual feedforward and visual feedback. Trans Beijing Inst Technol 29(11):983–987
8. Janez P, Marko M (2006) Robotassisted evaluation of coordination between grasp and load forces in a power grasp in humans. Adv Robot 20(8):933–951
9. Ude A (2004) Programming full-body movements for munanoid robots by observation. Robot Auton Syst 47(2–3):93–108

10. Ee SN, Yokoi K, Kajita S et al (2007) Whole-boy motion generation integrating operator's intention and robot's autonomy in controlling humanoid robots. IEEE Trans Robot 23(4):763–775
11. Wang HW, Yu SH (2010) Tracking control of robot manipulators based on orthogonal neural network. Int J Model Ident Control 11(1–2):130–135
12. Alexander GE, DeLong MR, Crutcher MD (1992) Do cortical and basal ganglionic motor areas use motor programs to control movement? Behav Brain Sci 15(1):656–665
13. Paulignan Y, MacKenzie C, Marteniuk R et al (1991) Selective perturbation of visual input during prehension movement. Exp Brain Res 83(1):502–512
14. Turetta A, Casalino G, Sorbara A (2008) Distribute control architecture for self-reconfigurable manipulators. Int J Robot Res 27(3–4):481–504
15. Cheng L, Xu WX, Wu HY et al (2012) A new procedure for multi-mode sequential flocking with application to multiple non-holonomic mobile robot motion control: mode description and integration principle. Int J Model Ident Control 15(1):39–47
16. Cheng L, Zheng XJ, Wu HY et al (2012) A new procedure for multi-mode sequential flocking with application to multiple non-holonomic mobile robot motion control: implementation and analysis. Int J Model Ident Control 16(1):50–59
17. Ge WM, Ye DF (2011) Sliding mode variable structure control of mobile manipulators. Int J Model Ident Control 12(1/2):166–172
18. Du QJ, Shi X, Dai B, Lu EQ (2013) Binocular stereo vision system for a humanoid robot. Int J Comput Appl Technol 46(4):316–322

Chapter 16
Low Intensity Laser Irradiation Influence Proliferation of Mesenchymal Stem Cells: Comparison of Experimental Data to Intelligent Agent-Based Model Predictions

Aya Sedky Adly, Mohamed H. Haggag and Mostafa-Sami M. Mostafa

Abstract Over the past several decades, evidences have shown that low intensity laser can stimulate a number of biological processes, including stem cell proliferation. In order to fully utilize stem cells in research and medical studies, understanding these processes is essential. However, for gaining this fundamental understanding in a rapid and cost-effective manner, model predictions and computer simulations are required as they may yield useful information and represent powerful supportive tools. This chapter provides some of the experiments employed to measure influence of low intensity laser on proliferation of mesenchymal stem cells which can vary considerably according to many parameters and biological conditions such as laser nature of emission, irradiation time, wavelength, and energy density. These experiments were compared to intelligent agent-based model predictions and detailed information about the model description and comparison results are provided. The model was capable of predicting the data for the scenarios fairly well although a few were somewhat problematic. This study recommends a wave length ranging from 600 to 680 nm, and an energy density ranging from 0.3 to 4.0 J/cm^2 for enhancing proliferation of mesenchymal stem cells.

Keywords Mesenchymal stem cells · Intelligent agent-based model · Low intensity laser irradiation · Proliferation · Biological systems

A. S. Adly (✉) · M. H. Haggag · M.-S. M. Mostafa
Department of Computer Science, Helwan University, Helwan, Egypt
e-mail: ayasedky@helwan.edu.eg

M. H. Haggag
e-mail: mohamed.haggag@fci.helwan.edu.eg

M.-S. M. Mostafa
e-mail: mostafa.sami@fci.helwan.edu.eg

L. Liu et al. (eds.), *Applied Methods and Techniques for Mechatronic Systems*, Lecture Notes in Control and Information Sciences 452,
DOI: 10.1007/978-3-642-36385-6_16,

16.1 Introduction

The effect of low intensity laser therapy on different fields has been known for over 40 years since the invention of lasers and has become a focus of recent research [1–3]. With respect to tissue engineering its effects have been reported in promoting healing [4–8], collagen synthesis [9, 10], relieving pain [11–14], decreasing inflammation [5, 15, 16], reducing edema [17, 18], stimulating cell proliferation [19–28], and facilitating cell differentiation [25, 27, 29]. Moreover, there are many evidences gathered concerning low intensity laser irradiation beneficial effects in the promotion of various cells survival and viability [30–32].

Low intensity laser irradiation promotes proliferation of multiple cell types, including mesenchymal stem cells (MSCs) [25–28], which is mainly through the activation of mitochondrial respiratory chain and the initiation of cellular signaling.

Stem cells are present throughout life, from the fertilized oocyte to the adult. They have remarkable regenerative ability, extensive proliferative potential, and the amazing ability to develop into many different cell types. Research on stem cells enables scientists to learn about the cells' essential properties and to advance their knowledge about their characteristics and behaviors in different situations. Stem cell research is one of the most fascinating areas of contemporary biology that rapidly generates new discoveries [33–38]. Today's scientific emerging interest in tissue engineering using stem cells stimulated us to develop a model to estimate, examine, and investigate the effects of low intensity laser irradiation on stem cells.

Mesenchymal stem cells are adult stem cells traditionally found in the bone marrow. However, they can also be isolated from other tissues including cord blood, peripheral blood, fallopian tube, and fetal liver and lung. Multipotent stem cells differentiate to form adipocytes, cartilage, bone, tendons, muscle, and skin. MSCs are a distinct entity to the mesenchyme, embryonic connective tissue which is derived from the mesoderm and differentiates to form hematopoietic stem cells. They are of importance for therapeutic use due to their special abilities. Morphologically, MSCs have long thin cell bodies with a large nucleus. As with other stem cell types, MSCs have a high capacity for self-renewal while maintaining multipotency. Thus, MSCs have enormous therapeutic potential for tissue repair and they have been shown to be capable of differentiating into multiple cell types including adipocytes, chondrocytes, osteocytes, and cardiomyocytes [39, 40].

16.2 Model Description

To conduct an experiment, a laser source, and a cell proliferation assay (or any suitable techniques to infer cell proliferation) are needed. However, to run the model, genomic sequences of the chromosomes along with their annotation databases are needed. The model may require setting some parameters and defining information on gene expression patterns according to different biological conditions along with gene expression changes measured according to cell type, organism part, developmental stage, disease state, and any other biological/experimental conditions of interest.

The model applies the idea of intelligent agents where its structure is defined as a specific environment in which agents are situated. The behavior of agents is influenced by states and types of agents that are situated in adjacent and at distant sites.

The capabilities of agents include both ability to adapt and ability to learn. Adaptation implies sensing the environment and reconfiguring in response. This can be achieved through the choice of alternative problem-solving rules or algorithms, or through the discovery of problem-solving strategies. Learning may proceed through observing actual data, and then it implies a capability of introspection and analysis of behavior and success. Alternatively, learning may proceed by example and generalization, and then it implies a capacity to abstract and generalize. We could group agents used in this work into five classes based on their degree of perceived intelligence and capability: (1) simple reflex agent (acts only on the basis of the current percept), (2) model-based reflex agent (keeps track of the current state of its world using an internal model, before choosing an action), (3) goal-based agent (store information regarding desirable situations, to select the one which reaches a goal state), (4) utility-based agent (distinguish between goal states and non-goal states and can define how desirable a particular state is), and (5) learning agent (can initially operate in unknown environments and to become more competent than its initial knowledge alone might allow). Agents' behavior is nondeterministic. Interactions and responding abilities are governed by a set of rules that define their behavior. Rules will often contradict and there must be some mechanism for selecting over them. Agents are not aware of the system macro-view although having the capability of responding to local environmental factors as well as external dynamics and signaling under various conditions. In addition, they are not able to know the state and current behavior of every other agent [41, 42].

The system development of this model is highly related to time and is considered at discrete time steps $t = \Delta t_m = m \times \Delta t$, $(m = 0, 1, 2, \ldots)$ with a fixed length Δt. The illustrations below show a very simple formula to calculate the population growth and estimations of proliferation parameters based on the initial number of cells:

$$\mathcal{N}\underline{o}_{\psi} = \mathcal{N}\underline{o}_{o}\left[(1-\mu)\sum_{i=0}^{\psi-1}\left[(1-\alpha)^{\psi-i}(2\mu(1-\rho)(1-D))^{i}\right] + (2\mu(1-\rho)(1-D))^{\psi}\right], \quad (16.1)$$

where $\psi = t/(\mathrm{CCT} + l_{G0})$ is the average number of divisions occurred during time t, l_{G0} is the time length that a cell resides in G_0 phase, CCT is the average cell cycle time, and is subdivided into check points l_{G1}, l_s, l_{G2}, l_M, finally $D = f(\overrightarrow{D}, \overleftarrow{D})$ is the differentiation index, the rest parameters are illustrated in Table 16.1.

The checkpoint l_{G1} which is located at the end of the cell cycle's G_1 phase, just before entry into S phase, is responsible for making the key decision of whether the

Table 16.1 Nomenclature

$\mathcal{N}_{\underline{\mathcal{O}}}(t)$	Number of cells at time t in population
$\mathcal{N}_{\underline{\mathcal{O}}_o}$	The initial number of cells in population
CCT	Average cell cycle time
Δt	Time step
ψ	The average number of divisions occurred during time t
l_{G1}	Checkpoint located at the end of the cell cycle's G_1 phase
l_S	Checkpoint located at the end of the cell cycle's S phase
l_{G2}	Checkpoint located at the end of the cell cycle's G_2 phase
l_M	Checkpoint located at the end of the cell cycle's M phase
l_{G0}	Checkpoint located at the end of the cell cycle's G_0 phase
μ	The mitotic index
D	The differentiation index
$\overrightarrow{D}$	The differentiation rate
$\overleftarrow{D}$	The dedifferentiation rate
ρ	The cell death index for normal cells
α	The cell death index for senescent cells
e	Number of available population crypts
E	Tendency matrix for population cells, $\mathcal{N}_{\underline{\mathcal{O}}\,\psi} \times e$ matrix valued functions of t
$E_{ij}(t)$	The tendency of the ith cell to the jth crypt at time t
ε	Number of available energy densities
S	Stimulation rates matrix of proliferation, $\varepsilon \times e$ matrix valued functions of t
$S_{kj}(t)$	Stimulation rate of proliferation at the kth energy intensity on the jth crypt at time t
F	Facilitation rates matrix of differentiation, $\varepsilon \times e$ matrix valued functions of t
$F_{kj}(t)$	Facilitation rate of differentiation at the kth energy intensity on the jth crypt at time t

cell should divide, delay division, or enter a resting stage and cease proliferation, i.e., become quiescent. Most cells stop at this stage and enter the resting state G_0 , in which case a cell does not act as an actual stem cell, but since it can reenter the cycle it has the potential to act as a stem cell.

G_1 phase, S phase, G_2 phase collectively known as *interphase* and M phase as *metaphase*. A newborn cell resides either in G_0, the nondividing state, or in G_1 until physiological parameters allow it to enter the S phase and to start replicating its genetic material. Metabolic activity, cell growth, and cell differentiation all occur during interphase and G_1 phase are considered the major period of cell growth during its lifespan [43, 44]

Therefore, a cell differentiation, which is possible only during the interphase, can occur with probability $d = 0$ when the cell is during the metaphase.

Although stem cells normally reside in a specific microenvironmental niche, they are also assumed to be able to reside in more than one environment and migrate from one crypt to another within their system according to their tendency to this crypt. Cell migration is a central process in the development and maintenance of multicellular organisms. Tissue formation during embryonic development, wound healing, and immune responses all require the movement of cells in a particular direction to a specific location [45, 46].

We can define the cell migration and cell tendency as:

e Number of available population crypts.
E Tendency matrix for population cells, $\mathcal{N}\underline{o}_{\psi} \times e$ matrix valued functions of t

where $E_{ij}(t)$ is the tendency ith of the cell to the jth crypt at time t. On every step a cell should take a decision whether to reside on its crypt or to migrate. This decision depends mainly on its current position, state, and its tendency matrix.

In addition, it is well known that low intensity laser irradiation significantly stimulates proliferation and facilitates differentiation of stem cells, which was confirmed by previous studies in this field [25–28].

We can define the effect of low intensity laser on population growth as:

ε Number of available energy densities.
S Stimulation rates matrix of proliferation, $\varepsilon \times e$ matrix valued functions of t.
F Facilitation rates matrix of differentiation, $\varepsilon \times e$ matrix valued functions of t.

where $S_{kj}(t)$ is the stimulation rate of proliferation at the kth energy intensity on the jth crypt at time t, and $F_{kj}(t)$ is the facilitation rate of differentiation at the kth energy intensity on the jth crypt at time t.

In fact, training the model by previous experimental data that have the same biological/experimental conditions may yield more accuracy in the model predictions, but it is not a necessity as long as all the required data are available. However, to ensure a meaningful modeling, we have conducted validation tests of the proposed model. By doing so, we can identify and compare the parameters of the model, according to a verified scenario.

To validate the model utility and performance, we have applied it to previously investigated and measured experiments. Each experiment investigated a different cell type. The first experiment was performed by Hou et al. [25] on proliferation of bone marrow derived MSC. The second was by Horvát-Karajz et al. [26] on proliferation of murine MSC. The third was by Leonida et al. [27] on proliferation of human MSC. The last was by Hana et al. [28] on proliferation of MSC.

16.3 Results

16.3.1 Effect of Low Intensity Laser on the Proliferation of Bone Marrow Derived Mesenchymal Stem Cells

Laser Irradiation

Mode of application	Continuous
Wavelength (nm)	635
Irradiation time (s)	75, 150, 300, 750
Power (mW)	60
Power density (mW/cm^2)	6.61
Energy (J)	4.5, 9, 18, 45

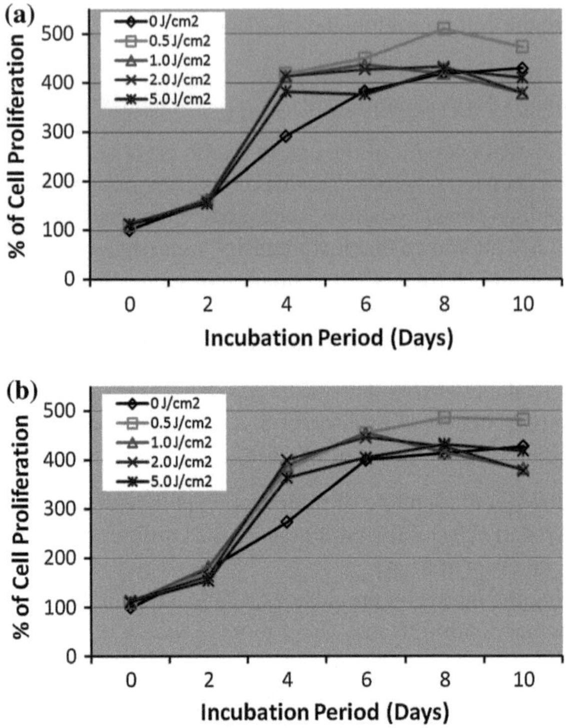

Fig. 16.1 Effect of low intensity laser irradiation on proliferation of bone marrow-derived MSC. **a** *Curves* of the actual experimental results of Jian-feng Hou et al. **b** *Curves* of the model results for the same experiment

Energy density (J/cm^2)	0.5, 1.0, 2.0, 5.0
Beam diameter (cm)	3.4
Beam area (cm^2)	9

Experimental Results

Hou et al. [25] measured the effect of laser irradiation on proliferation of bone marrow derived mesenchymal stem cells isolated from Sprague-Dawley rats. Cell numbers increased rapidly from day 0 to day 4 after low intensity laser treatment, and then reached a stationary phase by day 6. They also observed cell growth curves between the nonirradiated and irradiated groups throughout the cell-culture period. They indicated no significant differences between the nonirradiated and irradiated groups between day 0 and day 10. At day 2, groups irradiated at 0.5, 1.0, and 2.0 J/cm^2 showed significant higher numbers than that in the nonirradiated group. At day 4, numbers in all irradiated groups were higher than that in the nonirradiated group and there were no significant differences among the irradiated groups. At day 6, numbers in groups irradiated at 0.5, 1.0, and $0.2\,\text{J/cm}^2$ were significantly higher than that in the nonirradiated group. At day 8, only the group irradiated at $0.5\,\text{J/cm}^2$ showed significantly higher numbers than that in other groups including the nonirradiated group (Fig. 16.1a).

Model Predictions

The results computed after applying the model to their experiment showed no significant difference than the actual experimental results. However, in day 2, numbers in the nonirradiated and irradiated groups were slightly higher than the actual experimental results except for the group irradiated at $5.0\,\text{J/cm}^2$. At day 4, the model results were nearly the same for all groups except for groups irradiated at 0.5, and $1.0\,\text{J/cm}^2$ which has a lower number than that of the actual results. The overall results in all days seem to change more smoothly in the model results than that of the actual experimental results (Fig. 16.1b).

Observations

Low-level laser irradiation results in statistically significant increase in proliferation of bone marrow-derived mesenchymal stem cells and $0.5\,\text{J/cm}^2$ was found to be an optimal energy density in both the experimental results and the model predictions.

Conclusions

Low intensity laser can significantly enhance the proliferation of mesenchymal stem cells when appropriate laser irradiation parameters are used. This study recommends a wavelength ranging from 600 to 680 nm, and an energy density ranging from 0.3 to 6.0 J/cm^2

16.3.2 Effect of Low Intensity Laser on the Proliferation of Murine Mesenchymal Stem Cells

Laser Irradiation

Mode of application	Continuous
Wavelength (nm)	660
Irradiation time (s)	25, 50, 75, 25, 75
Power (mW)	60
Power density (mW/cm^2)	76, 76, 76, 94, 156
Energy (J)	1.5, 3, 4.5, 1.5, 4.5
Energy density (J/cm^2)	1.9, 3.8, 5.7, 2.35, 11.7

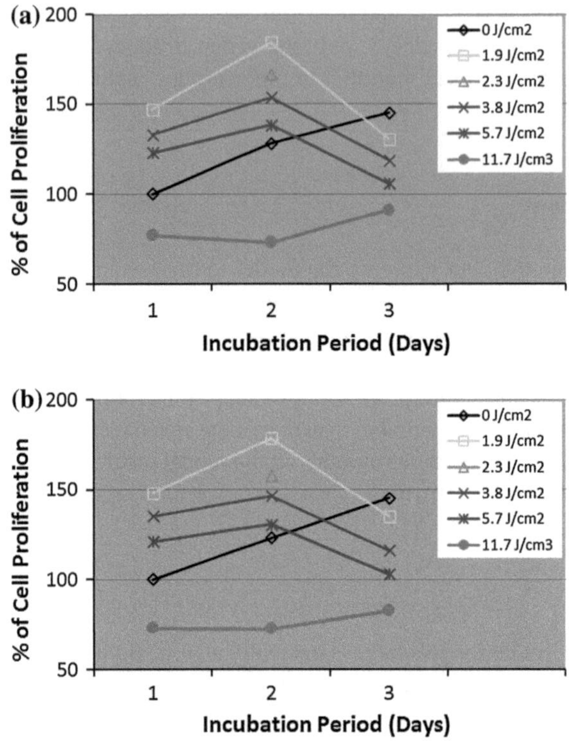

Fig. 16.2 Effect of low intensity laser irradiation on proliferation of human MSC. **a** *Curves* of the actual experimental results of Horvát-Karajz et al. **b** *Curves* of the model results for the same experiment

Beam diameter (cm)	1, 1, 1, 0.9, 0.7
Beam area (cm^2)	0.785, 0.785, 0.785, 0.636, 0.385

Experimental Results

Horvat-Karajz et al. [26] measured the effect of laser irradiation on proliferation of bone marrow derived mesenchymal stem cells isolated from mice. Cells were irradiated at different doses. The $1.9\,\mathrm{J/cm^2}$ dose resulted in the largest culture's growth during the first and second days of incubation. The same effect was observed with $3.8\,\mathrm{J/cm^2}$ dose after 1-day incubation and then with 2.3 and $3.8\,\mathrm{J/cm^2}$ dose after 2-day incubation. No significant changes were detected with 1.9 and $3.8\,\mathrm{J/cm^2}$ after 3-day incubation compared to nonirradiated group. Stimulatory activity was seen with $5.7\,\mathrm{J/cm^2}$ after 1 day and inhibition after 3-day incubation, while significant effect was not present after 2-day period. Higher dose $11.7\,\mathrm{J/cm^2}$ of laser achieved the inhibition of proliferation at 2- and 3-day incubation, but 2 days was not enough to present any significant effect with this dose (Fig. 16.2a).

Model Predictions

The results computed after applying the model to their experiment showed no significant difference than the actual experimental results. However, the higher dose $11.7\,\mathrm{J/cm^2}$ of laser achieved more inhibition of proliferation than that of the actual experimental results especially at 3-day incubation (Fig. 16.2b).

Observations

Low-level laser effectively increases the proliferation especially at the dose of $1.9\,\mathrm{J/cm^2}$ at 1- and 2-day incubation in both model predictions and actual experimental results.

16.3.3 Effect of Low Intensity Laser on the Proliferation of Human Mesenchymal Stem Cells

Laser Irradiation

Mode of application	Pulsed
wavelength (nm)	1064
Irradiation time (s)	0.135
Power (mW)	1500, 2250
Power density (mW/cm^2)	18750, 28100
Energy (J)	0.1, 0.15
Energy density (J/cm^2)	2.52, 3.81
Beam diameter (cm)	0.32, 0.226
Beam area (cm^2)	0.08, 0.04

Experimental Results

Leonida et al. [27] measured the effect of laser irradiation on proliferation of human mesenchymal stem cells. After the first week, proliferation of irradiated cells resulted in statistically significant increase in both groups compared to nonirradiated cells. After 2 weeks, irradiated cells showed the first signs of suffering as they did not record any increase in proliferation compared to nonirradiated cells although there was a significant increase of both nonirradiated cells proliferation and irradiated cells differentiation. These signs probably not directly related to laser treatment, since the process of cell proliferation decreases with increase in the process of cell differentiation (Fig. 16.3a).

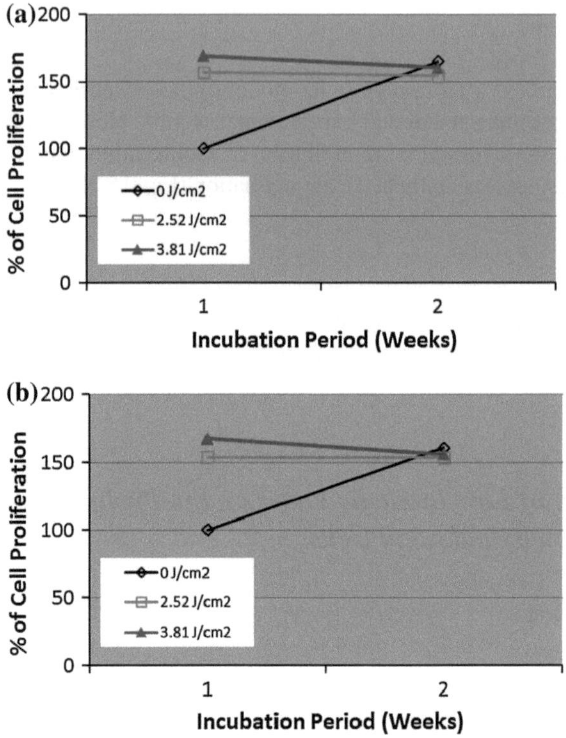

Fig. 16.3 Effect of low intensity laser irradiation on proliferation of human MSC. **a** *Curves* of the actual experimental results of Leonida et al. **b** *Curves* of the model results for the same experiment

Model Predictions

Comparing the effect of laser irradiation computed by the model to their experiment, it was noted that cell proliferation increased in both groups after a week of the actual experiment. After 2 weeks, cell numbers in all groups were somewhat similar. All the values were slightly less than that of the actual experimental results (Fig. 16.3b).

Observations

Statistically significant cell proliferation was observed 1 week after irradiation. There was no significant difference in proliferation between the two used power levels of irradiation. After 2 weeks, the number of cells contained in all groups was similar. There was no significant difference between the model results and the actual experimental results.

16.3.4 Effect of Low Intensity Laser on the Proliferation of Mesenchymal Stem Cells

Laser Irradiation

Mode of application	Pulsed
Wavelength (nm)	804
Irradiation time (s)	20, 60
Power (mW)	40
Power density (mW/cm^2)	50
Energy (J)	0.8, 2.4
Energy density (J/cm^2)	1, 3
Beam diameter (cm)	1.0
Beam area (cm^2)	0.785

Experimental Results

Hana et al. [28] measured the effect of laser irradiation on proliferation of mesenchymal stem cells isolated from charles river rats. The number of MSCs increased rapidly from the first week to the fourth week after low intensity laser treatment for both duration of 20 s at an energy density $1\,\mathrm{J/cm^2}$ and duration of 60 s at an energy density $3\,\mathrm{J/cm^2}$ as compared to nonirradiated cells. It would appear that the energy density of $1\,\mathrm{J/cm^2}$ caused a higher stimulation of cell proliferation than $3\,\mathrm{J/cm^2}$. The extent of enhancement of cell proliferation did not increase with time but leveled off or decreased after 3 weeks (Fig. 16.4a).

Model Predictions

The model results compared to their experiment demonstrate that numbers of MSC increased nearly the same as the experimental results although the enhancement increased more smoothly than that of the experimental results which can be noticed clearly in energy density of $1\,\mathrm{J/cm^2}$.

From the current findings, it would appear that the model accuracy is higher for short incubation periods than long ones (Fig. 16.4b).

Observations

Low-level laser irradiation significantly promotes the proliferation of both Mesenchymal and Cardiac stem cells. Enhancement in cell proliferation was found to be higher at 3 weeks post-culturing than at the rest for MSC and lower at 2 weeks than at 1 week for cardiac stem cells in both the experimental results and the model predictions.

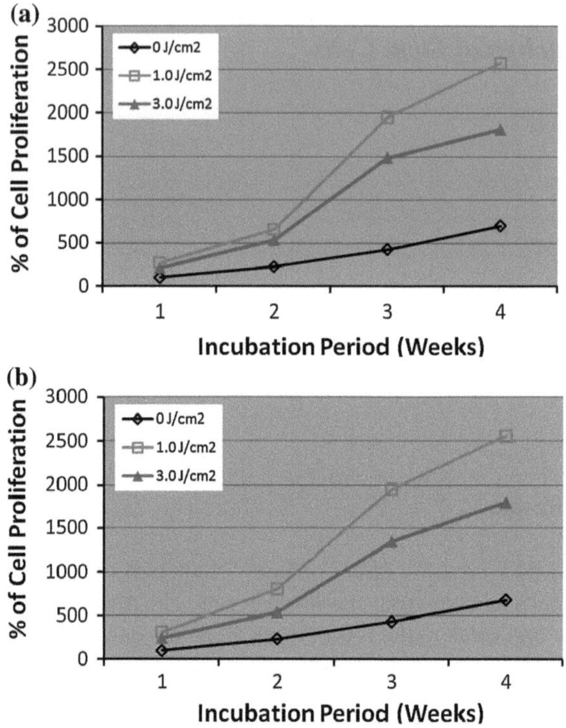

Fig. 16.4 Effect of low intensity laser irradiation on proliferation of MSC. **a** *Curves* of the actual experimental results of Hana Tuby et al. **b** *Curves* of the model results for the same experiment

Conclusion

Low intensity laser can significantly enhance the proliferation of mesenchymal stem cells when appropriate laser irradiation parameters are used. This study recommends a wavelength ranging from 600 to 680 nm, and an energy density ranging from 0.3 to 6.0 J/cm^2.

References

1. Splinter R (2006) An introduction to biomedical optics. CRC Press, Boca Raton
2. Tuner J, Hode L (2002) Laser therapy—clinical practice and scientific background. Prima Books, Grängesberg
3. Karu T (2007) Ten lectures on basic science of laser phototherapy. Prima Books, Grangesberg
4. Arany PR, Nayak RS, Hallikerimath S, Limaye AM, Kale AD et al (2007) Activation of latent TGF-beta1 by low-power laser in vitro correlates with increased TGF-beta1 levels in laser-enhanced oral wound healing. Wound Repair Regen 15:866–874. doi:10.1111/j.1524-475x.2007.00306.x

5. Arora H, Pai KM, Maiya A,Vidyasagar MS, Rajeev A (2008) Efficacy of He-Ne laser in the prevention and treatment of radiotherapy-induced oral mucositis in oral cancer patients. Oral Surg Oral Med Oral Pathol Oral Radiol Endod 105:180–186
6. Gal P, Mokry M, Vidinsky B, Kilik R, Depta F et al (2009) Effect of equal daily doses achieved by different power densities of low-level laser therapy at 635 nm on open skin wound healing in normal and corticosteroid-treated rats. Lasers Med Sci 24:539–547. doi:10.1007/s10103-008-0604-9
7. Woodruff L, Bounkeo J, Brannon W, Dawes K, Barham C, Waddell D, Enwemeka C (2004) The efficacy of laser therapy in wound repair: a meta-analysis of the literature. Photomed Laser Surg 22:241–247
8. Hawkins D, Abrahamse H (2007) Phototherapy - a treatment modality for wound healing and pain relief. Afr J Biomed Res 10:99–109
9. Lam TS, Abergel RP, Meeker CA, Castel JC, Dwyer RM, Uitto J (1986) Laser stimulation of collagen synthesis in human skin fibroblast cultures. Lasers Life Sci 1:61–77
10. Yu W, Naim JO, Lanzafame RJ (1994) The effect of laser irradiation on the release of bFGF from 3T3 fibroblasts. Photochem Photobiol 59(2):70–167
11. Djavid GE, Mehrdad R, Ghasemi M, Hasan-Zadeh H, Sotoodehmanesh A, Pouryaghoub G (2007) In chronic low back pain, low level laser therapy combined with exercise is more beneficial than exercise alone in long term: a randomized trial. Aust J Physiother 53:155–160
12. Karu T (1998) The science of low power laser therapy. Gordon and Breach Science Publishers, London
13. Iijima K, Shimoyama N, Shimoyama M, Yamamoto T, Shimizu T et al (1989) Effect of repeated irradiation of low-power He-Ne laser in pain relief from postherpetic neuralgia. Clin J Pain 5:271–274. doi:10.1097/00002508-198909000-00013. CrossRef PubMed/NCBI
14. Kemmotsu O, Sato K, Furomido H, Harada K, Takigawa C, Kaseno S (1991) Efficacy of low reactive-level laser therapy for pain attenuation of postherpetic neuralgia. Laser Ther 3:1–75
15. Qadri T, Bohdanecka P, Tunér J, Miranda L, Altamash M, Gustafsson A (2007) The importance of coherence length in laser phototherapy of gingival inflammation: a pilot study. Lasers Med Sci 22(4):245–251
16. Martin R (2003) Laser-accelerated inflammation/pain reduction and healing practical. Pain Manag 3:20–25
17. Stergioulas A (2004) Low-level laser treatment can reduce edema in second degree ankle sprains. J Clin Laser Med Surg 22(2):125–128
18. Enwemeka CS, Parker JC, Dowdy DS, Harkness EE, Sanford LE, Woodruff LD (2004) The efficacy of low power lasers in tissue repair and pain control: a meta analysis study. Photomed Laser Surg 22:323–329
19. Zhang L, Xing D, Gao X, Wu S (2009) Low-power laser irradiation promotes cell proliferation by activating PI3K/Akt pathway. J Cell Physiol 219:553–562
20. Gao X, Xing D (2009) Molecular mechanisms of cell proliferation induced by low-power laser irradiation. J Biomed Sci 16:4
21. Hu WP, Wang JJ, Yu CL, Lan CCE, Chen GS, Yu HS (2007) Helium-Neon laser irradiation stimulates cell proliferation through photostimulatory effects in mitochondria. J Inv Dermatol 127:2048–2057
22. Moore P, Ridgway TD, Higbee RG, Howard EW, Lucroy MD (2005) Effect of wavelength on low-intensity laser irradiationstimulated cell proliferation in vitro. Lasers Surg Med 36:8–12
23. Fujihara NA, Hiraki KRN, Marque MM (2006) Irradiation at 780 nm increases proliferation rate of osteoblasts independently of dexamethasone presence. Lasers Surg Med 38:332–336
24. Taniguchi D, Dai P, Hojo T, Yamaoka Y, Kubo T, Takamatsu T (2009) Low-energy laser irradiation promotes synovial fibroblast proliferation by modulating p15 subcellular localization. Lasers Surg Med 41(3):232–239
25. Hou JF, Zhang H, Yuan X et al (2008) In vitro effects of low-level laser irradiation for bone marrow mesenchymal stem cells: proliferation, growth factors secretion and myogenic differentiation. Lasers Surg Med 40(10):726–733

26. Horvát-Karajz K, Kovács V, Balogh Z, Sréter L, Uher F (2009) In vitro effect of carboplatin, cytarabine, paclitaxel, vincistine and low power laser irradiation on murine mesenchymal stem cells. Onkologie 32/4:216–217, Lasers Surg Med 41:463–469
27. Leonida A, Paiusco A, Rossi G et al (2013) Effects of low-level laser irradiation on proliferation and osteoblastic differentiation of human mesenchymal stem cells seeded on a three-dimensional biomatrix: in vitro pilot study. Lasers Med Sci 28(1):125–132
28. Hana Tuby, Maltz Lidya, Oron Uri (2007) Low-level laser irradiation (LLLI) promotes proliferation of mesenchymal and cardiac stem cells in culture. Lasers Surg Med 39:373–378
29. Stein E, Koehn J, Sutter W, Wendtlandt G, Wanschitz F et al (2008) Initial effects of low-level laser therapy on growth and differentiation of human osteoblast-like cells. Wien Klin Wochenschr 120:112–117. doi:10.1007/s00508-008-0932-6
30. Hawkins DH, Abrahamse H (2006) The role of laser fluence in cell viability, proliferation, and membrane integrity of wounded human skin fibroblasts following helium-neon laser irradiation. Lasers Surg Med 38:74–83
31. Arisu HD, Turkoz E, Bala O (2006) Effects of Nd:Yag laser irradiation on osteoblast cell cultures. Lasers Med Sci 21:175–180. doi:10.1007/s10103-006-0398-6
32. Soares CP, Oliveira DAAP, de Oliveira RF, Zangaro RA (2008) Evaluation of low-level laser therapy of osteoblastic cells. Photomed Laser Surg 26:401–404. doi:10.1089/pho.2007.2101
33. Siniscalco D, Giordano A, Galderisi U (2012) Novel insights in basic and applied stem cell therapy. J Cell Physiol 227:2283–2286
34. Regenerative medicine glossary (2009) Regener Med 4(4 Suppl):S1–88. doi:10.2217/rme.09.s1. PMID 19604041
35. Mason C, Dunnill P (2008) A brief definition of regenerative medicine. Regener Med 3(1):1–5. doi:10.2217/17460751.3.1.1. PMID 18154457
36. Riazi AM, Kwon SY, Stanford WL (2009) Stem cell sources for regenerative medicine. Methods Mol Biol 482:55–90. doi:10.1007/978-1-59745-060-7_5. PMID 19089350
37. Lysaght MJ, Crager J (2009) Origins. Tissue Eng Part A 15(7):50–1449. doi:10.1089/ten.tea.2007.0412. PMID 19327019
38. Placzek MR, Chung IM, Macedo HM et al (2009) Stem cell bioprocessing: fundamentals and principles. J R Soc Interface 6(32):32–209. doi:10.1098/rsif.2008.0442. PMC 2659585. PMID 19033137
39. Waterman R, Betancourt A (2011) Outside the operating room: unlimited directions in research and beyond. Ochsner J (Spring) 11(1):14–16. PMCID: PMC3096162
40. Lodi D, Iannitti T, Palmieri B (2011) Stem cells in clinical practice: applications and warnings. J Exp Clin Cancer Res 30:9. doi:10.1186/1756-9966-30-9
41. Adly AS, Aboutabl AE, Ibrahim MS (2011) Modeling of gene therapy for regenerative cells using intelligent agents. Adv Exp Med Biol 696:317–25. doi:10.1007/978-1-4419-7046-6_32, PMID 21431572
42. Adly AS, Kandil OA, Ibrahim MS et al (2010) Computational and theoretical concepts for regulating stem cells using viral and physical methods. Mach Learn Syst Eng 68:533–546. doi:10.1007/978-90-481-9419-3_41
43. Qu Z, MacLellan WR, Weiss JN (2003) Dynamics of the cell cycle: checkpoints, sizers and timers. Biophys J 85(6):3600–3611
44. Burdon T, Smith A, Savatier P (2002) Signalling, cell cycle and pluripotency in embryonic stem cells. Trends Cell Biol 12(9):432–438
45. Watt FM, Hogan BLM (2000) Out of Eden: stem cells and their niches. Science 287(5457):1427
46. Fuchs E, Tumbar T, Guasch G (2004) Socializing with the neighbors: stem cells and their niche. Cell 116:769–778

Chapter 17
Motor Soft Starter Based on Variable Reactance

Xiang Shi and Qinjun Du

Abstract When an industrial induction motor is started directly, the stator current is generally 5–7 times greater than the rated one, which makes the voltage sag of the transmission lines increase rapidly, furthermore, affecting the power quality seriously. However, applying variable reactance type motor soft starter to start motors, the starting current of the motor can be reduced effectively. The mathematical model of the soft starter which is based on variable reactance is established, and the relationship between the thyristor conduction angle and the motor current is analyzed. The purpose of the motor and equipment's protection is achieved by selecting the appropriate conduction angle to control the starting current amplitude.

Keywords Variable reactor · Soft starter · Thyristor · Induction angle

17.1 Introduction

When a three-phase induction motor is started directly, the starting current is very huge, about 5–7 times greater than the rated one [1], that it will impact on the power grid. Meanwhile, the starting torque is about 2 times greater than the rated one; the excessive start torque will increase mechanical wear abrasion and may cause serious equipment damage under extreme conditions [2]. Among the starting current, the reactive current accounts for the major parts [3–5]. The excessive reactive current may result in distribution line bad waveform, which reduces the quality of power supply and causes additional expenditure. With the aim to solve the motor startup

X. Shi · Q. Du (✉)
School of Electrical and Electronic Engineering, Shandong University of Technology,
12 Zhangzhou road, Zibo, 255091, Shandong, China
e-mail: qinjundu@sohu.com

X. Shi
e-mail: 116629689@qq.com

L. Liu et al. (eds.), *Applied Methods and Techniques for Mechatronic Systems*,
Lecture Notes in Control and Information Sciences 452,
DOI: 10.1007/978-3-642-36385-6_17,

problem, especially for high-power motor in the startup procedure, a method using the soft start appliance is proposed [6, 7].

There are many soft start methods applied in the industry. The first common one is the soft starter with series thyristor technique. This method is superior to other traditional starting ones, e.g., series resistor and autotransformer. Therefore, there are three aspects of problems that have to be addressed. The first question involves the requirement for component parameters consistency, which is very high. The second relates to the component properties for balancing voltage which performs poorly, after being applied for a period of time. The third deals with the damage to all components which may easily be caused, leading to the reduced reliability of the device. The situation is extremely serious for the high voltage motor soft start device.

The second common method is the liquid resistance soft starting, which is an energy loss type motor starting method. When this method is used, a large amount of energy is consumed in water resistance, and then gradually transferred to the motor. The weaknesses are (1) during the evaporation period, the water resistance is very small, while the current value is much greater than the rated one, which makes the voltage sag of the distribution line obviously affect the normal operation of other equipments; (2) liquid resistance is related to vaporization, plate, ambient temperature, and other factors, so the precise starting control is difficult; (3) the generating heat warms the water quickly, resulting in cool down time before next starting, so there is limitation on starting times [8–11].

The third common method is the inverter soft starting. This method has good performance on starting current and torque control, as its starting voltage and frequency can be continuously adjusted from zero, keeping the motor with a small slip value. But the weaknesses are obvious: (1) complex inverter circuit and high demands of the maintenance technology; (2) expensive inverter and spare parts; and (3) the inverter output has a large amount of harmonics, which is harmful to the motor.

While there are many reactive power compensation methods that have been applied in the industry, the first method is the synchronous condenser compensation. The synchronous condenser is a rotating-type reactive power compensation device. As a rotating machinery, it is comparatively poorer than a new static reactive power compensation device on the power loss, response speed, operation, and maintenance. So it has gradually died out.

The second common method is capacitor compensation. Reactive power can be compensated by using shunt power capacitor group. This is a kind of traditional reactive compensation method. Because the capacitor value is fixed and cannot adjust smoothly for the reactive power requirement, it cannot easily reach the needed power factor range to avoid overcompensation.

The third common method is thyristor-controlled reactor compensation. This method uses triacs to control main circuit current. However, the method can generate a large mount of harmonics and can easily lead to instability in power grid.

After analyzing these advantages and disadvantages, further studies on this problem will be summarized in this chapter. A cheaper and simpler method for motor start and reactive compensation is urgent to be proposed. In this chapter, a method

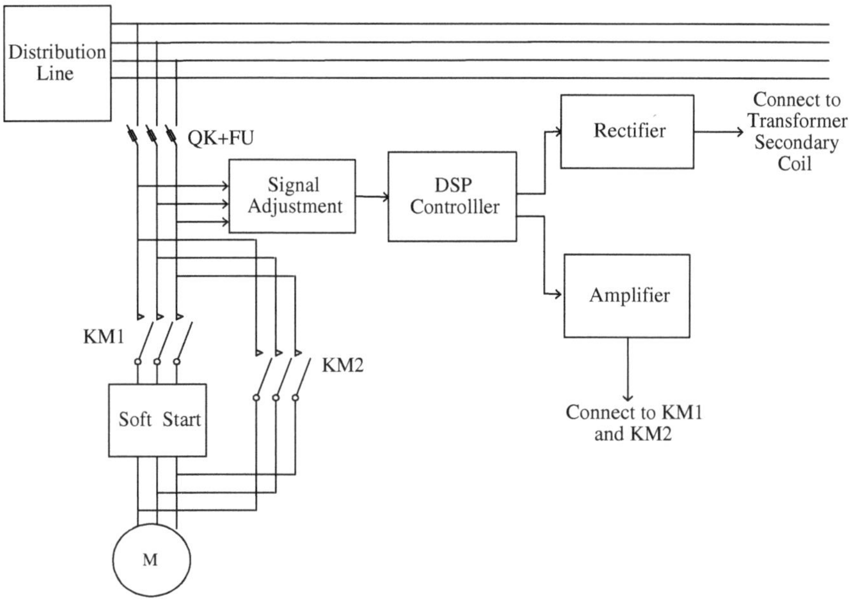

Fig. 17.1 Soft starter structure diagram

based on variable reactance is proposed. By using a combination of capacitors and magnetic controlled reactors, the excessive starting current can be reduced.

17.2 The Topology of the Induction Motor Soft Starter

The soft start device system structure is shown in Fig. 17.1. When we need to start the motor, the fuse switch closes at first, then the signal generated by the DSP controller is amplified, which drives the contactor KM1 to close and connects the soft start device and the induction motor. Under the control of the soft starter, the motor voltage increases gradually, and the motor speed increases smoothly. When the motor speed reaches the rated speed range, the DSP controller makes the contactor KM2 to close and then disconnects the contactor KM1. After that, the induction motor completes the soft start and runs at the rated speed.

17.2.1 Variable Reactance Transformation Principles

Magnetic control reactor soft starter is derived from the traditional reactor soft starter, using the controlled saturated reactor to buck the asynchronous motor stator voltage. The main circuit for motor starting consists of variable reactance converter's primary

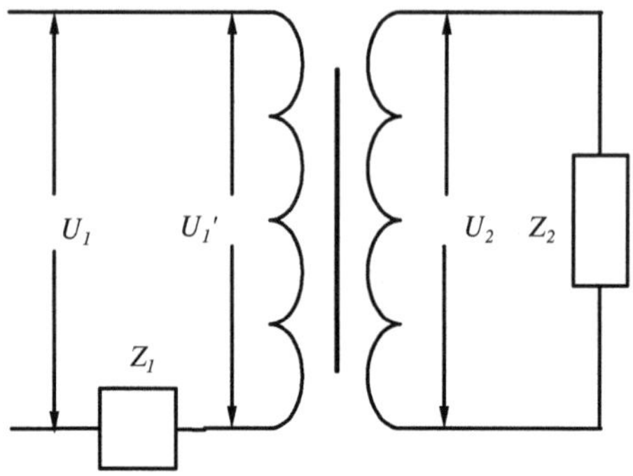

Fig. 17.2 Variable reactor equivalent circuit diagram

coil and induction motor, and they are connected in series. The variable reactance unit is constructed by the secondary coil and the rectifier bridge is controlled by DSP with load. The secondary reactance is changed by the DSP controller. Through the converter's reactance transformation, the secondary reactance change can be reflected on the primary side and can regulate the induction motor voltage and complete motor soft start [12–14].

17.2.2 The Equivalent Model of the Variable Reactor

The induction motor load is simplified as a reactor with the reactance value Z_1 [15], the obtained variable reactor as shown in Fig. 17.2.

According to the principle of reactance transformation, we can yield

$$\frac{U_1'}{I_1} = \frac{N_1}{N_2}U_2 \Big/ \left(\frac{N_2}{N_1}I_2\right) = \left(\frac{N_1}{N_2}\right)^2 \frac{U_2}{I_2} \tag{17.1}$$

where,

$$\frac{U_1'}{I_1} = Z_1',$$

$$\frac{U_2}{I_2} = Z_2,$$

$$\frac{U_1}{U_2} = K.$$

Hence,

$$\left|Z_1'\right| = \frac{U_1'}{I_1} = \left(\frac{N_1}{N_2}\right)^2 |Z_2| = K^2 |Z_2| . \tag{17.2}$$

Fig. 17.3 Variable reactor simplified model

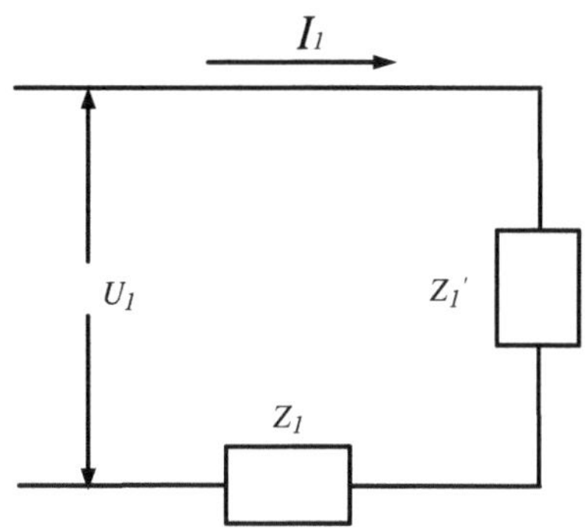

According to Kirchhoff's voltage law, we can yield

$$U_1 = I_1(Z_1 + Z_1'). \tag{17.3}$$

We can deduct,

$$I_1 = \frac{U_1}{(Z_1 + Z_1')}. \tag{17.4}$$

Formulas (17.3) and (17.4) are discussed as follows:

Therefore, the original circuit model can be equivalent as the simplified equivalent circuit as shown in Fig. 17.3.

Obviously, when the thyristors are full conducted, the impedance value $Z_1 + Z_1'$ gets its maximum value, and the conduct current value I_1 gets its minimum value; while the thyristors are full cut-off, the impedance value $Z_1 + Z_1'$ gets its minimum value, and the conduct current value I_1 gets its maximum value. Therefore, the adjustment of the thyristor conduct angle α can change the magnitude of the conduct current I_1, control the induction motor starting current, and complete the soft starting of induction motors.

17.3 The Mathematical Model of the Induction Motor Soft Starter

With the assumption of neglecting the three-phase bridge rectifier circuit resistance, inductance load commutation process, current ripple, the three-phase balance power supply is shown as follows [16–19]:

$$\begin{cases} e_a = E_m \sin(\omega t + \alpha) = \sqrt{2}E \sin(\omega t + \alpha) \\ e_b = E_m \sin(\omega t + \alpha - \frac{2}{3}\pi) = \sqrt{2}E \sin(\omega t + \alpha - \frac{2}{3}\pi) \\ e_c = E_m \sin(\omega t + \alpha + \frac{2}{3}\pi) = \sqrt{2}E \sin(\omega t + \alpha + \frac{2}{3}\pi) \end{cases} \tag{17.5}$$

where, E_m is the maximum value of the supply voltage;

E is the effective value of the supply voltage;

α is the trigger delay angle.

The current can be divided into two parts: the positive and negative half-cycle parts. Both parts are the square wave currents, three phase current wave are the same, and in turn differ by 120°, the relationship between rms value and the dc current is [20]:

$$I = \sqrt{\frac{2}{3}} I_d. \tag{17.6}$$

Also, the current waveform can be decomposed into Fourier series. Taking the phase A current for example, we can choose the midpoint between the current positive part and negative half wave as zero, the time point then yields [21, 22]:

$$\begin{aligned} i_a &= \frac{2\sqrt{3}}{\pi} I_d \left[\sin \omega t - \frac{1}{5} \sin 5\omega t - \frac{1}{7} \sin 7\omega t + \frac{1}{11} \sin 11\omega t + \cdots \right] \\ &= \frac{2\sqrt{3}}{\pi} I_d \sin \omega t + \frac{2\sqrt{3}}{\pi} I_d \sum_{\substack{n=6k\pm1 \\ k=1,2,3,\ldots}}^{\infty} (-1)^k I_n \frac{1}{n} \sin n\omega t \\ &= \sqrt{2} I_1 \sin \omega t + \sqrt{2} I_d \sum_{\substack{n=6k\pm1 \\ k=1,2,3,\ldots}}^{\infty} (-1)^k I_n \sin n\omega t. \end{aligned} \tag{17.7}$$

From formula (17.7) we can get the current effective value of fundamental and harmonics, respectively:

$$\begin{cases} I_1 = \frac{\sqrt{6}}{\pi} I_d \\ I_n = \frac{\sqrt{6}}{n\pi} I_d \end{cases} \\ (n = 6k \pm 1, k = 1, 2, 3, \ldots). \tag{17.8}$$

From formulas (17.6) and (17.8), the fundamental wave factor can be obtained as follows:

$$\upsilon = \frac{I_1}{I} = \frac{3}{\pi} \approx 0.955. \tag{17.9}$$

Therefore, the fundamental current in the secondary coil current is the main factor affecting the circuit current magnitude [23].

According to the Fourier decomposition formula the transformer secondary coil current can be decomposed as follows:

$$i_2(\omega t) = \sum_{\substack{n=6k\pm1 \\ k=1,2,3,\ldots}}^{\infty} (a_n \cos n\omega t + b_n \sin n\omega t) \tag{17.10}$$

So the fundamental current factor can be obtained as

$$\begin{aligned} a_1 &= \frac{1}{\pi}\int_{-\pi}^{\pi} i(\omega t)\cos\omega t \mathrm{d}wt \\ &= \frac{2}{\pi}\int_{0}^{\pi} i(\omega t)\cos\omega t \mathrm{d}wt \\ &= \frac{2}{\pi}\int_{\alpha}^{\pi} \frac{\sqrt{2}U_2}{\left|Z_1'\right|}\sin\omega t\cos\omega t \mathrm{d}wt \\ &= \frac{\sqrt{2}U_2}{2\pi\left|Z_1'\right|}(\cos 2\alpha - 1) \end{aligned} \tag{17.11}$$

$$\begin{aligned} b_1 &= \frac{1}{\pi}\int_{-\pi}^{\pi} i(\omega t)\sin\omega t \mathrm{d}wt \\ &= \frac{2}{\pi}\int_{0}^{\pi} i(\omega t)\sin\omega t \mathrm{d}wt \\ &= \frac{2}{\pi}\int_{\alpha}^{\pi} \frac{\sqrt{2}U_2}{\left|Z_1'\right|}\sin\omega t\sin\omega t \mathrm{d}wt \\ &= \frac{\sqrt{2}U_2}{2\pi\left|Z_1'\right|}[(\sin 2\alpha + 2(\pi - \alpha)] \end{aligned} \tag{17.12}$$

Yields

$$i_2(\omega t) \approx a_1\cos\omega t + b_1\sin\omega t. \tag{17.13}$$

According to formula (17.13), we can obtain

$$|i_2| \approx \sqrt{\frac{a_1^2 + b_1^2}{2}}. \tag{17.14}$$

And also yields

$$\begin{aligned} |i_2| &\approx \frac{U_2}{2\pi\left|Z_1'\right|}\sqrt{(\cos 2\alpha - 1)^2 + [\sin 2\alpha + 2(\pi - \alpha)]^2} \\ &= \frac{U_2}{\pi\left|Z_1'\right|}\sqrt{\sin^2\alpha + (\pi - \alpha)\sin 2\alpha + (\pi - \alpha)^2} \end{aligned} \tag{17.15}$$

Yields

$$\left|Z_1'\right| = \frac{U_2}{I_2} = \frac{\pi\left|Z_1'\right|}{\sqrt{\sin^2\alpha + (\pi - \alpha)\sin 2\alpha + (\pi - \alpha)^2}}. \tag{17.16}$$

Formula (17.16) shows the equivalent impedance of reactance converter changes as well as the thyristor conduction angle changes; so the reactance converter can be equivalent to the variable impedance regulator.

17.4 Instantaneous Reactive Power Detection

17.4.1 The Theory of Instantaneous Reactive Power

The instantaneous reactive power theory was put forward by Hirofumi Akagi, who is a Japanese scholar. The theory is improved step-by-step after being verified by Quade and Akagi and other deep researches made on it. Now, the theory of instantaneous reactive power is applied well in the field of reactive current and harmonic detection.

The core of the theory can be stated as follows briefly: Using the 3/2 coordinate transformation (Clark transformation), the expression of voltage and current in abc three-phase coordinate system can be re-expressed in $\alpha\beta$ two-phase coordinate system, and the concepts of the instantaneous active power P and instantaneous reactive power Q can be established relatively [24].

Using Clark transformation, the instantaneous voltage and current in three-phase circuits can be transformed into $\alpha\beta$ two-phase coordinate system.

$$\begin{bmatrix} u_\alpha \\ u_\beta \end{bmatrix} = C_{32} \begin{bmatrix} u_a \\ u_b \\ u_c \end{bmatrix} \tag{17.17}$$

where u_a, u_b, u_c are instantaneous voltages in three-phase circuit;

u_α, u_β are instantaneous voltages in $\alpha\beta$ coordinate system.

$$\begin{bmatrix} i_\alpha \\ i_\beta \end{bmatrix} = C_{32} \begin{bmatrix} i_a \\ i_b \\ i_c \end{bmatrix}. \tag{17.18}$$

Relatively, i_a, i_b, i_c are instantaneous currents in three-phase circuit;

i_α, i_β are instantaneous currents in $\alpha\beta$ coordinate system.

C_{32} is Clarke transform and expressed as follows:

$$C_{32} = \sqrt{\frac{2}{3}} \begin{bmatrix} 1 & -1/2 & -1/2 \\ 0 & \sqrt{3}/2 & -\sqrt{3}/2 \end{bmatrix} \tag{17.19}$$

17.4.2 The Implementation of Instantaneous Reactive Power Theory

The relationship between three-phase abc coordinate system and $\alpha\beta$ coordinate system has been given in Fig. 17.4. Then, the specific implementation of reactive current detection in variable reactor system is analyzed as follows:

In the instantaneous reactive power theory, the grid voltage can be expressed as

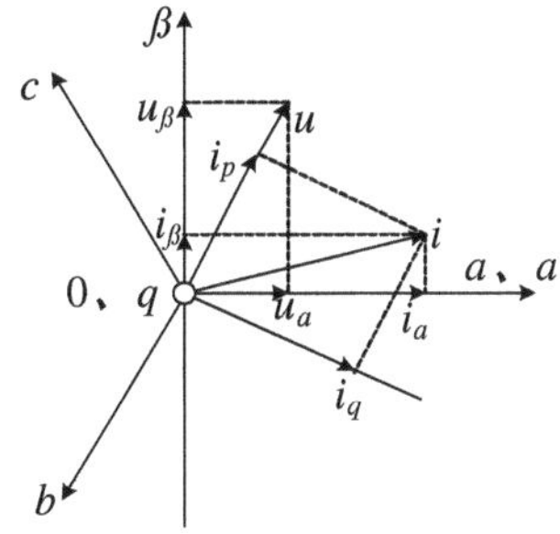

Fig. 17.4 The relationship between three-phase abc and $\alpha\beta$ coordinate system

$$\begin{cases} U_a = U_m \sin \omega t \\ U_b = U_m \sin (\omega t - 2\pi/3) \\ U_c = U_m \sin (\omega t + 2\pi/3) \end{cases} \tag{17.20}$$

where, U_m is the voltage magnitude of power grid;

U_a, U_b, and U_c are three-phase power grid voltages.

Load currents are presented as

$$\begin{cases} i_{La} = \sum I_{Lan} \sin [n(\omega t) - \theta_{an}] \\ i_{Lb} = \sum I_{Lbn} \sin [n(\omega t - 2\pi/3) - \theta_{bn}] \\ i_{Lc} = \sum I_{Lcn} \sin [n(\omega t + 2\pi/3) - \theta_{cn}] \end{cases} \tag{17.21}$$

where, i_{La}, i_{Lb}, and i_{Lc} are the load currents.

Using Clark transform, U_a, U_b, and U_c can be transformed into $\alpha\beta$ coordinate system as U_α, U_β:

$$\begin{bmatrix} U_\alpha \\ U_\beta \end{bmatrix} = \sqrt{\frac{2}{3}} \begin{bmatrix} 1 & -1/2 & -1/2 \\ 0 & \sqrt{3}/2 & -\sqrt{3}/2 \end{bmatrix} \begin{bmatrix} U_a \\ U_b \\ U_c \end{bmatrix}. \tag{17.22}$$

Relatively, i_{La}, i_{Lb}, and i_{Lc} can be transformed as i_α, i_β:

$$\begin{bmatrix} i_\alpha \\ i_\beta \end{bmatrix} = \sqrt{\frac{2}{3}} \begin{bmatrix} 1 & -1/2 & 1/2 \\ 0 & \sqrt{3}/2 & -\sqrt{3}/2 \end{bmatrix} \begin{bmatrix} i_{La} \\ i_{Lb} \\ i_{Lc} \end{bmatrix} \tag{17.23}$$

In the $\alpha\beta$ coordinate system, instantaneous active power of variable reactor system is defined as

$$p = u_\alpha i_\alpha + u_\beta i_\beta. \tag{17.24}$$

Similarly, instantaneous reactive power of the system is defined as

$$q = u_\alpha i_\alpha - u_\beta i_\beta. \tag{17.25}$$

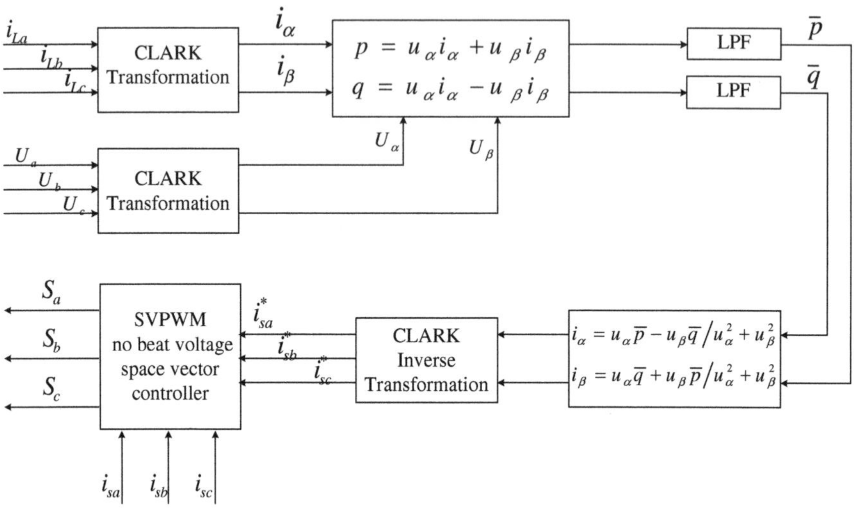

Fig. 17.5 Detection of the instantaneous reactive power based on theory

So we can yield the expression in the $\alpha\beta$ coordinate system

$$\begin{bmatrix} i_\alpha \\ i_\beta \end{bmatrix} = \frac{1}{u_\alpha^2 + u_\beta^2} \begin{bmatrix} u_\alpha & -u_\beta \\ u_\beta & u_\alpha \end{bmatrix} \begin{bmatrix} p \\ q \end{bmatrix} \tag{17.26}$$

The instantaneous active power p and reactive power q can be decomposed into DC and AC components

$$p = \bar{p} + \tilde{p} \tag{17.27}$$

$$q = \bar{q} + \tilde{q} \tag{17.28}$$

where, $\bar{p}$ and $\bar{q}$ are DC components;

$\tilde{p}$ and $\tilde{q}$ are AC components.

Hence, $i_{s\alpha}^*$ and $i_{s\beta}^*$ can be calculated in the $\alpha\beta$ coordinate system. And then, the reference current in abc coordinate system can be calculated by Clark inverse transformation.

$$\begin{bmatrix} i_{sa}^* \\ i_{sb}^* \\ i_{sc}^* \end{bmatrix} = \sqrt{\frac{2}{3}} \begin{bmatrix} 1/\sqrt{2} & 1 & 0 \\ 1/\sqrt{2} & -1/2 & \sqrt{3}/2 \\ 1/\sqrt{2} & -1/2 & -\sqrt{3}/2 \end{bmatrix} \begin{bmatrix} 0 \\ i_{s\alpha}^* \\ {}_{s\beta}^* \end{bmatrix}. \tag{17.29}$$

The diagram of instantaneous reactive power detection is shown in Fig. 17.5. u_a, u_b, u_c and i_{La}, i_{Lb}, i_{Lc} are the inputs of the controller. We can get the instruction currents of compensation currents i_{sa}^*, i_{sb}^*, and i_{sc}^* in the abc coordinate system after using Clark transformation, LPF low-pass filter, and then the Clark inverse transformation.

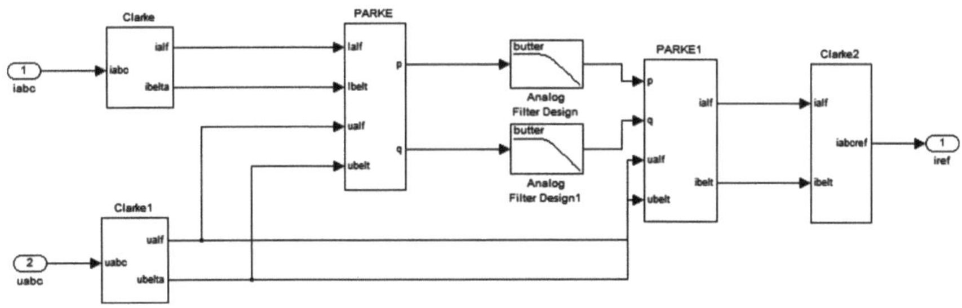

Fig. 17.6 Instantaneous reactive current detection module

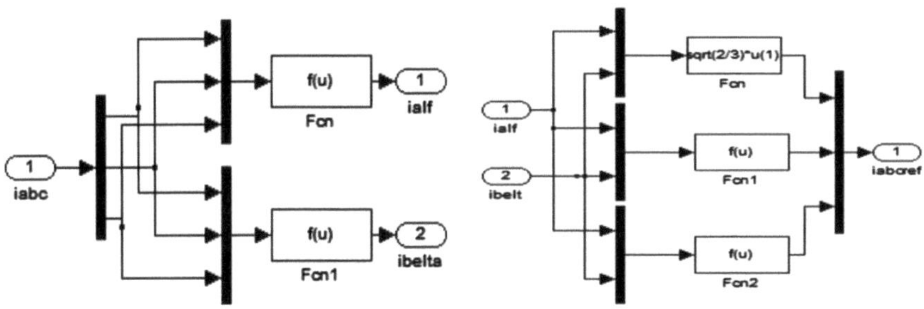

Fig. 17.7 C_{32} transformation and inverse transformation calculation module

The no beat voltage space vector controller uses these instruction currents to generate output PWM pulse signal for variable reactance control.

After using the Clark transformation, LPF low-pass filter, and Clark inverse transformation, the inputs i_{La}, i_{Lb}, and i_{Lc} are transformed into the instruction current i_{sa}^*, i_{sb}^*, and i_{sc}^* in abc coordinates, which is shown in Figs. 17.6 and 17.7. The voltage space vector (SVPWM) deadbeat current controller uses instruction currents to control the variable reactor.

Under the situation of symmetrical three-phase voltage and high-order harmonic, using the reactive current detection algorithm, some simulations are gained as follows:

(1) Under the situation of three-phase voltage symmetry without high-order harmonic, and phase A reactive current waveform is shown in Fig. 17.8.

(2) Under the situation of three-phase voltage symmetry without high-order harmonic is shown in Fig. 17.9, and phase A reactive current waveform is shown in Fig. 17.10.

(3) Under the situation of three-phase voltage symmetry without high-order harmonic is shown in Fig. 17.11, and phase A reactive current waveform is shown in Fig. 17.12.

From these simulation results, the instantaneous reactive current detection method can detect current easily and accurately.

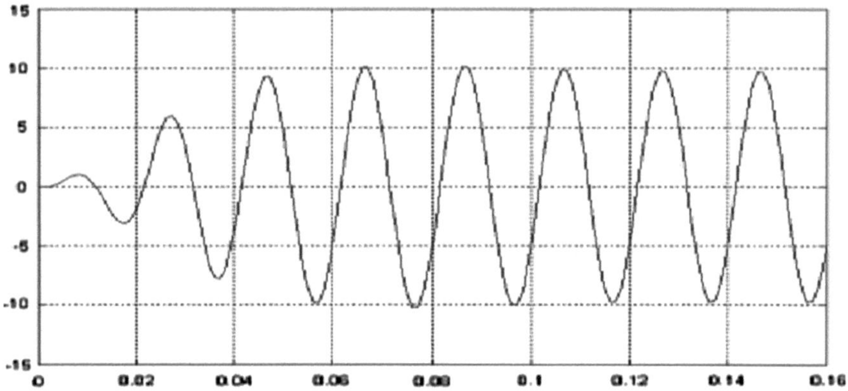

Fig. 17.8 Phase A reactive current wave without high-order harmonic

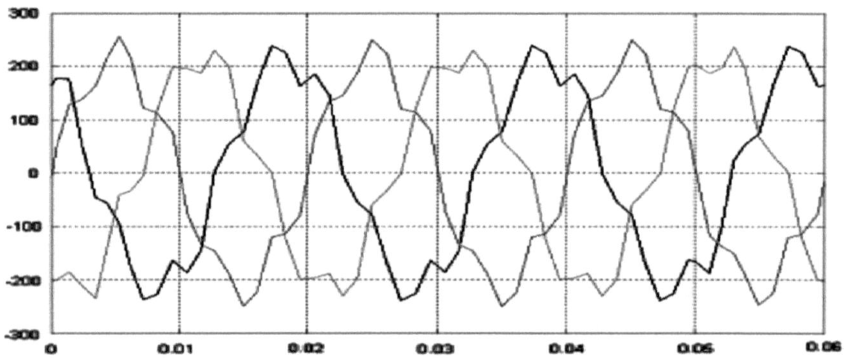

Fig. 17.9 Three-phase reactive current wave without high-order harmonic

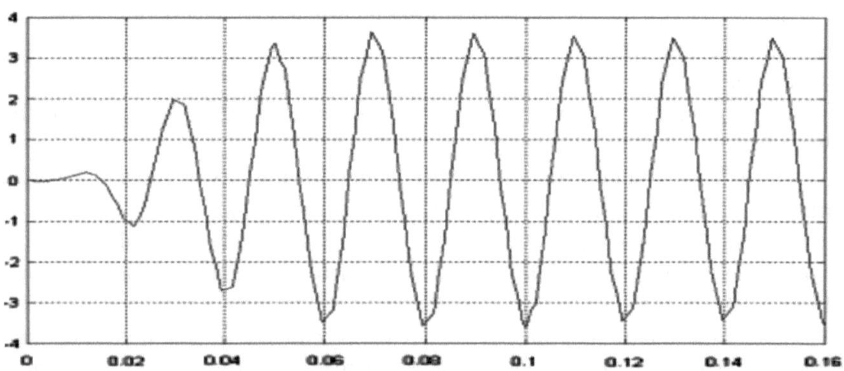

Fig. 17.10 Phase A reactive current wave without high-order harmonic

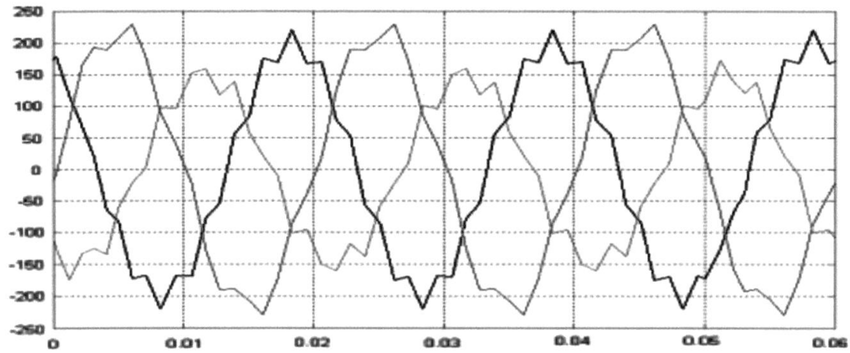

Fig. 17.11 Three-phase reactive current wave with high-order harmonic

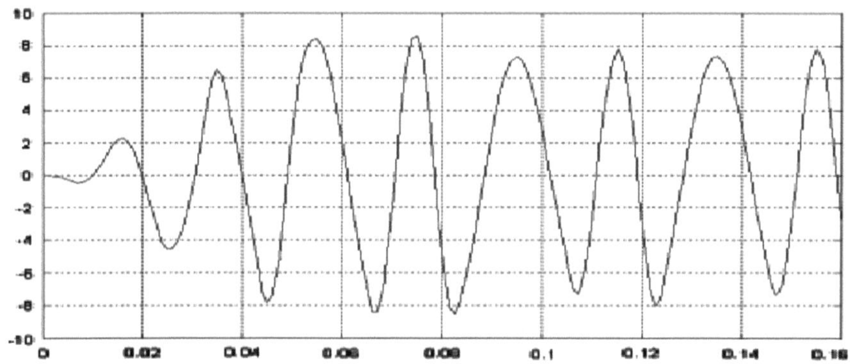

Fig. 17.12 Phase A reactive current wave with high-order harmonic

17.5 Fuzzy Self-Adaptive PID Control Algorithm

Traditional converters were realized by analog control technique and linear system control method. In recent years, with the development of high frequency digital control technology, digital control has entered the field of application of high frequency switch converter, and gradually replaces the analog control technique. Along with this trend, more and more control methods have been proposed, but among many intelligent control methods, fuzzy control method is recognized as the most effective control method for switch converter.

At present, all kinds of application complex situations raise more and more high requirements on the dynamic properties of the converter, especially on the most important properties of output overshoot and recovery time.

Because the converter system is a nonlinear time-varying system, it is difficult to establish an accurate mathematical model. While using the traditional control strategy, the system is approximated as a linear system model and implemented by adopting the PID control algorithm.

The traditional PID algorithm is based on a precise mathematical model, so the system performs best only under certain conditions. So scholars have put forward a number of methods on how to improve the dynamic performance of the variable reactor. A fuzzy adaptive PID controller is designed for control circuit in this chapter which can improve output dynamic performance.

In the traditional PID control, the control parameters that directly affect the system dynamic performance and steady-state performance are based on the mathematics models of the specific control object, mainly theoretical optimum parameters gained after the analysis of transfer function. There are a lot of PID parameter setting methods, e.g.:

(1) the trial and error method;
(2) the ISTE optimal setting method;
(3) the Ziegler–Nichols setting method;
(4) the root locus method;
(5) the pole assignment method;
(6) the engineering tuning method.

Because the converter is a time-varying, nonlinear system, it is difficult to establish a precise mathematical model according to the traditional PID control algorithm. In recent years, some advanced control methods have been developed, which provide new design ideas and broad strategies for converters control; parameters fuzzy self-adaptive PID control is included. This control principle is proposed on the following view. The sudden changes on the converter input or output, such as voltage surge or load surge, are considered to be converter structure changes. In this case, if the parameters are not changed with the control object, it will affect the control system response speed.

Considering such a situation, if the controller can adjust PID parameters according to the actual converter operation situation, the control system response speed will be increased. This controller combines traditional PID control algorithm and expert decision system in order to achieve the optimal control results. This implementation requires a large amount of experimental data accumulated and the fuzzy reasoning method. The controller can adjust online PID parameters according to the actual operation situation to make the control system have good dynamic and steady properties. The fuzzy self-adaptive PID control procedure is shown in Fig. 17.13.

17.6 System Simulation

The soft starter system simulation model built in the Matlab is composed of a three-phase power, a three-phase transformer, a three-phase universal rectifier bridge, an induction motor, and other components.

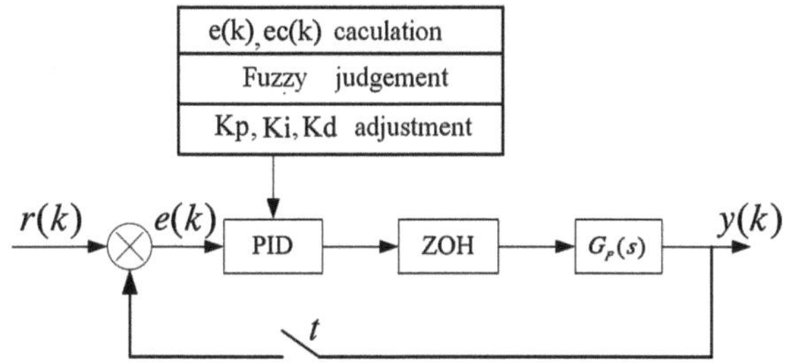

Fig. 17.13 Diagram of fuzzy self-adaptive PID control

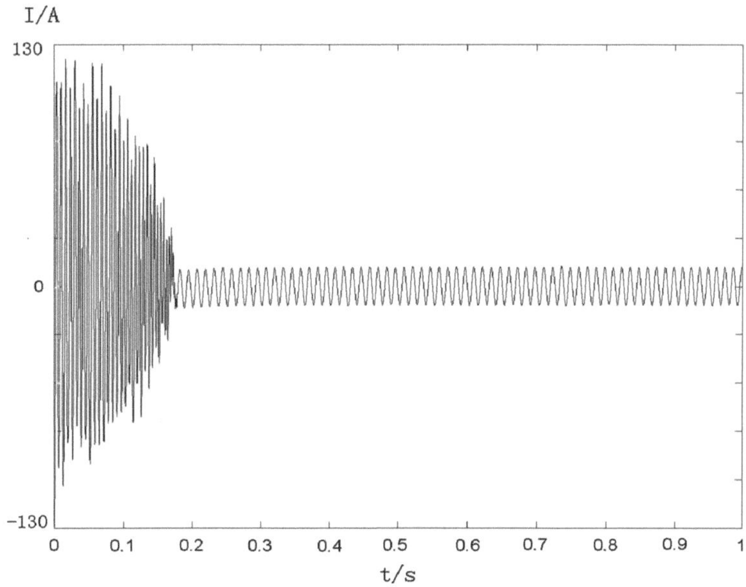

Fig. 17.14 Starting process current waveform diagram (start directly)

17.6.1 No-Load Full-Voltage Starting

Applying this simulation system on the 7.5 kW three-phase induction motor under the condition of no-load startup process simulation directly, we can get current waveform shown as Fig. 17.14.

Figure 17.14 shows the induction motor full voltage starting process current waveform diagram during 1 s. The diagram shows that the motor starting current increased hugely and reached the maximum value of about 130a. After about 0.2 s, the induction motor current amplitude drops rapidly to the no-load current amplitude, while the induction motor completes the start procedure.

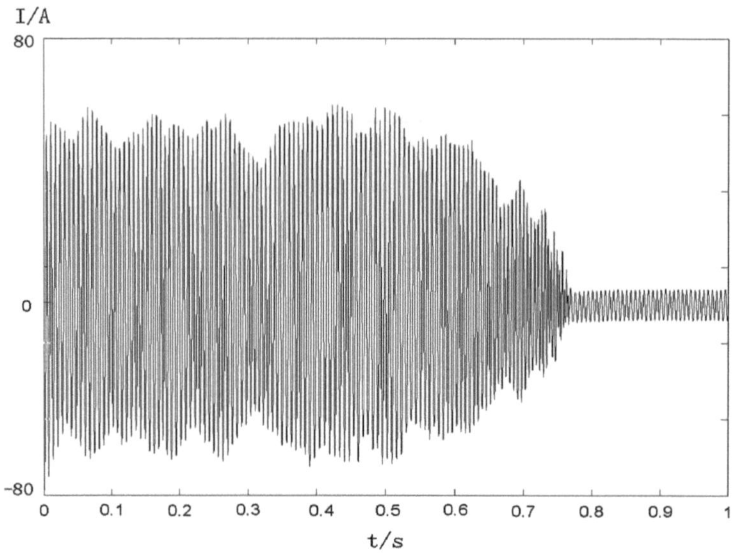

Fig. 17.15 Starting process current waveform diagram (soft starter)

17.6.2 Soft Start Mode Starting

To use the soft start mode on the same induction motor, the current amplitude decreased greatly during the starting period and the current waveform is shown in Fig. 17.15.

Figure 17.15 shows that after applying the soft starter, the starting current reduces significantly and reaches the maximum value of about 50a, while the starting time extended to almost 0.75 s. Therefore, the impact current amplitude decreases obviously during the induction motor starting process, and the performance of starting process improves significantly.

Comparing the two sets of simulation data, we can get that applying the soft starter in the starting of AC induction motor can reduce the starting current and the impact of related equipment.

17.7 Conclusions

The study clearly demonstrates that, the mathematical model of the soft starter is established, the relationship between thyristor conduction angle and the motor current is analyzed. Based on the relationship, the mathematical analytic expressions, which is used to characterize the relationship between them is obtained. According to the induction motor capacity, the appropriate conduction angle is chosen, the simulation results verified that it can effectively reduce the motor starting current and achieve the preset purpose.

References

1. Yuan YX, Zhao YW, Ding Y et al (2006) Study on AC motor soft starter based on fuzzy control. Electr Drive 36(5):13–15
2. Ding YZ, Yuan YX, Xiao YP et al (2008) Research on variable reactor based on power convert. J Wuhan Univ Tech 30(3):136–138
3. Bedi P, Qayumi K, Kaur T (2012) Home security surveillance system using multi-robot system. Int J Comput Appl Technol 45(4):272–279
4. Henmi T, Deng MC, Lnoue A (2009) Swing-up controller for the acrobot using skill of human movements. Int J Model Ident Control 6(3):222–229
5. Du QJ, Shi X, Lu EQ et al (2012) Motion control system analysis and design for a humanoid robot. Int J Model Ident Control 17(2):158–165
6. Jia DY, Huang Q, Tian Y et al (2009) Object manipulatin of a humanoid robot based on visual feedforward and visual feedback. Trans Beijing Inst Technol 29(11):983–987
7. Wang HW, Yu SH (2010) Tacking control of robot manipulators based on orthogonal neural network. Int J Model Ident Control 11(1–2):130–135
8. Alexander GE, DeLong MR, Crutcher MD (1992) Do cortical and basal ganglionic motor areas use motor programs to control movement? Behav Brain Sci 15(1):656–665
9. Cheng L, Xu WX, Wu HY et al (2012) A new procedure for multi-mode sequential flocking with application to multiple non-holonomic mobile robot motion control: mode description and integration principle. Int J Model Ident Control 15(1):39–47
10. Cheng L, Zheng XJ, Wu HY et al (2012) A new procedure for multi-mode sequential flocking with application to multiple non-holonomic mobile robot motion control: implementation and analysis. Int J Model Ident Control 16(1):50–59
11. Ge WM, Ye DF (2011) Sliding mode variable structure control of mobile manipulators. Int J Model Ident Control 12(1/2):166–172
12. Du QJ, Shi X, Dai B, Lu EQ (2013) Binocular stereo vision system for a humanoid robot. Int J Comput Appl Technol 46(4):316–322
13. Tian MX, Li QF, Wang SH (2003) An equivalent physical model and a mathematical model of the controlled saturable reactor. Trans China Electrotech Soc 18(16):18–21
14. Sheng JK, Chen QF, Xiong LY et al (2006) Techniques and experiment for high-voltage and high-power adjustable reactor based on magnetic flux controllable. High Voltage Eng 30(3):91–94
15. Zhou K (2010) Design of intelligent magnetic-controlled soft starter for motor. Mech Eng Autom 17(5):185–186
16. Wang YW, Liu J, Yu HF (2012) Optimal design of iron-core reactor based on genetic algorithm. Heilongjiang Electr Power 34(1):74–77
17. Zhu BS, Guan Y, Chen QG et al (2012) Design and characteristics analysis of orthogonal magnetization controllable reactor electric. Mach Control 16(5):26–32
18. Huang MH, Cheng HX, Zhang JH et al (2012) Commutation study on magnetically-controlled reactor. Jiangsu Electrical Apparatus 1:7–12
19. Luo LF, Chen B, Xu JZ (2011) Application of equivalent physical model of magnetically saturated controllable reactor in the design. Proc Chinese Soc Univ Electr Power Sys Autom 32(5):70–74
20. Liu QF, Dang HG, Liang YC et al (2011) Optimization algorithm of the reactor design. Power Capacitor React Power Compensat 32(5):46–50
21. Wang L (2011) A simple designed method for saturable reactor. Manufact Autom 33(1):31–33
22. Yang GP, Luo JT, Peng B et al (2010) Magnetic field research and circulating current calculation of power reactor with air core. Transformer 32(5):70–74
23. Lin ZK, Liu HP, Jiang Y (2010) The design of control system for the reactor of transformer type with controllable load. Power Electron 44(8):71–74
24. Chen X, Tong Z, Ouyang G (2009) Operational characteristics and simulation of the magnetic-saturated controllable reactor. High Voltage Apparatus 45(2):53–56

Chapter 18
Mixed H_2/H_∞ Robust Controller Design Based LMI Techniques

Yuan Peng, Quanmin Zhu and Hassan Nouri

Abstract H_2 and H_∞ optimal control are two important breakthroughs in modern robust control. In recent years, the H_2 and H_∞ controller design techniques have gained a lot of research attention. Both have strong theoretical basis and are efficient algorithms for synthesizing optimal controllers. The performance of H_2 robust controller is useful to handle stochastic aspects such as measurement noise and capture the control cost. In the robust H_2 approach, the controller is designed to minimize an upper bound on the worst case H_2 norm for a range of admissible plant perturbations. When a model under analysis is applied with H_2 robust controller, the closed loop system will have a good dynamic system performance, but it has poor robustness for the external disturbances of the uncertain system model. The performance of H_∞ is convenient to enforce robustness to model uncertainty, but it is based on compromising system performance. Their combination, the mixed H_2/H_∞ allows intuitive quadratic performance specifications of the H_2 synthesis with robust stability requirements specifications expressed by the H_∞ synthesis. In time domain aspects, satisfactory time response and closed loop damping can often be achieved by enforcing the closed loop poles into a predetermined region of the left half plane. Combining these requirements to form so-called mixed H_2/H_∞ design with regional pole placement constraints allows for more flexible and accurate specification of closed-loop behavior. This chapter introduces the work of design of improved LMI-based robust output feedback controller and related simulations.

Keywords Linear matrix inequality (LMI) · Power system stabilizers (PSS) · Robust control

---

Y. Peng (✉) · Q. Zhu · H. Nouri
Department of Engineering Design and Mathematics, University of the West of England, Coldharbour Lane, Bristol, BS16 1QY, UK
e-mail: Yuan2.Peng@uwe.ac.uk; Yuan.Peng@ukpowersolutions.co.uk

L. Liu et al. (eds.), *Applied Methods and Techniques for Mechatronic Systems*, Lecture Notes in Control and Information Sciences 452,
DOI: 10.1007/978-3-642-36385-6_18, © Springer-Verlag Berlin Heidelberg 2014

18.1 Introduction

Despite the popularity of the above methods among researchers and the many successfully solved control problems, the actual application of those approaches is quite limited. This is mainly due to two reasons listed below.

1. The order of the synthesized controllers is the same as the order of the generalized plant, which in many cases is very high for practical applications.
2. It is normally convenient or necessary to use a certain controller structure, for example, PID control or decentralized control, which is impossible with the standard methods.

LMI technique is often considered for multiobjective synthesis [1, 2], it provides more flexibility for combining various design objectives in a numerically tractable manner, and can even cope with those problems to which analytical solution is out of question. LMI-based robust controller design techniques have been used for PSS design for many years. The robust PSSs can be classified into three categories: H_2 robust PSS [3, 4], H_∞ robust PSS [5, 6], and mixed H_2/H_∞ robust PSS [7–9]. It has been observed that most LMI robust PSSs design are based on either H_2 or H_∞ norm. Furthermore, over half of them are state feedback controller design method, which is not available for practical PSS validation. LMI-based mixed H_2/H_∞ robust PSS design [7–10] has been studied in recent years, but all these design methods do not provide a comprehensive way for solving stability problems and they also have their individual drawbacks. Due to the complexity of modern power systems, low frequency oscillations caused by uncertain external and internal disturbances lead to complicated cases for PSS damping ability analysis [11, 12]. Although several LMI robust H_2/H_∞ PSS design approaches have been developed to enhance the performance of PSSs, unfortunately there still exist a number of unsatisfactory issues, typically as listed below.

1. The mixed H_2/H_∞ robust PSSs design is mainly based on state feedback controller design approach [13, 14], which in many cases is very difficult for practical applications.
2. Most H-norm-based robust PSSs are designed without concerning the parameter optimizations and the maximum stability region of the uncertain parameters in the target power system.
3. The system uncertainty and unknown disturbances are not clearly defined in most previous studies [15–18], for robust PSS design. Simulation results obtained from them might be conservative because the controllers are designed with unclear defined system uncertainty. Some mixed H_2/H_∞ robust PSSs have an acceptable damping ability for disturbances in a certain range, but for other operating point far away from the reference one, they are no longer available.
4. The weighting function of mixed sensitivity in power system stability analysis [19, 20], known as one of the significant components in H_∞ performance [21], are rarely discussed in robust PSSs design approaches. They can directly affect the damping performance of robust PSS suffered by unknown disturbances.

5. Most of mixed H_2/H_∞ robust PSSs are applied to SMIB system, some of the methods are not able to indicate whether the proposed PSSs can still be available for multi-machine power systems.
6. Disturbances for testing power system stability are normally selected as step input signals. However, because of the various faults in modern power system, step inputs are not enough to prove the PSS have a strong guarantee for system stability [22].

Research in this area related to wind turbine has been undertaken, particularly, during the last several years. However, due to the difficulty of making further progress in applying robust PSS to power system with large wind parks, the studies have been developed with a low speed, and consequently there have been no closely related publications issued for about 2 years. It is hoped this study will provide a new impetus to the research progress in the field of H_2/H_∞ robust PSS for DFIG wind turbine involved power systems.

In this study, the design process of H_2/H_∞ robust output feedback controller as a PSS for power systems is specified as four parts to overcome the problems mentioned above. Based on this robust PSS, a set validation tests is developed on both single machine infinite bus (SMIB) and multi-machine power systems. These tests provide a comprehensive and effective method of damping LFOs in power systems for various types of faults. Some simulation examples are provided to illustrate the efficiency and effectiveness of the damping ability of the proposed robust PSS.

This chapter is organised as follows. In Sect. 18.2, the design process of the improved H_2/H_∞ robust output feedback controller is explained to lay a basis for the controller design approach. In Sect. 18.3, the robust PSS validation for a single machine infinite bus system is introduced with simulation results. In Sect. 18.4, Robust PSS validation for a multi-machine power system is also introduced with simulation results. All the simulations are presented to demonstrate the effect of the proposed PSS in power systems while under small and large disturbances. In Sect. 18.5, conclusions are drawn to summarize this chapter.

18.2 Improved H_2/H_∞ Output Feedback Controller Design Methodologies

18.2.1 Power System Mathematical Model Selection and Comparison

Power system stability has a close relationship with the elements in power systems, especially the generator and related control systems. Dynamic performance of synchronous generators can be described by fundamental equations in d, q axis and they are named as Park Equations [23]. Figure 18.1 shows the relationship between the traditional a, b, c axis and the d, q axis after Park transform.

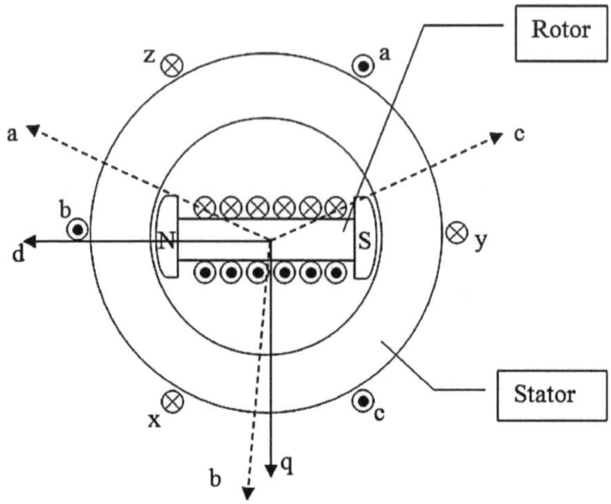

Fig. 18.1 Simple structure of a synchronous generator

In developing equations of a synchronous machine, the following assumptions need to be made

(1) The stator windings are sinusoidally distributed along the air gap as far as the mutual effects with the rotor are concerned.
(2) The stator slots cause no appreciable variation of the inductances with rotor position.
(3) Magnetic hysteresis and saturation effects are negligible.

Equations to describe synchronous generators can be obtained based on the mathematical model [24] and assumptions and normally they are made up of five basic windings. They are stator winding in d and q axis, field winding, and damper winding in d and q coordinate. Different mathematical models can be obtained by choosing different windings and simplify some of them. For example, small signal oscillatory in power system is one kind of LFOs (low frequency oscillations) with low rotor velocity. However, the excitation control system affects such oscillatory a lot. In this case, the damper winding can be ignored because the field winding (in d axis) can be regarded as the damper winding in d axis. After that, a four-winding mathematical model [25] of synchronous generator can be formed for stability analysis.

In power system stability analysis, voltages, currents, and flux linkages in rotor side are normally expressed by some defined electromotive in stator side (such as E_{fd} and E'_q) not the form of original Park Equations. The reason for doing this is to regard one generator as an electromotive with internal impedance. After that, it can be solved by approaches for normal electrical circuits.

The four-winding model only considers the effect of damper winding in q axis, however, the three-winding model just neglect all the impact of damper winding which can be acceptable for small signal stability analysis. Because the time constant of damper winding (around 0.03–0.04 s) is much smaller than the swing period of

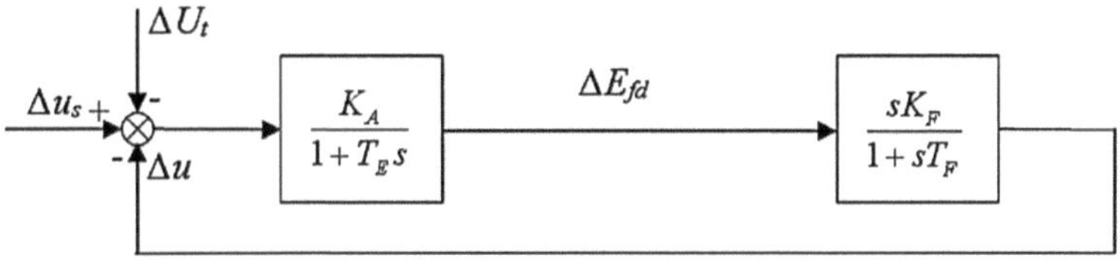

Fig. 18.2 The mathematical model of a fast excitation system

rotor (about 1 s). In this case, the synchronous generator can be expressed by stator winding in d and q axis and the field winding as following [25].

$$\frac{\mathrm{d}}{\mathrm{d}t}E'_q = \frac{1}{T'_{d0}}\left(E_{fd} - E_q\right) \tag{18.1}$$

$$\frac{\mathrm{d}}{\mathrm{d}t}\omega = \frac{1}{T_J}\left(M_m - M_e\right) - \frac{1}{T_J}D \tag{18.2}$$

$$\frac{\mathrm{d}}{\mathrm{d}t}\delta = \omega_0\left(\omega - 1\right), \tag{18.3}$$

where D is a compensation constant for damper winding. If the damper winding is taken into consideration, the four-winding model can be transferred to a five-winding model [25] which can be expressed by Eqs. (18.2), (18.3) and following

$$\frac{\mathrm{d}}{\mathrm{d}t}E'_q = \frac{1}{T'_{d0}}\left[E_{fd} - E'_q - \left(x_d - x'_d\right) i_d\right] \tag{18.4}$$

$$\frac{\mathrm{d}}{\mathrm{d}t}E''_q = -\frac{1}{T''_{d0}}\left[E'_q - E''_q - \left(x'_d - x''_d\right) i_d\right] \tag{18.5}$$

$$\frac{\mathrm{d}}{\mathrm{d}t}E'_d = -\frac{1}{T'_{q0}}E_d = \frac{1}{T'_{q0}}\left[-E'_d + \left(x_q - x'_q\right) i_d\right], \tag{18.6}$$

where E''_q is the post-transient electromotive, T''_{d0} is the special time constant used in five-winding model, and x''_d is the post-transient reactance in d axis.

In another aspect, the excitation system can be classified into two large categories, the fast excitation system and the regular excitation system. The mathematical models of them are shown in Figs. 18.2 and 18.3. In these figures, u_s is the output voltage of PSS and u_R is the reference voltage input to the exciter.

A complete mathematical model of a generator for stability analysis should include both generator model and the model of excitation system. If the generator model is chosen as three-winding model and five-winding model (with or without damper winding), as mentioned above, there are four different integrated mathematical models listed as following

(1) The three-winding model with the fast excitation system (Heffron-Philips model or the $K_1 - K_6$ model).
(2) The five-winding model with the fast excitation system ($C_1 - C_{12}$ model [26]).

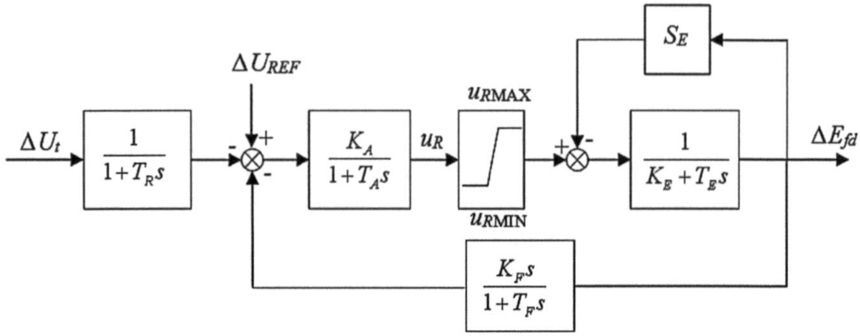

Fig. 18.3 The mathematical model of a regular excitation system

Table 18.1 Comparison between eigenvalues of the two models

Q p.u.	P p.u.	C_1-C_{12} Model		K_1-K_6 Model		
		Swing mode	Damping ratio	D	Swing mode	Damping ratio
−0.8	0	−2.2717 ± j8.2406	0.2658	8.1057	−1.5088 ± j5.2474	0.2763
	0.5	−1.2112 ± j8.7881	0.1365	12.0863	−1.0840 ± j6.5534	0.1632
	1.0	0.6862 ± j8.9663	−0.0763	9.7634	0.5017 ± j8.2560	−0.0607
0	0	−1.0714 ± j7.8385	0.1354	10.1279	−0.8440 ± j6.6635	0.1257
	0.5	−0.7406 ± j8.0347	0.0918	8.9484	−0.6324 ± j7.0046	0.0899
	1.0	0.2461 ± j7.0130	−0.0351	3.5739	0.2203 ± j6.8329	−0.0322
0.8	0	−0.4096 ± j6.7923	0.0602	4.3871	−0.3656 ± j6.4209	0.0568
	0.5	−0.2429 ± j6.8671	0.0354	4.0996	−0.2253 ± j6.5436	0.0344
	1.0	0.2461 ± j7.0130	−0.0351	3.5739	0.2203 ± j6.8329	−0.0322

(3) The three-winding model with the regular excitation system.
(4) The five-winding model with the regular excitation system.

The first comparison is made between the K_1-K_6 model and the C_1-C_{12} model. The most significant difference between the two models is that whether the damper winding is taken into consideration. When generators are suffered from LFOs, the damper winding only increase the damping torque of generators, i.e., increase the positive real part of the eigenvalues of the mathematical model, which can be equivalently transferred to the increase of damping torque factor D in Eq. (18.2).

Simulations for a randomly selected SMIB system are taken to compare eigenvalues between the two models, electrical parameters of the SMIB system can be found in related works [26]. Table 18.1 shows the calculation results of eigenvalues of the K_1-K_6 model and the C_1-C_{12} model.

It can be drawn from Table 18.1 (P stands for active power and Q stands for reactive power) that the damping torque factor D can be used to replace the damper winding in various swing modes and the damping ratios between the two models are

Table 18.2 Synchronizing and damping torque of different excitation systems

Exciter	Torque			
	Low frequency ($f < 1$)		High frequency ($f \gg 1$)	
	ΔM_s	ΔM_D	ΔM_s	ΔM_D
No excitation system	$K_1 - K_2K_3K_4$	$T'_dK_2K_3K_4$	$K_1 - \frac{K_2K_3K_4}{\omega_d{}^2T'^2_d}$	$\frac{K_2K_3K_4}{\omega_d{}^2T'^2_d}$
Fast excitation system	$K_1 - \frac{K_2K_5}{K_6}$	$\frac{T'_dK_2K_5}{K_EK_6^2K_3}$	$K_1 - \frac{K_2K_3^2K_5K_6K_E^2}{\omega_d{}^2T'^2_d}$	$\frac{K_2K_3K_5K_E}{\omega_d{}^2T'^2_d}$
Regular excitation system	$K_1 - \frac{K_2K_5}{K_6}$	MFK_5	$K_1 - \frac{K_2K_3K_4b_3}{T_AT_FT_E\omega_d{}^2T'^2_d}$	$\frac{K_2K_3K_4}{\omega_d{}^2T'^2_d}$

nearly the same. The three-winding model can be treated as a model reduction to the five-winding model and the system performance can be well retained. It is obvious that the $K_1 - K_6$ model is more suitable for power system stability analysis for its low system order and simple model structure.

Related to the excitation system, another comparison should be made between the three-winding model with the fast excitation system and the regular excitation system. Synchronizing torque ΔM_s and damping torque ΔM_D are two very important parameters for stability analysis [25]. Synchronizing torque is generated by the demagnetizing effect due to stator current and the system will be gradually unstable if $\Delta M_s \leq 0$. In another aspect, damping torque is generated by damper winding itself and the system will be oscillatory unstable if $\Delta M_D \leq 0$.

Table 18.2 shows the comparison of synchronizing and damping torque between generators with fast excitation systems and regular excitation systems. For a comprehensive comparison, cases of both low frequency and high frequency are taken into considerations.

In Table 18.2, $T'_d = K_3T'_{d0}$ is the time constant of the field winding, ω_d is the frequency domain expression of Laplace operator s, and symbols b_3, M, and F are defined as following:

$$b_3 = T_AT_FT_E + T'_dmT_AT_F + T'_dT_E\left(T_A + T_F\right), F = \frac{K_A^2K_2K_3K_4}{\left(b_0^3 + b_1^2\omega^2\right)}$$

$$M = \frac{K_F\left(K_5 - K_3K_4K_6\right)}{K_4K_5},$$

where $m = K_E + S_E$, $b_0 = m + K_3K_6K_A$ and

$$b_1 = T_E + m\left(T_A + T_F\right) + mT'_d + K_3K_6K_AT_F + K_AK_F.$$

Some previous works [27, 28] have already tested the value of synchronizing and damping torques for regular power systems. Simulation results show that the related system may lose its stability because of the negative value of the damping torque. However, fast excitation systems are more reliable compared with regular excitation systems. In other aspects, regular excitation systems have rotation mechanical parts

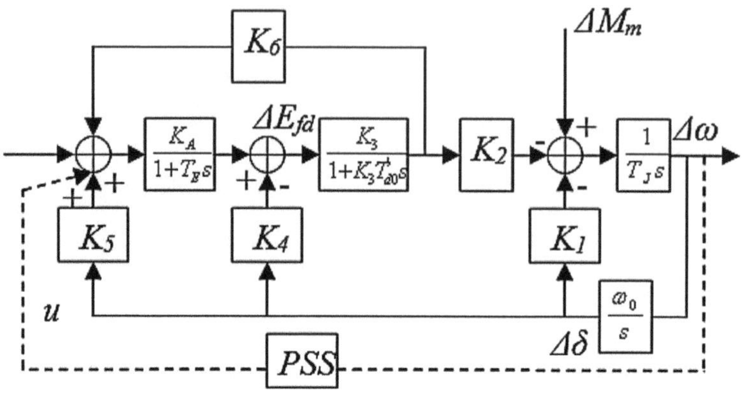

Fig. 18.4 Small signal model of SMIB system

which are time-varying systems and they are hard to be analyzed in some cases [25]. Because of their structures, regular excitation systems do not directly affected by the system operating conditions, i.e., there is a time delay between the excitation system and the generator. As shown in above equations, the expressions of synchronizing and damping torque of regular excitation systems are more complicated than fast excitation systems, which lead to inconvenience for stability analysis. Furthermore, fast excitation systems are widely applied in modern power systems for a large number of generators without capacity constraints. Therefore, the three-winding model with fast excitation systems is better for power system stability study compared with the regular excitation systems.

Finally, the three-winding model with fast excitation system (Heffron-Philips Model) is selected for the present study in both SMIB and multi-machine systems.

18.2.2 State Space Description

The single machine infinite bus systems, the structure of the power system mathematical model is shown in Fig. 18.4.

The state space matrices of the SMIB system [22] with system uncertainty can be obtained by using $\Delta\omega$, $\Delta\delta$, $\Delta E'_q$ and ΔE_{fd}.

$$A = \begin{bmatrix} -\frac{MD}{T_J} & -\frac{K_1}{T_J} & -\frac{K_2}{T_J} & 0 \\ \omega_0 & 0 & 0 & 0 \\ 0 & -\frac{K_4}{T'_{d0}} & -\frac{1}{K_3 T'_{d0}} & \frac{1}{T'_{d0}} \\ 0 & -\frac{K_5 K_A}{T_E} & -\frac{K_6 K_A}{T_E} & -\frac{1}{T_E} \end{bmatrix}, B = \begin{bmatrix} 0 \\ 0 \\ 0 \\ \frac{K_A}{T_E} \end{bmatrix}, \quad C = \begin{bmatrix} 1\,0\,0\,0 \end{bmatrix}, \quad D = 0.$$

The input signal for PSS in this chapter is selected as $\Delta\omega$. The parameters $K_1, K_2, K_3, K_4, K_5, K_6$ of the model depend on the active power, reactive power, and synchronous reactance as stated in Chap. 2. Direct analytical expressions that relate parameters $K_1, K_2, K_3, K_4, K_5, K_6$ to power and reactance can be found in previous works [25]. The nonlinear system matrix A is affinely dependent on different value of $K_1, K_2, K_3, K_4, K_5, K_6$, i.e., matrix A can be written in the form of

$$A = A_0 + A_1K_1 + A_2K_2 + A_3K_3 + A_4K_4 + A_5K_5 + A_6K_6,$$

where $A_0, A_1, A_2, A_3, A_4, A_5, A_6$ are all constant matrices. In another aspect, each parameter of $K_1, K_2, K_3, K_4, K_5, K_6$ varies within a certain range under different loading conditions and such concept can be expressed as

$$K_1 \in \left[K_{1-}, K_{1+}\right], \quad K_2 \in \left[K_{2-}, K_{2+}\right], \quad K_3 \in \left[K_{3-}, K_{3+}\right]$$

$$K_4 \in \left[K_{4-}, K_{4+}\right], \quad K_5 \in \left[K_{5-}, K_{5+}\right], \quad K_6 \in \left[K_{6-}, K_{6+}\right],$$

where $\left[K_{\eta-}, K_{\eta+}\right]$ denotes the lower and upper bound of $K_1, \ldots, K_6$ for all loading conditions. These bounds can be easily calculated by using active and reactive powers in different loading conditions [25, 29].

The state space matrices of the multi-machine system with system uncertainty can be expressed as [25]:

$$A_M = \begin{bmatrix}
-\frac{MD}{T_{J1}} & 0 & 0 & -\frac{K_{111}}{T_{J1}} & \cdots & -\frac{K_{11n}}{T_{J1}} & -\frac{K_{211}}{T_{J1}} & \cdots & -\frac{K_{21n}}{T_{J1}} & 0 & 0 & 0 \\
0 & \ddots & 0 & \vdots & \vdots & \vdots & \vdots & \vdots & \vdots & 0 & 0 & 0 \\
0 & 0 & -\frac{MD}{T_{Jn}} & -\frac{K_{1n1}}{T_{Jn}} & \cdots & -\frac{K_{1nn}}{T_{Jn}} & -\frac{K_{2n1}}{T_{Jn}} & \cdots & -\frac{K_{2nn}}{T_{Jn}} & 0 & 0 & 0 \\
\omega_0 & 0 & 0 & 0 & 0 & 0 & 0 & 0 & 0 & 0 & 0 & 0 \\
0 & \ddots & 0 & 0 & 0 & 0 & 0 & 0 & 0 & 0 & 0 & 0 \\
0 & 0 & \omega_0 & 0 & 0 & 0 & 0 & 0 & 0 & 0 & 0 & 0 \\
0 & 0 & 0 & -\frac{K_{411}}{T'_{d01}} & \cdots & -\frac{K_{41n}}{T'_{d01}} & -\frac{1}{T'_{d01}K_{311}} & \cdots & -\frac{1}{T'_{d01}K_{31n}} & -\frac{1}{T'_{d01}} & 0 & 0 \\
0 & 0 & 0 & \vdots & \vdots & \vdots & \vdots & \vdots & \vdots & 0 & \ddots & 0 \\
0 & 0 & 0 & -\frac{K_{4n1}}{T'_{d0n}} & \cdots & -\frac{K_{4nn}}{T'_{d0n}} & -\frac{1}{T'_{d0n}K_{3n1}} & \cdots & -\frac{1}{T'_{d0n}K_{3nn}} & 0 & 0 & -\frac{1}{T'_{d0n}} \\
0 & 0 & 0 & -\frac{K_{A1}K_{511}}{T_{E1}} & \cdots & -\frac{K_{A1}K_{51n}}{T_{E1}} & -\frac{K_{A1}K_{6n1}}{T_{E1}} & \cdots & -\frac{K_{A1}K_{61n}}{T_{E1}} & -\frac{1}{T_{E1}} & 0 & 0 \\
0 & 0 & 0 & \vdots & \vdots & \vdots & \vdots & \vdots & \vdots & 0 & 0 & 0 \\
0 & 0 & 0 & -\frac{K_{An}K_{5n1}}{T_{En}} & \cdots & -\frac{K_{An}K_{5nn}}{T_{En}} & -\frac{K_{An}K_{6n1}}{T_{En}} & \cdots & -\frac{K_{An}K_{6nn}}{T_{En}} & 0 & 0 & -\frac{1}{T_{En}}
\end{bmatrix}$$

$$B_M = \begin{bmatrix} 0 \cdots 0 & \frac{K_{A1}}{T_{E1}} & 0 & 0 \\ \vdots \; \vdots \; \vdots & 0 & \ddots & 0 \\ 0 \cdots 0 & 0 & 0 & \frac{K_{An}}{T_{En}} \end{bmatrix}^T, \quad C_M = \begin{bmatrix} 1 & 0 & 0 & 0 \cdots 0 \\ 0 & \ddots & 0 & \vdots \; \vdots \; \vdots \\ 0 & 0 & 1 & 0 \cdots 0 \end{bmatrix}, \quad D_M = 0,$$

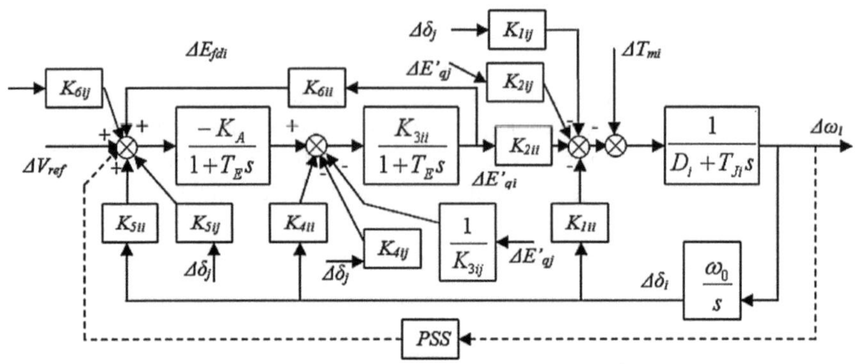

Fig. 18.5 A mathematical model of multi-machine system

where the subscript M means multi-machine system and the output is the speed deviation of each synchronous generator. Similarly, the lower and upper bound of $K_{1ij}, \ldots, K_{6ij}$ can be calculated by using active and reactive powers in different loading conditions.

For multi-machine system, the structure of the mechanical model is shown in Fig. 18.5. This model [30] is the extension of the Hafferon-Phillips model and the $K_1, \ldots, K_6$ is replaced by $K_{1ij}, \ldots, K_{6ij}$, where the subscribes i, j denote the relationship between different synchronous generators in the multi-machine system. Damper windings are negligible because they are always related to high frequency modes and have little relationship with LFO modes in power systems. In the simulation, such model without damper windings is adequate to deal with dynamic performance of LFOs in power systems and the location selection for PSSs installation.

One of the most important advantages of this model is to divide the multi-machine system into n subsystems (n is the number of synchronous generators in the system). In this way, the interaction between synchronous generators can be clearly defined and used for stability analysis. Once locations for PSSs application are selected, the proposed output feedback robust controller design algorithm can be directly applied for the PSSs design. The reason is that, for each synchronous generator in the multi-machine system, it can be treated as a single machine system with related parameters $K_{1ij}, \ldots K_{6ij}$.

The H_2/H_∞ state space model adopted for the power system stability study can be expressed as [31]:

$$\begin{aligned} \dot{x} &= Ax + B_1 w + B_2 u \\ z_{\text{inf}} &= C_{\text{inf}} x + D_{\text{inf}1} w + D_{\text{inf}2} u \\ z_2 &= C_2 x + D_{21} w + D_{22} u \\ y &= C_o x + D_{o1} w + D_{o2} u, \end{aligned} \tag{18.7}$$

where u, w, y are the robust PSS output, system disturbance input, and system output. Whereas, A is the uncertain state matrix of the system, B_1, B_2 are the disturbance

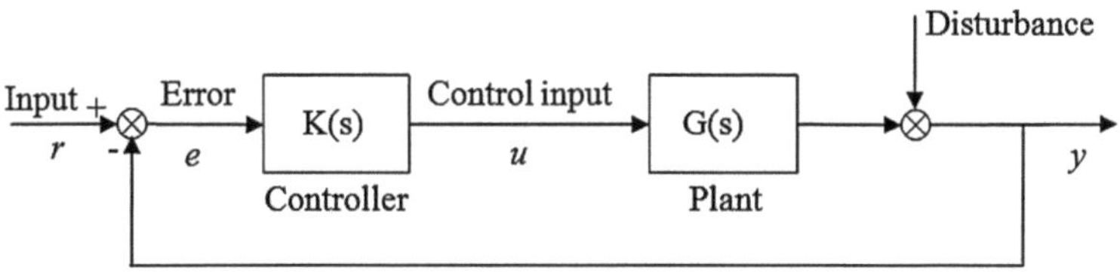

Fig. 18.6 A mathematical model of multi-machine system small

input matrices and regulated input matrix, C_{inf}, C_2, C_o are the H_∞ output, H_2 output and system output. Matrices D with different subscribes are real matrices with proper dimension for the system.

18.2.3 Mixed Sensitivity and Weighting Functions Selection

The mixed H_2/H_∞ robust controller design includes both H_2 controller performance and H_∞ controller performance, which can provide the system robustness for unknown disturbances and a good dynamic performance. Mixed sensitivity [20, 32], the guarantee of system robustness, is a significant part in H_∞ controller design. It can build augment matrix of the selected system by using different weighting functions. In this way, the robust controller can be designed to cope with the worst case and to keep the system to be stable.

In practical control systems, there are external disturbances and system uncertainties. Mixed sensitivity can be used to control the disturbances and uncertainties and the stability and performance of the target system can be measured by following functions [21]:

(1) Sensitivity function: $S(s) = (I + L(s))^{-1}$
(2) Complementary sensitivity function: $T(s) = L(s)(I + L(s))^{-1}$
(3) Input sensitivity function: $R(s) = K(s)(I + L(s))^{-1}$

where I is the identity matrix, $L(s) = G(s)K(s)$ is the loop transfer matrix, and $K(s)$ is the transfer function of controller.

As shown in Fig. 18.6, the transfer function of tracking error signal $e(s)$ can be expressed as

$$e(s) = (I + L(s))^{-1} r(s) = S(s) r(s),$$

where $r(s)$ is the transfer function of the input signal, $G(s)$ is the plant transfer function, and y is the output signal.

To obtain an acceptable tracking error performance, the tracking error $\|e(j\omega)\|$ should be as small as possible within the bandwidth, i.e., the maximum gain of the error signal $\bar{\sigma}(S(j\omega))$ should be small enough. In order to keep the robust tracking performance, $\bar{\sigma}(S(j\omega))$ should be restricted by

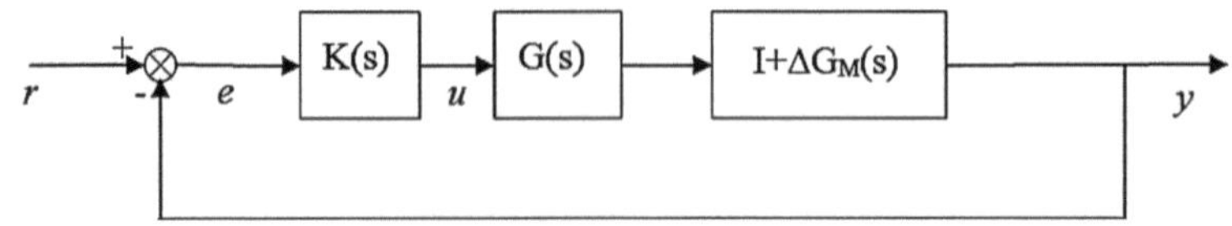

Fig. 18.7 Block diagram of system with multiplicative uncertainty

$$\bar{\sigma}\left(S\left(j\omega\right)\right) \leq \left|W_1^{-1}\left(j\omega\right)\right| \quad \forall\omega, \tag{18.8}$$

where $W_1\,(j\omega)$ is defined as the error envelop. Inequality (18.8) indicates that if the maximum gain reaches the requirement in (18.8), gains in all different cases will be smaller than $\left|W_1^{-1}\left(j\omega\right)\right|$. In this case, (18.8) can be rewritten as

$$\bar{\sigma}\left(S\left(j\omega\right)\right) \leq \left|W_1^{-1}\left(j\omega\right)\right| \quad \forall\omega \tag{18.9}$$

$$\Leftrightarrow \underset{\omega}{Sup}\,\bar{\sigma}\left(S\left(j\omega\right)W_1^{-1}\left(j\omega\right)\right) \leq 1 \Leftrightarrow \|SW_1\|_\infty \leq 1 \Leftrightarrow \left\|\left(I+GF\right)^{-1}W_1\right\|_\infty \leq 1.$$

The complementary sensitivity function $T\,(s)$ is used to describe systems with multiplicative uncertainties [33] and they need to be set within related stability margins. As shown in Fig. 18.7, the system can be expressed as (18.10) after small signal disturbances.

$$Y(s)/U(s) = \left(I + \Delta G_M\left(s\right)\right)G\left(s\right). \tag{18.10}$$

$\Delta G_M\,(s)$ in (18.10) is the transfer function of multiplicative uncertainty.

In order to find the requirements for the controller $K\,(s)$ to keep the system stable with the uncertainty $\Delta G_M\,(s)$, (18.11) can be obtained from the small gain theorem [21].

$$\bar{\sigma}\left(\Delta G_M\left(j\omega\right)\right) \leq \frac{1}{\bar{\sigma}\left(T\left(j\omega\right)\right)} \quad \forall\omega. \tag{18.11}$$

Define $|W_3\,(j\omega)| = \bar{\sigma}\left(\Delta G_M\left(j\omega\right)\right)$ as the perturbation envelop, (18.11) can be rewritten as [33]:

$$\begin{aligned}&\bar{\sigma}\left(T\left(j\omega\right)\right) \leq \left|W_3\left(j\omega\right)^{-1}\right| \quad \forall\omega\\ &\Leftrightarrow \bar{\sigma}\left(T\left(j\omega\right)\right)\left|W_3\left(j\omega\right)\right| \leq 1 \Leftrightarrow \underset{\omega}{Sup}\,\bar{\sigma}\left(T\left(j\omega\right)W_3\left(j\omega\right)\right) \leq 1 \Leftrightarrow \|TW_3\|_\infty \leq 1.\end{aligned} \tag{18.12}$$

The input sensitivity $R\,(s)$ is the transfer function between the reference input and the control input. In order to avoid the saturation of the machine [34, 35], the gain of $R\,(s)$ in high frequency part should be restricted. In another aspect, the

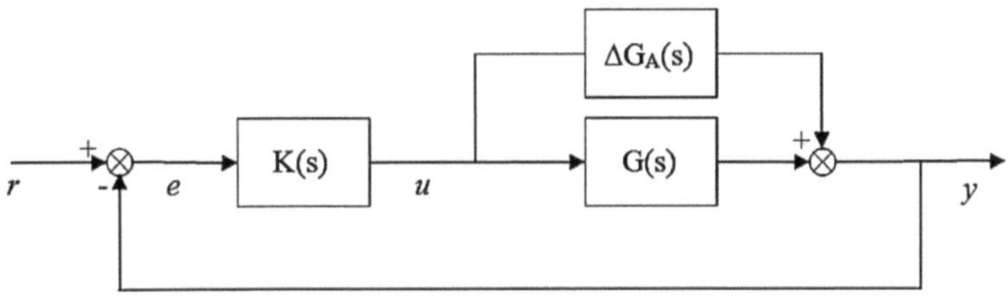

Fig. 18.8 Block diagram of system with additive uncertainty

input sensitivity can affect system stability while the system has additive uncertainty $\Delta G_A(s)$ as shown in Fig. 18.8.

It can be drawn from Fig. 18.7 that the transfer function of the system with small signal disturbance can be expressed as

$$Y(s)/U(s) = G(s) + \Delta G_A(s).$$

Similarly, obtained from the small gain theorem, the maximum additive uncertainty for the robust controller $K(s)$ to keep the system stable can be expressed as [21]

$$\bar{\sigma}(\Delta G_A(j\omega)) \leq \frac{1}{\bar{\sigma}(R(j\omega))} \quad \forall\omega. \tag{18.13}$$

Define $|W_2(j\omega)| = \bar{\sigma}(\Delta G_A(j\omega))$ as the limit input value of the actuator, (18.13) can be rewritten as [21]

$$\bar{\sigma}(R(j\omega)) \leq \left|W_2(j\omega)^{-1}\right| \quad \forall\omega \tag{18.14}$$

$$\Leftrightarrow \bar{\sigma}(R(j\omega))\,|W_2(j\omega)| \leq 1 \Leftrightarrow \underset{\omega}{Sup}\,\bar{\sigma}(R(j\omega)\,W_2(j\omega)) \leq 1 \Leftrightarrow \|RW_2\|_\infty \leq 1.$$

As a result, in order to meet the criterion of the error-tracking ability, avoid the saturation of the machine, and keep stable with additive and multiplicative disturbances, the open loop transfer function $L(s)$ should be constrained as [21]

$$\left\|\begin{bmatrix} SW_1 \\ RW_2 \\ TW_3 \end{bmatrix}\right\|_\infty = \left\|\begin{bmatrix} z_{w1} \\ z_{w2} \\ z_{w3} \end{bmatrix}\right\|_\infty \leq 1, \tag{18.15}$$

where z_{w1} is the tracking error signal controlled by weighting function W_1, z_{w2} is the control input signal restricted by weighting function W_2, and z_{w3} is the system output signal controlled by weighting function W_3. The structure of the augment system is given in Fig. 18.9, and it contains the system transfer function $G(s)$ and all three weighting functions, where M_i is the measurement input to the system

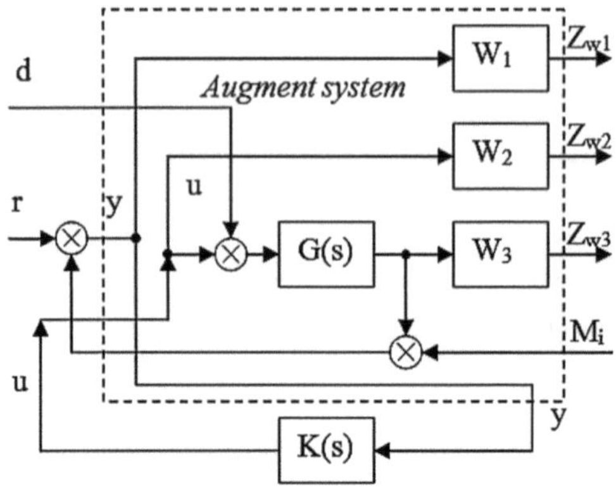

Fig. 18.9 Block diagram of the augment system

and d, r, u are disturbance signal, system reference input signal, and controller input signal, respectively.

Finally, the transfer function of the augment system can be used as the model for the H_2/H_∞ robust controller design.

18.2.4 LMI Regional Pole Placement

In most practical cases, however, a LMI-based robust PSS should deliver sufficient fast and well-damped time responses to the power system. The most popular way to guarantee satisfactory transients is to place the closed loop poles in a suitable region of the complex plane. This technique is referred as LMI regional pole placement, where the poles are assigned to specific locations in the complex plane. Regional pole assignment has also been considered in conjunction with other design objectives, such as H_∞ or H_2 performance [21, 36]. In previous regional pole placement research, the most important problem is the precision of the pole placement, i.e., place closed loop poles to the desired locations. However, because of the system uncertainty and disturbances, the precision of pole placement cannot be implemented easily. In fact, the system stability and dynamic performance can be guaranteed if the poles of the closed loop system are placed in proper areas of the complex plane, which are named as LMI regions.

An LMI region is used to describe any subset D_{LMI} of the complex plane [37] that can be defined as

$$D_{\text{LMI}} = \left\{ z \in C : L + zM + \bar{z}M^T < 0 \right\}, \tag{18.16}$$

where L and M are real matrices such that $L^T = L$. The matrix-valued function (18.17) is called the characteristic function of D_{LMI}.

$$f_{D_{\mathrm{LMI}}}(z) = L + zM + \bar{z}M^T, \tag{18.17}$$

where z is a Hermitian matrix and $f_{D_{\mathrm{LMI}}}(z) < 0$ represents the matrix is negative definite.

There are several key facts about LMI regions need to be presented. The first is that intersections of LMI regions are LMI regions. Second, any convex region symmetric with respect to the real axis can be approximated by an LMI region to any desired accuracy. A real matrix A is D-stable [21], i.e., has all its eigenvalues in the LMI region D_{LMI}, if and only if a symmetric matrix X exists such that

$$M_D(A, X) = L \otimes X + M \otimes (XA) + M^T \otimes \left(A^T X\right) < 0, \quad X > 0, \tag{18.18}$$

where $\otimes$ is the Kronecker product [37] which is an important tool for the subsequent analysis in LMI regional pole placement.

Consider a linear system with uncertainty which can be expressed as [37]

$$\begin{aligned} \dot{x} &= A(\Delta)\,x \\ A(\Delta) &= A + B(I - \Delta D)^{-1}\Delta C, \end{aligned} \tag{18.19}$$

where the state matrix $A(\Delta)$ depends fractionally on the norm bounded uncertainty matrix $\Delta \in R$, C, $\sigma_{\max}(\Delta) \leq \gamma^{-1}$. A sufficient LMI-based condition for quadratic D-stability [37] is that the system (18.19) with its uncertainty is quadratically D-stable if matrix $X > 0$ and $P > 0$ exist such that

$$\begin{bmatrix} M_D(A, X) & M_1 \otimes (XB) & \left(M_2^T P\right) \otimes C^T \\ M_1 \otimes \left(B^T X\right) & -\gamma P \otimes I & P \otimes D^T \\ (PM_2) \otimes C & P \otimes D & -\gamma P \otimes I \end{bmatrix} < 0. \tag{18.20}$$

The related proof can be found in [37] and the concept of quadratic D-stability is very significant in uncertain system pole placement.

In practical applications, LMI regions are often specified as the intersection of elementary regions, such as conic sectors, disks, or vertical half-planes. Given LMI regions$D_{\mathrm{LMI}1}, \ldots, D_{\mathrm{LMI}n}$, the intersection $D_{\mathrm{LMI}} = D_{\mathrm{LMI}1} \cap \ldots \cap D_{\mathrm{LMI}n}$ has characteristic function [37]

$$f_{D_{\mathrm{LMI}}}(z) = \operatorname*{diag}_{i=1}^{n} f_{D_{\mathrm{LMI}i}}(z), \quad i = 1, \ldots, n. \tag{18.21}$$

The LMI feasibility problems should be solved independently for each region $D_{\mathrm{LMI}i}$ because no coupling exists between the constraints for each region. By contrast, (18.19) can be directly applied to (18.20) to jointly solving all LMIs with $X = X_1 = \cdots = X_n$ and P_i, $i = 1, \ldots, n$ as variables [37]. This approach is clearly more costly and more conservative because of the requirement that a single X satisfy for all regions $D_{\mathrm{LMI}i}$.

In modern power systems, the system uncertainties are always the time-varying electrical parameters. In order to reduce the conservative mentioned above, some previous works [21] have been taken and proven to be significantly less conservative than quadratic stability for time-invatiant parameter uncertainty. The analysis below deals with the same basic uncertain model (18.19), but now assumes

$$\Delta = \Delta(\delta) = \operatorname*{diag}_{i=1}^{q} \delta_i I_{r_i}, \quad \delta_i \in R, \quad |\delta_i| \leq 1, \tag{18.22}$$

where δ_i denotes the uncertain parameter.

To stress the dependence on the parameter vector δ, the uncertain state matrix can be written as

$$A(\delta) = A + B(I - \Delta(\delta) D)^{-1} \Delta(\delta) C. \tag{18.23}$$

For such parameter uncertainty, robust pole clustering in the LMI region is equivalent to the existence of symmetric matrices $X(\delta) > 0$ parameterized by δ such that

$$L \otimes X(\delta) + M \otimes X(\delta) A(\delta) + M^T \otimes A(\delta)^T X(\delta) < 0. \tag{18.24}$$

In order to enforce the tractability of (18.24), the search of functions $X(\delta)$ can be restricted to matrices with affine dependence [38] on δ

$$X(\delta) = X_0 + \sum_{i=1}^{q} \delta_i X_i = X_0 + J^T \hat{\Delta}(\delta) \hat{X} J, \tag{18.25}$$

where $J = [I_n, \ldots, I_n]^T$ with q times I_n, $\hat{\Delta}(\delta) = \operatorname*{diag}_{i=1}^{q} \delta_i I_n$, and $\hat{X} = \operatorname*{diag}_{i=1}^{q} X_i$.

Two robust D-stability tests are derived using such affine $X(\delta)$. The first one applies to general linear fractional dependence of $A(\delta)$ on δ, whereas the other one is restricted to affine dependence. Such tests can be found in [38] and they will not be further introduced in this study.

18.2.5 Design of Improved Mixed H_2/H_∞ Robust Output Feedback Controller

In this section, the complete design process of the improved LMI-based H_2/H_∞ output feedback controller (IMOFC) design for PSS is comprehensively introduced with a case study of a SMIB system. First, as explained in previous sections, the mathematical model of power system and excitation system should be chosen as Heffron-Philips model and static fast exciter for both SMIB and multi-machine system. Second, the robust controller design process will be presented with full algorithm descriptions. Finally, a step-by-step procedure for the robust PSS design will be presented.

18.2.5.1 Optimal Location and Set Point Selection for PSS Design

Location selection for PSSs is the problem of controller design for multi-machine systems. Once LFOs are detected in a large power system, it is not possible to damp LFOs by installing PSSs to all the generators in the system. The inappropriate location selection may bring little help for the damping abilities of synchronous generators.

The technique of participation factors [39] for selecting PSS optimal locations has been widely used and regarded as one of the most efficient way. For any swing mode, PSS need to be installed on the generator with the largest participation factor. If participation factors of several machines are nearly the same and they are very large in one certain swing mode, each of these machines should be applied with a PSS.

Set point selection should take all loading conditions into considerations as well as the oscillation frequencies. The system may be unstable in some loading conditions and the set point for controller design should be selected within them. Normally, the loading condition with the largest (f_{up})/lowest (f_{low}) oscillation frequency (all swing modes should be concerned for multi-machine system) can be selected as set point and the one with lowest/largest oscillation frequency can be selected as the "worst case" for mixed sensitivity analysis.

18.2.5.2 Weighting Function Selection Criterions

It is notable that the success of mixed sensitivity designs reported in some previous works [40, 41] depends largely on the appropriate selection of the weights used in the optimization process. While for typical applications appropriate weights are often easily chosen after several trials and errors, the stringent performance requirements imposed for the high performance applications makes the selection of appropriate weights difficult, or at least time-consuming.

It is necessary to mention that the selection of the optimal weights for H norm-based robust control has received attention in the last decade [42–44], where attempts have been made to formulate the problem in a setting by which one obtains the controller and the weights simultaneously and in an iterative manner. However, the question of the suitability of the weights obtained and the complexity of the algorithm employed are yet to be judged. On the other hand, the development of weighting function selection for PSS design has been undertaken and it is important for the performance of the mixed H_2/H_∞ robust controller. Because of system uncertainties, robust controllers should be able to keep the system stable in the case of both small disturbances and large disturbances. The robust PSS can damp oscillations easily even if the system operating point changes a lot. In this study, a certain process for selecting weighting functions for robust PSS design is obtained based on mixed sensitivity analysis mentioned above.

Step 1. Select the rated operating point of the power system as the dynamic mode of nominal working condition for controller design and record as $G_0(s)$. Decide the worst case with related operating point as the maximum uncertainty and record as $G_x(s)$.

Step 2. The multiplicative uncertainty of the system can be expressed as $G_x(s) - G_0(s) = \Delta G_M(s)\, G_0(s)$, for which $\bar{\sigma}(\Delta G_M(s)\, G_0(s)) \leq \bar{\sigma}(G_0(s)\, W_3(s))$. The weighting function of $W_3(s)$ should be selected as a high-bandwidth transfer function for two reasons. First, it will reduce any unnecessary restriction on the performance specification. Second, it will allow a possible reduction in the complexity for the controller by removing high-frequency controller poles induced by $W_3(s)$. In this case, function $W_3(s)$ can be selected to make the curve for $|G_0(s)\, W_3(s)|$ above that for $|\Delta G_M(s)\, G_0(s)|$ in bode diagram.

Step 3. Weighting function $W_1(s)$ should be selected to have a strong error-tracking ability. Normally, $W_1(s)$ is in the form of a first order low bandwidth transfer function because of similar reasons for $W_3(s)$ but in low frequency domain. Define $G_{w1}(s) = 1/(1 + G_0(s))$, according to the relationship between $S(s)$ and $T(s)$, weighting function of $W_1(s)$ can be selected to make the curve for $|W_1(s)|$ above that for $|G_{w1}(s)|$ in bode diagram.

Step 4. Weighting function $W_2(s)$ should be selected according to the frequency distribution of LFOs in power systems. In such bandwidth, $W_2(s)$ should provide enough lagged phase to guarantee the robust PSS enough leading phase, i.e., it can restrict the poles of the designed controller and the bandwidth of the actuator. Because of the additive uncertainties as mentioned, define $G_{w2}(s) = G_x(s) - G_0(s)$, weighting function $W_2(s)$ can be selected to make the curve for $|W_2(s)|$ above that for $|G_{w2}(s)|$.

The system external input signals, shown in Fig. 18.10, can be classified as reference input signal (r), measurement input signal (n), and disturbance input signal (d). Define $w = \begin{bmatrix} d\ n\ r \end{bmatrix}^T$ and $z = \begin{bmatrix} z_1\ z_2\ z_3 \end{bmatrix}^T$, relationships between all the inputs and output signals in Fig. 18.8 can be expressed as:

$$\begin{bmatrix} z_1 \\ z_2 \\ z_3 \\ y \end{bmatrix} = \begin{bmatrix} -W_1(s)\, G_0(s) & -W_1(s) & W_1(s) & -W_1(s)\, G_0(s) \\ 0 & 0 & 0 & W_2(s) \\ W_3(s)\, G_0(s) & 0 & 0 & W_3(s)\, G_0(s) \\ -G_0(s) & -1 & 1 & -G_0(s) \end{bmatrix} \begin{bmatrix} d \\ n \\ r \\ u \end{bmatrix} = M_{\mathrm{aug}} \begin{bmatrix} d \\ n \\ r \\ u \end{bmatrix}, \tag{18.26}$$

where u is the controller input signal and M_{aug} indicates that the matrix found of $W_1(s)$, $W_2(s)$, $W_3(s)$ and $G_0(s)$ is the augment matrix of the system.

The detailed information about bode diagrams of weighting functions and system uncertainties will be further stated in case study.

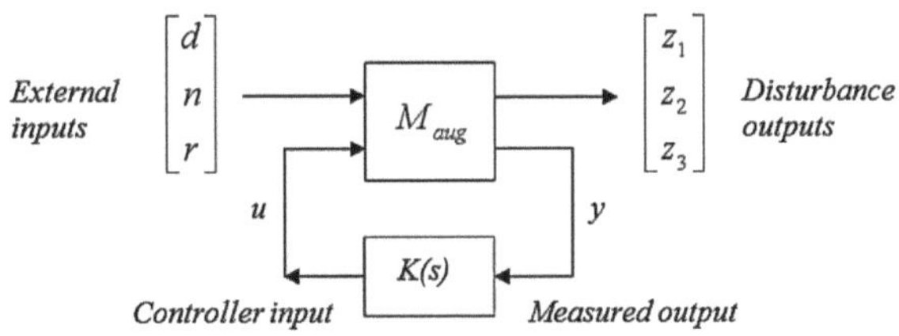

Fig. 18.10 Block diagram for the integrated system

18.2.5.3 Mixed H_2/H_∞ Output Feedback Controller Design Algorithm

Before introducing the controller design algorithm, the system which is made up of A, B_2, C_o, D_{o2} should be confirmed to be controllable and observable. Because it is the sufficient but necessary condition for output feedback controllers design.

As mentioned by equation group (18.25), a mixed H_2/H_∞ output feedback controller $K(s)$ should be designed forcing the closed loop system to own the performance of asymptotic stability. The state space expression for $K(s)$ can be present as:

$$\begin{aligned} \dot{x}_k &= A_k x_k + B_k u \\ u &= C_k x_k + D_k y, \end{aligned} \tag{18.27}$$

where x_k is the state variable and A_k, B_k, C_k, D_k are unknown mixed H_2/H_∞ output feedback controller matrices. Combining (18.25) with (18.27), the closed loop system can be expressed as:

$$\begin{aligned} \dot{x}_{\text{ct}} &= A_{\text{ct}} x_{\text{ct}} + B_{\text{ct}} w \\ z_{\text{inf ct}} &= C_{\text{infct}} x_{\text{ct}} + D_{\text{inf ct}} w \\ z_{2\text{ct}} &= C_{2\text{ct}} x_{\text{ct}} + D_{2\text{ct}} w, \end{aligned} \tag{18.28}$$

where

$$x_{\text{ct}} = \begin{bmatrix} x \\ x_k \end{bmatrix}, \quad A_{\text{ct}} = \begin{bmatrix} A + B_2 D_k C_o & B_2 C_k \\ B_k C_o & B_k D_{o1} \end{bmatrix}, \quad B_{\text{ct}} = \begin{bmatrix} B_1 + B_2 D_k D_{o1} \\ B_k D_{o1} \end{bmatrix}$$

$$C_{\text{inf ct}} = \begin{bmatrix} C_{\text{inf}} + D_{\text{inf2}} D_k C_o & D_{\text{inf2}} C_k \end{bmatrix}, \quad D_{\text{infct}} = D_{\text{inf1}} + D_{\text{inf2}} D_k D_{o1}$$

$$C_{2\text{ct}} = \begin{bmatrix} C_2 + D_{22} D_k C_o & D_{22} C_k \end{bmatrix}, \quad D_{2\text{ct}} = D_{21} + D_{22} D_k D_{o1}.$$

The mixed H_2/H_∞ output feedback controller should be designed to take both mixed H_∞ performance ($\|T_{\text{cl}\infty}(s)\|_\infty < \gamma_1$) and H_2 performance ($\|T_{\text{cl}2}(s)\|_2 < \gamma_2$) into consideration, where $\|T_{\text{cl}\infty}(s)\|_\infty$ is the H_∞ norm of the transfer function from w to $z_{\text{inf ct}}$, $\|T_{\text{cl}2}(s)\|_2$ is the H_2 norm of the transfer function from w to $z_{2\text{ct}}$, and γ_1, γ_2 are the upper bound of $\|T_{\text{cl}\infty}(s)\|_\infty$ and $\|T_{\text{cl}2}(s)\|_2$. Such output feedback controller ought to be designed to make the system to have an acceptable H_∞ norm form w to $z_{\text{inf ct}}$ to keep the system robustness and a suitable H_2 upper bound to keep the system dynamic performance. Lemma 1 is obtained for the H_2/H_∞ performance of the closed loop system and it is defined as:

Lemma 1: If there are symmetric positive matrices X, Y and matrices W, Q, J, V with proper dimension, the H_2/H_∞ performance of a closed loop system with mixed H_2/H_∞ output feedback controller can be expressed as:

$$\begin{bmatrix} AX + B_2J + XA^T + (B_2J)^T & W^T + A + B_2D_kC_o & B_1 + B_2D_kC_o & (C_{\inf}X + D_{\inf 2}J)^T \\ * & AY + QC_o + YA^T + (QC_o)^T & YB_1 + QD_{o1} & (C_{\inf} + D_{\inf 2}D_kC_o)^T \\ * & * & -\gamma_1 I & D_{o1} + D_{o2}D_kD_{o1} \\ * & * & * & -\gamma_1 I \end{bmatrix} < 0$$

$$\begin{bmatrix} AX + B_2J + XA^T + (B_2J)^T & W^T + A + B_2D_kC_o & B_1 + B_2D_kC_o \\ * & AY + QC_o + YA^T + (QC_o)^T & YB_1 + QD_{o1} \\ * & * & -I \end{bmatrix} < 0$$

$$\begin{bmatrix} X & I & (C_2X + D_{22}J)^T \\ * & Y & C_2 + D_{22}D_kC_o \\ * & * & V \end{bmatrix} < 0$$

$$Trace\,(V) < \gamma_2,$$

where * means the symmetric part of the matrix. The proof of Lemma 1 will be given as following.

Proof for Lemma 1: The H_2/H_∞ performance of the closed loop system can be expressed as [45] :

$$\begin{bmatrix} A_{ct}X_{ct} + X_{ct}A_{ct}{}^T & X_{ct}B_{ct} & C_{infct}^T \\ B_{ct}^TX_{ct} & -\gamma_1 I & D_{infct}{}^T \\ C_{infct} & D_{infct} & -\gamma_1 I \end{bmatrix} < 0 \quad (18.29)$$

$$\begin{bmatrix} A_{ct}X_{ct} + X_{ct}A_{ct}^T & B_{ct} \\ B_{ct}^T & -I \end{bmatrix} < 0 \quad (18.30)$$

$$\begin{bmatrix} X_{ct} & X_{ct}C_{2ct}^T \\ C_{2ct}X_{ct} & V \end{bmatrix} < 0 \quad (18.31)$$

$$Trace\,(V) < \gamma_2^2. \quad (18.32)$$

where X_{ct} is a symmetric positive matrix with proper dimension. However, the matrices $A_{ct}, \ldots, D_{infct}$ are all made up of the unknown controller matrices A_k, B_k, C_k, D_k, and the matrix X_{ct} is also a matrix need to be calculated. The nonlinear expression brings huge trouble for solving the LMIs in Lemma 1. That's also the reason for not using the method (directly use the unknown matrix $A\,(\eta)$ for controller design) mentioned at the beginning of this subsection.

In order to solve the nonlinearity problem for LMIs, define

$$X_{ct} = \begin{bmatrix} X & N^T \\ N & \Gamma \end{bmatrix}, \quad X_{ct}^{-1} = \begin{bmatrix} Y & E^T \\ E & \Gamma \end{bmatrix},$$

where X, Y are symmetric positive matrices and N, E, Γ are matrices with proper dimensions. It can be easily found that

$$X_{ct}^{-1} \cdot X_{ct} = \begin{bmatrix} Y & E^T \\ E & \Lambda \end{bmatrix} \begin{bmatrix} X & N^T \\ N & \Gamma \end{bmatrix} = \begin{bmatrix} I & 0 \\ 0 & I \end{bmatrix} \tag{18.33}$$

and $\begin{bmatrix} Y & E^T \end{bmatrix} X_{ct} = \begin{bmatrix} I & 0 \end{bmatrix}$. From Eq. (18.33), it can also be found that

$$\begin{bmatrix} I & 0 \\ Y & E^T \end{bmatrix} X_{ct} = \begin{bmatrix} X & N^T \\ I & 0 \end{bmatrix}. \tag{18.34}$$

Define

$$\Psi_1 = \begin{bmatrix} I & 0 \\ Y & E^T \end{bmatrix}, \Psi_2 = \begin{bmatrix} X & N^T \\ I & 0 \end{bmatrix} \tag{18.35}$$

and (18.34) can be rewritten as $\Psi_1 X_{ct} = \Psi_2$. In another aspect, LMIs (18.29) to (18.31) arė equivalent to LMIs

$$\Psi_1 \begin{bmatrix} A_{ct}X_{ct} + X_{ct}A_{ct}{}^T & X_{ct}B_{ct} & C_{infct}^T \\ B_{ct}^T X_{ct} & -\gamma_1 I & D_{infct}{}^T \\ C_{infct} & D_{infct} & -\gamma_1 I \end{bmatrix} \Psi_1^T < 0 \tag{18.36}$$

$$\Psi_1 \begin{bmatrix} A_{ct}X_{ct} + X_{ct}A_{ct}{}^T & B_{ct} \\ B_{ct}{}^T & -I \end{bmatrix} \Psi_1^T < 0 \tag{18.37}$$

$$\Psi_1 \begin{bmatrix} X_{ct} & X_{ct}C_{2ct}^T \\ C_{2ct}X_{ct} & V \end{bmatrix} \Psi_1^T < 0. \tag{18.38}$$

Note that in the first element of (18.36), it can be found that

$$\Psi_1 A_{ct} X_{ct} \Psi_1^T = \Psi_1 A_{ct} \left(\Psi_1 X_{ct}\right)^T = \Psi_1 A_{ct} \Psi_2^T.$$

In this way, the nonlinear expression of $A_{ct}X_{ct}$ can be transferred to a bilinear expression by matrices $X, Y$and similarly for other nonlinear expressions in (18.36) to (18.38). Following inequalities can be yielded from (18.36) to (18.38)

$$\begin{bmatrix} \mathrm{INF11} & \mathrm{INF12} & B_1 + B_2 D_k D_{o1} & \left(C_{inf}X + D_{o2}\left(D_k C_{inf} X + C_k N^T\right)\right)^T \\ \mathrm{INF21} & \mathrm{INF22} & YB_1 + \left(YB_2 D_k + E^T B_k\right) D_{o1} & \left(C_{inf} + D_{o2} D_k C_o\right)^T \\ * & * & -\gamma_1 I & D_{o1} + D_{inf\,1} D_k D_{inf\,2} \\ * & * & * & -\gamma_1 I \end{bmatrix} < 0 \tag{18.39}$$

$$\begin{bmatrix} \mathrm{INF11} & A + B_2 D_k C_o & B_1 + B_2 D_k D_{o1} \\ \mathrm{INF21} & \mathrm{INF22} & YB_1 + \left(YB_2 D_k + E^T B_k\right) D_{o1} \\ * & * & -I \end{bmatrix} < 0 \tag{18.40}$$

$$\begin{bmatrix} X & I & \left(C_{\inf}X + D_{o2}\left(D_k C_{\inf} X + C_k N^T\right)\right)^T \\ I & Y & \left(C_{\inf} + D_{o2} D_k C_o\right)^T \\ * & * & V \end{bmatrix} < 0, \tag{18.41}$$

where

$$\begin{aligned} \mathrm{INF11} &= AX + B_2\left(D_k C_o X + C_k N^T\right) + \left(AX + B_2\left(D_k C_o X + C_k N^T\right)\right)^T \\ \mathrm{INF12} &= \left(Y\left(A + B_2 D_k C_o\right)X + E^T B_k C_o X + Y B_2 C_k N^T + E^T A_k N^T\right)^T + A + B_2 D_k C_o \\ \mathrm{INF21} &= Y\left(A + B_2 D_k C_o\right)X + E^T B_k C_o X + Y B_2 C_k N^T + E^T A_k N^T \\ \mathrm{INF22} &= YA + \left(Y B_2 D_k + E^T B_k\right) C_o. \end{aligned}$$

If define $W = \mathrm{INF21}$, $Q = YB_2D_k + E^T B_k$, and $J = D_k C_o X + C_k N^T$ in (18.36) to (18.38), Lemma 1 can be proved to be available.

As introduced in previous sections, all the poles of the closed loop system should be restricted in a certain LMI region. Based on some related studies [46], the H_2/H_∞ performance of the closed loop system after LMI regional pole placement is applied can be expressed as:

$$\begin{bmatrix} L \otimes \begin{bmatrix} X & I \\ I & Y \end{bmatrix} + M \otimes \Phi_A + M^T \otimes {\Phi_A}^T & {M_1}^T \otimes \Phi_B & {M_2}^T \otimes {\Phi_{C\infty}}^T \\ M_1 \otimes \Phi_B^T & -\gamma_1 I & I \otimes {\Phi_D}^T \\ M_2 \otimes \Phi_{C\infty} & I \otimes \Phi_D & -\gamma_1 I \end{bmatrix} < 0 \tag{18.42}$$

$$\begin{bmatrix} L \otimes \begin{bmatrix} X & I \\ I & Y \end{bmatrix} + M \otimes \Phi_A + M^T \otimes {\Phi_A}^T & {M_1}^T \otimes \Phi_B \\ M_1 \otimes \Phi_B^T & -I \end{bmatrix} < 0 \tag{18.43}$$

$$\begin{bmatrix} L \otimes \begin{bmatrix} X & I \\ I & Y \end{bmatrix} + M \otimes \begin{bmatrix} X & I \\ I & Y \end{bmatrix} + M^T \otimes \begin{bmatrix} X & I \\ I & Y \end{bmatrix}^T & M_2^T \otimes \Phi_{C2}^T \\ M_2 \otimes \Phi_{C2} & I \otimes V \end{bmatrix} < 0, \tag{18.44}$$

where M_1, M_2 is one of the full rank factorization of $M(M_1^T M_2 = M)$ and $\Phi_A, \Phi_B, \Phi_C, \Phi_D$ are listed as following [38]:

$$\Phi_A = \begin{bmatrix} AX + B_2 J & A + B_2 D_k C_o \\ W & YA + QC_o \end{bmatrix}, \quad \Phi_B = \begin{bmatrix} B_1 + B_2 D_k D_{o1} \\ YB_1 + QD_{o1} \end{bmatrix}, \quad \Phi_D = D_{o1} + D_{\inf 1} D_k D_{\inf 2}$$

$$\Phi_{C\infty} = \begin{bmatrix} C_{\inf} X + D_{\inf 2} J & C_{\inf} + D_{\inf 2} D_k C_o \end{bmatrix}, \quad \Phi_{C2} = \begin{bmatrix} C_2 X + D_{22} J & C_2 + D_{22} D_k C_o \end{bmatrix}.$$

The controller matrices A_k, B_k, C_k, D_k can be obtained by solving BMIs (18.32), (18.38)–(18.44).

18.2.5.4 Parameter Tunings for the Robust PSS

The mixed H_2/H_∞ robust output feedback controller can be successfully designed after the work mentioned above. However, the controller parameters should be tuned considering the practical performance of the power system. In another aspect, the maximum stability region of the selected power system should also be discussed for taking a comprehensive analysis for the proposed PSS.

The feasibility radius saturation (FRS) is the key criterion for judging whether the parameters of the robust controller need to be optimized. It is the norm of the LMIs solution as a percentage of the feasibility radius [47] and the ideal value should be 100 %. However, considering the PSS practical usage and the performance of the target system, the FRS should set as close as 100 %. Compared with normal H_2/H_∞ robust controllers, IMOFC is designed after tuning γ_1, γ_2, i.e., set the H_2 upper bound as small as possible within the acceptable range H_∞ norm. Controller performance and simulation results will be presented in following sections.

In another aspect, the maximum stability margin is necessary to be concerned with. After the IMOFC is designed, the uncertain closed loop system can stay stable for all given $K_1, \ldots, K_6$ in different loading conditions. However, more loading conditions are unknown and the designed IMOFC cannot cover all the cases. If the uncertain system is stable for all values of uncertain elements ($K_1, \ldots, K_6$) within the allowable ranges, the uncertain system is robustly stable. Conversely, if there is a combination of element values that cause instability, and all lie within their allowable ranges, then the uncertain system is not robustly stable.

Known from Sect. 18.2.2, the model of the uncertain closed loop system for stability analysis can be expressed as [25]:

$$G_{\text{cl}} = \frac{G_{\text{op}}(s)\left(1 + \Delta G_{\text{op}}(s)\right)}{1 + G_k(s)\, G_{\text{op}}(s)\left(1 + \Delta G_{\text{op}}(s)\right)}, \tag{18.45}$$

where $G_{\text{op}}(s)$ is the transfer function of the open loop system, $G_k(s)$ is the transfer function of the designed IMOFC, and $\Delta G_{\text{op}}(s)$ is the system uncertainty.

While implementing a rigorous robust stability analysis, two steps need to be followed. The first is to verify the stability of the nominal system. The second is to verify that no poles cross the stability boundary as the uncertain elements vary within their ranges. Because the stability boundary is also associated with the frequency response, the second step can be interpreted as a frequency domain calculation. This amounts to a classical μ-analysis problem [25] which is not going to be stated in this study. Detailed information for the stability margin will be given in the case study.

18.2.5.5 A Step-by-Step Procedure to Design IMOFC

Step 1. Choose the optimal location for PSS installations by the approach of participation factors (multi-machine system only).

Table 18.3 Lower and upper bounds of K_1–K_6 in different loading conditions

	K_1	K_2	K_3	K_4	K_5	K_6
Upper bound	1.6342	1.9174	0.3600	2.4543	0.0643	0.3608
Plant number	(6)	(12)	(16)	(12)	(1)	(16)
Lowe bound	1.0170	1.1624	0.2889	1.4385	−0.1878	0.1805
Plant number	(1)	(1)	(1)	(1)	(12)	(12)

Step 2. Select the set point and the "worst case" of the target synchronous generator by eigenvalues and oscillation frequencies in all loading conditions. For multi-machine systems, all swing modes should be taken into consideration.

Step 3. Based on the result of step 2, choose proper mixed sensitivity weighting functions by the proposed criterions and build the augment system.

Step 4. Solve BMIs in Lemma 1 as well as (18.42)–(18.44) to calculated the controller matrices A_k, B_k, C_k, D_k.

Step 5. Regulate γ_1, γ_2 (balance the H_2 and H_∞ performance) and go to step 4 until the FRS is close enough to 100 %.

Step 6. Apply the optimal controller parameters (obtained in step 5) to (18.45) for robust stability analysis. Calculated the stability margin and the critical value of frequency at which instability occurs, with uncertain elements closest to their nominal values.

18.2.5.6 Design Tutorial: Step-by-Step Implementation for a SMIB Power System

IMOFC-based robust PSS design for a SMIB power system is selected to demonstrate the design procedure. In the system, each plant number indicates one loading condition and all related data are given in Appendix A. The IMOFC design can be started in the following order:

1. Select the proper set point based on the analysis for all loading conditions.

In the 16 different loading conditions [6], the per-unit value of active power P_G and reactive power Q_G varies from 0.4 to 2.0 and 0.2 to 0.4. Additionally, the lower and upper bounds of $K_1, K_2, K_3, K_4, K_5, K_6$ is given in Table 18.3.

The dominant poles of the eigenvalues of the SMIB system in the 16 loading conditions are listed in Table 18.4 and are displayed in Fig. 18.11.

It can be easily drawn from Table 18.2 that the open loop system will be stable in loading conditions 1–4 and it does not have the ability to damp LFOs in loading conditions 5–16. Hence, the set point of the target system should be selected among loading condition 5–16.

The oscillation frequency varies from 0.8367 to 1.2070 Hz. All of them are set within the range of 0.1 to 2.0 Hz, which indicates that all the oscillations belong to LFOs. Among plant 5 to plant 16, the set point should be selected as the plant with

Table 18.4 Dominant poles and of the open loop system in different loading condition

Loading conditions	Plant 1	Plant 2	Plant 3	Plant 4
Dominant poles	−0.4242±j5.2542	−0.4624±j5.6740	−0.2753±j6.5306	−0.4303±j5.6671
Oscillation frequency (Hz)	0.8367	0.9035	1.0399	0.9024
Loading conditions	Plant 5[a]	Plant 6[a]	Plant 7[a]	Plant 8[a]
Dominant poles	0.0739±j7.0455	0.6384±j7.4500	0.7684±j7.5048	0.89444±j7.5476
Oscillation frequency (Hz)	1.1219	1.1863	1.1950	1.2018
Loading conditions	Plant 9[a]	Plant 10[a]	Plant 11[a]	Plant 12[a]
Dominant poles	0.7702±j7.5034	0.8964±j7.5461	1.0156±j7.5799	1.1502±j7.4728
Oscillation frequency (Hz)	1.1948	1.2016	1.2070	1.1899
Loading conditions	Plant 13[a]	Plant 14[a]	Plant 15[a]	Plant 16[a]
Dominant poles	0.1817±j6.8170	0.2673±j6.6182	0.3354±j6.4372	0.4638±j6.8600
Oscillation frequency (Hz)	1.0855	1.0539	1.0250	1.0924

[a] The plants with unstable eigenvalues

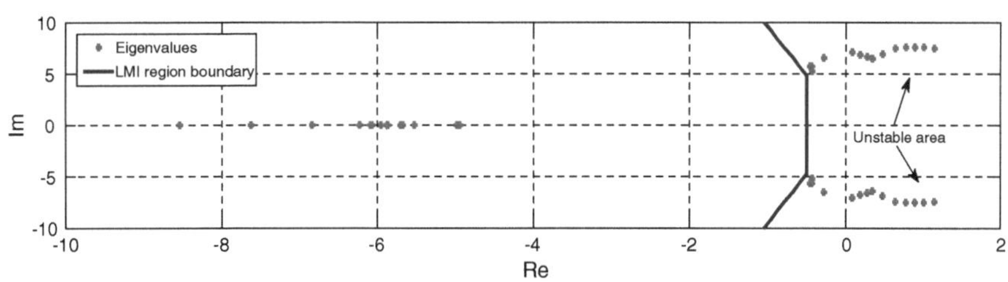

Fig. 18.11 Block diagram for the integrated system

the highest oscillation frequency. The reason is that, the robust controller should be designed to cover all the loading condition. At last, the set point for the H_2/H_∞ output feedback robust controller design is selected as plant 11.

Before using mixed sensitivity technique to build augment system for robust controller design, the loading condition of the so called "worst case" should be decided. In the target SMIB power system, it can be drawn from Table 18.1 that plant 1 got the extreme value of K_1–K_6. It is clearly to see from Table 18.2 that plant 1 also got the lower bound of the oscillation frequency. In this case, the transfer function of the operation point of loading condition 1 (plant 1) can be selected as $G_x(s)$.

2. Select the proper set point based on the analysis for all loading conditions.

Once the transfer function of the set point $G_0(s)$ and the "worst case" $G_x(s)$ are selected, the weighting functions $W_1(s)$, $W_2(s)$, $W_3(s)$ can be found by following

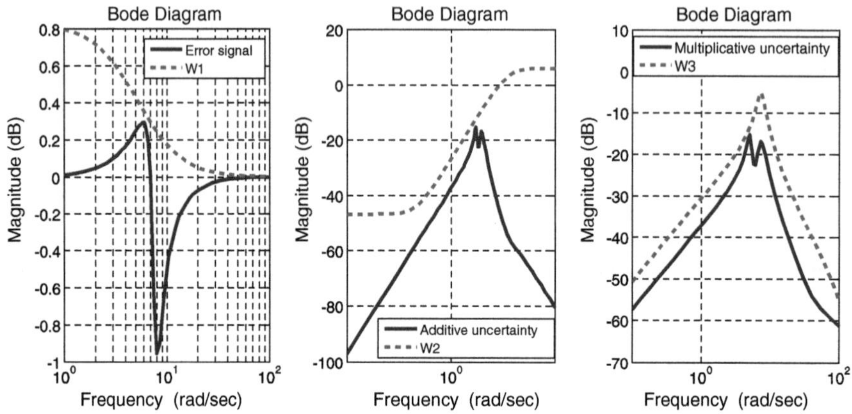

Fig. 18.12 Bode diagram of $W_1(s)$, $W_2(s)$, $W_3(s)$ and related uncertainties

the steps of weighting function selection criterions. After several times of regulation, the expressions of the weighting functions can be obtained as following:

$$W_1(s) = \frac{s+5.5}{s+5}, \quad W_2(s) = \frac{6s+0.6}{3s+130}, \quad W_3(s) = \frac{20s+400}{s+130}.$$

Bode diagrams for presenting the selected weighting function performances are shown in Fig. 18.12.

3. Using (18.26) to build the augment system and obtain the related matrices.

The transfer function of the augment system can be obtained via (18.26) and all matrices mentioned in (18.7) can be calculated and they are listed as following:

$$A = \begin{bmatrix} -50.5769 & -14.6441 & -9.7762 & -8.0728 & 0 & 0 & 0 \\ 16.0000 & 0 & 0 & 0 & 0 & 0 & 0 \\ 0 & 16.0000 & 0 & 0 & 0 & 0 & 0 \\ 0 & 0 & 8.0000 & 0 & 0 & 0 & 0 \\ 0 & 0 & 0.6181 & 0 & -5.0000 & 0 & 0 \\ 0 & 0 & 0 & 0 & 0 & -13.3333 & 0 \\ 0 & 0 & -0.6181 & 0 & 0 & 0 & -130.0000 \end{bmatrix}$$

$$B_1 = \left[0\,0\,0\,0\,1\,0\,0\right]^T, \quad B_2 = \left[1\,0\,0\,0\,0\,1\,0\right]^T$$

$$C_{\mathrm{inf}} = \left[0\,0\,0.6181\,0\,0.5\,0\,0\right], \quad D_{\mathrm{inf}\,1} = 1, \quad D_{\mathrm{inf}\,2} = 0$$

$$C_2 = \left[0\,0\,0\,0\,0\,-86.4667\,0\right], \quad D_{21} = 0, \quad D_{22} = 2$$

$$C_o = \left[0\,0\,-12.3613\,0\,0\,0\,-2200\right], \quad D_{o1} = 0, \quad D_{o2} = 0.$$

4. Using (18.17) to create proper LMI region

In low frequency oscillation swing modes, the damping ratio of power systems is normally required as larger than 10 % and the real part of eigenvalues of power systems after PSSs installed should be smaller than −0.5 [17, 20]. Hence, the matrix for describing such LMI region can be expressed as

$$\text{Region} = \begin{bmatrix} L & M \end{bmatrix} = \begin{bmatrix} 1+j & 0 & 0 & 1 & 0 & 0 \\ 0 & 2j & 0 & 0 & 0.9945 & -0.1045 \\ 0 & 0 & 0 & 0 & 0.1045 & 0.9945 \end{bmatrix}, \tag{18.46}$$

where L, M are defined in (18.16) and the related LMI region can be expressed as

$$\begin{aligned} D_{\text{LMI}} &= \left\{ z \in C : L + zM + \bar{z}M^T < 0 \right\} \\ &= \left\{ z \in C : \begin{bmatrix} 1+j & 0 & 0 \\ 0 & 2j & 0 \\ 0 & 0 & 0 \end{bmatrix} + z \begin{bmatrix} 1 & 0 & 0 \\ 0 & 0.9945 & -0.1045 \\ 0 & 0.1045 & 0.9945 \end{bmatrix} + \bar{z} \begin{bmatrix} 1 & 0 & 0 \\ 0 & 0.9945 & -0.1045 \\ 0 & 0.1045 & 0.9945 \end{bmatrix}^T < 0 \right\}. \end{aligned}$$

5. Solving bilinear matrix inequalities and tuning parameters of the designed controller

The improved LMI-based H2/H∞ output feedback controller (IMOFC) can be obtained by solving related BMIs, however, the feasibility radius saturation of the initially designed IMOFC is 38.559 % which is not close enough to 100 %. It can be found that the initially designed IMOFC is available for all loading conditions (with the FRS of 38.559 %) which indicate that the system has a strong robustness. In this way, system dynamic performance can be further improved by sacrificing the robustness a little bit. After changing the value of γ_1, γ_2 for several times, the optimal FRS can be obtained as 97.087 %. The transfer function of the accepted IMOFC after order reduction can be expressed as:

$$G_k(s) = \frac{54.422\,(s+5567)\,(s+182)\,(s+9.727)\,(s-1.989)}{(s+4712)\,(s+151.6)\,(s+71.54)\,(s+0.8991)}. \tag{18.47}$$

After the proposed robust PSS has been installed for the power system, the dominant poles of the eigenvalues under the 16 loading conditions are also displayed in Fig. 18.13.

6. Using (18.45) to implement robust stability analysis in frequency domain

The stability robust margin is calculated as 1.7738 which implies that the uncertain system remains stable for all values of uncertain elements up to 77.38 % outside their modeled uncertain ranges. The destabilizing frequency is 8.3085 Hz which means the poles migrate across the stability boundary (imaginary axis in continuous system) at the exact value.

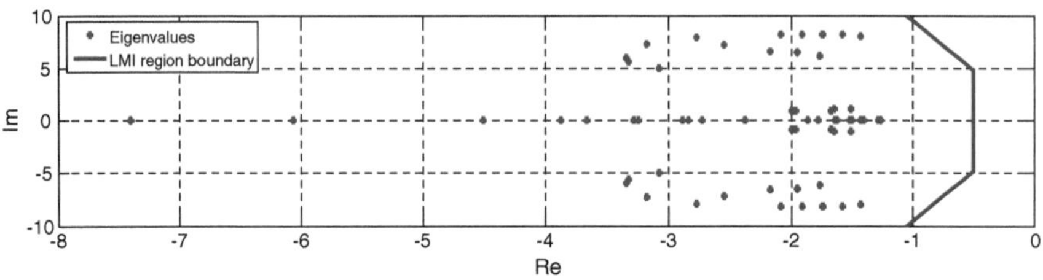

Fig. 18.13 Dominant poles distribution for the SMIB system with PSS under different loading conditions

18.3 Robust PSS Validations for a Single Machine Infinite Bus Power System

All the simulation results for SMIB system are discussed in this section. The performance of IMOFC is compared with normal mixed H_2/H_∞ robust controllers and traditional H_2 robust controllers. For practical usage, the proposed IMORC-based robust PSS is used to compare with conventional PSS under all loading conditions. Additionally, different kinds of disturbances are also applied to the model to test the performance of the proposed IMOFC-based PSS. Per-unit (p.u.) is applied for the unit of speed deviation in this study. Here is the list of the experiments.

1. Comparison of IMOFC and normal mixed H_2/H_∞ robust controller (test 1).
2. Comparison of IMOFC and conventional H_2 robust controller (test 1).
3. Comparison of IMOFC and CPSS (test 2).
4. IMOFC performances tests with different types of external disturbances (tests 1, 2, and 3).

18.4 SMIB System Tests 1

Compared with traditional mixed H_2/H_∞ robust controllers, the IMOFC is applied with clear weighting function selection criterions and optimal controller parameters. In the first simulations, a 10 % step input in reference voltage is selected as disturbances to test the performances of closed loop SMIB system. The simulation results under all loading conditions are shown in Figs. 18.14, 18.15, 18.16 and 18.17.

The simulation results from the 16 loading conditions show that the performance of IMOFC is better than traditional mixed H_2/H_∞ robust controllers. System robustness can be guaranteed by both types of controller. However, with lower H_2 bound (γ_2), the damping ability of IMOFC is stronger than traditional mixed robust controllers.

Instead of merely comparing the improvement of system dynamic performance by IMOFC, the key point here is to demonstrate the importance of FRS and the

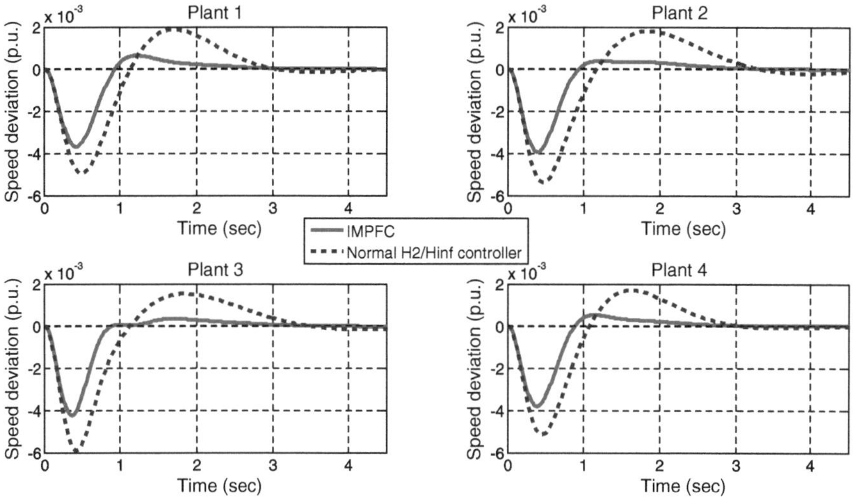

Fig. 18.14 System performances of plant 1–4 after a 10 % step input disturbance in reference voltage

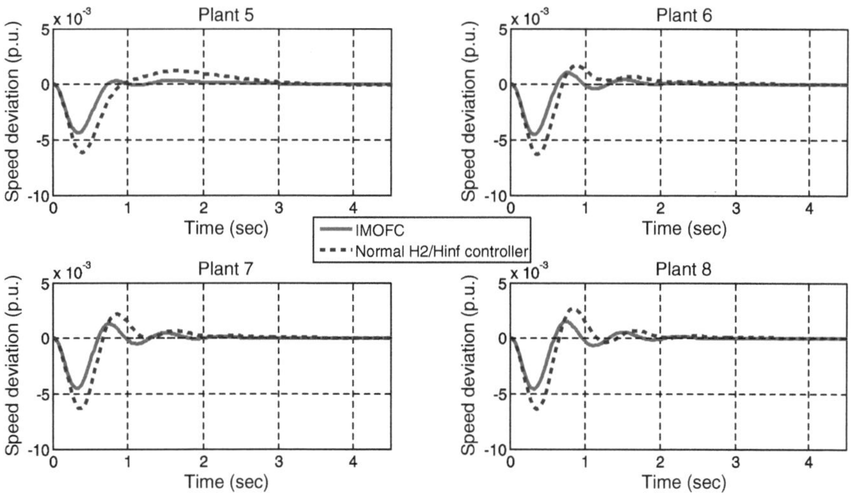

Fig. 18.15 System performances of plant 5–8 after a 10 % step input disturbance in reference voltage

weighting between H_2 and H_∞ for robust for controller design. The balance of these two features can improve the performance of controllers to target systems.

The robust stability margin of the traditional mixed H_2/H_∞ robust controller is 1.5315 and the destabilizing frequency is 7.2478 Hz. It shows that the traditional mixed H_2/H_∞ robust controller have less conservative robustness for the target system than IMOFC with the given loading conditions. These are consistent with

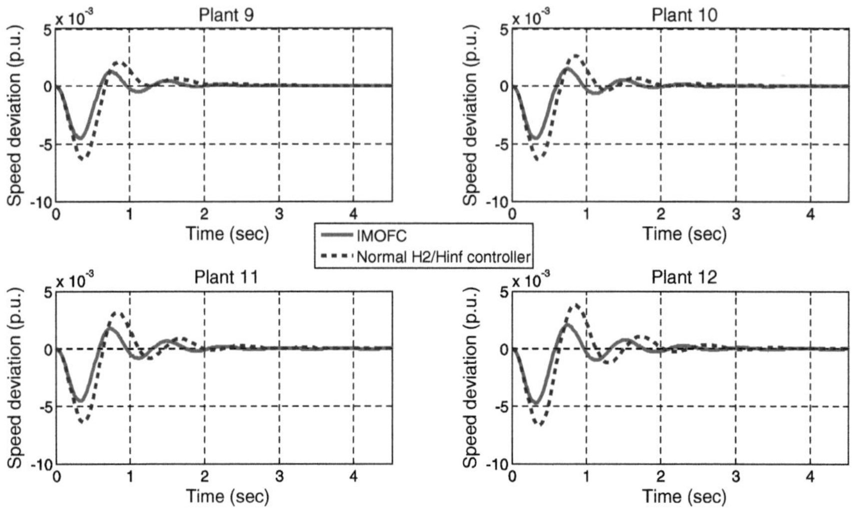

Fig. 18.16 System performances of plant 9–12 after a 10 % step input disturbance in reference voltage

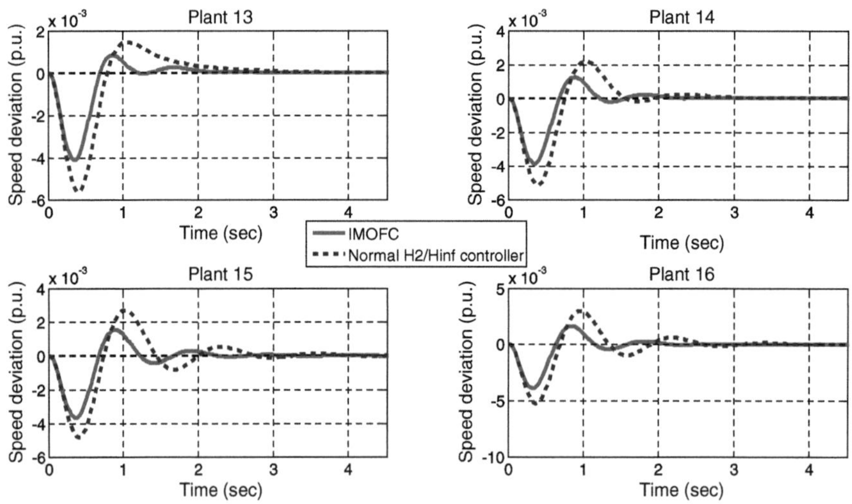

Fig. 18.17 System performances of plant 9–12 after a 10 % step input disturbance in reference voltage

the parameter tuning conception of improving H_2 performance by sacrificing H_∞ performance in Sect. 18.2.5.6. The simulation results can also indicate that within the certain range of robustness, the system performance can be improved by tuning the weighting of H_2 and H_∞ performance according to the actual situation of the target system.

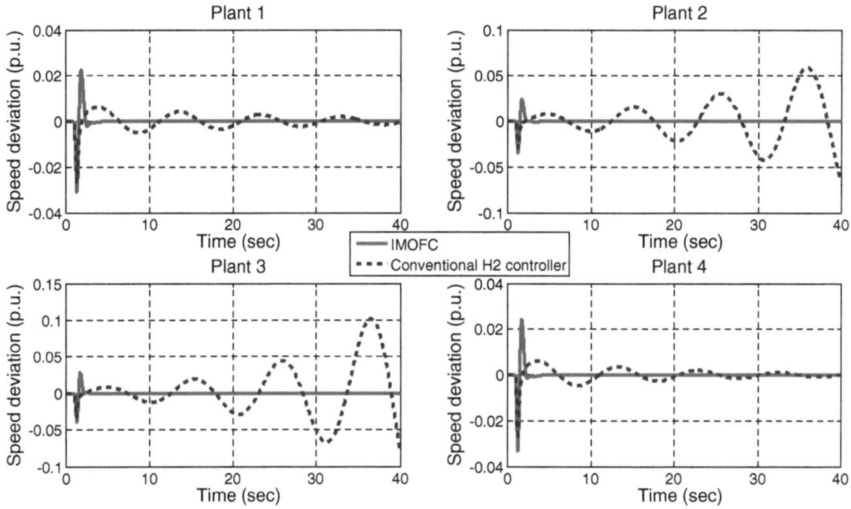

Fig. 18.18 System robustness of plant 1–4 after a short circuit fault

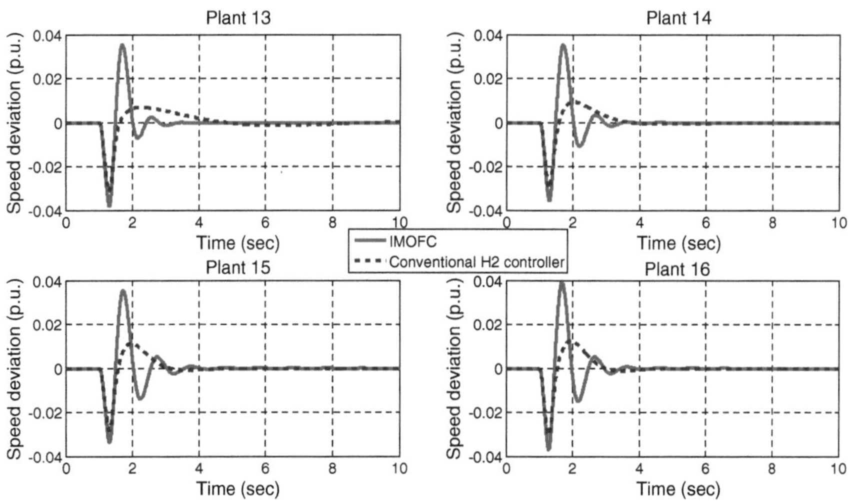

Fig. 18.19 System robustness of plant 13–16 after a short circuit fault

For controller robustness analysis, IMOFC is also used to compare with the traditional H_2 robust controllers. In the second simulations, a three-phase short circuit fault (resembles 100 % square wave input of the voltage) is used as disturbance to test the robustness of closed loop SMIB system. The fault is applied at 1 s on the SMIB system near the terminals of the transformer of the synchronous generator and it is cleared 350 ms later. Simulation results are shown in Figs. 18.18 and 18.19.

Inspection of the simulation results show that although the conventional H_2 controller has a better system performance in some loading conditions, it still lacks of robustness for all loading conditions. In plant 14, 15, and 16, the H_2 robust controller has a better system performance than IMOFC which indicates that the conventional H_2 controller can eliminate LFOs within a certain range of uncertain elements. In plant 13, it can be found that it takes about 10 s for conventional H_2 controller to damp LFOs compared with 3 s for IMOFC. The reason is that the poles of the closed loop system start to migrate to the stability boundary with the change of uncertain elements. In plant 1 and 4, it takes about 60 s for conventional H_2 controller to damp LFOs which show that such controller is not adequate to the related loading conditions. Finally, in plants 2 and 3 the LFOs cannot be damped by conventional H_2 controller because of the lack of robustness.

The robust stability margin of the traditional H_2 robust controller is 0.3431 which implies two aspects. The first is that the uncertain system remains stable for all values of uncertain elements that are less than 0.3431 normalized units away from their nominal values. The second is that there is a collection of uncertain elements that are less than or equal to 0.3431 normalized units away from their nominal values that results in instability.

18.4.1 SMIB System Tests 2

Widely used CPSS is normally based on the phase compensation method [25] which has a close relationship with excitation system parameters. The transfer function of CPSS ($G_{\mathrm{cp}}(s)$) can be expressed as [25]:

$$G_{\mathrm{cp}}(s) = \frac{K_P s}{1 + T_w s}\left(\frac{1 + \alpha T s}{1 + T s}\right)^2, \tag{18.48}$$

where K_P is the gain of CPSS, T_w is washout time constant, α is a constant, and T is the compensation time constant. According to the CPSS design procedure, the transfer function of the related conventional PSS can be expressed as:

$$G_{\mathrm{cp}}(s) = \frac{159.1s}{1 + 4s}\left(\frac{s + 8.9}{s + 12.8}\right)^2. \tag{18.49}$$

For evaluation purposes, the performance of the system with IMOFC was compared to the CPSS. Simulation was carried out by the disturbance of a short circuit fault mentioned in Fig. 18.14 . Simulation results are shown in Figs. 18.20 and 18.21.

From inspection of the simulation results, the closed loop system will be unstable in loading condition 2. It shows that CPSS does not have a capacity for damping LFOs in all the loading conditions compared with IMOFC-based PSS. In all other plants, CPSS needs long time to damp LFOs compared with IMOPC-based PSS. It also indicates that the robustness and dynamic performance of the CPSS-based system are worse than IMOFC-based system.

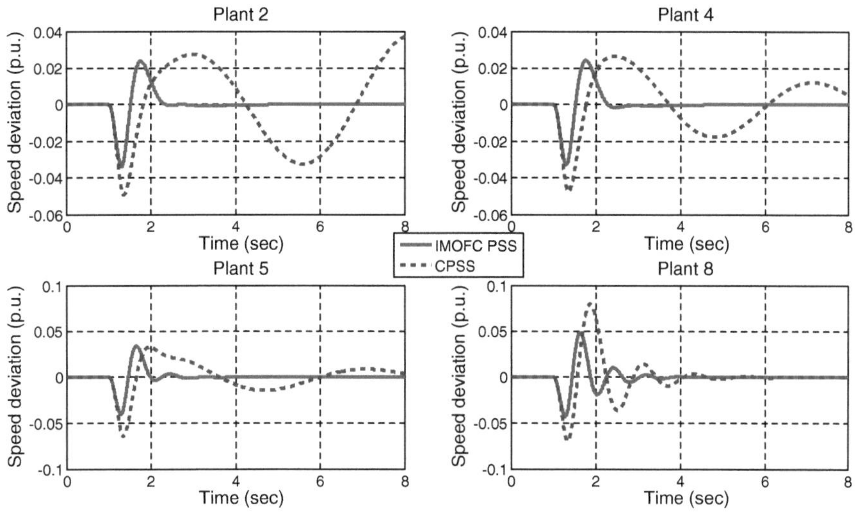

Fig. 18.20 System performances after a short circuit fault (compared with CPSS) for plant 2, 4, 5, 8

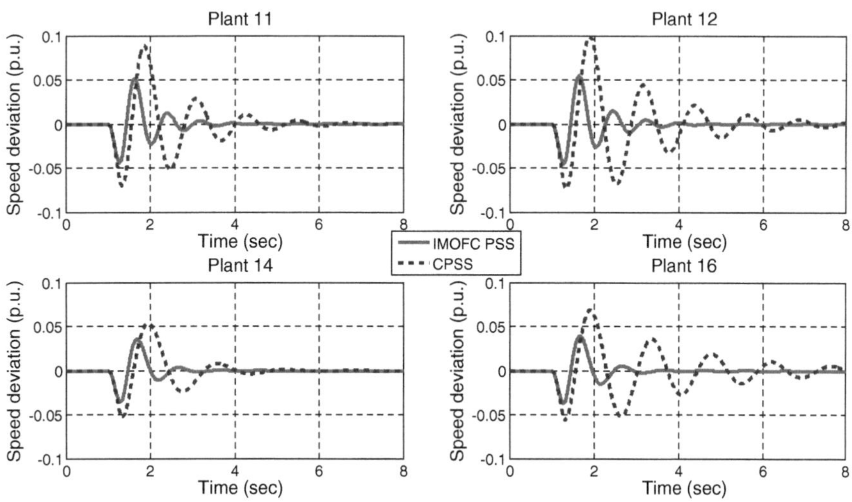

Fig. 18.21 System performances after a short circuit fault (compared with CPSS) for plant 11, 12, 14, 16

Table 18.5 shows the comparisons obtained from simulation results for all the mentioned controllers. The IMOFC-based PSS should be the first choice for power system PSS design.

Table 18.5 Controller tests results for damping LFOs

	IMOFC	Normal H_2/H_∞ controller	Conventional H_2 controller	CPSS
Stability for all plants	Yes	Yes	No	No
Stability margin	1.7738	1.5315	0.3431	0.1737

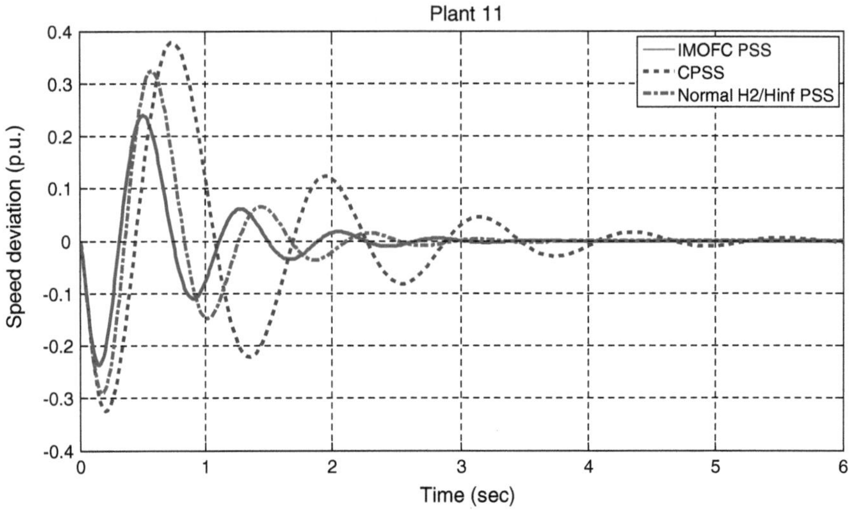

Fig. 18.22 System performances after an impulse input disturbance (resembles 100 % voltage input)

18.4.2 SMIB System Tests 3

Different types of disturbances are used to test the performance of IMOFC in following simulations. An impulse input disturbance is applied to the system at $t = 0$ and the test result is shown in Fig. 18.22.

Another simulation by applying a square wave input as disturbance is given in Fig. 18.23 to show the performance of mentioned controllers.

At last, a stochastic input is selected as the disturbance to test whether the propose controller is still available for the stability for the closed loop system (shown in Fig. 18.24).

To comprehensively and systematically investigate the performance of IMOFC, five different types of disturbances (impulse signal, impulse square wave, step signal, square wave, and stochastic signal) are applied in this study. Based on the results, the designed IMOFC is capable for all loading conditions of the target SMIB power system with high system performance. The IMOFC implementation to multi-machine system will be introduced in the following section.

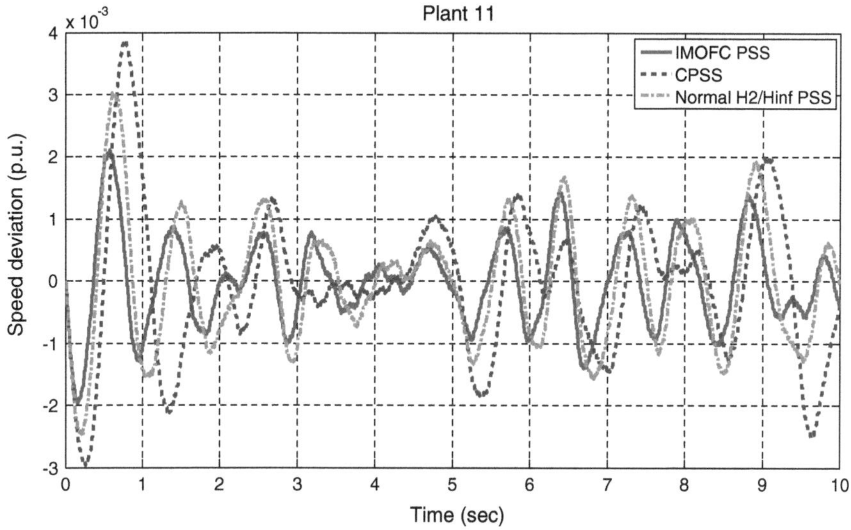

Fig. 18.23 System performances after a square wave input disturbance (resembles 100 % voltage input)

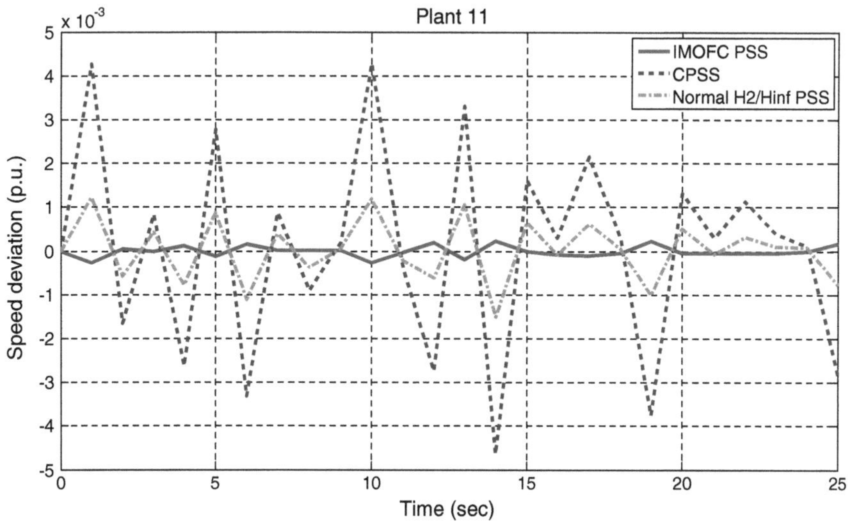

Fig. 18.24 System performances after a stochastic input disturbance (resembles 100 % voltage input)

Table 18.6 Loading conditions (in p.u.) for the multi-machine system

Generator	Heavy		Medium		Light	
	P (p.u.)	Q (p.u.)	P (p.u.)	Q (p.u.)	P (p.u.)	Q (p.u.)
G1	1.330	0.630	0.816	0.270	0.289	0.110
G2	1.900	0.361	1.600	0.162	0.849	0.035
G3	1.200	0.120	0.893	0.058	0.664	−0.085

Table 18.7 Participation factors for each generator

	G1	G2	G3
Swing mode 1	0.7545	**1.5997**	1.1555
Swing mode 2	0.0200	0.7418	**1.2799**

18.5 Robust PSS Validations for a Multi-Machine Power System

All the simulation results for multi-machine system are discussed in this section. Although the implementation of IMOFC is quite successful for SMIB system, its application for multi-machine power system is more significant because of the interactions among all synchronous generators. The model of a standard IEEE three-machine nine-bus system [48] is selected for simulations and related parameters can be found in Appendix A. To design the proposed robust controller, three loading conditions [13], i.e., a heavy loading, a medium loading condition, and a light loading condition, are considered, as shown in Table 18.6.

Simulated studies for interactions among generators in all loading conditions are presented and here is the list of the tests.

1. Design tutorial: step-by-step design procedure of IMOFC for the three-machine nine-bus system (test 1).
2. Comparison of IMOFC and CPSS in light loading condition (test 2).
3. Comparison of IMOFC and CPSS in medium loading condition (test 3).
4. Comparison of IMOFC and CPSS in heavy loading condition (test 4).

18.6 Multi-Machine System Test 1 (Step-by-Step Tutorial)

Design tutorial of the implementation of IMOFC for multi-machine system is introduced in test 1. Upon the model in Fig. 18.5, the IMOFC design is started in the following order.

1. Identify the optimal locations for PSSs installation via the participation factor method. The results (shown in Table 18.7) indicate that G2 (generator No. 2) and G3 (generator No. 3) are the optimum locations for installing PSSs to damp out the electromechanical modes of oscillations.

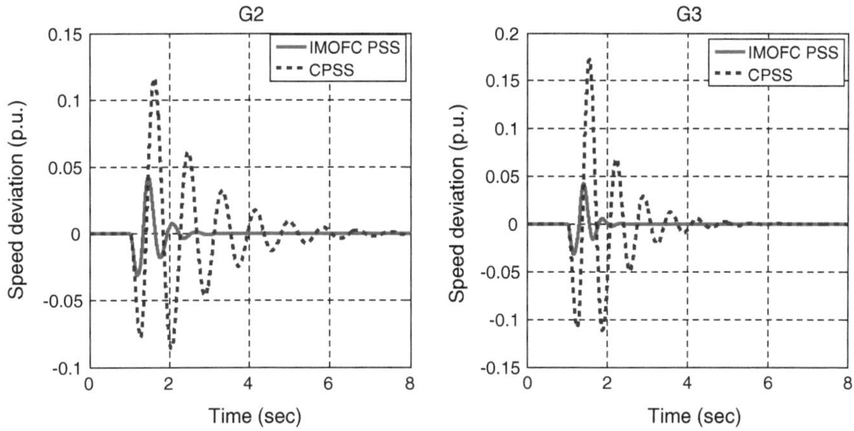

Fig. 18.25 PSS performance for G2 and G3 with a short circuit fault disturbance

2. Select the light loading condition as the set point and the heavy loading condition as the "worst case." Specify the sets of weighting functions $W_1(s), W_2(s), W_3(s)$ for G2 and G3 and build the augment systems respectively by related criterions.
3. Create the same LMI region with (18.46) and solve related BMIs for the controller matrices.
4. Check the FRS for each PSS and tune the parameters for the optimal controller. For the initial designed IMOFC, the FRS for G2 is 89.434 % and for G3 is 27.438 %.
5. Finally, obtain the optimized IMOFC and start simulations.

Simulation results are shown in Fig. 18.25 to present the speed deviation of G2 and G3 after a short circuit fault at the terminal of each of them respectively.

It can be drawn from Fig. 18.25 that generators installed with IMOFC-based PSS have a stronger damping ability compared with CPSS. However, the most important point in multi-machine power system is the interactions within all machines. The simulation results will be discussed in following tests.

18.6.1 Multi-Machine System Test 2 (Light Loading Condition)

Interactions among G1, G2, and G3 in the light loading condition are discussed in this section. The same three-phase short circuit fault (resembles 100 % square wave input of the voltage) is applied at the terminal of each generator to simulate the performance of other two generators. First, the fault is assumed to occur at the terminal of G1 and the response of G2 and G3 are shown in Fig. 18.26. Second, the fault is assumed to occur at the terminal of G2 and the response of G1 and G3 are shown in Fig. 18.27. Finally, the fault is assumed to occur at the terminal of G3 and the response of G1 and G2 are shown in Fig. 18.28.

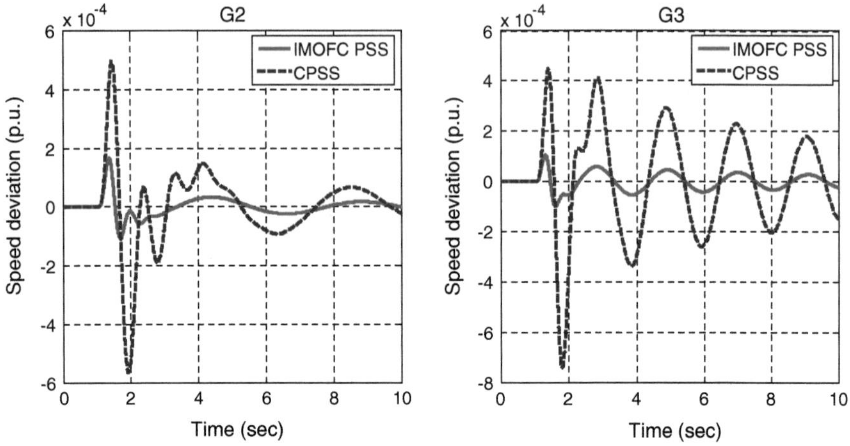

Fig. 18.26 Response of G2 and G3 after a three-phase fault at G1 (light load)

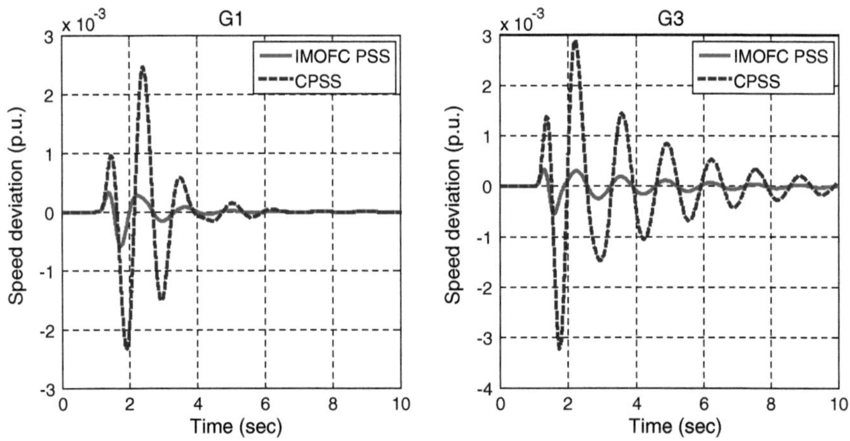

Fig. 18.27 Response of G1 and G3 after a three-phase fault at G2 (light load)

From inspection of the simulation results, the IMOFC-based PSS has a better performance for damping oscillations in multi-machine systems. Instead of merely comparing the damping ability between IMOFC and CPSS, the performance of the oscillations should also be noticed. It can be easily found that the magnitude in Fig. 18.26 is quite smaller than the ones in Figs. 18.27 and 18.28. The reason is that, the participation factor of G1 is the smallest one in both swing mode 1 and 2. If the fault is occurred in G1, the impact is not as large as in G2 and G3. In another aspect, if the fault is occurred in G2/G3, because it has the largest participation factor in swing mode 1/2, the response of the other two generators will be affected and the magnitudes of oscillations in Figs. 18.27, 18.28 are also consistent with it.

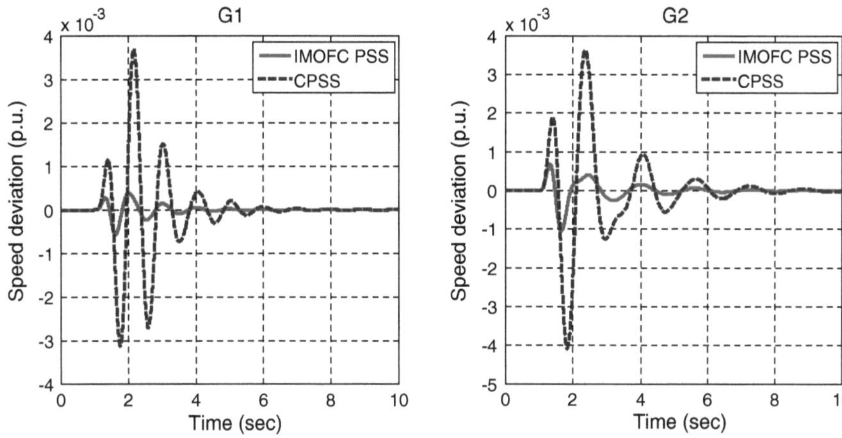

Fig. 18.28 Response of G1 and G2 after a three-phase fault at G3 (light load)

Another observation from the simulation is that oscillations can be damped within about 20s when the fault is applied at G1 compared with 8s at G2 and G3. The reason is because there is no PSS installed for G1 because it is not the optimal location for PSS installation. Without a PSS, it will take G1 longer time to damp oscillations by itself which definitely affects the response of other two generators.

18.6.2 Multi-Machine System Test 3 (Medium Loading Condition)

Interactions among G1, G2, and G3 in the medium loading condition are discussed in this section. The same three-phase short circuit fault (resembles 100% square wave input of the voltage) is applied at the terminal of each generator to simulate the performance of other two generators. First, the fault is assumed to occur at the terminal of G1 and the response of G2 and G3 are shown in Fig. 18.29. Second, the fault is assumed to occur at the terminal of G2 and the response of G1 and G3 are shown in Fig. 18.30. Finally, the fault is assumed to occur at the terminal of G3 and the response of G1 and G2 are shown in Fig. 18.31.

From inspection of the results, IMOFC-based PSS can guarantee a strong damping ability for the whole system in medium loading conditions. Because CPSSs are designed in the light loading condition and they have lack of robustness, the damping effects are reduced a lot especially for the CPSS installed at G3. Known from Fig. 18.29, when a fault occurs at G1, the CPSS can hardly damp the related oscillations at G3 because the loading condition is totally changed. However, with a strong robustness, the IMOFC-based CPSS can still guarantee the damping performance. Similarly in Fig. 18.30, the CPSS takes longer time than to damp the oscillations

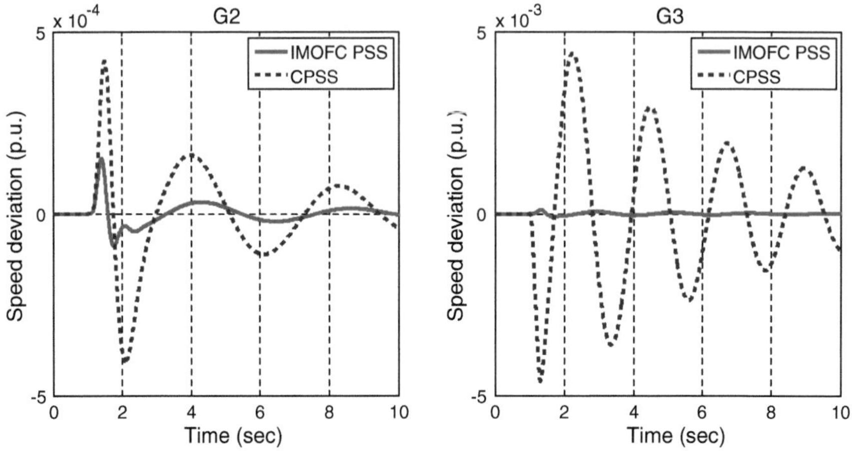

Fig. 18.29 Response of G2 and G3 after a three-phase fault at G1 (medium load)

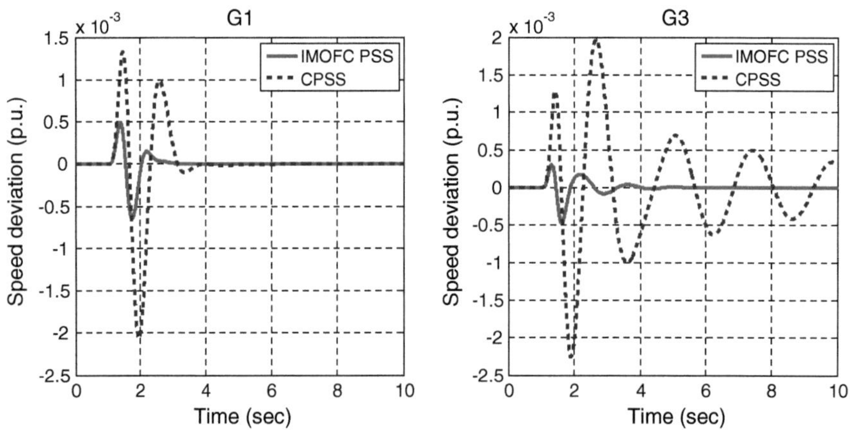

Fig. 18.30 Response of G1 and G3 after a three-phase fault at G2 (medium load)

(brought by the fault at G2) than in light loading condition. G1 is also affected by the weakness of CPSS at G3, it take G1 longer time to damp the oscillations (when fault occurred at G3) than in the light loading condition.

18.6.3 Multi-Machine System Test 4 (Heavy Loading Condition)

Interactions among G1, G2, and G3 in the heavy loading condition are discussed in this section. The same three-phase short circuit fault (resembles 100 % square wave input of the voltage) is applied at the terminal of each generator to simulate

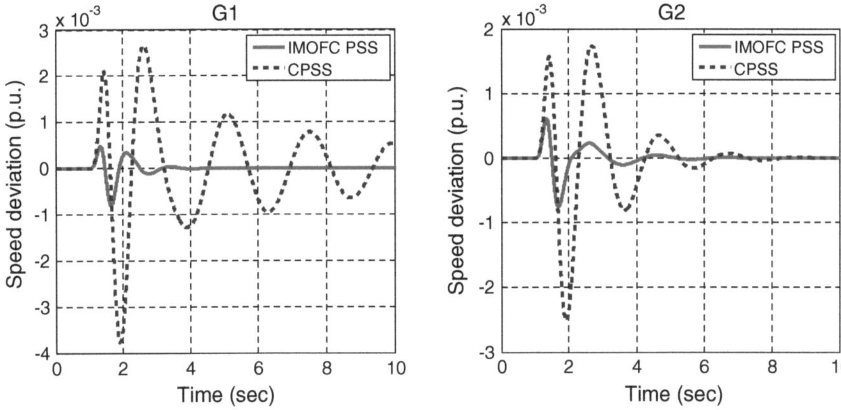

Fig. 18.31 Response of G1 and G2 after a three-phase fault at G3 (medium load)

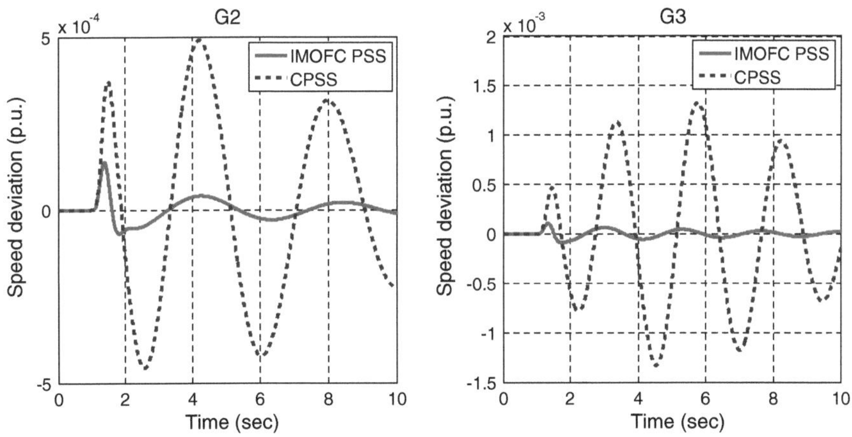

Fig. 18.32 Response of G2 and G3 after a three-phase fault at G1 (heavy load)

the performance of other two generators. First, the fault is assumed to occur at the terminal of G1 and the response of G2 and G3 are shown in Fig. 18.32. Second, the fault is assumed to occur at the terminal of G2 and the response of G1 and G3 are shown in Fig. 18.33. Finally, the fault is assumed to occur at the terminal of G3 and the response of G1 and G2 are shown in Fig. 18.34.

From inspection of the response of G1 when fault at G2 and G3, the self-damping ability of G1 is reduced a lot in the heavy loading condition. In another aspect, CPSS for G3 is not available in heavy loading condition and the parameter is slightly changed to make the LFOs to be damped in this test. Furthermore, the damping ability of CPSS in G2 is also reduced significantly compared in medium loading condition and the response of G2 when fault at G1 (shown in Fig. 18.32). However, the IMOFC-based PSS can still guarantee a strong robustness for the whole three-

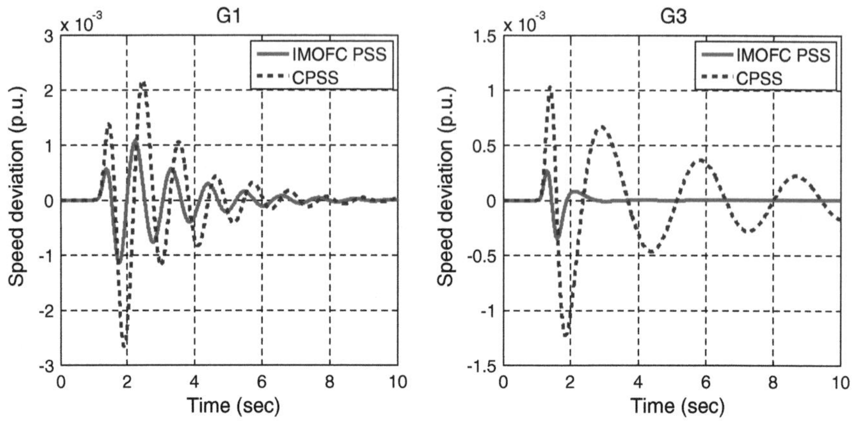

Fig. 18.33 Response of G1 and G3 after a three-phase fault at G2 (heavy load)

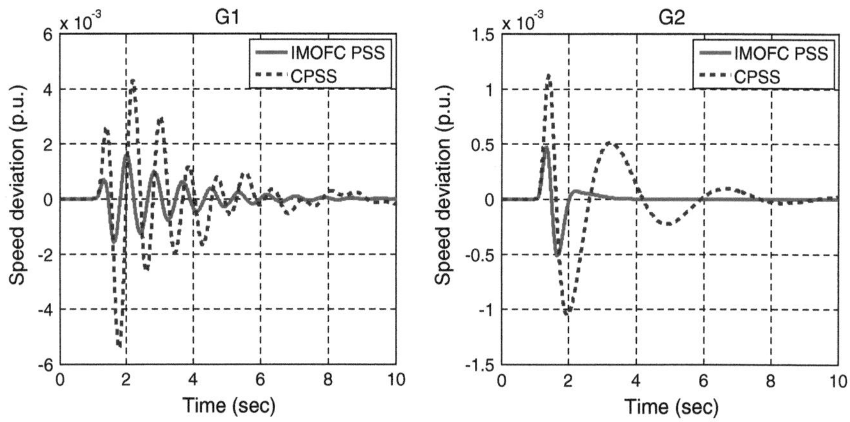

Fig. 18.34 Response of G1 and G2 after a three-phase fault at G3 (heavy load)

Table 18.8 Open loop dominant eigenvalues of the study system

	G1	G2	G3
Swing mode 1	2.266 ± j10.446	1.257 ± j9.937	0.064 ± j9.471
Swing mode 2	3.767 ± j12.856	2.771 ± j11.575	1.834 ± j11.396

machine nine-bus system in the heavy loading condition. Interactions among all generators will not suffer from disturbances with the help of IMOFC-based PSS.

The open loop and closed loop dominant eigenvalues of the system under study for three loading conditions are given in Tables 18.8 and 18.9.

It is clear that the system eigenvalues associated with the electromechanical modes have been successfully shifted to the LMI region with the IMOFC PSS. This demonstrates that the system damping with the proposed PSS is greatly enhanced.

Table 18.9 Closed loop dominant eigenvalues of the study system

	G1	G2	G3
Swing mode 1	$-2.845 \pm j10.513$	$-3.605 \pm j9.334$	$-4.919 \pm j7.969$
Swing mode 2	$-4.161 \pm j13.219$	$-4.882 \pm j11.315$	$-6.787 \pm j10.161$

18.7 Conclusions

An improved mixed H_2/H_∞ output feedback robust controller, particularly efficient in dealing with system robustness and dynamic performance, has been proposed in this Chapter. The interpretation and performance of the methodology have been proved in theory and demonstrated through numerical simulations. The integrated controller design procedure is given for both SMIB and multi-machine system. It provides a more comprehensive and effective detection of system robustness and performance for power system analysis. For applications the proposed controller can be widely used to provide PSS solutions for modern power systems. Furthermore, in the multi-machine case, the control is decentralized and only locally measured output signals are feedback at each generator. Simulation results show that the IMOFC PSS can effectively enhance the damping of LFOs. It has better system dynamic performance that traditional H_2/H_∞ controllers, better robustness than conventional H_2 robust controllers and it also performs better than CPSS.

References

1. Soliman M, Bendary F, Mansour W, Elshafei AL (2011) LMI static output feedback design of fuzzy power system stabilisers. Int J Model Ident Control 12(4):328–340
2. Chesi G (2011) LMI conditions for time-varing uncertain systems can be non-conservative. Automatica 47(3):621–624
3. Mendonca A, Lopes JAP (2003) Robust tuning of PSS in power systems with different operating conditions. Proc IEEE Power Tech Conf 1:1–4
4. Yuan SQ, Fang DZ (2009) Robust PSS parameters design using a trajectory sensitivity approach. IEEE Trans Power Syst 24(2):1011–1018
5. Xu GK, Wang J, Zhang T (2012) A moving horizon H_∞ output feedback control scheme for constrained uncertain linear systems. Int J Model Ident Control 16(1):79–85
6. Gupta R, Bandyopadhyay B, Kulkami AM (2005) Design of power system stabilizer for single-machine system using robust periodic output feedback controller. IEE Proc Gener Transm Disturb 150(2):211–216
7. Chowdhury A, Datta KB, Duttagupta PB (2002) A genetic based mixed H_2/H_∞ PID PSS: sensitivity and control sensitivity minimization. In: International conference on electrical and computer engineering, vol. 28, p 31
8. Werner H, Korba P, Yang TS (2003) Robust tuning of power system stabilizers using LMI-Techniques. IEEE Trans Control Syst Technol 11(1):147–152
9. Abdellatif H, Heimann B (2008) Modelling, identification and robust control of 6-DOF parrallel manipulators. Int J Model Ident Control 4(3):213–225
10. Huang H, Li DW, Xi YG (2010) Synthesis of robust model predictive control based on mixed H_2/H_∞ control approach. Control Decis 25(8):1269–1272

11. Lee S, Park Y (2010) Evaluation of PSS concepts for successful shift from product to PSS: an approach based on AHP and niche theory. In: IEEE International conference on industrial engineering and, engineering management, pp 453–457
12. Yan R, Dong ZY, Saha TK, Majumder R (2010) A power system nonlinear adaptive decentralized controller design. Automatica 46(2):330–336
13. Hardiansyah H, Furuya S, Irisawa J (2004) LMI based robust H_2 control design with regional pole constraints for damping power system oscillations. Euro Trans Power 15:13–29
14. Hardiansyah H, Junaidi J (2012) Multiobjective H_2/H_∞ control design with regional pole constraints. Telkomnika 10(1):103–112
15. Swarcewicz A, Wroblewska K (2001) Robust power system stabilizer. IEEE Trans Power Technol Proc 2:1–6
16. Chen H, Chen J, Bai H, Guo ZZ (2006) Wide-area robust control for damping multiple inter-area oscillations. In: International conference on, power system technology, pp 1–7
17. Cai CH (2006) Design of power system stabilizer for damped low-frequency oscillation. Electr Power Sci Eng 1:22–26
18. Soliman M, Emara H, Elshafei A, Bahgat A, Malik OP (2008) Robust output feedback power stabilizer design: an LMI approach. IEEE power and energy society general meeting, pp 1–8
19. Chen H, Carsten WS (2006) Moving horizon H_∞ control with $H\infty$ performance adaptation for constrained linear systems. Automatica 42(6):1033–1040
20. Cai CH, Liu F (2007) H_2/H_∞ mixed control for power system stabilizer under regional pole constraint. Guangdong Electr Power 20(4):46–49
21. Skogestad S, Postlethwaite I (2007) Multivariable feedback control. Wiley, New York
22. Peng Y, Zhu QM, Nouri H (2011) Robust H_2 power system stabilizer design using LMI techniques. In: Proceedings of international conference on modelling, identification and control, pp 405–410
23. Zhang X, Chen YK, Calay RK (2013) Modelling and analysis of a novel wind turbine structure. Int J Model Ident Control 19(2):142–149
24. Anderson PM, Fouad AA (2002) Power system control and stability, 2nd edn. Wiley-IEEE Press, Hoboken
25. Liu Q (2007) Power system stability and generator excitation system control. China Electrical Power Publisher, Beijing
26. Soliman HM, Bayoumi EHE, Soliman M (2012) Robust guaranteed-cost sliding control for brushless DC motors by LMI. Int J Model Ident Control 17(3):251–260
27. Ghasemi H, Canizares C (2007) On-line damping torque estimation and oscillatory stability margin prediction. IEEE Trans Power Syst 22(2):667–674
28. Gurrala G, Sen I (2011) Synchronizing and damping torques analysis of nonlinear voltage regulators. IEEE Trans Power Syst 26(3):1175–1185
29. Malik OP (2009) Adaptive and intelligent control applications to power system stabiliser. Int J Model Ident Control 6(1):51–61
30. Huang Y (2005) Study on dynamic analysis methods for AC/DC power systems. Dissertation for the Doctoral Degree of Engineering, Zhejiang University
31. Peng Y, Zhu QM, Nouri H (2012) LMI based H_2/H_∞ power system stabilizers fir large disturbances in power systems with wind plant.In: The proceedings of 2012 international conference on modelling identification and control (ICMIC), pp 939–944
32. Zhao AD, Xi ML, Sun J (2010) Method for designing $H\infty$ structured specified controllers. Comput Eng Appl 46(19):67–70
33. Mei SW, Shen TL, Liu KZ (2003) Modern robust control applications. Tsing Hua University Publisher, Beijing
34. Xia H, Majecki P, Ordys A, Grimble M (2005) Calculating setpoint range under actuator saturation and stochastic disturbance. In: IEEE conference on control applications, pp 1069–1074
35. Akasaka N (2008) Design of piecewise linear LQ control for linear systems with actuator rate saturations using LMIs optimization. In: International conference on control, automation and systems, pp 2517–2522

36. Gierusz W (2009) The H_2 and robust H_∞ regulators applied to multivariable ship steering. Int J Marine Navig Saf Sea Transp 3(4):431–440
37. Yu L (2002) Robust control: Linear matrix inequality techniques. Tsing Hua University Publisher, Beijing
38. Wu M, Gui WH (2005) Advanced robust control. Zhongnan University Publisher, Wuhan
39. Peng Y, Nouri H, Zhu QM (2012) Robust power system stabilizers for multi-machine systems with large wind parks. In: The 47th international universities power, engineering conference, pp 1–6
40. Oloomi H, Shafai B (2003) Weight selection in mixed sensitivity robust control for improving the sinusoidal tracking performance. In: Proceedings of the IEEE conference on decision and, control, pp 300–305
41. Yao Y, Fu SW, He FH, Wang XC (2009) Servo system performance optimization design method based on weight function selection. ACTA Automatica SINICA 35(11):1470–1475
42. Zhong ZZ, Wang JC (2009) A comparison and simulation study of robust excitation control strategies for single-machine infinite bus power system. Int J Model Ident Control 7(4):351–356
43. Chung KML, Joo S, Kim J (2004) Design of frequency-dependent weighting functions for H_2 control of seismic-excited structures. J Vib Control 11:137–157
44. Xu ZP, Chang SQ (2009) Modelling and control of an internal combusion linear generator integrated power system. Int J Model Ident Control 7(4):398–404
45. Ali RS (2010) Design of robust mixed H_2/H_∞ PID controller using particle swarm optimization. Int J Adv Comput Technol 2(5):53–60
46. Bayon B, Scorletti G, Blanco E (2012) An LMI solution for a class of robust open-loop problems. In: American control conference, pp 5234–5239
47. Alamo T, Tempo R, Camacho EF (2009) Randomized strategies for probabilistic solutions of uncertain feasibility and optimization. IEEE Trans Autom Control 54(11):2545–2599
48. Anderson PM, Fouad AA (1977) Power Syst Control Stab. The Iowa State University Press, Ames, Iowa

Chapter 19
Advanced Control of Atomic Force Microscope for Faster Image Scanning

M. S. Rana, H. R. Pota and I. R. Petersen

Abstract In atomic force microscopy (AFM), the dynamics and nonlinearities of its nanopositioning stage are major sources of image distortion, especially when imaging at high scanning speed. This chapter discusses the design and experimental implementation of an observer-based model predictive control (OMPC) scheme which aims to compensate for the effects of creep, hysteresis, cross-coupling, and vibration in piezoactuators in order to improve the nanopositioning of an AFM. The controller design is based on an identified model of the piezoelectric tube scanner (PTS) for which the control scheme achieves significant compensation of its creep, hysteresis, cross-coupling, and vibration effects and ensures better tracking of the reference signal. A Kalman filter is used to obtain full-state information about the plant. The experimental results illustrate the use of this proposed control scheme.

19.1 Introduction

Nanotechnology is an area of modern science which deals with the control of matter at dimensions of 100 nm or less. In recent years, of all the available microscopy techniques, atomic force microscopy (AFM) has proved itself extremely versatile as an investigative tool in this field. It is becoming a driving technology in nanomanipulation and nanoassembly [1, 2].

M. S. Rana (✉) · H. R. Pota · I. R. Petersen
School of Engineering and Information Technology, The University of New South Wales, Canberra, ACT, 2600, Australia
e-mail: md.rana@student.adfa.edu.au

H. R. Pota
e-mail: h.pota@adfa.edu.au

I. R. Petersen
e-mail: i.petersen@adfa.edu.au

L. Liu et al. (eds.), *Applied Methods and Techniques for Mechatronic Systems*,
Lecture Notes in Control and Information Sciences 452,
DOI: 10.1007/978-3-642-36385-6_19, © Springer-Verlag Berlin Heidelberg 2014

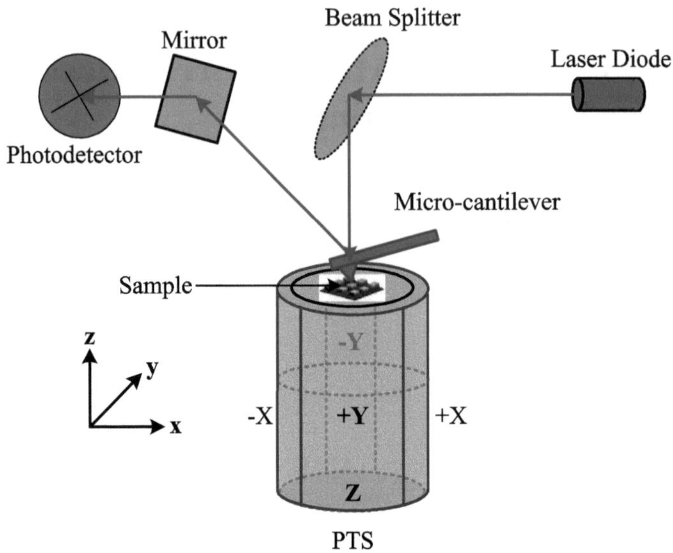

Fig. 19.1 Schematic view of an AFM

The AFM's attractive features are its fast and easy sample preparation, relatively low cost, and ability to operate in various environments. It provides a 3D profile of a surface on a nanoscale by measuring forces, such as the van der Waals, capillary, electrostatic, and magnetic, between a sharp probe (<10 nm) and a surface at a very short distance (0.2–10 nm probe-sample separation) [3].

For control researchers, this instrument provides interesting challenges since its ability to image the surface of a sample is entirely dependent upon the use of a feedback loop. As there appears to be some scope to improve the AFM's performance in terms of its image quality and scanning speed, a variety of different control methods have been developed to control its positioning in order to generate controlled motion for fast positioning. A basic schematic diagram of the AFM is shown in Fig. 19.1.

Imaging the topography of a sample with the AFM depends upon the lateral and vertical positioning precision of the piezoelectric tube scanner (PTS). The high precision lateral positioning of the scanner stage depends on the reference signal tracking performance of the system. So high quality imaging of the AFM depends on the high precision positioning of the scanner. However, the PTS suffers from different problems that degrade its positioning performance, like (i) nonlinear behavior due to hysteresis and creep, (ii) resonant modes due to its mechanical construction [4], and (iii) cross-couplings between the axes [5]. Because of these limitations, AFMs suffer in terms of tracking accuracy. Research work continues to model the dynamics of PTSs [6, 7]. Also, the use of feedback control techniques other than PI controllers to improve the tracking accuracy and scanning speed of AFMs has been undertaken by a number of researchers [4, 8–10]. In reference [8], a model predictive control (MPC) technique is presented, which achieves a significant improvement in the tracking of the reference signal.

During slow-speed scanning and scanning over an extended period of time, the precision positioning of the PTS is seriously affected by creep, the reason for which is that, when a voltage signal is applied to the PTS to move it along a certain area, it continues to displace in the same direction for a certain period of time. During the time between a step change and the beginning of the next scan line, a stretching or compression is seen in the image. Creep has an adverse effect on the vertical positioning of a sample and can contribute to distortions of generated images. A few methods for dealing with this effect have been proposed [11–13].

A loss in the PTS's precision positioning is caused by its nondifferentiable nonlinearity, i.e., hysteresis due to long-range positioning applications as, under applied voltage, piezoelectric materials exhibit hysteretic behaviors due to their ferroelectric properties. The effect of hysteresis varies according to the applied voltage amplitudes, and the hysteresis loop becomes wider with increasing amplitudes and frequencies of the voltage signal. Feedback control schemes have been used to compensate for hysteresis in [14–17]. A sector-bounded H^{∞} controller is implemented in [16] to improve the scanning speed and image quality, and this achieves a scan rate of 125 Hz, but it has distortions at the edges of the images at high-frequency scanning, because it cannot achieve much damping of the resonant modes.

Mechanical vibration, which is caused by the resonant modes in the piezoelectric tube actuator, is also responsible for poor AFM image quality. To avoid this problem, the scanning speed of AFMs is often limited to 1 % [18] of the scanner's first resonance frequency. Since, in most AFMs, the first resonant frequency of the PTS is approximately 1 kHz, this means that the scanning speed is limited to about 10 Hz. Research has been conducted by several authors on this issue [19–23].

In addition to these factors, there are some other important issues that cause distorted images, including the effect of a triangular reference signal. To generate the raster pattern, the fast axis of the AFM nanopositioner is driven by a triangular signal and its slow axis by a synchronized staircase or ramp signal [5]. The triangular signal contains odd harmonics of the fundamental frequency due to which a fast triangular signal excites the resonance of the PTS. Therefore, a major concern is to keep the tracking triangular signal undistorted so that it does not excite the tube resonance. In reference [9], a signal transformation method is implemented to track a triangular reference signal which achieves good results.

Improved mechanical designs such as stiff flexure-guided stages also offer significant improvement in scan speed in an AFM as discussed in [24–26]. In [24], a high-bandwidth, short-range vertical positioning stage integrated with a commercial scanning probe microscope (SPM) for dual-stage actuation has been introduced and achieves significant improvement in high-speed scanning. In [27], although shunt control is implemented in the AFM to improve its scanning rate, this approach struggles to find an effective Q factor and the scanned images it obtains are tilted due to the cross-coupling effect between the axes.

Capacitive sensors are commonly used in nanopositioning systems because of their high-resolution measurement capability. However, they have self noise which degrades an AFM's scanning performance. In our proposed control scheme, to deal with this noise, we introduce a Kalman filter.

In this work, we propose an observer-based model predictive control (OMPC) which reduces the tracking error and significantly damps the resonant modes of the AFM PTS. An MPC control scheme is commonly used when the tracking of a reference trajectory is the primary goal and it makes possible to constrain the control signal [28–30] within the allowed range of the system. Since MPC is designed according to the linear PTS model, the generation of the control action is less complicated than for nonlinear MPC. On the other hand, augmented integral action reduces the nonlinear behavior of the PTS, which in turn solves the tracking error problem [31]. It is also effective for nonminimum phase systems, such as the AFM. A Kalman filter is used to obtain full-state information and to filter the output sensor noise.

The organization of this chapter is as follows: Sect. 19.2 presents descriptions of the AFM and the experimental setup used in this work. Section 19.3 provides the modeling and identification of the system transfer functions. In Sect. 19.4, the control scheme for the AFM scanner is presented. Section 19.5 discusses the experimental results which illustrate the significant improvements in accuracy and imaging speed which can be achieved using the proposed control scheme. Finally, in Sect. 19.6, conclusions are drawn and suggestions regarding future work, which could further improve control of the AFM are discussed.

19.2 Experimental Setup

Figure 19.2 shows the experimental setup at the AFM laboratory, UNSW, Canberra, Australia, which is used to implement the proposed control scheme.

The experimental setup consists of an NT-MDT Ntegra scanning probe microscope (SPM), configured to operate as an AFM. This includes some important accessories such as a signal access module (SAM), control electronics, a noise isolator, and a computer. Other parts used for control implementation are a signal analyzer, a DSP dSPACE RT-1103 board, and a high-voltage amplifier (HVA).

The most important part of an AFM positioning system is the PTS. In this experiment, a NT-MDT z50313cl scanner is used to perform 3D positioning in the AFM. It is a "scan by sample" type of scanner with a scanning range of $100\,\mu m \times 100\,\mu m \times 12\,\mu m$, and resonance frequencies in both the X and Y directions of approximately 900 Hz and in the Z direction of about 5 kHz.

The existing PI controller uses the AFM's internal HVA and generates a triangular reference signal for scanning which can be used as a reference signal while implementing the external controller for scanning. Also, a reference signal for scanning externally can be generated using the dSPACE system.

Three capacitive position sensors, in the X, Y, and Z axes are incorporated into the scanner apparatus to allow for the direct measurement of the tube displacements in the X, Y, and Z directions. A block diagram of the experimental setup is shown in Fig. 19.3. The sensor displacement is observed in the CONTROLDESK software,

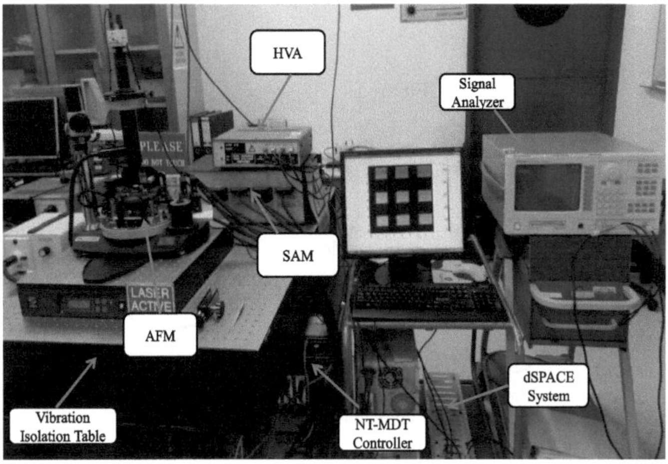

Fig. 19.2 AFM system and experimental setup

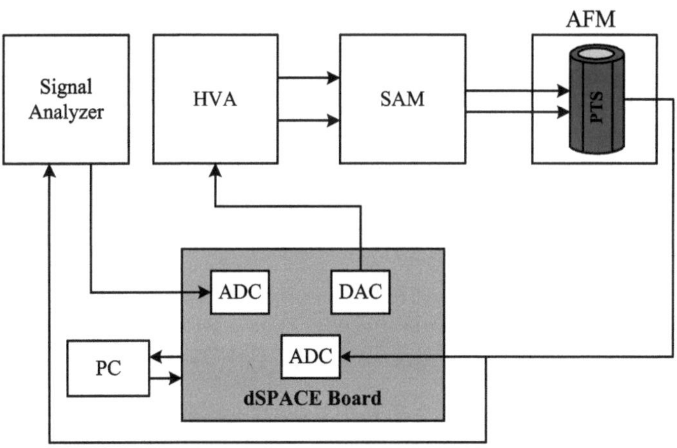

Fig. 19.3 Schematic diagram of the experimental setup

which is part of the dSPACE rapid prototyping system that allows direct access to the capacitive position sensor signals.

19.3 Modeling of the Piezoelectric Tube Scanner

The PTS is a useful actuator in nanopositioning applications, e.g., microscopes and is made of ceramic lead zirconate titanate (PZT). As illustrated in Fig. 19.4, it consists of a tube of radially poled piezoelectric material, four external electrodes, and a grounded internal electrode. The external electrodes are divided into four parts: $+X$

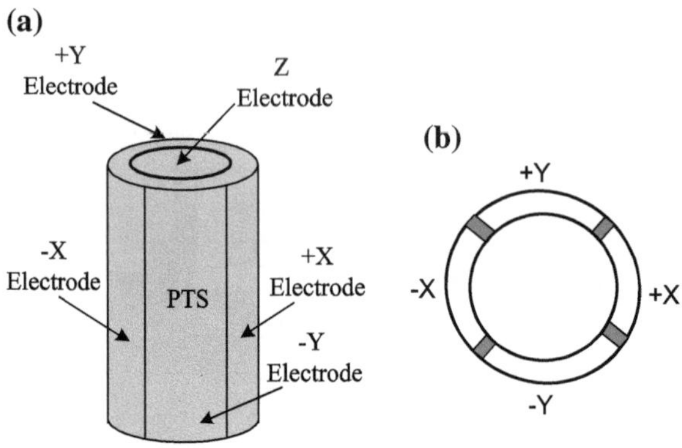

Fig. 19.4 Illustration of the PTS **a** side view and **b** top view

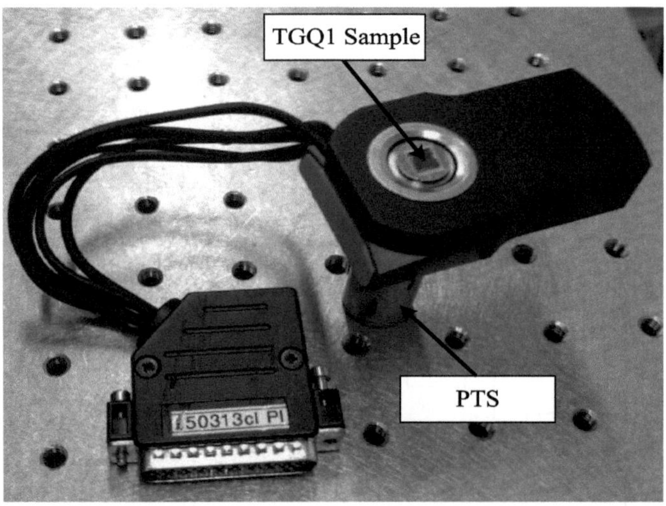

Fig. 19.5 The PTS with internally mounted capacitive position sensors

and $-X$ form the X electrode and $+Y$ and $-Y$ form the Y electrode pairs. The internal Z-electrode is continuous. The reason for using such a configuration is that it halves the input voltage requirements and results in more power being provide to the electrodes. When a differential voltage is applied to the external electrodes with respect to the internal one, the tube contracts or expands depending on the matching of the field and polarized directions, which leads to the tube bending in the x and y directions, respectively. A change in the internal electrode's potential with respect to all external sections results in a lengthening or reduction of the PTS along the z direction. Thus, it is possible to move the PTS in three dimensions. A more detailed illustration of the PTS can be found in [22, 32]. A PTS (NT-MDT z50313cl) mounted with capacitive sensors is shown in Fig. 19.5.

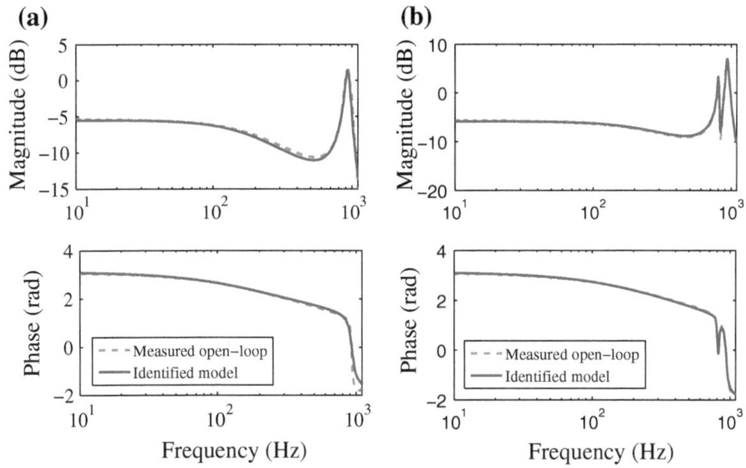

Fig. 19.6 Experimental and identified model frequency responses: **a** $G_{d_x v_x}(s)$ and **b** $G_{d_y v_y}(s)$

To simplify the control design, the AFM scanner is treated as two single-input single-output (SISO) systems. In this process, the plant is identified using a bandlimited random noise signal within the frequency range from 10 to 1100 Hz, using a dual channel HP35665A dynamic signal analyzer (DSA). This signal is supplied to the HVA (with a gain of 15) as an input and the corresponding amplified voltage is supplied to the SAM of the AFM from which there is a direct connection to the PTS.

The output displacement of the PTS is taken from the capacitive position sensor as a corresponding voltage and it is recorded in the DSA. The following frequency response functions (FRFs) were obtained using the DSA:

$$G_{d_x v_x}(i\omega) = \frac{D_x(i\omega)}{V_x(i\omega)}; \tag{19.1}$$

and

$$G_{d_y v_y}(i\omega) = \frac{D_y(i\omega)}{V_y(i\omega)}; \tag{19.2}$$

where V_x and V_y are the voltage signals applied in the corresponding X and Y axes in the piezo, and D_x and D_y the measured displacements from the corresponding capacitive position sensors. From these FRFs, using the prediction error method (PEM), the best fit with the measured data as shown in Fig. 19.6 are chosen as described in references [33–35].

The system behavior is characterized by processing the data from the identified model. The transfer functions in Eqs. (19.3) and (19.4) are found to be a good fit to the experimentally observed frequency responses, as shown in Fig. 19.6a and b. It should be noted here that there is about a 180° phase shift between the inputs and

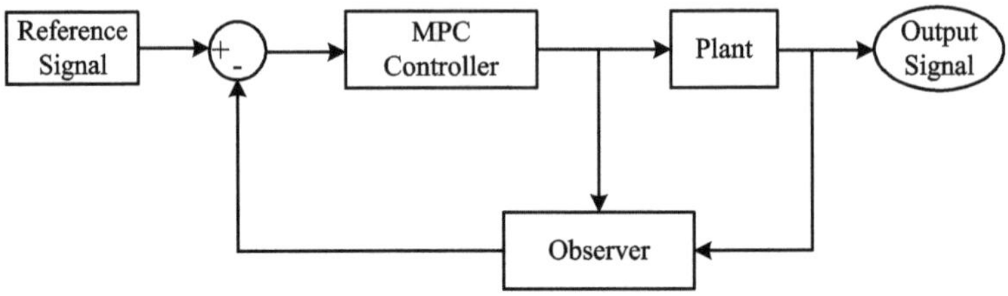

Fig. 19.7 Closed-loop system with the proposed control scheme

outputs at low frequencies in both the X and Y position sensors.

$$G_{d_x v_x}(s) = \frac{-16.2s^2 + 3.108 \times 10^4 s - 9.791 \times 10^7}{s^3 + 295.3s^2 + 8.897 \times 10^5 s + 1.843 \times 10^8}; \tag{19.3}$$

and

$$G_{d_y v_y}(s) = \frac{-14.81s^4 + 6.555 \times 10^4 s^3 - 1.282 \times 10^8 s^2 + 4.172 \times 10^{10} s - 8.19 \times 10^{13}}{s^5 + 355.5s^4 + 1.547 \times 10^6 s^3 + 4.832 \times 10^8 s^2 + 5.835 \times 10^{11} s + 1.604 \times 10^{14}}. \tag{19.4}$$

19.4 Controller and Observer Design

19.4.1 Controller Design

An MPC controller for reducing the tracking error of PTSs in the fast axis (X) and slow axis (Y) is designed. The construction of the corresponding closed-loop system is shown in Fig. 19.7.

The MPC scheme is employed to design a tracking controller, the output from which tracks a reference input. Its basic structure is shown in Fig. 19.8. It consists of a reference block, a predictor block, and an optimizer block. The reference block provides the future reference trajectory which is used to compute the cost function using the predictions from the predictor block. This cost function is then optimized in the optimizer block.

The plant is expressed as the following state-space model;

$$x_m(k+1) = A_m x_m(k) + B_m u(k); \tag{19.5}$$

$$y(k) = C_m x_m(k); \tag{19.6}$$

where A_m, B_m, and C_m define the discrete state-space plant model, derived from the identified FRFs in Eqs. (19.3) and (19.4), u is the manipulated variable or input variable, y is the measured output, and x_m is the state variable vector with dimension

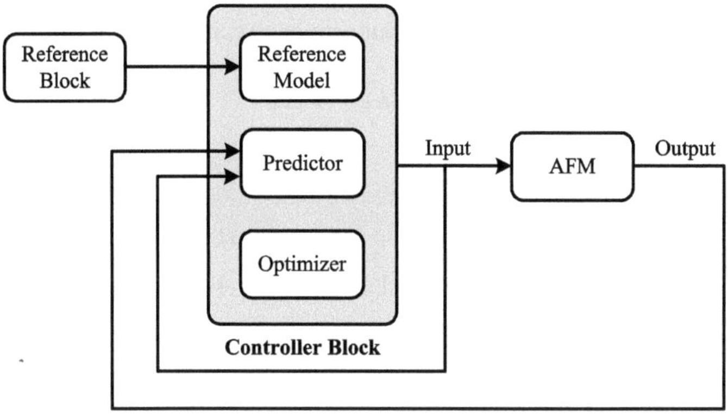

Fig. 19.8 Structure of the model predictive controller

n. In order to incorporate integral action for disturbance rejection and tracking a reference signal in the MPC algorithm, the plant can be augmented in the following way [31, 36]:

$$\begin{bmatrix} \Delta x_m(k+1) \\ y(k+1) \end{bmatrix} = A \begin{bmatrix} \Delta x_m(k) \\ y(k) \end{bmatrix} + B\Delta u(k); \tag{19.7}$$

$$y(k) = C \begin{bmatrix} \Delta x_m(k) \\ y(k) \end{bmatrix}; \tag{19.8}$$

where

$$A = \begin{bmatrix} A_m & 0 \\ C_m A_m & I \end{bmatrix}; B = \begin{bmatrix} B_m \\ C_m B_m \end{bmatrix}; C = \begin{bmatrix} 0 \ I \end{bmatrix};$$

$$\begin{aligned} \Delta u(k) &= u(k) - u(k-1); \\ \Delta x_m(k+1) &= x_m(k+1) - x_m(k); \\ \Delta x_m(k) &= x_m(k) - x_m(k-1); \end{aligned}$$

and A, B, and C are augmented system matrices. Based on this model, for the prediction horizon N_p and control horizon N_c, the future state variables and output sequence can be written as

$$x(k+N_p|k) = A^{N_p}x(k) + A^{N_p-1}B\Delta u(k) + \ldots + A^{N_p-N_c}B\Delta u(k+N_c-1); \tag{19.9}$$

and

$$y(k+N_p|k) = CA^{N_p}x(k) + CA^{N_p-1}B\Delta u(k) + \ldots + CA^{N_p-N_c}B\Delta u(k+N_c-1). \tag{19.10}$$

For a prediction horizon N_p, the output sequence can be written as

$$Y = Fx(k) + \Phi \Delta U; \tag{19.11}$$

in which

$$Y = [y(k+1|k), y(k+2|k), \ldots, y(k+N_p|k)]^T;$$
$$\Delta U = [\Delta u(k), \Delta u(k+1), \ldots, \Delta u(k+N_c-1)]^T.$$

It is assumed that there is no control action after time $k+N_c-1$, i.e., $\Delta u(k+i)=0$ for $i > N_c - 1$. The F matrix with dimensions of (N_P, n) and the Φ matrix with dimensions of (N_P, N_c) are:

$$F = \begin{bmatrix} CA \\ CA^2 \\ CA^3 \\ \vdots \\ CA^{N_p} \end{bmatrix};$$

$$\Phi = \begin{bmatrix} CB & 0 & \cdots\cdots & 0 \\ CAB & CB & \cdots\cdots & 0 \\ CA^2B & CAB & \cdots\cdots & 0 \\ \vdots & \vdots & \ddots \; \vdots & \vdots \\ CA^{N_p-1}B & CA^{N_p-2}B & \cdots\cdots & CA^{N_p-N_c}B \end{bmatrix}.$$

The control law is derived based on the minimization of the cost function

$$J = \sum_{m=1}^{N_p} Q(y(k+m|k) - R_s(k+m))^2 + \sum_{m=1}^{N_c} R(\Delta u(k+m-1))^2; \tag{19.12}$$

subject to the linear inequality constraints on the system inputs

$$u_{\min} \le u(k+i-1) \le u_{\max}, \quad i = 1, \ldots, N_c; \tag{19.13a}$$

$$\Delta u_{\min} \le \Delta u(k+i-1) \le \Delta u_{\max}, \quad i = 1, \ldots, N_c; \tag{19.13b}$$

where N_p is the prediction horizon, N_c is the control horizon, Q is the state weighting matrix, R is the control weighting, R_s is the reference signal, $u_{\min}$ and $u_{\max}$ are the low and high levels of the control action, respectively, and $\Delta u_{\min}$ and $\Delta u_{\max}$ are the low and high levels of the control increments, respectively.

By considering the above equations, the constrained MPC problem can be expressed as a quadratic programming (QP) problem:

$$\min(\frac{1}{2}\Delta U^T E \Delta U + \Delta U^T f); \tag{19.14}$$

s.t.

$$M \Delta U \leq \gamma;$$

where

$$E = \Phi^T Q \Phi + R;$$
$$f = \Phi^T Q F x(k+1|k) - \Phi^T Q R_s;$$

$M \in \mathbb{R}^{m_c \times N_c}$ and $\gamma \in \mathbb{R}^{N_c \times 1}$ are computed using Eq. (19.13), m_c is the number of constraints and $R_s \in \mathbb{R}^{N_p \times 1}$ is the reference signal.

19.4.2 Observer Design

Kalman observer can be used as a state observer and noise filter [37]. The displacements of the PTSs are taken from the capacitive position sensors. Capacitive sensors are commonly used in nanopositioning systems because of their high-resolution measurement capability. However, it adds unwanted noise and disturbances to the output displacement which degrades an AFM's scanning performance. To remove of which we design a Kalman state observer as a noise filter. The Kalman state observer estimates the states from the measured output. The Kalman observer dynamics is:

$$\hat{x}(k+1) = (A - LC)\hat{x}(k) + Bu(k) + Ly(k); \tag{19.15}$$
$$\hat{y}(k) = \hat{C}\hat{x}(k); \tag{19.16}$$

where $\hat{x}(k)$ is the estimated states, $\hat{y}(k)$ is the estimated state output, $\hat{C}$ is the identity matrix of dimension $n \times n$, and L is the observer gain which depends on the Gaussian white noise, process noise covariance, and measurement noise covariance.

19.5 Experimental Results

19.5.1 Creep Effect Compensation

During the slow operation of the AFM in open-loop, the scanned image shows a creep effect as shown in Fig. 19.9a. Due to this effect the right side bottom edge of the vertical axis of the scanned image is rolled-off. In Fig. 19.9b, the creep effect is reduced significantly using the proposed controller.

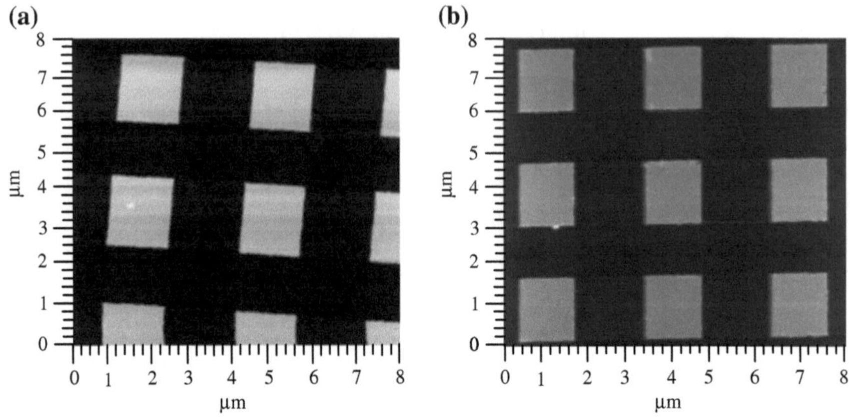

Fig. 19.9 Creep effect in the scanned image at open-loop (**a**) and compensation of creep effect using the proposed controller (**b**)

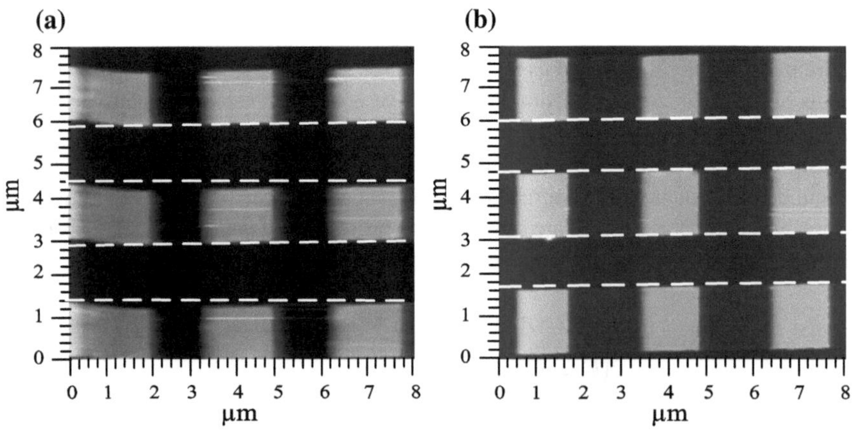

Fig. 19.10 Compensation of hysteresis effect at 31.25 Hz **a** using the AFM PI controller and **b** using the proposed controller

19.5.2 Hysteresis Effect Compensation

The reduction of hysteresis effect at 31.25 Hz using the AFM PI controller and the proposed controller is shown in Fig. 19.10a and b, respectively. The image of the grating has a distinct curvature for the image taken by using the AFM PI controller. In the case of proposed controller, there have no hysteresis effect in the scanned image as shown in Fig. 19.10b.

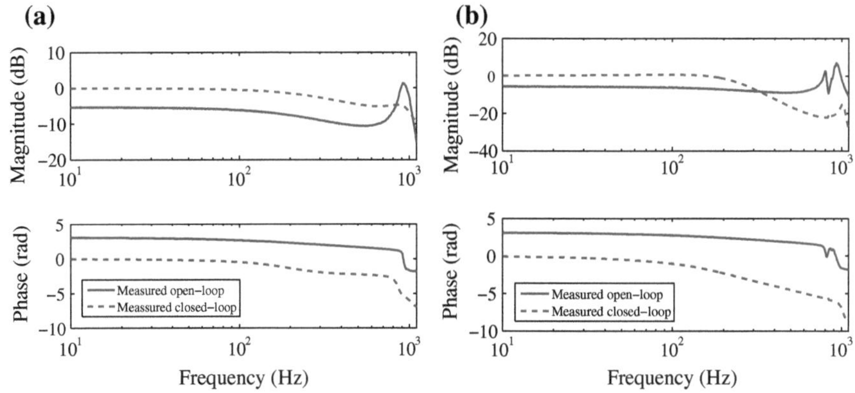

Fig. 19.11 Comparison of measured open-loop and closed-loop frequency responses of the *X* piezo (**a**) and *Y* piezo (**b**)

19.5.3 Vibration Effect Compensation

The performance of the OMPC control scheme is evaluated by measuring its closed-loop frequency responses. In Fig. 19.11a and b, the measured closed-loop frequency responses are plotted along with the open-loop frequency responses of the *X* and *Y* piezo displacements, respectively. By implementing this OMPC control scheme, we have achieved a reasonable damping in the closed-loop system for both axes, which in turn reduces the vibration significantly. Therefore, the closed-loop system has higher bandwidths for both the *X* and *Y* axes.

The reduction of vibration effect by using the proposed controller is shown in Fig. 19.12. Figure 19.12 illustrates the results obtained by implementing the OMPC scheme which show significantly different images from those taken in the open-loop condition. In the open-loop condition, the scanner displacement is quite poor because of the uncontrolled tube resonance and results in a sluggish image as shown in Fig. 19.12a, c, and e at 31.25, 62.50, and 125 Hz scanning speed and this problem is compensated by using the proposed controller as shown in Fig. 19.12b, d, and f. Fig. 19.12f demonstrates that the proposed controller can compensate the vibration effect at the higher scanning speed.

19.5.4 Cross-coupling Effect Compensation

Figure 19.13a and b show comparisons of the open-loop and closed-loop cross-couplings for the 5.21 and 31.25 Hz input reference signals, respectively. They illustrate that there is significant improvement in cross-coupling in the closed-loop. To

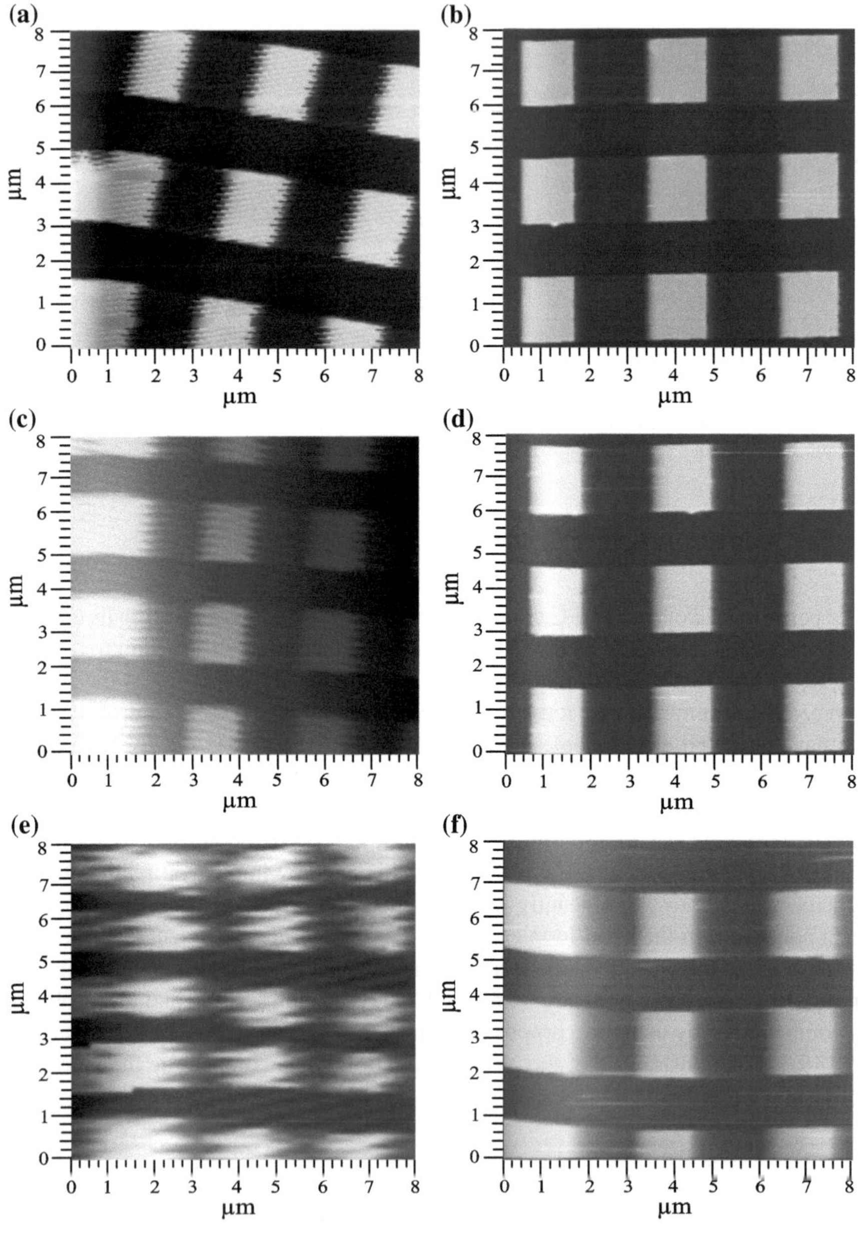

Fig. 19.12 Compensation of vibration effect at 31.25, 62.50, and 125 Hz **a**,**c**, and **e** open-loop and **b**, **d**, and **f** using the proposed controller

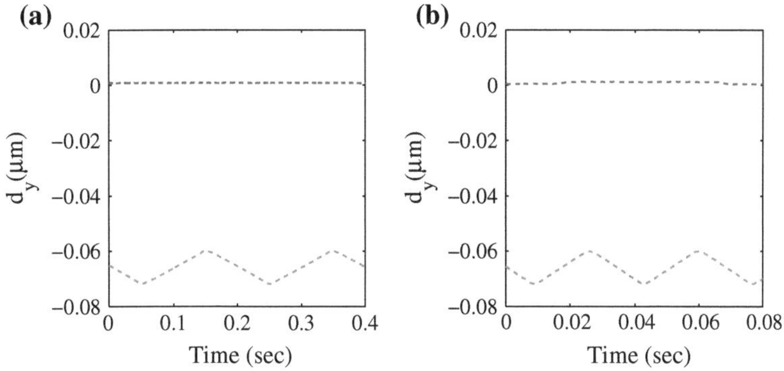

Fig. 19.13 Cross-coupling properties of the scanner at 5.21 and 31.25 Hz, respectively [closed-loop (*red*) and open-loop (*blue*)]

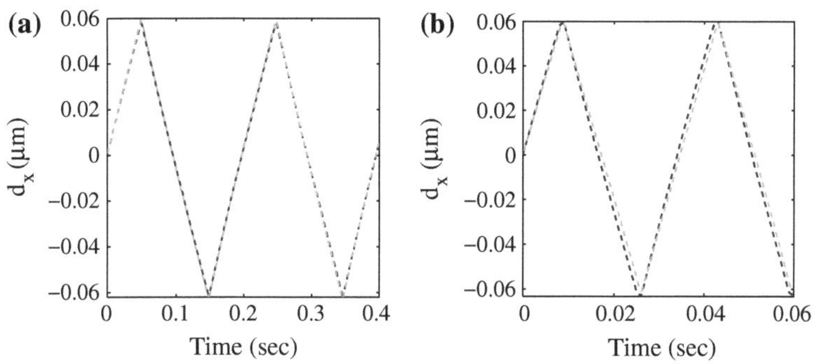

Fig. 19.14 Comparison of tracking performance of the proposed controller at 5.21 and 31.25 Hz, respectively, [reference signal (*blue*) and output signal (*red*)]

measure this cross-coupling, a reference triangular signal is applied to the X axis of the piezo and output taken from the Y position sensor.

19.5.5 Experimental Tracking Performance

The experimental tracking performance of the proposed controller is shown in Fig. 19.14. In Fig. 19.14a and b the closed-loop tracking is presented at 5.21 and 31.25 Hz, respectively, which reflect the tracking performance of the proposed controller. The closed-loop tracking error are summarized in Table 19.1.

Table 19.1 RMS values of tracking error in closed-loop

Scan frequency (Hz)	RMS tracking error (nm)
5.21	0.369
31.25	0.689

19.6 Conclusion and Future Work

The main contribution of this chapter was to compensate the nonlinear effect of the PTS by using the proposed controller for proper tracking and improved imaging at higher scanning speed of an AFM. The proposed OMPC controller was implemented to compensate the effect of creep, hysteresis, cross-coupling, and vibration of the PTS. The experimental results demonstrate that the tracking control performance is greatly improved in the high-speed application using the proposed controller.

In this work, the plant considered was a single-input single-output (SISO) system. However, the author is interested in working with a multi-input multi-output (MIMO) system in future.

Acknowledgments The authors would like to thank Mr. Shane Brandon, SEIT, UNSW, Canberra, Australia for his technical support during the experimental tests.

References

1. Yong YK, Ahmed B, Moheimani SOR (2010) Atomic force microscopy with a 12-electrode piezoelectric tube scanner. Rev Sci Instrum 81(3):033 701–10
2. Meyer E, Hug HJ, Bennewitz R (2004) Scanning probe microscopy. Springer, Berlin
3. Sarid D (1994) Scanning force microscopy: with applications to electric, magnetic and atomic forces. Oxford University Press, Oxford
4. Fleming AJ, Aphale SS, Moheimani SOR (2010) A new method for robust damping and tracking control of scanning probe microscope positioning stages. IEEE Trans Nanotechnol 9(4):438–448
5. Yong YK, Liu K, Moheimani SOR (2010) Reducing cross-coupling in a compliant XY nanopositioner for fast and accurate raster scanning. IEEE Trans Control Syst Technol 18(5):1172–1179
6. Taylor ME (1993) Dynamics of piezoelectric tube scanners for scanning probe microscopy. Rev Sci Instrum 64(1):154–158
7. Adriaens H, De Koning W, Banning R (2000) Modeling piezoelectric actuators. IEEE/ASME Trans Mechatron 5(4):331–341
8. Rana MS, Pota HR, Petersen IR (2012) Improved control of atomic force microscope for high-speed image scanning. In: Australian control conference (AUCC). Sydney, pp 470–475
9. Bazaei A, Yong YK, Moheimani SOR, Sebastian A (2012) Tracking of triangular references using signal transformation for control of a novel AFM scanner stage. IEEE Trans Control Syst Technol 20(2):453–464
10. Jung H, Shim JY, Gweon D (2001) Tracking control of piezoelectric actuators. Nanotechnology 12(1):14–20
11. Croft D, Shedd G, Devasia S (2000) Creep, hysteresis, and vibration compensation for piezoactuators: atomic force microscopy application. Proc Am Control Conf 3:2123–2128

12. Jung H, Shim JY, Gweon D (2000) New open-loop actuating method of piezoelectric actuators for removing hysteresis and creep. Rev Sci Instrum 71(9):3436–3440
13. Croft D, Shedd G, Devasia S (2001) Creep, hysteresis, and vibration compensation for piezoactuators: atomic force microscopy application. J Dyn Syst Meas Control Trans ASME 123(1):35–43
14. Leang K, Devasia S (2007) Feedback-linearized inverse feedforward for creep, hysteresis, and vibration compensation in afm piezoactuators. IEEE Trans Control Syst Technol 15(5):927–935
15. Yi KA, Veillette RJ (2005) A charge controller for linear operation of a piezoelectric stack actuator. IEEE Trans Control Syst Technol 13(4):517–526
16. Chuang N, Petersen IR, Pota HR (2013) Robust H^{∞} control in fast atomic force microscopy. Asian J Control 15(4):1–15
17. Cruz-Hernandez JM, Hayward V (2001) Phase control approach to hysteresis reduction. IEEE Trans Control Syst Technol 9(1):17–26
18. Mahmood IA, Moheimani SOR (2009) Making a commercial atomic force microscope more accurate and faster using positive position feedback control. Rev Sci Instrum 80(6):063 705-063–705-8
19. Moheimani SOR, Vautier BJG (2005) Resonant control of structural vibration using charge-driven piezoelectric actuators. IEEE Trans Control Syst Technol 13(6):1021–1035
20. Aphale SS, Bhikkaji B, Moheimani SOR (2008) Minimizing scanning errors in piezoelectric stack-actuated nanopositioning platforms. IEEE Trans Nanotechnol 7(1):79–90
21. Pota HR, Moheimani SOR, Smith M (2002) Resonant controller for smart structures. Smart Mater Struct 11:1–8
22. Bhikkaji B, Ratnam M, Fleming AJ, Moheimani SOR (2007) High-performance control of piezoelectric tube scanners. IEEE Trans Control Syst Technol 15(5):853–866
23. Moheimani SOR, Vautier BJG, Bhikkaji B (2006) Experimental implementation of extended multivariable PPF control on an active structure. IEEE Trans Control Syst Technol 14(3):443–455
24. Kenton BJ, Fleming AJ, Leang KK (2011) Compact ultra-fast vertical nanopositioner for improving scanning probe microscope scan speed. Rev Sci Instrum 82(12):123 703-123–703-8
25. Schitter G, Astrom K, DeMartini B, Thurner P, Turner K, Hansma P (2007) Design and modeling of a high-speed AFM-scanner. IEEE Trans Control Syst Technol 15(5):906–915
26. Kenton B, Leang K (2012) Design and control of a three-axis serial-kinematic high-bandwidth nanopositioner. IEEE/ASME Trans Mechatron 17(2):356–369
27. Fairbairn MW, Moheimani SOR, Fleming AJ (2011) Improving the scan rate and image quality in tapping mode atomic force microscopy with piezoelectric shunt control. In: Australian control conference (AUCC). pp 26–31
28. Grosswindhager S, Kozek M, Voigt A, Haffner L (2013) Fuzzy predictive control of district heating network. Int J Model Ident Control 19(2):161–170
29. Su B, Qi G, Van Wyk BJ (2012) Output feedback predictive control for uncertain non-linear switched systems. Int J Model Identif Control 17(3):195–205
30. Li D, Xi Y (2011) The synthesis of robust model predictive control with QP formulation. Int J Model Identif Control 13(1/2):1–8
31. Rana MS, Pota HR, Petersen IR (2012) Model predictive control of atomic force microscope for fast image scanning. In: 51st conference on decision and control (CDC). Hawaii, USA, pp 2477–2482
32. Devasia S, Eleftheriou E, Moheimani SOR (2007) A survey of control issues in nanopositioning. IEEE Trans Control Syst Technol 15(5):802–823
33. Privara S, Cigler J, Vana Z, Ferkl L (2012) Incorporation of system steady state properties into subspace identification algorithm. Int J Model Identif Control 16(2):159–167
34. Ljung L (2002) Prediction error estimation methods. Circ Syst Signal Process 21:11–21
35. Kabaila P (1983) On output-error methods for system identification. IEEE Trans Autom Control 28(1):12–23

36. Wang L (2009) Model predictive control system design and implementation using MATLAB. Springer, London
37. Ray P, Panda G (2012) Harmonics estimation using KF-Adaline algorithm and elimination with hybrid active power filter in distorted power system signals. Int J Model Identif Control 16(2):149–158

Chapter 20
Design of the State Estimation System in the Advanced Driver Assistance System

Wei Huang, Xiaoxin Su and David Bevly

Abstract The development of Advanced Driver Assistance System (ADAS) is emphasized in road transportation research. Reliability and safety is one of the most important issues in the design process of ADAS. In general, the system must fuse information from multiple sensors to obtain more complete and accurate information about the world. In this chapter, a novel loose coupling sensor fusion strategy is designed, which uses the Extended Kalman filtering (EKF) to fuse the sensor measurements from odometers, accelerometers, gyroscope, and GPS. By using a novel four-wheel vehicle model, the EKF is able to conduct a multi-output rate sensor fusion, compensate the latency for GPS signals, and increase the accuracy of vehicle state estimation even if there exist sensor errors, such as GPS outage, odometer reading error due to wheel slippage. From the road test, it is proved that the designed EKF has achieved good results for vehicle state estimation.

20.1 Introduction

Today, the development of Advanced Driver Assistance System (ADAS), which aids the driver by controlling the vehicle, is emphasized in road transportation research. These automotive mechatronic systems use sensors (e.g. radar, Global Positioning System (GPS), Inertial Navigation Sensors (INS)), and electronic control functions to recognize critical traffic situations, and give an appropriate warning to the driver. For

D. Bevly
Mechanical Engineering, Auburn University, Auburn, AL, USA
e-mail: bevlydm@auburn.edu

W. Huang (✉) · X. Su
Navistar Inc, Lisle, IL, USA
e-mail: huangjwcn@gmail.com

X.Su
e-mail: sxx1203@gmail.com

L. Liu et al. (eds.), *Applied Methods and Techniques for Mechatronic Systems*, Lecture Notes in Control and Information Sciences 452,
DOI: 10.1007/978-3-642-36385-6_20, © Springer-Verlag Berlin Heidelberg 2014

ADAS, reliability and safety is one of the most important issues in the design process. In general, the data from only one sensor is not deemed reliable enough for ADAS, and the system must fuse multiple sensors data to obtain more accurate information about the world. For example, the estimator should interpolate the position of the vehicle by fusing INS signals, i.e., odometer, accelerometer, and gyroscope, with GPS measurements. Here, the term "fusion" refers generally to the process of combining two sets of data to produce a better output.

Kalman filtering (KF) is the basis of the majority of navigational sensor fusion system in use today. In its complex form, the KF is able to fuse data from different sensors to estimate process states. For instance, to estimate the position of a vehicle, one could use a GPS system, or determine the traveled distance with a velocity sensor or possibly other sensors. A KF can then be built to combine the data from all of these sensors and knowledge of the system dynamics to generate an overall best position estimate. When the system model is not linear the KF cannot be used, instead an Extended Kalman filter (EKF) can be applied. In [1], sensor fusion using a KF is used in an autonomous vehicle. In [2] GPS, gyroscopes and an accelerometer have been used to estimate the sideslip and the roll of a vehicle. In [3], a vehicle state estimator is implemented by using a KF, which uses two plant models because of the different measurement frequency of GPS and the gyroscope. In [4], an EKF is used for integrating GPS with differential wheel speed sensors and a gyroscope. Tire slip has not been taken into account in the model used for the EKF. In [5], sensor fusion by KF for computation of the yaw rate and the absolute longitudinal velocity is used. EKFs are not only used for state estimation in road vehicles but in many autonomous systems, [6–11].

In ADAS implementation, GPS and INS measurements have complementary characteristics, which formulate the problems we need to solve in this chapter:

1. The output rate of the GPS is 1 Hz, that is not enough for ADAS application. Therefore, INS sensors (50 Hz output rate) can be used to achieve an overall rate of desired data.
2. The time delay, with which the data becomes available in the vehicle, exists in almost all GPS systems. This delay is called latency. Since the data from INS sensors are obtained in real-time, the latency of the GPS can introduce severe inaccuracies when it is not compensated.
3. GPS signals are not always available (GPS outage) when it is blocked by buildings, trees, bridges, etc. Therefore, there is a need for INS sensors that have their measurements always available.
4. INS systems accuracy can experience unbounded error growth over time. Therefore GPS, an external source, is used to correct the measurements from INS sensors from time to time.

Therefore, a novel GPS/INS loose coupling EKF is designed. Unlike previous studies, this work, by developing special vehicle models and fusion strategy, will allow the EKF to conduct a multi-output rate sensor fusion, compensate the latency for GPS signals, and increase the accuracy of vehicle state estimation even with the sensor errors. This chapter is organized as follows. Section 20.2 describes the sensor

Table 20.1 Sensor measurements in Smart 2

Sensor	Signals	Definition	Noise variance	Bias
DGPS	$\mathrm{DGPS}_{\mathrm{long}}$	x position	$10\,\mathrm{m}^2$	–
	$\mathrm{DGPS}_{\mathrm{lat}}$	y position	$10\,\mathrm{m}^2$	–
	DGPS_{ψ}	Heading	$0.4\,\mathrm{rad}^2$	–
Accelerometer	a_{long}	x acceleration	$0.02\,(\mathrm{m/s}^2)^2$	$-0.5\,\mathrm{m/s}^2$
	a_{lat}	y acceleration	$0.03\,(\mathrm{m/s}^2)^2$	$0.35\,\mathrm{m/s}^2$
Gyroscope	$\dot{\psi}$	Yaw rate	$1.2{\cdot}10^{-6}\,(\mathrm{rad/s})^2$	$-4\cdot 10^{-4}\,\mathrm{rad/s}$
Wheel speed sensor	$v_{\mathrm{f/r,l/r}}$	wheel speed	$3{\cdot}10^{-4}\,(\mathrm{m/s})^2$	$0.05\,\mathrm{m/s}$

measurements. In Sect. 20.3, sensor fusion design strategy is provided. Section 20.4 deals with the GPS system latency compensation. Section 20.5 shows the design of the EKF for the Smart car. In Sect. 20.6, experimental results are shown. The conclusion and future work are discussed in Sect. 20.7.

20.2 Sensors Measurements Specification

The test vehicle is a Smart vehicle that is a small 2-door vehicle with an automatic gearbox. The Smart is rear wheel driven and the engine is placed in the middle of the vehicle. In this work, the Smart is equipped with necessary sensors, such as wheel speed sensors, accelerometers, odometer, gyroscope, and GPS. Interfacing of the sensors of the Smart is done using a combination of a Controller Area network (CAN) and a laptop. Table 20.1 describes the variance, bias, and the drift of each sensor measurement, obtained from the product datasheet or calculated from the raw data.

For global position estimation a Trimble Lassen $SK8$ Differential Global Positioning System (DGPS) receiver has been used. In our experiment, the following GPS signals are applied: quality, Horizontal Dilution Precision (HDOP), number of satellites, time, longitudinal/lateral position, heading, and velocity. Based on the product datasheet, we know the DGPS update rate is 1 Hz, the accuracy is ± 5 m, and the resolution is following the NMEA-0183 message structure.

Two accelerometers have been used to record the lateral and longitudinal acceleration of the vehicle. The accelerometer signals have an update rate of 50 Hz and are biased. This bias is a low frequency signal in the accelerometer that is always present. This bias signal must be taken into account and subtracted from the accelerometer signal. Ideally, only the component of the acceleration signal that is caused by vehicle movements is desired.

The yaw rate measured from the gyroscope has the range of 100°/s and its bias is less than 2°/h. In our experiment, odometers are used to measure the wheel speeds of the vehicle. The odometer is a device that counts pulses that are generated when the wheel (where the sensor is mounted) is moving. Therefore, the signal from the odometer is an ascending signal when the vehicle is moving and is stationary when the vehicle is standing still.

20.3 Sensor Fusion Design Strategy

In this work, two EKFs are designed for state estimation of the Smart. The basic strategies behind these designs are the same, but the vehicle models are different.

20.3.1 Extended Kalman Filter Equations

For nonlinear systems the general stochastic difference equation for discrete systems is defined as

$$x_k = f(x_{k-1}, u_{k-1}, w_{k-1}), \; z_k = h(x_x, v_k), \tag{20.1}$$

where f is the nonlinear function, which relates the state x_k at the current time step k to the state at the previous time step $k-1$, the control input u_k, and the noise w_{k-1}. h is the nonlinear function, which relates the measurement z_k to the state x_k and the measurement noise v_k. In practice, the values of the noise w_k and v_k are not known and u_k is also set to zero because in the proposed system there is no control input, therefore, the state and measurement vectors have to be approximated by:

$$\widetilde{x}_k = f(\widehat{x}_{k-1}, 0, 0), \; \widetilde{z}_k = h(\widetilde{x}_k, 0), \tag{20.2}$$

where $\widetilde{x}_k$ is the estimate of the state based on the measurement (a posteriori).

The variables w_k and v_k are assumed to be Gaussian with zero mean and are represented by their covariance matrices Q and R. N represents a normal distribution.

$$p(w) \sim N(0, Q), \; p(v) \sim N(0, R). \tag{20.3}$$

Two kinds of estimation errors can be defined, the *a priori* estimation error e_{k^-} and the a *posteriori* estimation error e_k. They are described by

$$e_k^- \equiv x_k - \widehat{x}^-, \; e_k = x_k - \widehat{x}_k, \tag{20.4}$$

where $\widehat{x}^-$ is the *a priori* state estimate based on knowledge of the process prior to step k and $\widehat{x}$ is the *a posteriori* state based on the measurement z_k. The covariances of the estimation errors are defined by

$$P_k^- = E[e_k^-, (e_k^-)^T], \; P_k = E[e_k, e_k^T], \tag{20.5}$$

p_k^- is the *a priori* estimate error covariance and p_k is the *a posteriori* estimate error covariance.

The EKF algorithm consists of two stages, time update and measurement update [12]. The different steps in the **time update stage** are:

$$x_k^- = f(\widehat{x}_{k-1}) \tag{20.6}$$
$$P_k^- = A_k P_{k-1} A_k^T + Q_{k-1} \tag{20.7}$$

where A_k is the Jacobian of the system model equations described by $f(\cdot)$:

$$A_{k[i,j]} = \frac{\partial f_{[i]}}{\partial x_{[j]}}(\widehat{x}_{k-1}, 0). \tag{20.8}$$

The different steps in the **measurement update stage** are:

$$K_k = P_k^- C_k^T (C_k P_k^- C_k^T + R_k)^{-1} \tag{20.9}$$
$$\widehat{x}_k = \widehat{x}_k^- + K_k(z_k - h(\widehat{x}_k^-, 0)) \tag{20.10}$$
$$P_k = (I - K_k C_K) P_k^-, \tag{20.11}$$

where K_k is the Kalman gain that minimizes the posteriori estimate error covariance P_k, and C_k is the Jacobian of the measurement function $h(\cdot)$:

$$C_{k[i,j]} = \frac{\partial h_{[i]}}{\partial x_{[j]}}(\widetilde{x}_{k-1}, 0). \tag{20.12}$$

20.3.2 Sensor Fusion Strategy

Based on the sensor measurements, as listed in Table 20.1, and a vehicle kinematic model, an EKF can be designed to estimate some important states of the Smart. An overview of the sensor configuration for the Smart is shown in Fig. 20.1. In this configuration, INS sensor measurements are also used to update the vehicle estimated position and heading when DGPS update is not available.

Additionally, DGPS update pulse signals, u_{DGPS}, are used in this work: one pulse indicates an update for position signals available from the NMEA $GPGGA sentence; and the other indicates an update for heading and velocity signals from the $GPRMC sentence. The update value is either 0 or 1, where 1 means that the DGPS signal is available.

Since the INS and DGPS measurements have output rate of 50 Hz and of 1 Hz, respectively, a multi-rate EKF [13] is developed, which performs sensor fusion of DGPS and INS data at 1 Hz (whenever DGPS data are received) and relies on INS data alone between DGPS samples. The scheme for this method is plotted in Fig. 20.2. It can be seen that after the estimated state and error covariance is initialized, the time update is proceeded. There are three measurement updates: one INS and two DGPS updates (one for position and one for heading). If there is no DGPS available ($u_{\text{DGPS}} = 0$), then the estimated state and error covariance resulted from the INS measurement update will be fed back to the next time update. Otherwise, they will be fed forward to the DGPS position or heading measurement update. After DGPS

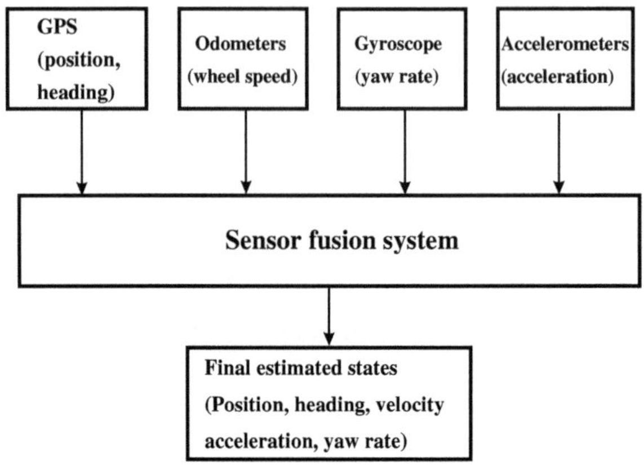

Fig. 20.1 Sensor configuration block diagram of the Smart

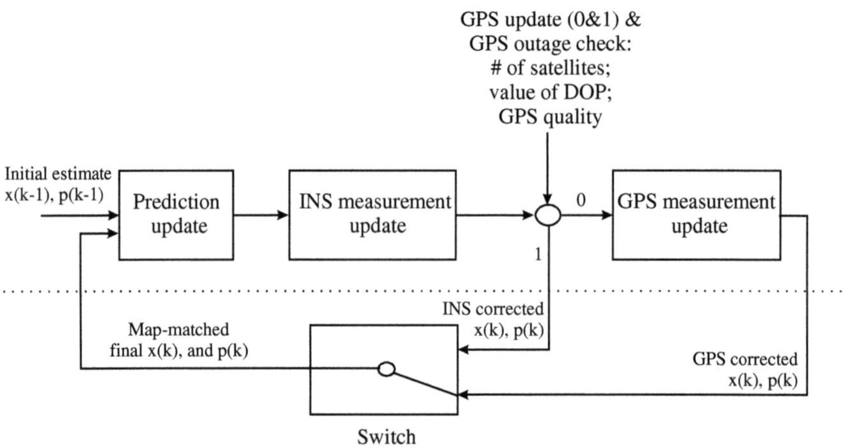

Fig. 20.2 Scheme for the fusion strategy

measurement update, the estimated state and error covariance will be fed back to the next time update. Meanwhile, the latency compensation for DGPS signals is conducted during DGPS measurement update as well.

20.4 Latency Compensation for DGPS System

A fundamental limitation of the DGPS system is the delay with which the data become available in the vehicle. This delay is called latency. There are several origins of this latency [14]. One is the time it takes to calculate the DGPS data, plus the time it takes to transmit these data to the vehicle. Another origin of latency is the delay, which occurs in the data transmission between the base station and moving receivers in a DGPS receiver. Since the data from the INS is obtained in real-time, the latency of the DGPS can introduce severe inaccuracies when it is not compensated.

The data received from DGPS at time instant t represent data belonging to a previous time instant $t - \tau$. This can be written as:

$$ {}^{N}y_x(t) = {}^{N}y_x(t - \tau) \tag{20.13} $$

$$ {}^{N}y_y(t) = {}^{N}y_y(t - \tau) \tag{20.14} $$

$$ {}^{N}y_\psi(t) = {}^{N}y_\psi(t - \tau) \tag{20.15} $$

where τ is the time in seconds the DGPS data is delayed and ${}^{N}y_x(t)$, ${}^{N}y_y(t)$, ${}^{N}y_\psi(t)$ are, respectively, the DGPS longitudinal, lateral position, and the heading.

When a DGPS update is received, it is known that this update represents data from τ seconds ago and not real-time data. To fix this latency issue, we compare several methods:

1. The easiest method is to delay the INS data until it is synchronized with the DGPS data. A major drawback of this method is that the whole state estimation is delayed with the latency of DGPS.
2. The second method is to use EKF to compensate the latency only during the final fusion step in the measurement update, where the estimated states are the real-time states buffered from ($N = \tau/T$) samples ago. It has been shown that this method can achieve good compensation performance.

The DGPS observations for the second method therefore can be written as:

$$ {}^{N}y_x(k) = {}^{N}x(k - N) + v_{{}^{N}y_x(k)} \tag{20.16} $$

$$ {}^{N}y_y(k) = {}^{N}y(k - N) + v_{{}^{N}y_y(k)} \tag{20.17} $$

$$ {}^{N}y_\psi(k) = {}^{N}\psi(k - N) + v_{{}^{N}y_\psi(k)}, \tag{20.18} $$

where $v_{(k)}$ is the measurement noise of the DGPS receiver. And the equations for measurement update can be given as follows:

$$ {}^{N}x(k+1) = {}^{N}x(k) + K({}^{N}y_x(k) - {}^{N}x(k - N)) \tag{20.19} $$

$$ {}^{N}y(k+1) = {}^{N}y(k) + K({}^{N}y_y(k) - {}^{N}y(k - N)) \tag{20.20} $$

$$ {}^{N}\psi(k+1) = {}^{N}\psi(k) + K({}^{N}y_\psi(k) - {}^{N}\psi(k - N)). \tag{20.21} $$

${}^{N}y_{x/y/\psi}(k)$ are the data received from DGPS at time instant k which represent data belonging to a previous time instant $k - N$. ${}^{N}x/y/\psi(k - N)$ are the lagged state estimates from N samples ago and K is the Kalman gain. Subsequently, the proposed method will be implemented in the EKF.

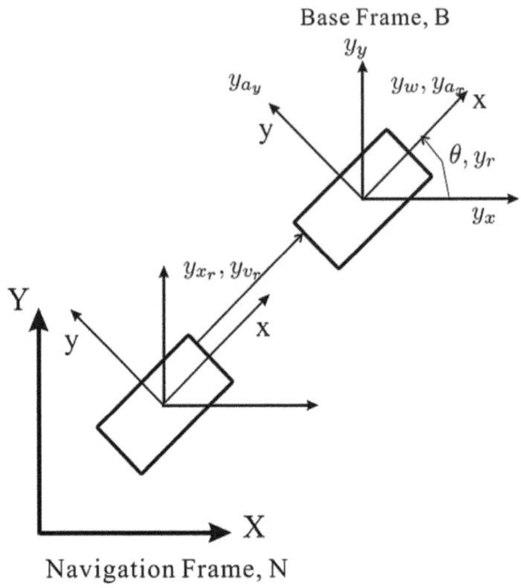

Fig. 20.3 One-track bicycle model: two tyres

20.5 Extended Kalman Filters Design for the Smart Car

EKFs based on both bicycle and four-wheel vehicle models are designed for the Smart in this section. The motion of a vehicle will be referred to a navigation frame N which is described by a right-handed orthogonal axis system (X, Y, Z) on the Earth. A second reference frame (base frame) B fixed in the vehicle is described by axes (x, y, z) fixed along the central principal axes of the vehicle. The relation between navigation and base frame can be seen in Fig. 20.3.

The sensor measurements will be assigned into different frames. DGPS longitudinal, lateral positions, and heading are in the navigation frame. The acceleration and wheel speed measured from INS sensors are in the base frame. To obtain a trajectory in the navigation frame it is necessary to convert the inertial signals from the base frame into the navigation frame, where we assume that the road is flat and then neglect the Z and z axes. This can be done by the following matrix (a simplified form of the direct cosine matrix):

$$R = \begin{bmatrix} \cos(\theta(k)) & -\sin(\theta(k)) \\ \sin(\theta(k)) & \cos(\theta(k)) \end{bmatrix}.$$

20.5.1 Vehicle Bicycle Model

In this section, the EKF based on a kinematic vehicle bicycle model is designed, and two assumptions are made in advance:

1. the wheel slippage in longitudinal direction is assumed to be zero;
2. the test track, on which the Smart is driven, is flat.

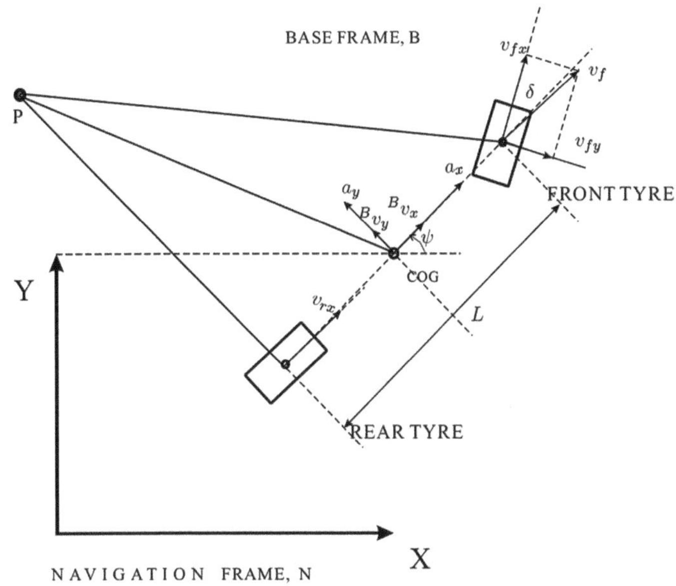

Fig. 20.4 One-track bicycle model: two tyres

In order to reproduce the trajectory followed by the vehicle, kinematic rules for a bicycle model have been used to describe the motion of a vehicle, as shown in Fig. 20.4. The kinematic vehicle model can be developed as:

$$^N\dot{x} = {}^Bv_x \cdot \cos\psi - {}^Bv_y \cdot \sin\psi \tag{20.22}$$

$$^N\dot{y} = {}^Bv_x \cdot \sin\psi + {}^Bv_y \cdot \cos\psi \tag{20.23}$$

$$^B\dot{v}_x = a_x \tag{20.24}$$

$$^B\dot{v}_y = a_y, \tag{20.25}$$

where ^Nx and ^Ny are longitudinal and lateral positions in the navigation frame, and $^Bv_{x/y}$ indicate the longitudinal and lateral velocities in the base frame. Additionally, for the front and rear wheel speed we have:

$$v_{enc_f} = \sqrt{{}^Bv_x{}^2 + ({}^Bv_y + \dot{\psi}L_1)^2} \tag{20.26}$$

$$v_{enc_r} = \sqrt{{}^Bv_x{}^2 + ({}^Bv_y - \dot{\psi}L_2)^2}, \tag{20.27}$$

where $v_{enc_{f/r}}$ are the front and rear wheel speeds, $\dot{\psi}$ is the yaw rate, and $L_{1/2}$ are the lengths from the Smart center of gravity (COG) to the front and rear axle, respectively.

In order to identify the errors of the wheel speed sensors, accelerometers, and gyroscope in real-time the biases are also estimated. The resulting error states are

Table 20.2 States of the bicycle model EKF of Smart 2

In the navigation frame	In the base frame
${}^{N}x_2$: Longitudinal position (m)	${}^{B}a_{x2}$: Longitudinal acceleration (m/s^2)
${}^{N}y_2$: Lateral position (m)	${}^{B}a_{y2}$: Lateral acceleration (m/s^2)
${}^{N}\psi_2$: Yaw angle (rad)	${}^{B}v_{x2}$: Vehicle longitudinal speed in the COG (m/s)
${}^{N}\dot{\psi}_2$: Yaw rate (rad/s)	${}^{B}v_{y2}$: Vehicle lateral speed in the COG (m/s)
	$b_{a_{x2}}$: Bias state for longitudinal accelerometer
	$b_{a_{y2}}$: Bias state for lateral accelerometer
	$b_{\dot{\psi}2}$: Bias state for gyroscope
	b_{f2}: Bias state for front odometer
	b_{r2}: Bias state for rear odometer (m/s)

Table 20.3 Measurement used by the bicycle model EKF of Smart 2

In the navigation frame	In the base frame
${}^{N}y_{x2}$: DGPS longitudinal position	${}^{B}y_{a_x2}$: Longitudinal accelerometer
${}^{N}y_{y2}$: DGPS lateral position	${}^{B}y_{a_y2}$: Lateral accelerometer
${}^{N}y_{\psi2}$: DGPS heading (yaw angle)	${}^{B}y_{w_f2}$: Front wheels mean speed
${}^{N}y_{\dot{\psi}2}$: Gyro yaw rate	${}^{B}y_{w_r2}$: Rear wheels mean speed

described in Table 20.2 and are modeled as random walks. By using the random walk model, growth of uncertainty of the true value of the errors and the rate at which it varies can be reflected [15]. Finally, the states and measurements of the EKF are identified in Tables 20.2 and 20.3.

Therefore, the discrete-time state equations of the EKF can be written as follows:

$$\begin{aligned} {}^{N}x_2(k) =& {}^{N}x_2(k-1) + T\cdot {}^{B}v_{x2}(k-1)\cdot\cos({}^{N}\psi_2(k-1)) - \\ & T\cdot {}^{B}v_{y2}(k-1)\cdot\sin({}^{N}\psi_2(k-1)) + w_{{}^{N}x_2(k)} \end{aligned} \tag{20.28}$$

$$\begin{aligned} {}^{N}y_2(k) =& {}^{N}y_2(k-1) + T\cdot {}^{B}v_{x2}(k-1)\cdot\sin({}^{N}\psi_2(k-1)) + \\ & T\cdot {}^{B}v_{y2}(k-1)\cdot\cos({}^{N}\psi_2(k-1)) + w_{{}^{N}y_2(k)} \end{aligned} \tag{20.29}$$

$${}^{B}v_{x2}(k) = {}^{B}v_{x2}(k-1) + T\cdot {}^{B}a_{x2}(k-1) + w_{{}^{B}v_{x2}(k)} \tag{20.30}$$

$${}^{B}v_{y2}(k) = {}^{B}v_{y2}(k-1) + T\cdot {}^{B}a_{y2}(k-1) + w_{{}^{B}v_{y2}(k)} \tag{20.31}$$

$${}^{B}a_{x2}(k) = {}^{B}a_{x2}(k-1) + w_{{}^{B}a_{x2}(k)} \tag{20.32}$$

$${}^{B}a_{y2}(k) = {}^{B}a_{y2}(k-1) + w_{{}^{B}a_{y2}(k)} \tag{20.33}$$

$${}^{N}\psi_2(k) = {}^{N}\psi_2(k-1) + T\cdot {}^{N}\dot{\psi}_2(k-1) + w_{{}^{N}\psi_2(k)} \tag{20.34}$$

$${}^{N}\dot{\psi}_2(k) = {}^{N}\dot{\psi}_2(k-1) + w_{{}^{N}\dot{\psi}_2(k)} \tag{20.35}$$

$$b_{f2}(k) = b_{f2}(k-1) + w_{b_{f2}} \tag{20.36}$$

$$b_{r2}(k) = b_{r2}(k-1) + w_{b_{r2}} \tag{20.37}$$

$$b_{a_{x2}}(k) = b_{a_{x2}}(k-1) + w_{b_{a_{x2}}} \tag{20.38}$$

$$b_{a_{y2}}(k) = b_{a_{y2}}(k-1) + w_{b_{a_{y2}}} \tag{20.39}$$

$$b_{\dot{\psi}2}(k) = b_{\dot{\psi}2}(k-1) + w_{b_{\dot{\psi}2}}. \tag{20.40}$$

Additionally, the measurement equations can be specified as follows:

$$^{N}y_{x2}(k) = {}^{N}x_2(k-N) + v_{^{N}y_{x2}(k)} \tag{20.41}$$

$$^{N}y_{y2}(k) = {}^{N}y_2(k-N) + v_{^{N}y_{y2}(k)} \tag{20.42}$$

$$^{N}y_{\psi 2}(k) = {}^{N}\psi_2(k-N) + v_{^{N}y_{\psi 2}(k)} \tag{20.43}$$

$$^{B}y_{a_x 2}(k) = {}^{B}a_{x2}(k) + b_{a_{x2}}(k) + v_{^{B}y_{a_x 2}(k)} \tag{20.44}$$

$$^{B}y_{a_y 2}(k) = {}^{B}a_{y2}(k) + b_{b_{y2}}(k) + v_{^{B}y_{a_y 2}(k)} \tag{20.45}$$

$$^{N}y_{\dot{\psi}2}(k) = {}^{N}\dot{\psi}_2(k) + b_{\dot{\psi}2}(k) + v_{^{N}y_{\dot{\psi}2}(k)} \tag{20.46}$$

$$\begin{aligned} ^{B}y_{w_f 2}(k) = &\sqrt{(^{B}v_{x2}(k))^2 + (^{B}v_{y2}(k) + L_1 \cdot {}^{N}\dot{\psi}_2(k))^2} + \\ &+ b_{f2}(k) + v_{^{B}y_{w_f 2}(k)} \end{aligned} \tag{20.47}$$

$$\begin{aligned} ^{B}y_{w_r 2}(k) = &\sqrt{(^{B}v_{x2}(k))^2 + (^{B}v_{y2}(k) - L_2 \cdot {}^{N}\dot{\psi}_2(k))^2} + \\ &+ b_{r2}(k) + v_{^{B}y_{w_r 2}(k)}. \end{aligned} \tag{20.48}$$

20.5.2 Vehicle Four-Wheel Model

Furthermore, a more accurate four-wheel vehicle model is developed for the EKF, as shown in Fig. 20.5. First, the longitudinal and lateral positions, and velocities can be developed similarly as for the bicycle model, as described in Eqs. (20.28–20.40). However, for each wheel speed, we have different equations as follows:

$$v_{fl} = \sqrt{(^{B}v_x - 0.5 \times \dot{\psi} W_1)^2 + (^{B}v_y + \dot{\psi} L_1)^2} \tag{20.49}$$

$$v_{fr} = \sqrt{(^{B}v_x + 0.5 \times \dot{\psi} W_1)^2 + (^{B}v_y + \dot{\psi} L_1)^2} \tag{20.50}$$

$$v_{rl} = {}^{B}v_x - 0.5 \times \dot{\psi} W_2 \tag{20.51}$$

$$v_{rr} = {}^{B}v_x + 0.5 \times \dot{\psi} W_2, \tag{20.52}$$

where $v_{(fl/fr/rl/rr)}$ are wheel speeds measured from the front left, front right, rear left, and rear right wheels, $\dot{\psi}$ is the yaw rate, $L_{1/2}$ are lengths from front and rear axles to the Smart COG, and $W_{1/2}$ are the length of front and rear track. Eqs. (20.51) and (20.52) are derived from the condition that rear wheels are not steering wheels and their wheel angles are equal to zero. Moreover, by adding up Eqs. (20.51) and (20.52), the rear wheel speeds can be finally calculated as:

$$v_{rl} + v_{rr} = 2 \cdot {}^{B}v_x. \tag{20.53}$$

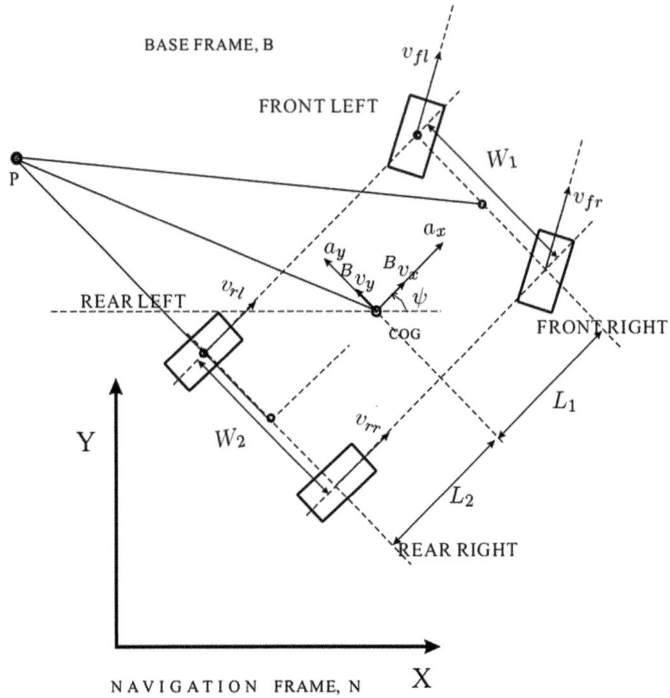

Fig. 20.5 Four-wheel bicycle model: four tires

Different from the bicycle model, the four-wheel model is able to take into account the occurrence of wheel slippage during vehicle road tests. As stated before, the Smart is a rear wheel driven vehicle and thus its measured wheel speed is largely affected by wheel slippage if it occurs. By using the four-wheel vehicle model, the rear wheel speeds, even if largely affected by wheel slippage, will not corrupt the calculation of lateral vehicle COG velocities, ${}^{B}v_{y}$, and therefore, the EKF will generate a robust state estimation.

The difference in the state equations between this model and the bicycle model is that two new states have been introduced, which are the bias state for each front wheel encoder, b_{fl} and b_{fr}. Two different state equations can be written as:

$$b_{\mathrm{fl}}(k) = b_{\mathrm{fl}}(k-1) + w_{b_{\mathrm{fl}}} \tag{20.54}$$

$$b_{\mathrm{fr}}(k) = b_{\mathrm{fr}}(k-1) + w_{b_{\mathrm{fr}}}. \tag{20.55}$$

Additionally, four new measurements are ${}^{B}y_{fl2}$, ${}^{B}y_{fr2}$, ${}^{B}y_{rl2}$ and ${}^{B}y_{rr2}$, which stand for each measured wheel speed. The measurement equations can be specified as follows:

$$^{N}y_{x2}(k) = {}^{N}x_2(k-N) + v_{^{N}y_{x2}(k)} \tag{20.56}$$

$$^{N}y_{y2}(k) = {}^{N}y_2(k-N) + v_{^{N}y_{y2}(k)} \tag{20.57}$$

$$^{N}y_{\psi2}(k) = {}^{N}\psi_2(k-N) + v_{^{N}y_{\psi2}(k)} \tag{20.58}$$

$$^{B}y_{a_x2}(k) = {}^{B}a_{x2}(k) + b_{a_{x2}}(k) + v_{^{B}y_{a_x2}(k)} \tag{20.59}$$

$$^{B}y_{a_y2}(k) = {}^{B}a_{y2}(k) + b_{b_{y2}}(k) + v_{^{B}y_{a_y2}(k)} \tag{20.60}$$

$$^{N}y_{\dot{\psi}2}(k) = {}^{N}\dot{\psi}_2(k) + b_{\dot{\psi}2}(k) + v_{^{N}y_{\dot{\psi}2}(k)} \tag{20.61}$$

$$^{B}y_{fl2}(k) = \sqrt{({}^{B}v_{x2}(k) - 0.5W_1{}^{N}\dot{\psi}_2(k))^2 + ({}^{B}v_{y2}(k) + L_1 \cdot {}^{N}\dot{\psi}_2(k))^2} + b_{\mathrm{fl}}(k) + v_{^{B}y_{fl2}(k)} \tag{20.62}$$

$$^{B}y_{fr2}(k) = \sqrt{({}^{B}v_{x2}(k) + 0.5W_1{}^{N}\dot{\psi}_2(k))^2 + ({}^{B}v_{y2}(k) + L_1 \cdot {}^{N}\dot{\psi}_2(k))^2} + b_{\mathrm{fr}}(k) + v_{^{B}y_{fr2}(k)} \tag{20.63}$$

$$^{B}y_{w_r2}(k) = {}^{B}v_{x2}(k) + b_{r2}(k) + v_{^{B}y_{w_r2}(k)}. \tag{20.64}$$

20.5.3 System Observability

A dynamic system is said to be observable if it is possible to uniquely reconstruct the state information based on the model of a system given the inputs and outputs of the system. The nonlinear model for the Smart car is:

$$\dot{x} = f(x) \tag{20.65}$$

$$z_i = h_i(x),\ i \in \{1, \ldots, p\}, \tag{20.66}$$

where h=col($h_1, \ldots h_p$): $R^n \rightarrow R^p$.

Although, this model is nonlinear, it is linearized each time for the EKF. The observability matrix, O, was calculated for each linearization in the run of the EKF. If O has full rank at each run, then the linearized model is locally observable. In this project, the observability matrix is accumulated over the entire time step from the initial to the final as a global check [16].

For the bicycle model EKF, the model is locally observable when the vehicle is driving; when the vehicle is standing still and yaw rate measurement is equal to zero, the estimated longitudinal/lateral COG velocities, ${}^{B}v_{x2}/{}^{B}v_{y2}$ and bias states for front/rear odometers, b_{f2} and b_{r2}, become unobservable. This is because when the vehicle is standing still no speed information is received from wheel encoders, so nothing can be said about the error in the speed calculation. Similarly, the four-wheel model used in the EKF is locally observable when the vehicle is driving. If the vehicle is standing still and yaw-rate measurement is equal to zero, then the state ${}^{B}v_{x2}$, ${}^{B}v_{y2}$, $By_{w_r,2}$, b_{fl}, b_{fr} and b_{r2} become unobservable.

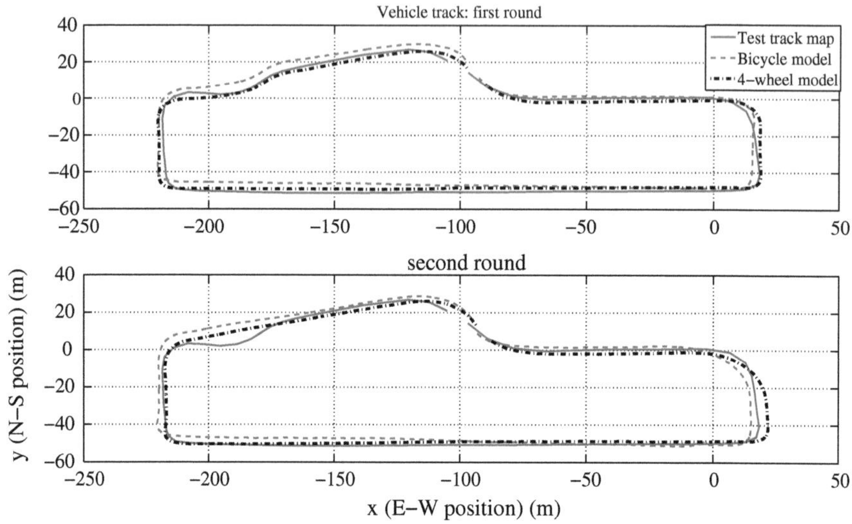

Fig. 20.6 Test track map versus INS-estimated track

20.5.4 Validation of Two Vehicle Models

The bicycle and four-wheel vehicle models are validated by evaluating the EKFs' performance for vehicle position estimation. To this end, two EKFs are applied to estimate vehicle position only based on their time update and INS measurements update, where all DGPS measurement updates, such as in Eqs. (20.41–20.43) and (20.56–20.58), are removed. By this way, if errors between the estimated and the real position are small enough, the designed vehicle models are validated. The real position is taken from the center line coordinator of the test track map converted into meters. The global position values of the vehicle and the track map are referenced to a point, which is located in the laboratory area.

Using this method, the vehicle estimated position is plotted in Fig. 20.6, where test track map and INS-estimated position are compared together. Meanwhile, the mean error and the variance of the error are given in Table 20.4. It can be seen that the estimated position follows the real position closely and consequently, the developed vehicle models are accurate enough for the ADAS EKF application. Additionally, we notice that the values shown in Table 20.4 for the four-wheel model are smaller than those for the bicycle model. Therefore, for the position estimation, the four-wheel vehicle model is more accurate than the bicycle model. This is because the four-wheel model represents the real vehicle kinematics more comprehensively and accurately. Therefore, in this work a four-wheel vehicle model is used for vehicle state estimation in the EKF.

In Fig. 20.7, raw speed measurements from four wheels, v_{fl}, v_{fr}, v_{rl}, and v_{rr}, are compared with the COG longitudinal velocity estimated from two EKFs. It is noticed

Table 20.4 Comparison of performances of the two EKFs

EKF model	Mean error (DGPS and INS)	Variance of error
Bicycle	3.713 m	8.599 m^2
Four-wheel	2.552 m	8.249 m^2

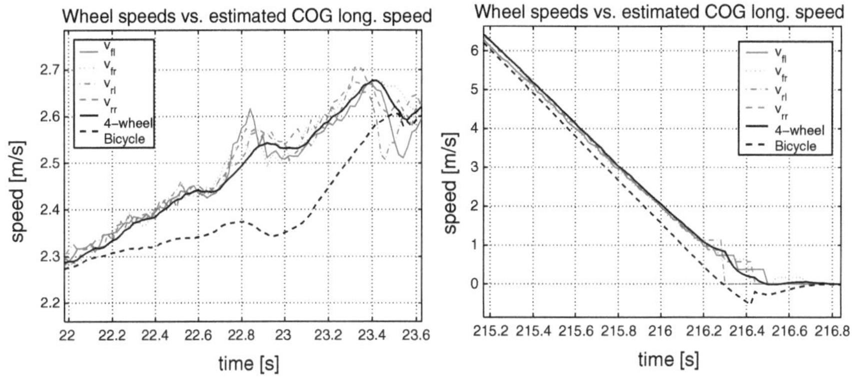

Fig. 20.7 (*left*) Raw wheel speeds versus EKF COG x velocity $t = 22 - 23.6$ s; (*right*) $t = 215.2 - 216.8$ s

that the estimated velocity from the bicycle model diverges largely from raw wheel speeds especially during time intervals, $t = [22, 23.6]$ s and $t = [215.2, 216.8]$ s. During this period, the Smart is either sharply accelerated or braked, which incurs big slippage in each wheel and makes bicycle model not accurate enough to calculate the COG velocity. This also explains why the error of the estimated position using the bicycle model is larger than the four-wheel model as shown in Table 20.4.

20.6 Vehicle Test for the EKF Evaluation

Finally, in order to evaluate the designed four-wheel model EKF for vehicle state estimation, a road test is conducted and the result is analyzed in this section. The road test is conducted on the test track as shown in Fig. 20.8. During the experiment, the Smart starts at the starting point P_1 and runs in the counterclockwise direction. With the CAN interface and data acquisition system installed on the Smart, the actual measurement data can be obtained from all sensors in vehicles, and inputted to the EKF built in a real-time Simulink environment on a Windows XP Laptop (Intel Core2 6300 1.86 Hz, 2 Gb of RAM) for the state estimation.

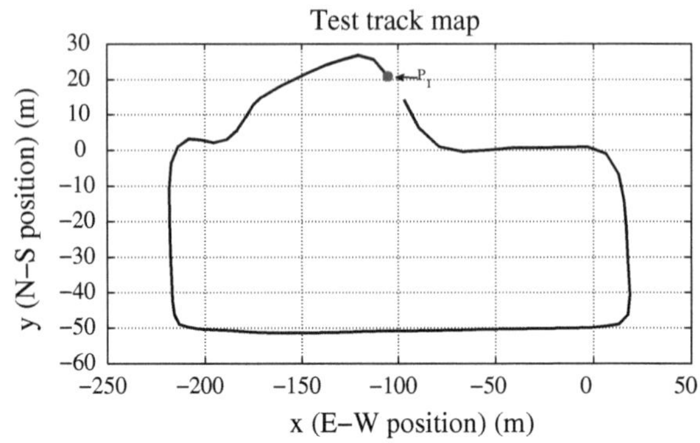

Fig. 20.8 Test track map

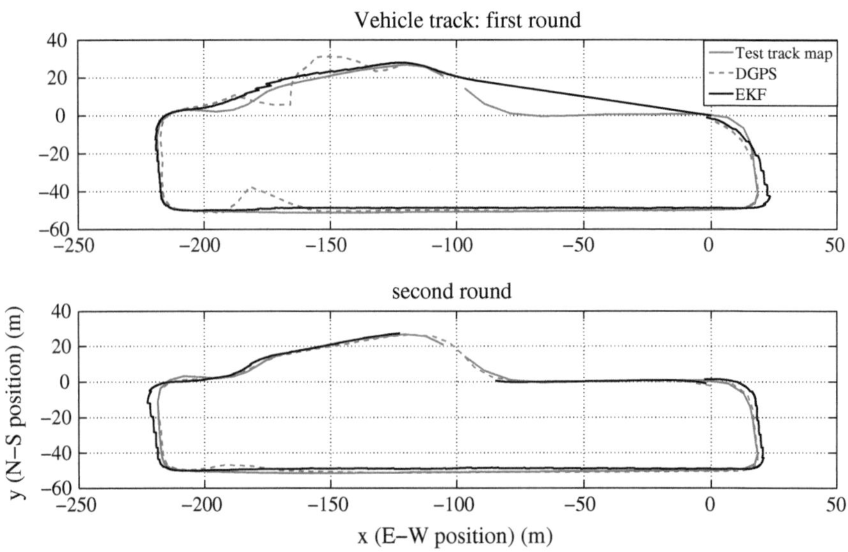

Fig. 20.9 Track map, DGPS, and EKF-estimated position

20.6.1 Evaluation of the EKF for State Estimation

The EKF is tested in this section to validate if all the problems, put forward in Sect. 20.1, have been solved. Important estimated states, i.e., longitudinal/lateral position, heading, estimated vehicle COG speed, acceleration, and yaw rate, are plotted together with the corresponding raw sensor data to show the performance of the state estimation strategy. During a 160 s test, the Smart is driven around the test track for two rounds. The estimated positions are plotted, along with the raw DGPS position and the track map, in Fig. 20.9.

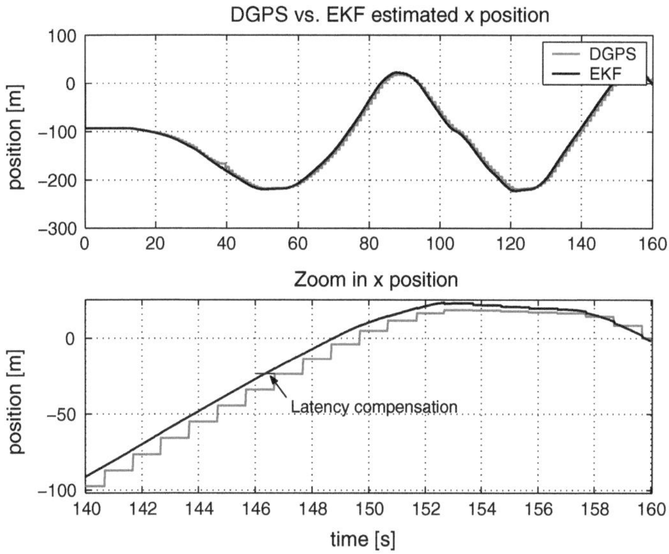

Fig. 20.10 DGPS versus EKF-estimated x position

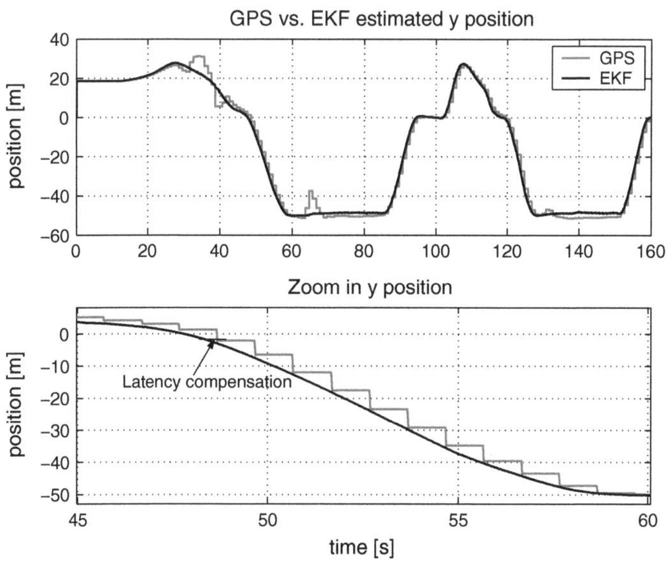

Fig. 20.11 DGPS versus EKF-estimated y position

The EKF-estimated longitudinal/lateral position and heading are plotted together with the corresponding DGPS measurements in Figs. 20.10, 20.11 and 20.12, where the zoomed-in figures are shown on the lower part. In the following, the performance of the designed EKF is analyzed:

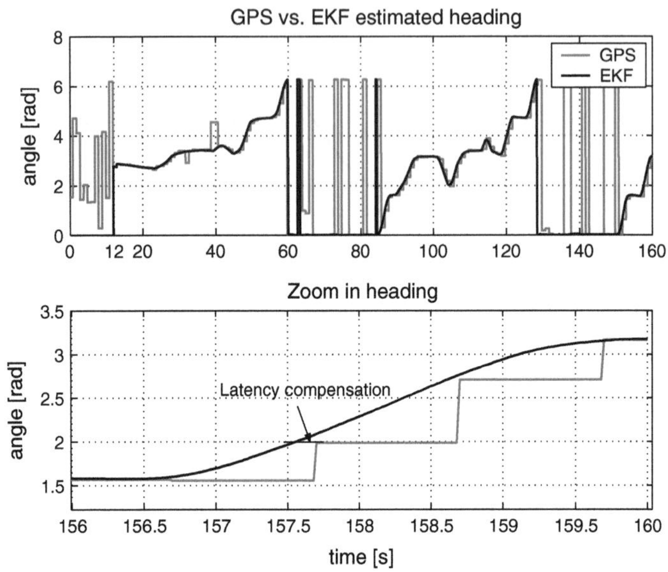

Fig. 20.12 DGPS versus EKF-estimated heading

1. Output rate of the system: the output rate of the DGPS is 1 Hz, which can be seen as the gray curves in Figs. 20.10, 20.11, and 20.12. The designed EKF fuses DGPS measurements together with the INS sensors data, and have an output rate at 50 Hz, which achieves an overall rate of desired data. It can be seen that the estimated position and heading, the black curve in those figures, have the same output rate at 50 Hz.
2. Latency of the DGPS system (0.58 s in this work): following with the strategy developed in Sect. 20.4, the position and heading estimated from the EKF are compensated with the latency from DGPS measurements. These latency compensations, around 0.5 s, are shown in the bottom plots of Figs. 20.10, 20.11, and 20.12.
3. DGPS data availability: DGPS signals can be blocked by buildings, trees, bridges, etc. This DGPS outage situation happens during $t = [20, 40]$ s in our test, as shown in Figs. 20.10 and 20.11. Since the number of satellites dropped below 4, there is big oscillation on DGPS measurements. During this time interval, the designed EKF estimates the vehicle states only based on INS sensors but not DGPS measurements. Therefore, the estimated positions shown in those figures are still accurate and located on the test track even if the DGPS data are not available.
4. The accuracy of INS navigation: INS systems can experience unbounded error growth over time. The mean error and variance of error from INS-estimated position are 2.552m and 8.249 m^2 separately, as shown in Sect. 20.5.4. The designed EKF corrects the INS states estimation every one second using the DGPS mea-

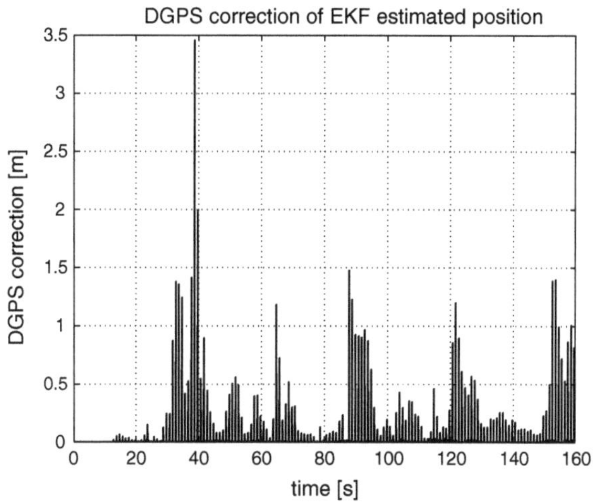

Fig. 20.13 DGPS correction of EKF-estimated position

surement. The mean error and the variance of error after such a DGPS correction are 1.150m and 6.152 m^2, respectively. It is clear that the accuracy of the navigation system is improved.

Additionally, the estimated heading is a fixed value for sections where the vehicle does not take a turn, as can be seen in Fig. 20.12. Moreover, during the time interval 0–12 s, when the vehicle is standing still, the estimated vehicle heading does not follow the oscillated DGPS heading measurement, but stays as zero.

In Fig. 20.13, the absolute DGPS position corrections are shown. These corrections are the difference of the absolute EKF position estimate of one sample before the update and the sample in which the data are corrected by the DGPS. The following equation has been used to calculate the difference:

$$d(k_{\text{upd}}) = \sqrt{(x(k_{\text{upd}}) - x(k_{\text{upd}} - 1))^2 + (y(k_{\text{upd}}) - y(k_{\text{upd}} - 1))^2} \qquad (20.67)$$

where k_{upd} is the time at which DGPS data arrives, x and y are the longitudinal and lateral positions. From Fig. 20.13, it can be seen that the most corrections are below 0.5 m, sometimes there are relatively large corrections of about 1 m. After analysis, we find that the big corrections occur during cornering of the vehicle. This is because the use of only accelerometer and gyroscope in lateral directions to reconstruct the vehicle track in corners is sometimes insufficient. Similarly, resulting from the occurrence of large DGPS corrections, there are staircases in the estimated vehicle track especially during the vehicle cornering, which are shown in Fig. 20.9.

The estimated Smart COG longitudinal and lateral velocity are shown in Fig. 20.14. Here, four-wheel speed signals, front left (v_{fl}), front right (v_{fr}), rear left (v_{rl}), and rear right (v_{rr}), are plotted together with the estimated COG longitudinal velocity,

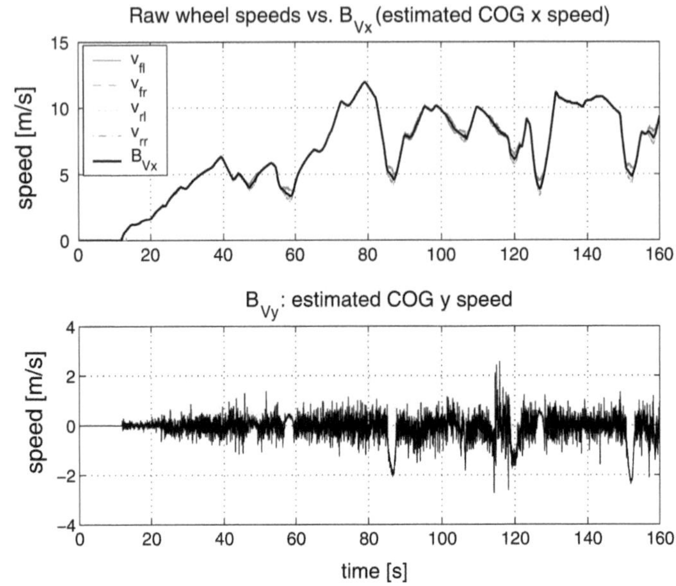

Fig. 20.14 EKF-estimated speed

${}^{B}v_{x2}$. We notice that ${}^{B}v_{x2}$ comes close to raw wheel speed signals only except for the periods when the Smart car is cornering, where, in 20.49, lateral kinematic components, i.e., yaw rate, lateral acceleration and lateral velocity are considerably large. Such a situation can be seen also from the lower plot of 20.14, where the lateral COG velocity, ${}^{B}v_{y2}$, is always around zero, except for the course of vehicle cornering.

Additionally, comparing to ${}^{B}v_{x2}$, ${}^{B}v_{y2}$ is more noisy. The reason is that ${}^{B}v_{x2}$ and ${}^{B}v_{y2}$ are updated together by the same measurement wheel speed signal, which can be seen in Eqs. (20.62) and (20.63). Therefore, a trade-off resides during the estimation of these two velocities; if one is estimated accurately and smoothly by assigning it a smaller process noise covariance, the other will end up with a noisy value. Some extra sensors for monitoring vehicle lateral kinematics must be applied in order to achieve good estimations both for ${}^{B}v_{x2}$ and ${}^{B}v_{y2}$. Therefore, under current experiment condition, the designed EKF can only accurately estimate ${}^{B}v_{x2}$ on the longitudinal direction and ${}^{B}v_{y2}$ during vehicle cornering.

The x-acceleration varies because of the continuous braking and accelerating of the vehicle, as shown in Fig. 20.15. The y-acceleration signal remains close to 0 when the vehicle is driving straightforward, but varies when the vehicle is driving through a corner. When the vehicle makes a left turn, the y-acceleration and yaw rate are positive, when the vehicle takes a right turn the measurements are negative, as shown in the two lower plots of Fig. 20.15.

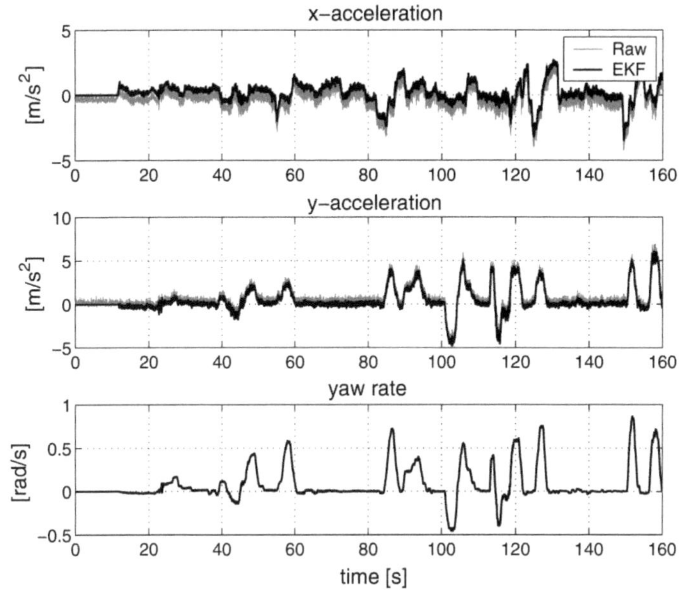

Fig. 20.15 Raw versus EKF-estimated acceleration and yaw rate

20.7 Conclusion and Future Work

In this work, a novel Extended Kalman filter (EKF) strategy has been designed to estimate states that are needed for the design of ADAS of the Smart car. Two EKFs, one based on a bicycle model and the other based on a four-wheel vehicle model, have been developed and their performances for state estimation compared. The four-wheel model based EKF is more accurate for vehicle state estimation. In this EKF, the estimated states include position, heading, acceleration, velocity, and yaw rate. The measurements are obtained from DGPS, accelerometers, gyroscope, and odometers. The designed EKF fuses multi-output rate sensor signals for achieving a desired overall rate of estimated states. Moreover, it is capable of dealing with latency compensation for DGPS signals and improving the accuracy of important vehicle state estimation even if there exist sensor errors. From the road test, it is proved that the designed EKF has achieved good results for vehicle state estimation.

The future work will consider several problems that have not been handled by the designed system in this chapter:

1. Design a sensor fusion system to handle the vehicle wheel slippage in a better way. The Smart car is a rear wheel driven vehicle. Actually, it is difficult to compute velocities using the rear (driven wheel) wheel speed, since they are affected by wheel slippage. Front (nondriven) wheel speed measurements, however, may be used to provide high accuracy velocity measurements. Therefore, in order to better estimate the vehicle velocity, the front and rear-wheel speed measurements may be applied with different weighting during the EKF measurement update.

2. Apply extra sensors to monitor the vehicle lateral dynamics and improve the accuracy of the vehicle lateral state estimation. For instance, it will smooth the estimated position when vehicle is cornering, and make the estimated vehicle lateral COG velocity more accurate.

References

1. Becker JC, Simon A (2000) Sensor and navigation data fusion for an autonomous vehicle. In: Proceedings of the IEEE intelligent vehicle symposium
2. Ryu J, Rossetter EJ, Gerdes JC (2002) Vehicle sideslip and roll parameter estimation using GPS. In: Proceedings of AVEC 2002, Japan, Sept 9–13 2002
3. Fukuba H, Adachi T, Yoshimoto A, Takahashi H, Yoshioka T (2002) Precise 6-DOF movement measurement for vehicle by GPS and angular rate sensor and Kalman filter. In: Proceedings of AVEC 2002, Japan, 9–13 Sept 2002
4. Carlson CR, Gerdes JC, Powell, JD (2002) Practical position and yaw rate estimation with GPS and differential wheelspeeds. In: Proceedings of AVEC 2002, Japan, 9–13 Sept 2002
5. Gustafson F, Persson N, Forsell U, Ahlqvist S (2001) Sensor fusion and accurate computation of yaw rate and absolute velocity. In: SAE 2001 World Congress Detroit, Michigan, 5–8 March 2001
6. Healy AJ, An EP, Marco DB (1998) On line compensation of heading sensor bias for low cost AUVs. In: Proceedings of the IEEE workshop on autonomous underwater vehicles, AUV98, Cambridge, Mass, pp 35–42, Aug 20–21, 1998
7. Kiriy E, Buehler M (2002) Three-state extended Kalman filter for mobile robot localization. Centre for Intelligent Machines (CIM), McGill University, April 12 2002
8. Qu S, Tian Y, Chen C, Ai L (2012) A small intelligent car system based on fuzzy control and CCD camera. Int J Model Ident Control 15(1):48–54
9. Ren W, Gu Q, He D, Zhao J (2013) Modelling and implementation of a car-like mobile robot for trajectory-tracking. Int J Model Ident Control 19(2):150–160
10. Shamsudin SS, Chen X (2012) Identification of an unmanned helicopter system using optimised neural network structure. Int J Model Ident Control 17(3):223–241
11. Thrapp R, Westbrook C, Subramanian D (2001) Robust localization algorithms for an autonomous campus tour guide. In: Proceedings of the 2001 IEEE international conference on robotics and automation, vol 2, pp 2065–2071
12. Grewal M, Andrews A (2000) Kalman filtering theory and practice using matlab, 2nd edn. Wiley, New York
13. Lee D, Tomizuka M (2003) Multirate optimal state estimation with sensor fusion. In: American control conference, Denver, June 4–6 2003
14. Bouvet D, Garcia G (2000) Improving the accuracy of dynamic localization systems using RTK GPS by identifying the GPS latency. In: Proceedings of the IEEE international conference on robotics and automation, San Francisco, pp 2525–2530 April 2000
15. Wada M, Yoon KS, Hashimoto H (2000) High accuracy multisensor road vehicle state estimation. In: IEEE international conference on industrial electronics, control and instrumentation, Nagoya, Japan, October 2000
16. Hermann R, Krener A (1977) Nonlinear controllability and observability. IEEE Trans Autom Control 22(5):728–740

Chapter 21
A Collaborative Learning Optimization Strategy for Shared Control of Walking-Aid Robot

Wenxia Xu, Jian Huang, Yongji Wang and Chunjing Tao

Abstract Elderly and disabled people often require assistance in getting about with maximum freedom and control while maintaining overall safety. In this chapter, we develop a collaborative learning optimization strategy for shared control of an intelligent walking-aid robot for the purpose of assisting elderly and disabled people. The proposed architecture can adjust two user control weights dynamically by a learning algorithm according to user control habit and walking environment, allowing both human and robot to maintain control of the walking-aid robot. Finally, the experiment results illustrate the validity of the collaborative learning optimization strategy as part of a shared control algorithm.

Keywords Walking-aid robot · Shared control · Collaborative learning · Sarsa-learning

21.1 Introduction

According to the World Health Organization's definition of an aging society: the proportion of the population of age 65 or over must exceed 7 %. In many countries, such as China and Japan, the proportion of the population aged 65 years or over was more than 7 %. Furthermore, the increase in the life span of many elderly people means that there are more demands for aids to support them in their daily lives [1]. A lack of nursing care accompanies this aging society. The elderly, therefore, must find ways to be independent from traditional human-based nursing care.

W. Xu · J. Huang (✉) · Y. Wang
School of Automation, Huazhong University of Science and Technology, Wuhan, 430074, China
e-mail: huang_jan@mail.hust.edu.cn

C. Tao
National Rehabilitation Center for Rehabilitation Technical Aids, Beijing, 100721, China

L. Liu et al. (eds.), *Applied Methods and Techniques for Mechatronic Systems*,
Lecture Notes in Control and Information Sciences 452,
DOI: 10.1007/978-3-642-36385-6_21,

Aged, frail, or disabled people often require a general quality-of-life commensurate to their activities: household chores, going shopping, visiting friends, enjoying the park, watching sporting events, or exercising. Moreover, those who suffer from muscular strength deterioration and amblyopia have a strong need to improve their mobility to maintain the viability of their daily lives. Traditional walking auxiliary equipment including walking sticks, and step assistance frames have simple functions along with limited safety and comfort. They do not greatly assist the elderly and disabled toward achieving a high quality and independent life. Therefore, there is an increasing need for walking-aid systems which directly lead to the improvement of life quality for elderly people.

In recent years, the walking-aid robot has combined robot technologies, such as motion control, sensors, and navigation devices, with traditional walking auxiliary equipment, as the basis for increasing research. A number of walking-aid robots are in various stages of development around the world. Omni RT Walker-II (ORTW-II) [2] is a passive-type omnidirectional walking-aid system, which uses MR brakes for its control. The mass and the grip resistance of this passive-type walker negatively influences performance in some environments, including upward slopes for instance. The active-type walking-aid robot Personal Aid for Mobility and Monitoring (PAMM) [3] uses motors to drive the system to address the load issue from which passive-type walking-aids suffer, provides mobility assistance and monitoring for the elderly. Walking Helper [4] relies on electric walking auxiliary structures, according to the user force measured by force sensors, to realize omnidirectional walking mobile support.

Physical defects such as hand trembling, gait disability, vision defects, and so on make it difficult for the elderly and disabled to manipulate the walking-aid robot to avoid obstacles or pass through the door using normal force control. The robot should provide intelligent and autonomous capabilities to properly assist the user. In such cases, both the robot and the user must join to complete a task smoothly, potentially interfering with each other. Our research is concerned with combining the control abilities of walking-aid robots and human operators to execute tasks collaboratively.

Situations where machines and humans cooperate to achieve a common goal fall within the field of collaborative control. There have been several studies investigating the level of autonomy a robot should properly achieve when interacting with a human and vice versa [5, 6]. Depending on the amount of autonomy given to the machine, collaborative approaches for human/machine control can be roughly categorized into (1) safeguarded operation; and (2) shared control. Safeguarded operation mainly uses the algorithm of the reaction formula. Shared control mainly proceeds through the negotiation method.

Shared control is a control framework that facilitates the sharing of control between a human user and an intelligent machine (or an intelligent agent) furnishing itself with the unique strengths of both the human and the robotic agent and aiding each in their areas of weaknesses[7]. Sawaragi [8] applied shared control to the design of a mobile robot teleoperation system. Shen [9] presented a method that combines shared control with a collaborative control scheme. A collaborative control component allows operator intervention when the robot is facing complex situations, while a shared control component provides an automatic control mechanism to assist and to monitor correct or irrational operator actions. However, this system is not intelligent enough to meet the demands of smooth function and suffers from low operability. Ivanisevic [10] augments the human performance in teleoperation tasks by shared control when transforming the task from the workspace to a configuration space. In this method, human and robot act together as partners rather than as master and slave. For example, the robot assesses changes in the environment by machine intelligence in spatial reasoning while the human relies on human intelligence. In a manner similar to human thinking, a system developed by Philips [11] combines the user's intention (by means of a brain control interface) and the current state as a shared control input and then based on the surrounding environment along with other information, chooses a behavior by comparing all behavior probabilities. Whether the user is able to restore the direction when activating a behavior, however, is not considered. Li [12] proposed a dynamic shared control method to adapt a wheelchair's assistance to variations in user control ability. Notable, however, is the fact that few researchers have so far considered the robot's ability to autonomously adapt to different user's skills. What is more, walking-aid robots aim to help elderly and disabled individual's motions and assist in their rehabilitation. It is significant that humans and robots can collaboratively learn different user operation habits and motor abilities to help their rehabilitation.

In this chapter, we introduce an intelligent walking-aid robot and propose a special arrangement of one-dimensional push–pull force sensors to estimate user motion intent. Our proposal is based on a shared control method, where both human and robot control the walking-aid robot. Our main novelty is the development of a collaborative learning optimization strategy in shared control: the human and robot control weight in shared control are dynamically optimized by collaborative learning between human and robot. This produces a synthetic motion command that combines both inputs in an efficient, safe, and smooth way and is dynamically adapted to different users. The algorithm was tested on the intelligent walking-aid robot in our laboratory. The cooperative avoid obstacle task is introduced in Sect. 21.2. The collaborative learning optimization strategy in our shared control algorithm is given in Sect. 21.3, and experiment results are presented in Sect. 21.4.

21.2 Cooperative Avoid Obstacle Task

A cooperative obstacle avoidance task is employed in an intelligent walking-aid robot and a human subject together within a given workspace. The cooperative obstacle avoidance task is shown in Fig. 21.1. The experimental area was composed of a room (5 m*5.5 m) and a door (88 cm) (Fig. 21.2). The intelligent walking-aid robot and subject cooperatively avoided obstacle and passed through a door. The subject was given a short tour of the room and was placed at the starting point of the experimental circuit and asked to move to the target point. We also informed the subject about the following three points: (1) the walking-aid robot moves automatically, (2) the subject

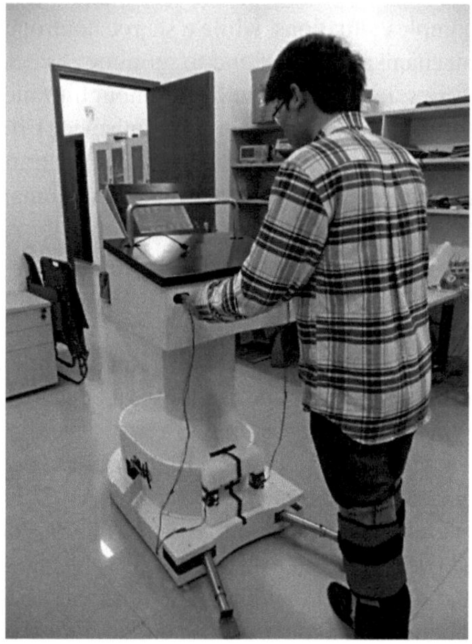

Fig. 21.1 The intelligent walking-aid robot

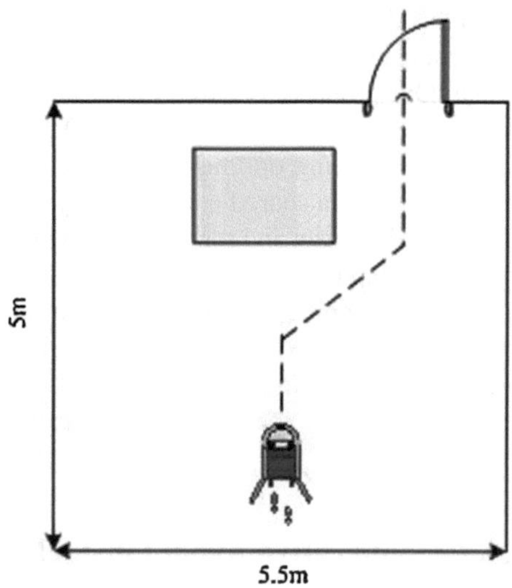

Fig. 21.2 The experiment environment

can control the robot motion by push/pull the handle of walking-aid robot, (3) the robot changes motion control weights by learning the environment and the subject's control ability.

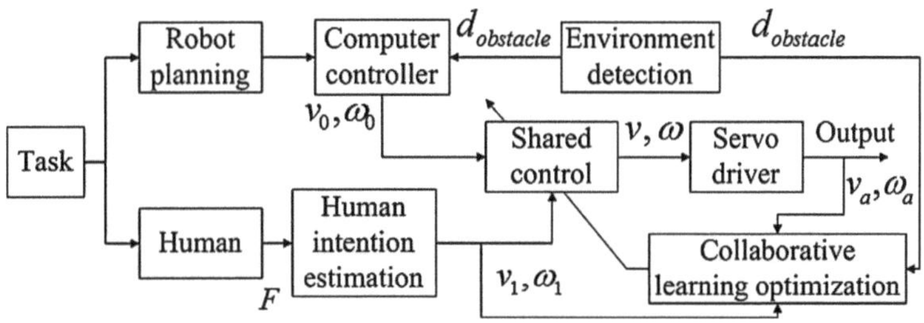

Fig. 21.3 The walking-aid robot shared control system structure

21.3 The Model and System

21.3.1 Shared Control Architecture

The main objective of our shared control architecture is to give users as much control as possible to ensure adequate performance and safety. For walking-aid robots, the architecture is used to assist the user rather than replace the user in a moving task. There are two fundamental requirements: first, the user should feel in charge of control of the walking-aid robot, which in turn should behave as the user expects. Second, the waking-aid robot should be safe and comfortable while being operated by the user.

However, due to elderly or disabled people's limited cognitive or physical capabilities, such as preserving body curvature, or due to irrational behavior, difficulty in perceiving surroundings, and manipulating the device flexibly may arise. A method that can combine human control ability and machine control ability effectively would greatly improve patient rehabilitation. Machine assistance, moreover, allows a device to adapt to differences in human control abilities.

To satisfy the requirement, a shared control architecture for walking-aid robot is proposed in this chapter, see Fig. 21.3. v_0 and ω_0 are the planned robot translational and rotational speed generated by computer controller. v_1 and ω_1 are the user-desired translational and rotational speed estimated by a human intention estimation algorithm. v_a and ω_a are the robot current translational and rotational speed of the intelligent walking-aid robot recorded by the odometer. d_{obstacle} is obstacle distance obtained from the ultrasonic sensor. The aim of computer controller is to generate planned translational and rotational speed to avoid obstacle. By perceiving the push/pull forces on the U-shape arm, human intention estimation algorithm can estimate user-desired translational and rotational speed by an array of push/pull sensors which are detailed in [13]. The collaborative learning optimization assigns control weights to user and the robot motion controller. Details of the collaborative learning optimization algorithm will be discussed in the following section.

The shared control outputs v and ω are synthesized from two control inputs to the shared controller by user translational speed control weight k_v and user rotational speed control weight k_ω.

This shared control algorithm includes six steps:

(1) Update obstacle distances d_{obstacle} from ultrasonic sensor.
(2) Generate robot-planned translational and rotational speed v_0, ω_0 through computer controller.
(3) Estimate user-desired translational and rotational speed v_1, ω_1 by a force sensors array and human intent estimation algorithm.
(4) Calculate user translational and rotational speed control weight k_v, k_ω by a collaborative learning optimization algorithm which will be discussed later.
(5) Calculate final translational and rotational speed v, ω by the shared control algorithm according to:

$$\begin{aligned} v &= k_v \cdot v_1 + (1 - k_v)\, v_0 \\ \omega &= k_\omega \cdot \omega_1 + (1 - k_\omega)\, \omega_0. \end{aligned} \tag{21.1}$$

(6) Send the final translational and rotational speed to servo driver.

The shared control architecture combines the merits of the human and the robot. Its novelty lies in a synthesis of the expected user motion velocity with the planned robot velocity according to the translational and rotational speed user control weights which are dynamically adjusted by a cooperative learning optimization algorithm. Weight adjustment is performed online based on various user control efficiencies, the robot walking state, and current environment information. Details are introduced in next section.

21.3.2 Collaborative Learning Optimization Strategy

So far, few researchers have considered the robot's ability to autonomously adapt to different user's skills and to different environments. It is generally observed that user rehabilitation benefits from human and robot collaboration and learning of different user operation habits and motor abilities. Therefore, we use a collaborative learning optimization strategy to automatically optimize user control weights for shared control by learning user operation habits and motor abilities collaboratively. Two Sarsa-learning modules optimize user translational speed control weight k_v and user rotational speed control weight k_ω, respectively, as seen in Fig. 21.4.

As reinforcement learning was used in robot widely [14]. To accomplish the above goal, the collaborative learning optimization strategy capitalizes on Sarsa-learning which is a kind of reinforcement learning capable of creating personalized adaption based on previous experiences of interaction, and is also capable of adapting and learning state-by-state. Here, the Sarsa-learning can be described as the interaction between user and the intelligent walking-aid robot and occurs through the generation of a sequence of states which include user-desired translational and rotational speeds

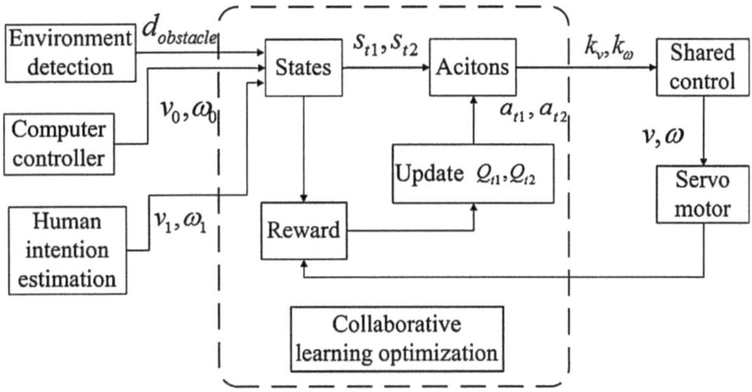

Fig. 21.4 Collaborative learning optimization strategy of shared control

as well as obstacle distance information mapped by the user to a series of actions of the robot. This action sequence moves and positions the walking-aid robot in the robot's workspace to achieve a goal. The robot and user must learn collaboratively (based on actions and observed states and rewards) which strategies will maximize the earned reward.

There are three important elements of Sarsa-learning: states, actions, and rewards. The robot's states are dynamic, and the states S_1, S_2 are obtained from user and robot, $S_1 = \{d_{\text{obstacle}}, v_s, \}S_2 = \{d_{\text{obstacle}}, \omega_s\}$. d_{obstacle} is the obstacle distances obtained by ultrasonic sensor. v_s is the differentials between user-desired translational speed and robot actual translational speed. ω_s is the differentials between user-desired rotational speed and robot actual rotational speed.

$$\begin{aligned} v_s &= |v_1 - v_0| \\ \omega_s &= |\omega_1 - \omega_0| \end{aligned} \tag{21.2}$$

The two actions A_1, A_2 are user translational speed control weights and user rotational speed control weights satisfying, $A_1 = \{0, 0.1, 0.2... 0.9, 1\}$, $A_2 = \{0, 0.1, 0.2... 0.9, 1\}$. The states are discretized to satisfy Markov Decision Process (MDP) which is characterized by a set of states and actions. After the action, a reward is presented. The reward function computes the immediate robot reward based on human motion intention, the robot's current states, and environment. The reward function provides a scalar reward value to the sequence of state-action pairs that caused the robot to reach its current location. Here, the control system operates within the discrepancy between the user intent and the actual motion of the human-walker system. The reward function should thus consider the safety and smoothness of the sequence pairs. For example, consider a human pushing a walker through a doorway which is narrow relative to the width of the walker frame. Slight errors in the control of the walker frame by the user may cause one of the wheels to impact the side of the doorway. This can result in a collision with the door. The user's control weight is thus influenced by the user's performance and walking environment. Therefore, the reward function can be represented as

$$\begin{aligned} r_{t1} &= k_1 d_{\text{obstacle}} - k_2 v_s \\ r_{t2} &= k_3 d_{\text{obstacle}} - k_4 \omega_s \end{aligned} \tag{21.3}$$

where k_1, k_2, k_3, k_4 are positive constants.

After a learning episode is completed, a data tuple s_t, a_t, r_t is collected. The collected data are compressed into a table $Q(s_t, a_t)$ which is defined as an estimated accumulated reward starting in state s_t when taking action at the following policy π. Here, we select actions by Soft-Max method which chooses actions according to current action values, using a Boltzmann distribution. The updated equation for the value function estimation is defined as

$$Q(s_t, a_t) = Q(s_t, a_t) + \alpha_t[r(s_t, a_t) + \gamma Q(s_{t+1}, a_{t+1}) - Q(s_t, a_t)] \tag{21.4}$$

where α_t is a learning factor which reflects learning speed, $\alpha_{t+1} = \alpha_t/(1 + \alpha_t)$. γ is the discount factor.

21.4 Experiments

The collaborative learning optimization strategy of shared control is evaluated and compared in obstacle avoiding experiments with two subjects. Each subject and robot walk forward together, proceed through a door, and finally reach a table for 25 trials. The computer controller helps the robot avoid obstacles to ensure the user's safety. In this experiment the condition of no obstacle, robot-planned translational, and rotational speed are $\{-0.3\,\text{m/s}, 0\,\text{m/s}, 0\,\text{rad/s}\}$. When the distance between robot and obstacle is less than 1m, the robot-planned translational and rotational speed will be $\{-0.3\,\text{m/s}, 0.3\,\text{m/s}, 0\,\text{rad/s}\}$ to avoid obstacles. To be safe, if the distance between robot and obstacle is less than 0.3 m, the robot will stop, and the servo motor speed will reach zero.

21.4.1 Comparison of the Performances

Figure 21.5 shows the comparison of robot translational and rotational speed performances of subjects in the 1st and 25th trials. The horizontal axis shows the trial time. The vertical axis shows velocity in the robot's coordinate system. The blue line represents the actual robot velocity in the first trial, the black line represents the actual robot's velocity in the last trial, and the red line presents the planned robot velocity. After 25 trials' learning process, the robot's velocity becomes relatively stable and the user feels more comfortable. The collaborative learning optimization strategy

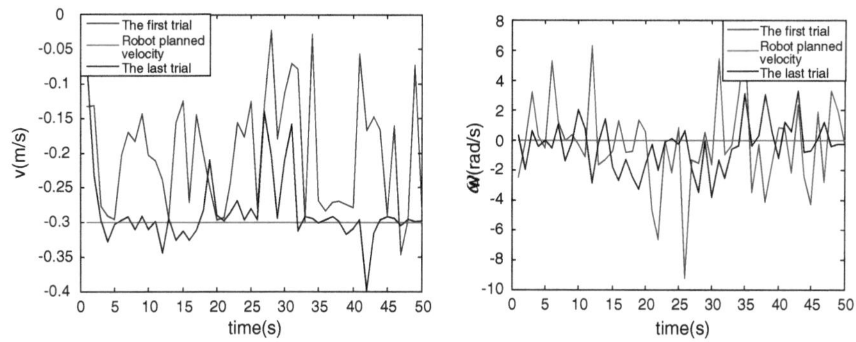

Fig. 21.5 Comparison of two user control weights to adjust robot translational and rotational speed

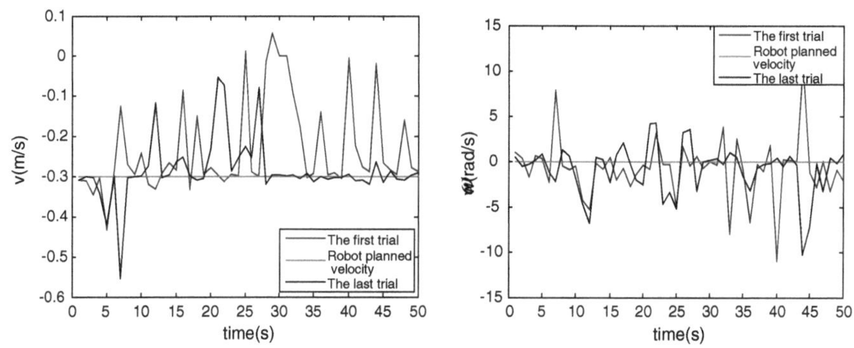

Fig. 21.6 Comparison of one user control weight to adjust robot translational and rotational speed

of shared control is therefore efficient. And compared to Fig. 21.6 which shows the performances of using one user translational speed control weight to adjust robot translational and rotational speed, the relative performance of using two user translational speed control weights and a rotational speed control weight is superior. This method also solves the problem of unstable rotational speed.

The transitions of the travel time of two subjects are shown in Fig. 21.7. All performances metrics improved in these figures. Although the transitions in travel time are not obvious in the first half of the learning process, differences appeared in the second half. After robot learning the subject control habit and environment, the robot makes comparatively better action choices than before.

21.5 Discussion

The differences between the first trial and last trial values of the fits for mean speed and peak speed are plotted in Figs. 21.8 and 21.9. Mean speed and peak speed are smoothness metrics. Mean speeds of robot translational and rotational speed in Fig. 21.8 are the average of the velocity of the robot during a trial. Filled circles represent the

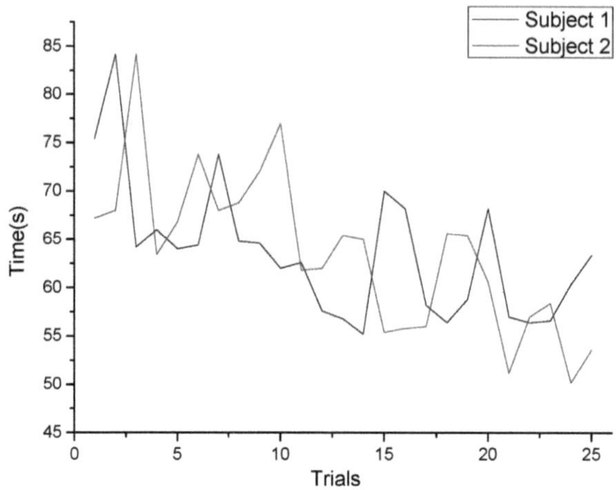

Fig. 21.7 Travel time of two subjects

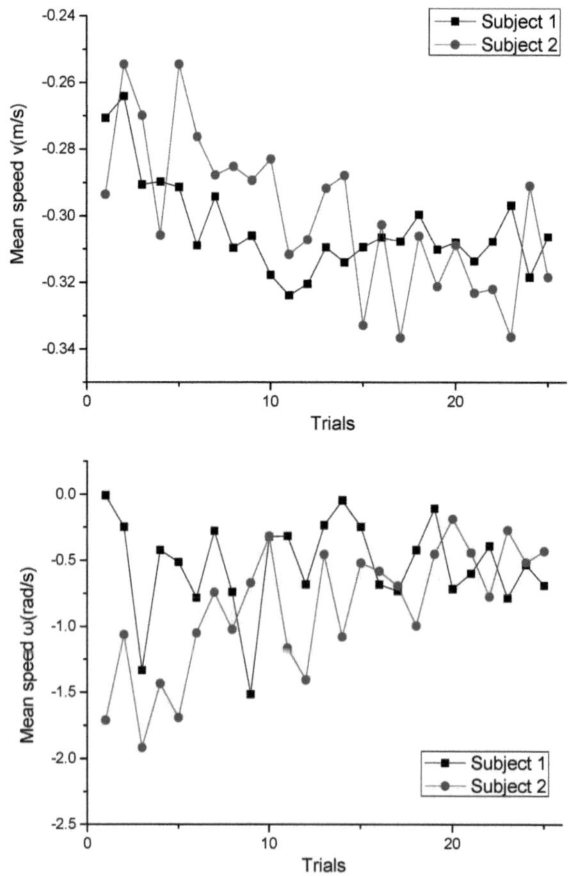

Fig. 21.8 Mean speed of two subjects

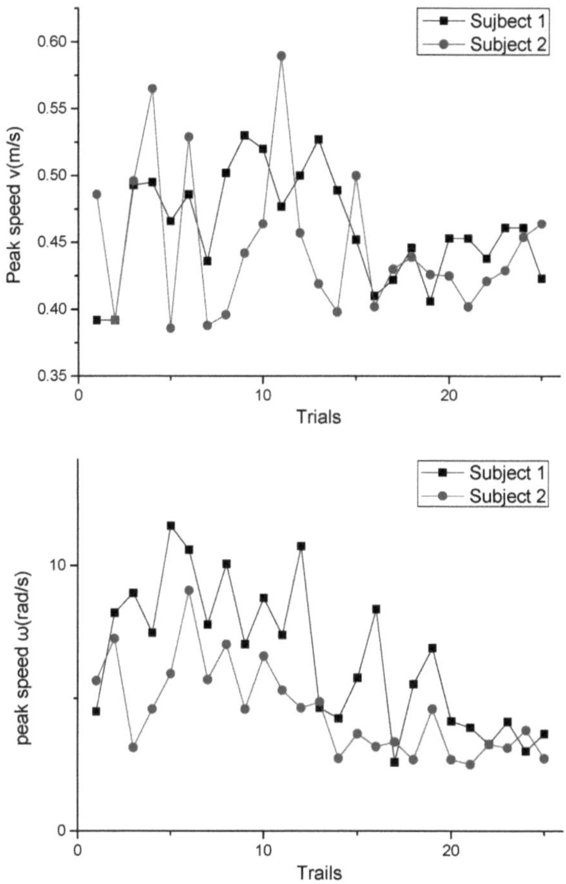

Fig. 21.9 Peak speed of two subjects

statistical significance of mean speed during the 25 trials. Subjects mean speeds have a slow trend in the direction of the change and tend to remain stable. It can been seen that the robots' movements become more steady and fluent, and that the task will be accomplished with greater speed after reinforcement learning. Figure 21.9 is the peak speed of the robot translational and rotational speed during 25 trials. It can be seen that after learning, the peak speed increased significantly in many cases.

21.6 Conclusions

In this chapter, we introduced a collaborative learning optimization strategy for shared control algorithm for intelligent walking-aid robot. The goal of this algorithm is to achieve maximum freedom and control for users. The proposed algorithm considers user personality characteristics and automatically adapts to user behaviors, in order to improve the user's degree of comfort. The proposed architecture can adjust two user control weights dynamically by a learning algorithm adapting to user control efficiency and walking environment, allowing both human and robot to maintain control of the walking-aid robot. Experimental data show that the proposed algorithm can ensure user movement performance and smoothness meets safety requirements while tangibly improving the quality of user control interactions. This algorithm may aid in the future development of more intuitive, people-friendly assistive walking devices.

Future work on collaborative learning as applied to shared control may proceed according to these three aspects:

(1) More detailed impact analysis of the different users' control characteristics.
(2) Analysis of the robustness of the collaborative learning in shared control algorithm.
(3) Further studies of human–robot mutual learning and human learning information according to robot actions.

Acknowledgments This work is supported by the Chinese national science and technology support project "Research and development of multi-functional walking-aid system" under Grant 2012BAI33B04, and the International Science & Technology Cooperation Program of China (Precision Manufacturing Technology and Equipment for Metal Parts under Grant No.2012DFG70640), and is also supported by International Science & Technology Cooperation Program of Hubei Province under Grant 2012IHA00601 "Joint Research on Green Smart Walking Assistance Rehabilitant Robot."

References

1. Xiao BJ, Su HM, Zhao YL, Chen X (2013) Ant colony optimisation algorithm-based multi-robot exploration. Int J Model Ident Control 18(1):41–46
2. Hashino S (1996) Daily life support robot. J Robot Soc Jpn 14(5):2–6
3. Nejatbakhsh N, Kosuge K (2005) A human adaptive Path tracking method for omnidirectional passive walking aid system. In: IEEE/RSJ international conference on intelligent robots and systems (IROS 2005), pp 1145–1150
4. Dubowsky S, Genot F, Godding S et al (2000) Pamm–a robotic aid to the elderly for mobility assistance and monitoring: a "helping-hand" for the elderly. In: Proceeding of IEEE international conference on robot and automation, CA, San Francisco, pp 570–576
5. Hirata Y, Baba T, Kosuge K (2003) Motion control of omnidirectional type walking support system "walking helper". In: Proceeding of the 12th international IEEE workshop on robot and human interactive communication, Millbrae, CA, pp 85–90
6. Kofman J, Wu X, Luu TJ et al (2005) Teleoperation of a robot manipulator using a vision-based human-robot interface. IEEE Trans Ind Electron 52(5):1206–1219
7. Tahboub KA (2001) Natural and manmade shared-control systems: an overview. In: Proceedings of the 2001 IEEE international conference on robotics and automation, Seoul, Korea, 21–26 May 2001, pp 2655–2660
8. Sawaragi T, Takayuki S, Akashi G (2000) Foundations for designing an ecological interface for mobile robot teleoperation. Robot Auton Syst 31:193–207

9. Shen J, Guzman JI, Chew TN et al (2004) A collaborative-shared control system with safe obstacle avoidance capability. In: Proceedings of the 2004 IEEE conference on robotics, automation and mechatronics, vol. 1. Singapore, pp 119–123
10. Ivanisevic I, Lumelsky VJ (2000) Configuration space as a means for augmenting human performance in teleoperation tasks. IEEE Trans Syst Man Cybern Part B 30(3):471–484
11. Philips J, del Millan JR, Vanacker G et al (2007) Adaptive shared control of a brain-actuated simulated wheelchair. In: Proceedings of IEEE 10th international conference on rehabilitation robotics, Noordwijk, The Netherland, pp 408–414
12. Li QN, Chen WD et al (2011) Dynamic shared control for human-wheelchair cooperation. In: Proceedings of IEEE international conference on robotics and automation (ICRA), Shanghai, China, pp 4278–4283
13. Ye JY, Huang J, He JP et al (2012) Development of a width-changeable intelligent walking-aid robot. In: Proceedings of IEEE international conference on micro-nano mechatronics and human science, Nagoya, Japan, pp 358–363
14. Yang EF, Gu DB (2009) Multi-robot systems with agent-based reinforcement learning: evolution, opportunities and challenges. Int J Model Ident Control 6(4):271–286

Chapter 22
Vibration Suppression of Deformable Linear Object Based on Vision Feedback

Jian Huang, Feng Ding, Huan Wang and Yongji Wang

Abstract Deformable Linear Objects, such as hose and cable, are widely used in daily life and industrial applications. Problems in robotic operation and precise control of DLOs, have limited the progression of automation in the electronics industry and automotive industry. In this chapter, we propose an approach based on robot vision to suppress the vibration of a DLO, by means of real-time image processing of high-speed visual feedback to obtain the geometric coordinates and posture of the DLO. A PID controller verifies the validity of the theory along with the expected posture of the DLO. The effectiveness of the proposed strategy is confirmed by experiment. Results show that the proposed method can damp vibrations effectively.

Keywords Deformable linear object · Robot vision · Vibration suppression · PID control

22.1 Introduction

Robotic manipulation research focuses mainly on rigid objects. Manipulation of deformable linear objects (DLOs) such as hoses, wires, cables are common in daily life, and many domains such as manufacture, medical surgery, and so on. DLOs are challenging to handle, as the uncertainty resulting from oscillation at the end of DLOs may cause failure during operation. Deformable linear objects suffer from an inadequate ability to adjust for vibration while undergoing movement. In the context of increasing demand for automatic control and robotic operation, research

J. Huang · H. Wang · Y. Wang
Department of Control Science and Engineering, Huazhong University of Science and Technology, Wuhan, Hubei, China

F. Ding (✉)
College of Computer Science, South-Central University for Nationalities, Wuhan, Hubei, China
e-mail: dingfeng@hust.edu.cn

L. Liu et al. (eds.), *Applied Methods and Techniques for Mechatronic Systems*, Lecture Notes in Control and Information Sciences 452,
DOI: 10.1007/978-3-642-36385-6_22,

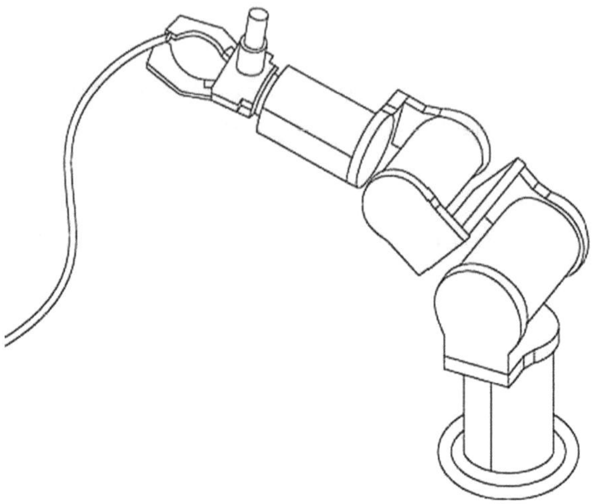

Fig. 22.1 Prototype of a robot manipulating a DLO with vibration

pertaining to this issue as applied to robot operational control has gradually become an important area in robotic control research.

In the past decades, much works have been done in the field of rigid objects. In general, the deformation of rigid objects can be analyzed by tension [1]. The method cannot suit deformable objects with huge deformation, such as paper, cotton material. In order to study the shape and characteristics of DLOs, Hirai et al [2] obtained the static model of DLO by a method based on differential geometry and minimum energy. In order to obtain a dynamic model, some methods such as differential geometry [3], finite element method (FEM) [4], combining FEM and Lagrange equations [5] have been applied to derive dynamic models of DLOs.

In light of the motion characteristics of DLOs, the reason that DLOs are challenging to handle is due to the vibration at the end of the DLO (as seen in Fig. 22.1), as the uncertainty resulting from their strong oscillation may cause failure during operation. Much works have been done in the field of vibration control. An adaptive controller is designed to reduce the vibration of a piezoelectric actuator-based flexible arm [6]. Yue [7] proposed a model-based adjustment motion for DLO vibration. He also presented a fast manipulation approach using force/torque sensor to eliminate the cute vibration [8]. With the goal of dampening vibrations, we designed a sliding mode controller to minimize the vibration at the end of a DLO [9].

For modeling, character information such as length, flexibility is important. As mode-based control methods offer poor portability in different robot operation situations, the scope of the application is limited. In order to accommodate the hazards of operational environment that hamper acquisition of the dynamic mode of a DLO, a control solution based on sensors has been proposed. A vibration reduction strategy inspired by human manipulation skill was proposed by Yue [10], which is force-based and implemented by fuzzy and P control. However, for lightweight DLOs

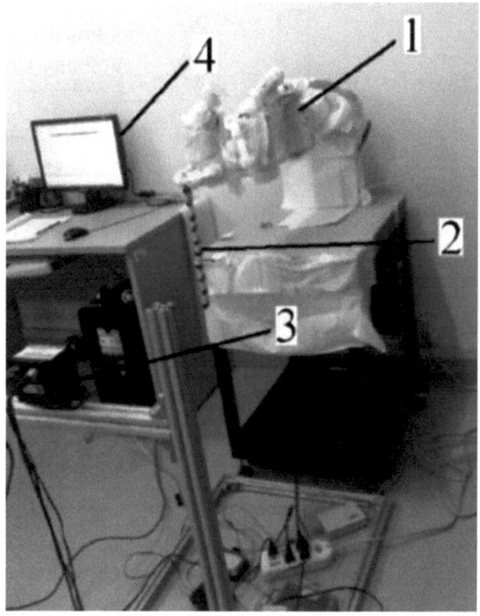

Fig. 22.2 The platform of DLO control system

such as ropes and strings, force-based manipulation strategies may not work well. This is because measurement of weak force signals is prone to noise pollution. Jiang [11] measured the end-effector position and link flexural behavior of flexible robotic manipulators by a CCD camera. A CCD camera was used to extract path information through image recognition [12]. In the view of these points, we propose a vibration reduction strategy based on robot vision feedback. We can obtain the velocity and position of the DLO though high-speed camera, and avoid to measurement of weak force signals. A PID controller is designed to suppress the vibration in the end of a DLO.

The rest of the chapter is organized as follows: the overall design and structure of experimental platform are presented in Sect. 22.2; the comparison of image processing algorithms and optimization is presented in Sect. 22.3; a PID controller is designed in Sect. 22.4. Finally, the effectiveness of the proposed strategy is conformed by experiment results. The experiment results are inspiring and show that the proposed method can damp vibration effectively.

22.2 Robot Operation System

The main content of this chapter is controller design and experiment research concerning vibration suppression of robot manipulating deformable linear objects. The experiment platform is shown in Fig. 22.2, where 1 is a robot manipulator, 2 is a DLO, which is fixed in the end of the manipulator, 3 is a high-speed camera which

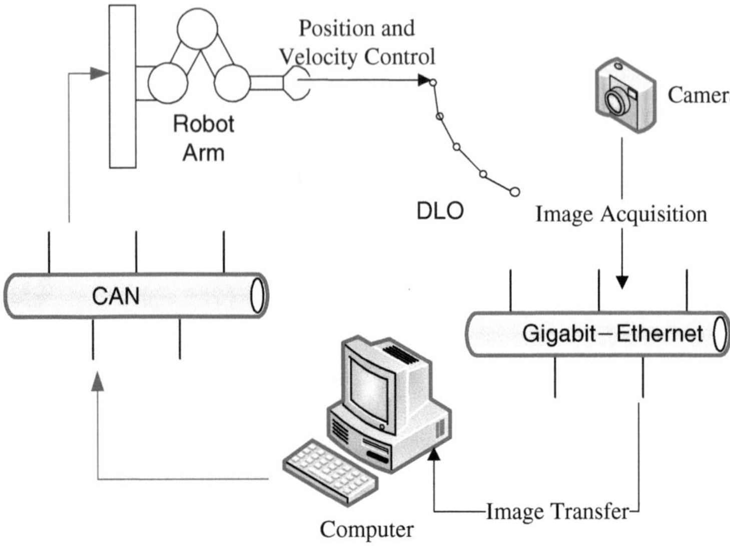

Fig. 22.3 The diagram of DLO control system

is fixed to the front of the DLO to detect the motion information of the DLO, and 4 is the host computer, which collects and processes the images captured by camera. Based on the information of the DLO obtaining from image processing, controller is designed which planes the motion of a robotic manipulator. The block diagram of the control system is shown in Fig. 22.3.

Four degrees of freedom manipulator is used for the experiments of vibration suppression at the end of a DLO. The joints of the manipulator are driven by Coreless brushless DC servo motors, and absolute encoders are used to measure speeds of motors. The robot manipulator system centers on a computer, establishes a small-scale distributed control system through CAN bus, and sends instructions to the inferior machine (DSP) to control a four joint coordinated operation.

Since the signals measured by force sensor are prone to noise pollution, we introduce vision system (high-speed camera) substitute force sensor to measure the movements of a DLO. MVC610DAM-GE110 is selected, which is an industrial CCD camera with progressive scan mode and high resolution. The frame rate is up to 110pfs, and the interface is simple. The electronic shutter can capture fast moving objects, and reduce the motion blur caused by image recognition errors. We developed it secondarily in the OpenCV (Open Source Computer Vision Library) environment.

The controller system requires a high real-time input, as the controller is designed by using online identification of the states of DLO. According to real-time requirements of the machine vision, the speed of image processing must be less than the reciprocal of the frame rate. That is, the image processing time is less than $1/110 = 0.0091\,\mathrm{s}$.

22.3 Image Processing Algorithms

Image processing is one of the most important parts in the vision-based system. According to the raw data collected by image sensor (camera), obtainment of the position and velocity information of the DLO is a key point to the experiment. Image acquisition is required before image processing. The block diagram of Gigabit Ethernet CCD vision sensor collecting image information is shown in Fig. 22.4.

The image processing method is developed in OpenCV environment [13], which is Intel's open source computer vision library containing by C functions and a few C++ classes, and implements some common algorithms with respect to image processing and computer vision. As OpenCV is written by C/C++, compared to MATLAB, the efficiency is much higher, especially in a loop. Meanwhile, the performance is not worse than some commercial solutions. Now, OpenCV is widely used in human–computer interaction, object recognition, image segmentation, face recognition, motion recognition, motion tracking and robot vision, and so no.

We study a control system based on a real-time visual feedback, the key and difficulty of which lies in how to solve the real-time accuracy of the image processing. Three algorithms utilizing dynamic background diminishing, edge detection based on canny operator and threshold-based moving target detection were studied and compared based on performance.

The main concept of a dynamic background diminishing algorithm is to obtain the spatial location of a moving target by detecting the differences between the foreground and background. The main idea of edge detection based on canny operator algorithm is to obtain the target profile by computing the differences between the moving target and background, and then extracting features. Threshold-based moving target detection algorithm functions by obtaining the feature points and by detecting the difference between the feature points and background, and then extrapolating the locations of the feature points by threshold. The effects of the three image processing algorithms detecting DLO are shown in Fig. 22.5.

From Fig. 22.5, we can see that all image processing algorithms can extract the features. But which one suits to the situation? We have to consider other factors such as time complexity, illumination robustness and motion blur robustness, and so on. After a few repeated experiments comparing, marked point detection success rate and the proportion of contour integrity statistic were carried out, we found which algorithm is more robustness with respect to illumination and motion blur. In order to obtain the time complexity of image processing algorithm, we have to obtain the machine time from beginning to end during image processing. Fortunately, Windows Operating System provides a hardware timer counter, implemented by the function *QueryPerformanceCounter()*, which returns a high-precision performance counter value with microsecond timing unit. However, the exact precise time is associated with the Windows System. We must query the system to get the returned frequency of ticking, which is provided by the function *QueryPerformanceCounter()*. Compared to other hardware time counter such as *GetTickCount()*, *QueryPerformanceCounter()* possesses more accuracy.

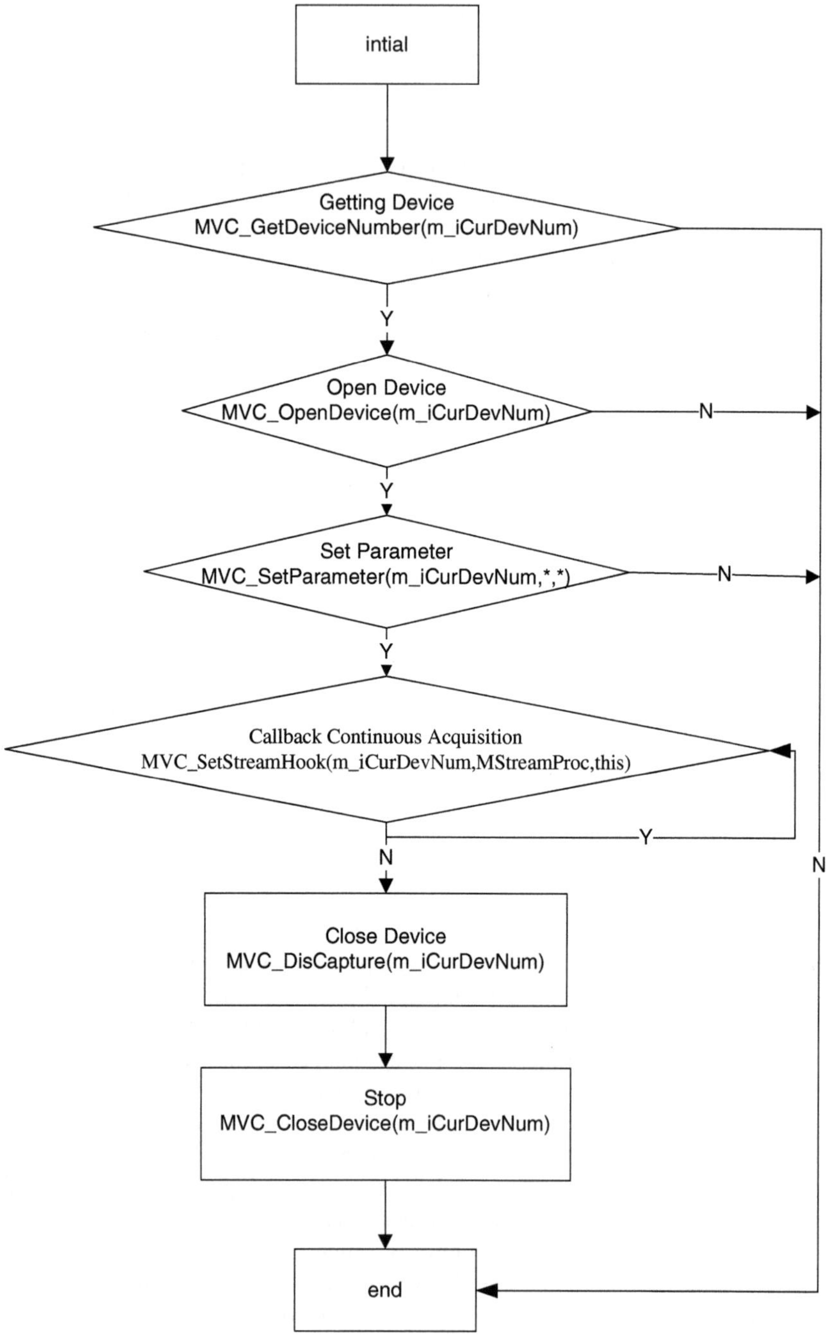

Fig. 22.4 The block of image collection

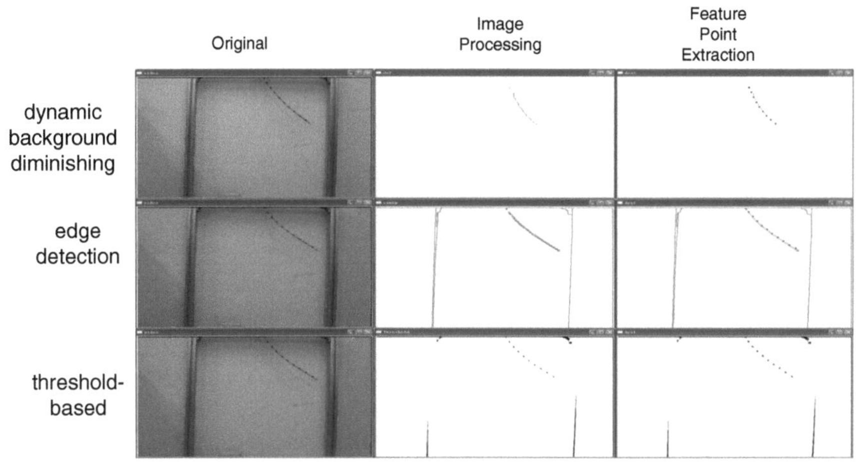

Fig. 22.5 The effects of the three image processing algorithms detecting DLO

The exact time is calculated from the first calling *QueryPerformanceCounter()* with returned value *nStartCounter*, and ends the function with *nStopCounter*. The machine time can be obtained by the difference between *nStopCounter* and *nStartCounter* dividing the frequency of *QueryPerformanceCounter()*. The calculated function as following:

$$T_e = \frac{C_{\text{stop}} - C_{\text{start}}}{f} - T_o, \tag{22.1}$$

where T_e is the cost time, C_{stop}, C_{start} are the stop time and initial time respectively, and T_0 is the time cost by the timer.

Theoretically, the timing function spends little time during timing T_0 which should be subtracted, but in engineering applications, this small time frame can be ignored. In general, the error is less than 1ms, and related to CPU. Equation (22.1) can be simplified as:

$$T_e = \frac{C_{\text{stop}} - C_{\text{start}}}{f}. \tag{22.2}$$

The image processing programs run in the computer with $f = 2.7$ GHz. According to some repeat experiments, time complexity of image processing is computed by the nonsingular means. The comparison on the performance of the image processing algorithms is shown in Table 22.1.

MVC610DAM-GE110 is used in the experiments, which is a high-speed camera with 110 fps. As mentioned in Sect. 22.2, the time complexity of image processing must be less than the reciprocal of the frame rate, that is less than $1/110 = 9.1$ ms. According to the Table 22.1, only the last column satisfies the condition. That is, threshold-based moving target detection is chosen for image processingalgorithm.

Table 22.1 Comparison on the performances of the image processing algorithms

	Dynamic background diminishing	Edge detection	Threshold-based
Time complexity (ms)	34.423	16.535	5.209
Illustration robustness	Better	Best	Worse
Motion blur robustness	Best	Better	Worse

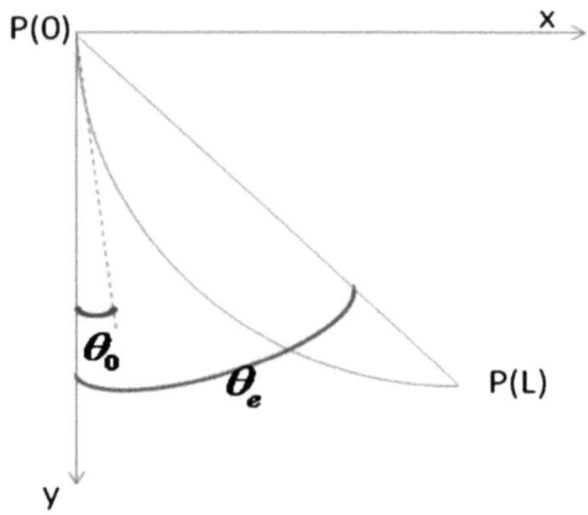

Fig. 22.6 The control of a DLO

Compared to other algorithms, threshold-based moving target detection is worse in illustration robustness and motion blur robustness, but it satisfies the experiment environment.

22.4 Control Strategy

In the experiments, we are concerned with suppressing the vibration at the end of a DLO quickly. Roughly, two angles (θ_0 and θ_e) can describe the deformation of the a DLO, where θ_0 is the angle between longitudinal level and the origin of a DLO, and θ_e is the angle between longitudinal level and the end of a DLO, as shown in Fig. 22.6.

A PID controller is widely used in industry applications and chosen to damp the vibration of a DLO. The most advantage is that PID controller is the simplest and most effective control strategy, and easy to apply to experiment. We don't need to model the DLO, and complex mathematical computation is avoided. The problem that model parameters such as flexural coefficient are difficult to measure is solved. There is enough experience to obtain PID parameters which are easy to experiment.

If a DLO deforms, θ_0 and θ_e exist. Our aim is to design a PID controller to eliminate θ_0 and θ_e. The controller includes three items [4].

$$u = u_{\mathrm{pi}} + u_{\mathrm{damp}} + u_{\mathrm{vsup}} \tag{22.3}$$

u_{pi} is a PI control item related to θ_e and its desired value, arming to damp the vibration at the end of a DLO.

$$u_{\text{pi}} = K_p \left(\theta_d - \theta_e\right) + K_i \int_0^t \left(\theta_d - \theta_e\right) \mathrm{d}t,$$

where K_p and K_i are proportional gains, and θ_d is the desired value.

u_{damp} is a damping item, designed as a D controller related to θ_0, arming to accelerate the controlled object approach to the desired value.

$$u_{\text{damp}} = K_d \left(\dot{\theta}_{d0} - \dot{\theta}_0\right),$$

where K_d is the differential gain, and $\dot{\theta}_{d0}$ is the desired value.

u_{vsup} is a P controller related the error between θ_0 and θ_e, aiming to inhibit the ripple oscillation caused by the error between θ_0 and θ_e.

$$u_{\text{vsup}} = K_v \left(\theta_e - \theta_0\right),$$

where K_v is the internal gain.

According to PID control theory, closed-loop control system based on visual feedback can suppress the vibration at the end of a DLO effectively. The PI controller can track the motion of a DLO quickly. The DLO and robot swinging with the same direction can destroy the energy of a DLO conversing between potential energy and kinetic energy. The total energy is reduced, and the vibration is limited.

22.5 Experiment

An experiment consisting of a robot manipulating DLO based on visual feedback has been constructed to demonstrate that the proposed image processing algorithm and motion control strategy can suppress the vibration in the end of a DLO effectively.

The manipulator is a four freedom robot arm designed by Shanghai Ingenious Automation Technology Co., Ltd. In the experiment, the first joint rotates 90° and the forth joint rotates −90°, and maintains constant angularity. Only the second and third joints are used. The four degrees of freedom robot can be transformed to a two degrees of freedom robot, as shown in Fig. 22.7.

The relationship of the position of end point(X, Y) and θ_1, θ_2 can be expressed as:

$$\begin{aligned} X &= l_1 \cos\left(\theta_1\right) + l_2 \cos\left(\theta_1 + \theta_2\right) \\ Y &= -l_1 \sin\left(\theta_1\right) - l_2 \sin\left(\theta_1 + \theta_2\right), \end{aligned} \tag{22.4}$$

Differential both side of (22.4).

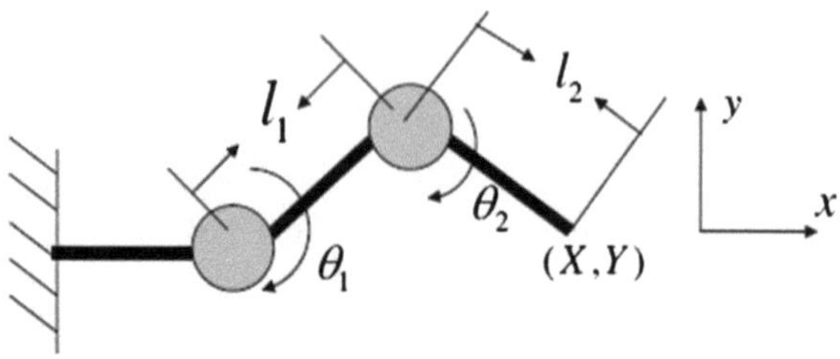

Fig. 22.7 The control of a DLO

$$\begin{aligned}\dot{X} &= -\left(l_1 \sin(\theta_1) + l_2 \sin(\theta_1 + \theta_2)\right)\dot{\theta}_1 - l_2 \sin(\theta_1 + \theta_2)\,\dot{\theta}_2 \\ \dot{Y} &= -\left(l_1 \cos(\theta_1) + l_2 \cos(\theta_1 + \theta_2)\right)\dot{\theta}_1 - l_2 \cos(\theta_1 + \theta_2)\,\dot{\theta}_2\end{aligned} \tag{22.5}$$

The angle and angle velocity are detected by encoders. According to constraint relationship, the manipulator can move in the X direction.

The camera (MVC610DAM-GE110) used in the experiment operates 110 fps and 659 × 494 pixel, and communicates with the PC by Gigabit Ethernet. Only planar motion can be detected by using a single camera. In order to prevent longitudinal movement, an elastic rubber strip of a certain width is chosen as the DLO, which meets the requirement of the swing on the plane, and is wrapped by white material. In order to highlight the target points, we mark the target points in back, as shown in Fig. 22.1.

The control period is 100 ms triggered by a timer. According to the position and velocity of the DLO feedback by vision system, compute the θ_e, θ_0, $\dot{\theta}_0$. According to PID parameter turning method [14], we used the proportional control only initially and obtained the proportional parameter when the system underwent undamped oscillation. We then added integral control and reduce proportional parameter. After obtaining proper PI parameters, we added the damping item and suppression item, and fine tuned the parameters. Because of the dead-zone characteristics and precision of joints, we obtained proper PID parameters after repeated adjustments. The parameters were chosen as $K_p = 1$, $K_i = 0.08$, $K_d = 0.01$, $K_v = 0.3$. The experiment results are shown in Figs. 22.8 and 22.9.

In Fig. 22.8, the red line represents the trajectory of θ_0 and blue line is the trajectory of θ_e without control, respectively. The vibrations of θ_0 and θ_e dye away gradually because of the air resistance. But relying on air resistance solely, it costs 8 s to eliminate the vibration, which is too slow in the industry application. The performance of DLO system controlled by PID controller is shown in Fig. 22.9, where red line is trajectory of θ_0 and blue line is the trajectory of θ_e. According to Fig. 22.9, we see that 2 s are required to suppress the vibration. Obviously, the proposed PID controller can suppress the vibration effectively.

Some overshoots during control are shown in Fig. 22.10.

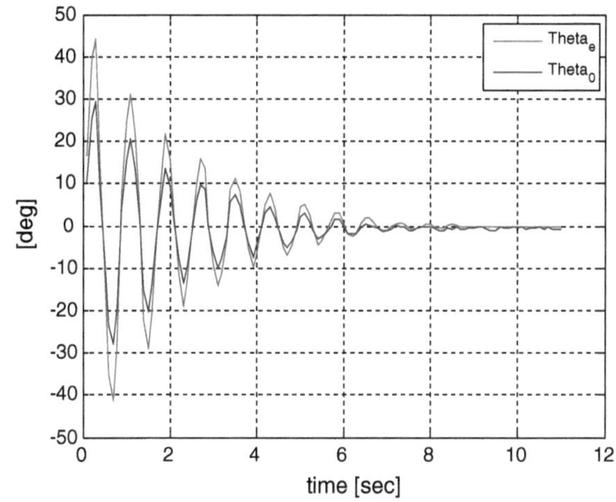

Fig. 22.8 The trajectories of θ_0 and θ_e without control

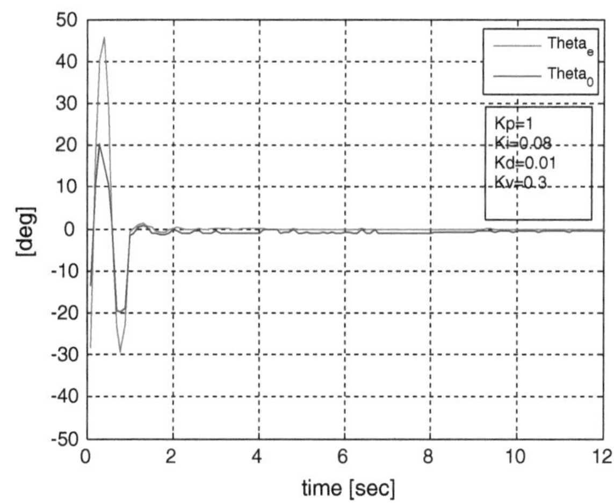

Fig. 22.9 The trajectories of θ_0 and θ_e controlled by PID controller

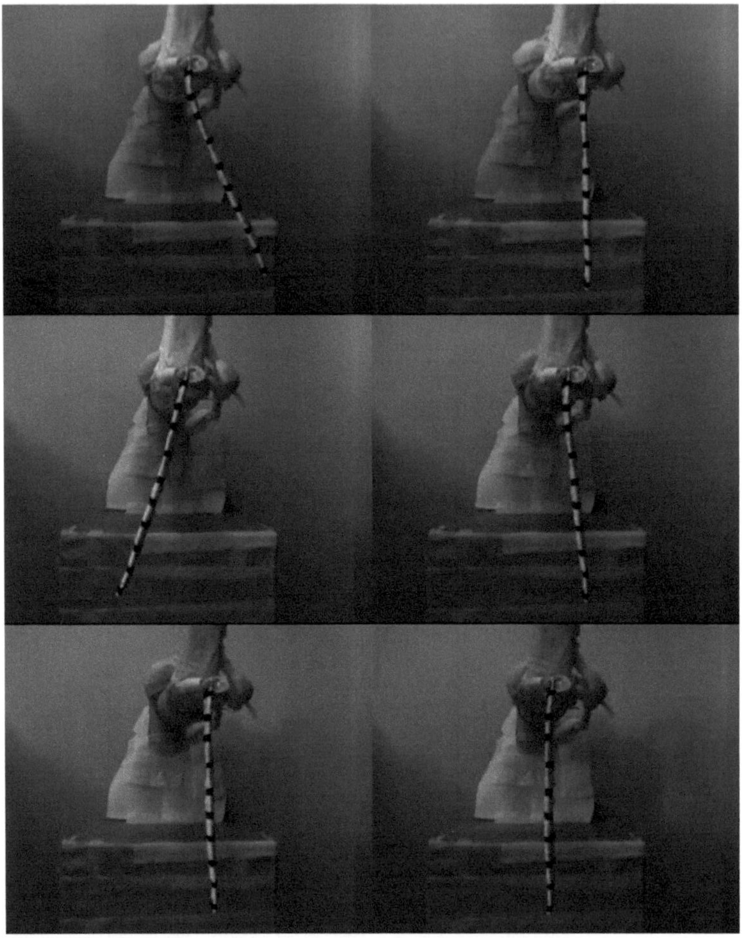

Fig. 22.10 Some overshoots during control

22.6 Conclusion

We proposed a vision-based control strategy for vibration suppression of a DLO manipulated by a robot. As image processing is the key work in the experiment, three image processing methods were studied and compared based on performance. From the perspective of real time, threshold-based moving target detection method was implemented for the detection of the movement of the DLO. Based on the errors between the detected states and desired value, a PID controller is designed to adjust the motion of the robot arm, and the vibration at the end of a DLO is suppressed rapidly. The proposed image processing method and control strategy are applied to experiment, and the results show that the theoretical analysis is accurate.

Acknowledgments This work is supported by the National Nature Science Foundation of China under Grant 60975058 and 61075095, and State Key Laboratory of Robotics and System (HIT) under Grant SKLRS-2010-MS-11, and also supported by International Science and Technology Cooperation Program of Hubei Province under Grant 2012IHA00601 "Joint Research on Green Smart Working Assistance Rehabilitant Robot."

References

1. Fung YC (1965) Foundations of Solid Mechanics. Prentice-Hall, America
2. Hirai S (2000) Energy-based modeling of deformable linear objects. In: Henrich D, Worn H (eds) Robot manipulation of deformable objects, advanced manufacturing series. Springer, London, pp 11–28
3. Yue SG, Henrich D (2001) Manipulating deformable linear objects: attachable adjustment motion for vibration reduction. J Robot Syst 18(7):375–389
4. Wakamatsu H, Takahashi K, Hirai S (2005) Dynamic modeling of linear objects deformation based on differential geometry coordinates. In: IEEE international conference on robotics and automation. IEEE press, USA, pp 1040–1045
5. Huang J, Di P, Fukuda T, et al (2008) Dynamic modeling and simulation of manipulating deformable linear objects. In: IEEE international conference on mechatronics and automation. IEEE Press, USA, pp 858–863
6. D Mingcong, I Akira (2009) Adaptive vibration control of piezoelectric actuator based flexible arm with unknown and unmeasurable disturbances. Int J Model Ident Control 6(3):215-221
7. Yue SG, Henrich D (2001) Manipulating deformable linear objects: model-based adjustment motion for vibration reduction. In: The 10th international conference on advanced robotic. Budapest Polytechnic press, Hungary, pp 503–508
8. Yue SG, Henrich D (2005) Manipulating deformable linear objects: sensor-based skills of adjustment motion for vibration reduction. J Robot Syst 22(2):67–85
9. F Ding, J Huang, Y Wang et al. (2012) Vibration damping in manipulation of deformable linear objects using sliding mode control. In: 31st Chinese control conference. IEEE Press, USA, pp 4924–4929
10. Yue SG, Henrich D (2006) Manipulating deformable linear objects: fuzzy-based active vibration damping skill. J Intel Robot Syst 46(3):201–219
11. Z Jiang (2008) Vision-based cartesian space motion control for flexible robotic manipulators. Int J Model Ident Control 4(4): 406–414 (Inderscience Enterprises Ltd press, Geneva)
12. Shaocheng Q, Tian Y, Chen C et al (2012) A small intelligent car system based on fuzzy control and CCD camera. Int J Model Ident Control 15(1):48–54
13. The home page of Open CV http://opencv.org/
14. Portillo J, Marcos M et al (1998) PID-ATC: a real-time tool for PID control and auto-tuning. In: Proceedings of 5th IFAC workshop on algorithms and architectures for real-time control. Elsevier press, UK, pp 15–17

About the Editors

Lei Liu is a Lecturer in Control Theory and Control Engineering at the Department of Control Science and Engineering, Huazhong University of Science and Technology, China. He received his Bachelor, Master, and Ph.D. degrees from Huazhong University of Science and Technology in 2003, 2005, and 2009, respectively. His main research interest is in the area of dynamic system modeling and control, such as high-speed flight vehicles and mobile robots. He has published over 30 papers on these topics. Currently, Dr. Liu is also a member of the Hubei Association of Automation, China.

Quanmin Zhu is a Professor of Control Systems at the Department of Engineering Design and Mathematics, University of the West of England, Bristol, UK. He completed his M.Sc. at the Harbin Institute of Technology, China in 1983 and his Ph.D. at the Faculty of Engineering, University of Warwick, UK in 1989. His main research interest is in the area of nonlinear system modeling, identification, and control. He has published over 160 papers on these topics and provided consultancy to various industries. Currently, Professor Zhu is also a member of the editorial boards of International Journal of Systems Science and of Chinese Journal of Scientific Instruments, editor (and founder) of International Journal of Modelling, Identification and Control, and editor of International Journal of Computer Applications in Technology.

Lei Cheng is an Associate Professor of Control Science and Engineering at the Engineering Research Center of Metallurgical Automation and Measurement Technology, Ministry of Education, Wuhan University of Science and Technology, Wuhan, China. He received his Bachelor, Master, and Ph.D. degrees from Huazhong University of Science and Technology, China, in 1999, 2002, and 2005, respectively. His main research interest is in Robotics and Complex Systems, a field in which he has published over 40 papers. Currently, he is also a member of the Chinese Association for Artificial Intelligence (CAAI) and of the Hubei Association of Automation, China. He was honored with the National Teaching Achievement Award of China (Second Prize) in 2009.

L. Liu et al. (eds.), *Applied Methods and Techniques for Mechatronic Systems*, 439
Lecture Notes in Control and Information Sciences 452,
DOI: 10.1007/978-3-642-36385-6, © Springer-Verlag Berlin Heidelberg 2014

Yongji Wang is a Professor of Control Theory and Control Engineering at the Department of Control Science and Engineering, Huazhong University of Science and Technology, China. He received his M.Sc. and Ph.D. degrees from Huazhong University of Science and Technology in 1984 and 1990, respectively. His main research interest is in the area of flight trajectory optimization, guidance and control, intelligent control system, biological information processing, and rehabilitation robots. He has published over 200 papers in these fields. Currently, Professor Wang is serving as head of the Hubei Association of Automation and as the Editor Asia-Pacific for International Journal of Modelling, Identification and Control.

Dongya Zhao received his B.Eng. from Shandong University, Jinan, China, in 1998, his M.Sc. from Tianhua Institute of Chemical Machinery and Automation, Lanzhou, China, in 2002, and his Ph.D. from Shanghai Jiao Tong University, Shanghai, China, in 2009. He was a research fellow at Nanyang Technological University from 2011 to 2012. Since 2002, he has been with the College of Chemical Engineering, China University of Petroleum, where he is currently an Associate Professor. His research interests include robot control, sliding mode control, process modeling and control, and nonlinear system control and analysis. He has published more than 30 papers in these areas. He was selected as the 2010 young key teacher of the China University of China.

MIX
Papier aus verantwortungsvollen Quellen
Paper from responsible sources
FSC® C105338
FSC
www.fsc.org

If you have any concerns about our products,
you can contact us on
ProductSafety@springernature.com

In case Publisher is established outside the EU,
the EU authorized representative is:
Springer Nature Customer Service Center GmbH
Europaplatz 3, 69115 Heidelberg, Germany

Printed by Libri Plureos GmbH
in Hamburg, Germany